STUDENT SOLUTIONS MANUAL

FOR

GUSTAFSON AND FRISK'S

COLLEGE ALGEBRA

Eighth Edition

Michael G. Welden
Rock Valley College

THOMSON

BROOKS/COLE

Australia • Canada • Mexico • Singapore • Spain • United Kingdom • United States

Printed in Canada
2   3   4   5   6   7   07   06   05   04   03

Printer: Webcom Limited

ISBN: 0-534-40069-8

For more information about our products, contact us at:
Thomson Learning Academic Resource Center
1-800-423-0563

For permission to use material from this text, contact us by:
Phone: 1-800-730-2214
Fax: 1-800-730-2215
Web: http://www.thomsonrights.com

Cover Design/Image: Larry Didona

Asia
Thomson Learning
5 Shenton Way #01-01
UIC Building
Singapore 068808

Australia/ New Zealand
Thomson Learning
102 Dodds Street
South Street
Southbank, Victoria 3006
Australia

Canada
Nelson
1120 Birchmount Road
Toronto, Ontario M1K 5G4
Canada

Europe/Middle East/South Africa
Thomson Learning
High Holborn House
50/51 Bedford Row
London WC1R 4LR
United Kingdom

Latin America
Thomson Learning
Seneca, 53
Colonia Polanco
11560 Mexico D.F.
Mexico

Spain/ Portugal
Paraninfo
Calle/Magallanes, 25
28015 Madrid, Spain

Contents

Preface

This manual contains detailed solutions to all of the odd exercises of the text *College Algebra*, eighth edition, by R. David Gustafson and Peter D. Frisk. It also contains solutions to all chapter summary, chapter test, and cumulative review exercises found in the text.

Many of the exercises in the text may be solved using more than one method, but it is not feasible to list all possible solutions in this manual. Also, some of the exercises may have been solved in this manual using a method that differs slightly from that presented in the text. There are a few exercises in the text whose solutions may vary from person to person. Some of these solutions may not have been included in this manual. For the solution to an exercise like this, the notation "answers may vary" has been included.

Please remember that only reading a solution does not teach you how to solve a problem. To repeat a commonly used phrase, mathematics is not a spectator sport. You MUST make an honest attempt to solve each exercise in the text without using this manual first. This manual should be viewed more or less as a last resort. Above all, DO NOT simply copy the solution from this manual onto your own paper. Doing so will not help you learn how to do the exercise, nor will it help you to do better on quizzes or tests.

I would like to thank the members of the mathematics faculty at Rock Valley College and Rachael Sturgeon of Brooks/Cole Publishing Company for their help and support. This solutions manual was prepared using EXP 5.0.

This book is dedicated to John, who helps me to realize that mathematics cannot describe everything in life.

May your study of this material be successful and rewarding.

Michael G. Welden

Exercise 0.1 (page 13)

1. decimal

3. 2

5. composite

7. decimals

9. negative

11. $x + (y + z)$

13. $5m + 5 \cdot 2$

15. interval

17. two

19. positive

21. Every natural number is a whole number, so $\mathbf{N} \subseteq \mathbf{W}$. $\boxed{\textbf{TRUE}}$

23. The rational number $\frac{1}{2}$ is not a natural number, so $\mathbf{Q} \nsubseteq \mathbf{N}$. $\boxed{\textbf{FALSE}}$

25. Every whole number is an integer, so $\mathbf{W} \subseteq \mathbf{Z}$. $\boxed{\textbf{TRUE}}$

27. natural: $1, 2, 6, 7$

29. integers: $-5, -4, 0, 1, 2, 6, 7$

31. irrational: $\sqrt{2}$

33. composite: 6

35. odd: $-5, 1, 7$

37.

39.

41.

43.

45. $x > 2 \to (2, \infty)$

47. $0 < x < 5 \to (0, 5)$

49. $x > -4 \to (-4, \infty)$

51. $-2 \le x < 2 \to [-2, 2)$

53. $x \le 5 \to (-\infty, 5]$

55. $-5 < x \le 0 \to (-5, 0]$

57. $-2 \le x \le 3 \to [-2, 3]$

59. $6 \ge x \ge 2 \to 2 \le x \le 6 \to [2, 6]$

61. $x > -5$ and $x < 4 \to (-5, \infty) \cap (-\infty, 4)$

$(-5, \infty)$

$(-\infty, 4)$

$(-5, \infty) \cap (-\infty, 4)$

63. $x \geq -8$ and $x \leq -3 \rightarrow [-8, \infty) \cap (-\infty, -3]$

$[-8, \infty)$

-8

$(-\infty, -3]$

-3

$[-8, \infty) \cap (-\infty, -3]$

$-8 \quad -3$

65. $x < -2$ or $x > 2 \rightarrow (-\infty, -2) \cup (2, \infty)$

$-2 \qquad 2$

67. $x \leq -1$ or $x \geq 3 \rightarrow (-\infty, -1] \cup [3, \infty)$

$-1 \qquad 3$

69. Since $13 \geq 0$, $|13| = 13$.

71. Since $0 \geq 0$, $|0| = 0$.

73. Since $-8 < 0$, $|-8| = -(-8) = 8$.
$-|-8| = -(8) = -8$

75. Since $32 \geq 0$, $|32| = 32$.
$-|32| = -(32) = -32$

77. Since $\pi - 5 < 0$,
$|\pi - 5| = -(\pi - 5) = -\pi + 5 = 5 - \pi$.

79. $|\pi - \pi| = |0| = 0$.

81. If $x \geq 2$, then $x + 1 \geq 0$. Then
$|x + 1| = x + 1$.

83. If $x < 0$, then $x - 4 < 0$. Then
$|x - 4| = -(x - 4)$.

85. distance $= |3 - 8| = |-5| = 5$

87. distance $= |-8 - (-3)| = |-5| = 5$

89. Since population must be positive and never has a fractional part, the set of **natural numbers** should be used.

91. Since temperatures are usually reported without fractional parts and may be either positive or negative (or zero), the set of **integers** should be used.

93. $-x$ will represent a positive number if x itself is negative. For instance, if $x = -3$, then $-x = -(-3) = 3$, which is a positive number.

95. The statement is always true.

97. The statement is not always true. (For example, let $a = 5$ and $b = -2$.)

99. The statement $a < b > c$ could be interpreted to mean that $a > c$, when this is not necessarily true.

Exercise 0.2   (page 24)

1. factor

3. $3, 2x$

5. scientific, integer

7. $x^m x^n = x^{m+n}$

9. $(xy)^n = x^n y^n$

11. $x^0 = 1$

13. $13^2 = 13 \cdot 13 = 169$

15. $-5^2 = -1 \cdot 5 \cdot 5 = -25$

17. $4x^3 = 4 \cdot x \cdot x \cdot x$

19. $(-5x)^4 = (-5x)(-5x)(-5x)(-5x)$

21. $7xxx = 7x^3$

23. $(-x)(-x) = (-1)(-1)x^2 = x^2$

25. $(3t)(3t)(-3t) = (3)(3)(-3)t^3 = -27t^3$

27. $xxxyy = x^3 y^2$

29. $2.2^3 = 10.648$

31. $-0.5^4 = -0.0625$

33. $x^2 x^3 = x^{2+3} = x^5$

35. $(z^2)^3 = z^{2 \cdot 3} = z^6$

37. $(y^5 y^2)^3 = (y^7)^3 = y^{21}$

39. $(z^2)^3 (z^4)^5 = z^6 z^{20} = z^{26}$

41. $(a^2)^3 (a^4)^2 = a^6 a^8 = a^{14}$

43. $(3x)^3 = 3^3 x^3 = 27x^3$

45. $(x^2 y)^3 = (x^2)^3 y^3 = x^6 y^3$

47. $\left(\dfrac{a^2}{b}\right)^3 = \dfrac{(a^2)^3}{b^3} = \dfrac{a^6}{b^3}$

49. $(-x)^0 = 1$

51. $(4x)^0 = 1$

53. $z^{-4} = \dfrac{1}{z^4}$

55. $y^{-2} y^{-3} = y^{-5} = \dfrac{1}{y^5}$

57. $\begin{aligned}(x^3 x^{-4})^{-2} &= (x^{-1})^{-2} \\ &= x^2\end{aligned}$

59. $\dfrac{x^7}{x^3} = x^{7-3} = x^4$

61. $\dfrac{a^{21}}{a^{17}} = a^{21-17} = a^4$

63. $\dfrac{(x^2)^2}{x^2 x} = \dfrac{x^4}{x^3} = x^{4-3} = x^1 = x$

65. $\left(\dfrac{m^3}{n^2}\right)^3 = \dfrac{(m^3)^3}{(n^2)^3} = \dfrac{m^9}{n^6}$

67. $\dfrac{(a^3)^{-2}}{aa^2} = \dfrac{a^{-6}}{a^3} = a^{-6-3} = a^{-9} = \dfrac{1}{a^9}$

69. $\left(\dfrac{a^{-3}}{b^{-1}}\right)^{-4} = \dfrac{(a^{-3})^{-4}}{(b^{-1})^{-4}} = \dfrac{a^{12}}{b^4}$

71. $\begin{aligned}\left(\dfrac{r^4 r^{-6}}{r^3 r^{-3}}\right)^2 &= \left(\dfrac{r^{-2}}{r^0}\right)^2 = (r^{-2})^2 = r^{-4} \\ &= \dfrac{1}{r^4}\end{aligned}$

73. $\left(\dfrac{x^5 y^{-2}}{x^{-3} y^2}\right)^4 = \left(\dfrac{x^5 x^3}{y^2 y^2}\right)^4 = \left(\dfrac{x^8}{y^4}\right)^4 = \dfrac{x^{32}}{y^{16}}$

75. $\left(\dfrac{5x^{-3} y^{-2}}{3x^2 y^{-3}}\right)^{-2} = \left(\dfrac{3x^2 y^{-3}}{5x^{-3} y^{-2}}\right)^2 = \left(\dfrac{3x^2 x^3 y^2}{5y^3}\right)^2 = \left(\dfrac{3x^5}{5y}\right)^2 = \dfrac{9x^{10}}{25y^2}$

77. $\left(\dfrac{3x^5 y^{-3}}{6x^{-5} y^3}\right)^{-2} = \left(\dfrac{6x^{-5} y^3}{3x^5 y^{-3}}\right)^2 = \left(\dfrac{2y^3 y^3}{1x^5 x^5}\right)^2 = \left(\dfrac{2y^6}{x^{10}}\right)^2 = \dfrac{4y^{12}}{x^{20}}$

79. $\dfrac{(8^{-2} z^{-3} y)^{-1}}{(5y^2 z^{-2})^3 (5yz^{-2})^{-1}} = \dfrac{8^2 z^3 y^{-1}}{5^3 y^6 z^{-6} \cdot 5^{-1} y^{-1} z^2} = \dfrac{64 z^3 y^{-1}}{5^2 y^5 z^{-4}} = \dfrac{64 z^3 z^4}{25 y^5 y^1} = \dfrac{64 z^7}{25 y^6}$

81. $-\dfrac{5[6^2 + (9 - 5)]}{4(2 - 3)} = -\dfrac{5[36 + 4]}{4(-1)} = -\dfrac{5[40]}{-4} = -\dfrac{200}{-4} = -(-50) = 50$

83. $x^2 = (-2)^2 = 4$ **85.** $x^3 = (-2)^3 = -8$

87. $(-xz)^3 = [-1 \cdot (-2) \cdot 3]^3 = 6^3 = 216$

89. $\dfrac{-(x^2 z^3)}{z^2 - y^2} = \dfrac{-[(-2)^2 \cdot 3^3]}{3^2 - 0^2} = \dfrac{-[4 \cdot 27]}{9 - 0} = \dfrac{-108}{9} = -12$

91. $5x^2 - 3y^3 z = 5(-2)^2 - 3(0)^3(3) = 5(4) - 3(0)(3) = 20 - 0 = 20$

93. $\dfrac{-3x^{-3}z^{-2}}{6x^2 z^{-3}} = \dfrac{-1z^3}{2x^2 x^3 z^2} = \dfrac{-z}{2x^5} = \dfrac{-3}{2(-2)^5} = \dfrac{-3}{2(-32)} = \dfrac{-3}{-64} = \dfrac{3}{64}$

95. $372{,}000 = 3.72 \times 10^5$ **97.** $-177{,}000{,}000 = -1.77 \times 10^8$

99. $0.007 = 7 \times 10^{-3}$ **101.** $-0.000000693 = -6.93 \times 10^{-7}$

103. $1{,}000{,}000{,}000{,}000 = 1 \times 10^{12}$ **105.** $9.37 \times 10^5 = 937{,}000$

107. $2.21 \times 10^{-5} = 0.0000221$ **109.** $0.00032 \times 10^4 = 3.2$ **111.** $-3.2 \times 10^{-3} = -0.0032$

113. $\dfrac{(65{,}000)(45{,}000)}{250{,}000} = \dfrac{(6.5 \times 10^4)(4.5 \times 10^4)}{2.5 \times 10^5} = \dfrac{(6.5)(4.5)}{2.5} \times 10^{4+4-5} = 11.7 \times 10^3$

$$= 1.17 \times 10^1 \times 10^3$$
$$= 1.17 \times 10^4$$

115. $\dfrac{(0.00000035)(170{,}000)}{0.00000085} = \dfrac{(3.5 \times 10^{-7})(1.7 \times 10^5)}{8.5 \times 10^{-7}} = \dfrac{(3.5)(1.7)}{8.5} \times 10^{(-7)+5-(-7)}$

$$= 0.7 \times 10^5$$
$$= 7 \times 10^{-1} \times 10^5 = 7 \times 10^4$$

117. $\dfrac{(45{,}000{,}000{,}000)(212{,}000)}{0.00018} = \dfrac{(4.5 \times 10^{10})(2.12 \times 10^5)}{1.8 \times 10^{-4}} = \dfrac{(4.5)(2.12)}{1.8} \times 10^{10+5-(-4)}$

$$= 5.3 \times 10^{19}$$

119. $3.31 \times 10^4 \text{ cm/sec} = \dfrac{3.31 \times 10^4 \text{ cm}}{1 \text{ sec}} \cdot \dfrac{1 \text{ m}}{100 \text{ cm}} \cdot \dfrac{60 \text{ sec}}{1 \text{ min}} = \dfrac{(3.31 \times 10^4)(6 \times 10^1)}{1 \times 10^2} \text{ m/min}$

$$= \dfrac{(3.31)(6)}{1} \times 10^{4+1-2} \text{ m/min}$$
$$= 19.86 \times 10^3 \text{ m/min}$$
$$= 1.986 \times 10^4 \text{ m/min}$$

121. mass $= 1,000,000,000(0.00000000000000000000000167248 \text{ g})$
$= (1 \times 10^9)(1.67248 \times 10^{-24} \text{ g}) = 1.67248 \times 10^{-15} \text{ g}$

123. $x^n x^2 = x^{n+2}$

125. $\dfrac{x^m x^2}{x^3} = \dfrac{x^{m+2}}{x^3} = x^{m+2-3} = x^{m-1}$

127. $x^{m+1} x^3 = x^{m+1+3} = x^{m+4}$

129. In the expression $-x^4$, the base of the exponent is x, while in the expression $(-x)^4$, the base of the exponent is $-x$.

131. $(-2, 4) \Rightarrow$

133. Since $\pi - 5 < 0$,
$|\pi - 5| = -(\pi - 5) = -\pi + 5 = 5 - \pi$.

Exercise 0.3 (page 36)

1. 0

3. not

5. $a^{1/n}$

7. $\sqrt[n]{ab}$

9. $\neq$

11. $9^{1/2} = (3^2)^{1/2} = 3$

13. $\left(\dfrac{1}{25}\right)^{1/2} = \left[\left(\dfrac{1}{5}\right)^2\right]^{1/2} = \dfrac{1}{5}$

15. $-81^{1/4} = -(3^4)^{1/4} = -3$

17. $(10,000)^{1/4} = (10^4)^{1/4} = 10$

19. $-64^{1/3} = -(4^3)^{1/3} = -4$

21. $64^{1/3} = (4^3)^{1/3} = 4$

23. $(16a^2)^{1/2} = [(4a)^2]^{1/2} = 4|a|$

25. $(16a^4)^{1/4} = [(2a)^4]^{1/4} = 2|a|$

27. $(-32a^5)^{1/5} = [(-2a)^5]^{1/5} = -2a$

29. $(-216b^6)^{1/3} = [(-6b^2)^3]^{1/3} = -6b^2$

31. $\left(\dfrac{16a^4}{25b^2}\right)^{1/2} = \left[\left(\dfrac{4a^2}{5b}\right)^2\right]^{1/2} = \left|\dfrac{4a^2}{5b}\right|$
$= \dfrac{4a^2}{5|b|}$

33. $\left(-\dfrac{1,000x^6}{27y^3}\right)^{1/3} = \left[\left(-\dfrac{10x^2}{3y}\right)^3\right]^{1/3}$
$= -\dfrac{10x^2}{3y}$

35. $4^{3/2} = (4^{1/2})^3 = 2^3 = 8$

37. $-16^{3/2} = -(16^{1/2})^3 = -(4)^3 = -64$

39. $-1000^{2/3} = -(1000^{1/3})^2 = -(10)^2$
$= -100$

41. $64^{-1/2} = \dfrac{1}{64^{1/2}} = \dfrac{1}{8}$

43. $64^{-3/2} = \dfrac{1}{64^{3/2}} = \dfrac{1}{\left(64^{1/2}\right)^3} = \dfrac{1}{8^3} = \dfrac{1}{512}$

45. $-9^{-3/2} = -\dfrac{1}{9^{3/2}} = -\dfrac{1}{\left(9^{1/2}\right)^3} = -\dfrac{1}{3^3}$
$$= -\dfrac{1}{27}$$

47. $\left(\dfrac{4}{9}\right)^{5/2} = \left[\left(\dfrac{4}{9}\right)^{1/2}\right]^5 = \left(\dfrac{2}{3}\right)^5 = \dfrac{32}{243}$

49. $\left(-\dfrac{27}{64}\right)^{-2/3} = \left(-\dfrac{64}{27}\right)^{2/3} = \left[\left(-\dfrac{64}{27}\right)^{1/3}\right]^2 = \left(-\dfrac{4}{3}\right)^2 = \dfrac{16}{9}$

51. $\left(100s^4\right)^{1/2} = 100^{1/2}\left(s^4\right)^{1/2} = 10s^2$

53. $\left(32y^{10}z^5\right)^{-1/5} = \dfrac{1}{\left(32y^{10}z^5\right)^{1/5}} = \dfrac{1}{32^{1/5}\left(y^{10}\right)^{1/5}\left(z^5\right)^{1/5}} = \dfrac{1}{2y^2z}$

55. $\left(x^{10}y^5\right)^{3/5} = x^{30/5}y^{15/5} = x^6y^3$

57. $\left(r^8s^{16}\right)^{-3/4} = r^{-24/4}s^{-48/4} = r^{-6}s^{-12}$
$$= \dfrac{1}{r^6s^{12}}$$

59. $\left(-\dfrac{8a^6}{125b^9}\right)^{2/3} = \dfrac{(-8)^{2/3}a^{12/3}}{125^{2/3}b^{18/3}} = \dfrac{(-2)^2a^4}{5^2b^6}$
$$= \dfrac{4a^4}{25b^6}$$

61. $\left(\dfrac{27r^6}{1000s^{12}}\right)^{-2/3} = \left(\dfrac{1000s^{12}}{27r^6}\right)^{2/3} = \dfrac{1000^{2/3}s^{24/3}}{27^{2/3}r^{12/3}} = \dfrac{10^2s^8}{3^2r^4} = \dfrac{100s^8}{9r^4}$

63. $\dfrac{a^{2/5}a^{4/5}}{a^{1/5}} = \dfrac{a^{6/5}}{a^{1/5}} = a^{5/5} = a$

65. $\sqrt{49} = \sqrt{7^2} = 7$

67. $\sqrt[3]{125} = \sqrt[3]{5^3} = 5$

69. $-\sqrt[4]{81} = -\sqrt[4]{3^4} = -3$

71. $\sqrt[5]{-\dfrac{32}{100,000}} = \sqrt[5]{\left(-\dfrac{2}{10}\right)^5} = -\dfrac{2}{10} = -\dfrac{1}{5}$

73. $\sqrt{36x^2} = \sqrt{(6x)^2} = |6x| = 6|x|$

75. $\sqrt{9y^4} = \sqrt{(3y^2)^2} = |3y^2| = 3y^2$

77. $\sqrt[3]{8y^3} = \sqrt[3]{(2y)^3} = 2y$

79. $\sqrt[4]{\dfrac{x^4y^8}{z^{12}}} = \sqrt[4]{\left(\dfrac{xy^2}{z^3}\right)^4} = \left|\dfrac{xy^2}{z^3}\right| = \dfrac{|x|y^2}{|z^3|}$

81. $\sqrt{8} - \sqrt{2} = \sqrt{4}\sqrt{2} - \sqrt{2} = 2\sqrt{2} - \sqrt{2}$
$$= \sqrt{2}$$

83. $\sqrt{200x^2} + \sqrt{98x^2} = \sqrt{100x^2}\sqrt{2} + \sqrt{49x^2}\sqrt{2} = 10x\sqrt{2} + 7x\sqrt{2} = 17x\sqrt{2}$

85. $2\sqrt{48y^5} - 3y\sqrt{12y^3} = 2\sqrt{16y^4}\sqrt{3y} - 3y\sqrt{4y^2}\sqrt{3y} = 2(4y^2)\sqrt{3y} - 3y(2y)\sqrt{3y}$
$$= 8y^2\sqrt{3y} - 6y^2\sqrt{3y} = 2y^2\sqrt{3y}$$

87. $2\sqrt[3]{81} + 3\sqrt[3]{24} = 2\sqrt[3]{27}\sqrt[3]{3} + 3\sqrt[3]{8}\sqrt[3]{3} = 2(3)\sqrt[3]{3} + 3(2)\sqrt[3]{3} = 6\sqrt[3]{3} + 6\sqrt[3]{3} = 12\sqrt[3]{3}$

89. $\sqrt[4]{768z^5} + \sqrt[4]{48z^5} = \sqrt[4]{256z^4}\sqrt[4]{3z} + \sqrt[4]{16z^4}\sqrt[4]{3z} = 4z\sqrt[4]{3z} + 2z\sqrt[4]{3z} = 6z\sqrt[4]{3z}$

91. $\sqrt{8x^2y} - x\sqrt{2y} + \sqrt{50x^2y} = \sqrt{4x^2}\sqrt{2y} - x\sqrt{2y} + \sqrt{25x^2}\sqrt{2y}$
$$= 2x\sqrt{2y} - x\sqrt{2y} + 5x\sqrt{2y} = 6x\sqrt{2y}$$

93. $\sqrt[3]{16xy^4} + y\sqrt[3]{2xy} - \sqrt[3]{54xy^4} = \sqrt[3]{8y^3}\sqrt[3]{2xy} + y\sqrt[3]{2xy} - \sqrt[3]{27y^3}\sqrt[3]{2xy}$
$$= 2y\sqrt[3]{2xy} + y\sqrt[3]{2xy} - 3y\sqrt[3]{2xy} = 0$$

95. $\dfrac{3}{\sqrt{3}} = \dfrac{3}{\sqrt{3}} \cdot \dfrac{\sqrt{3}}{\sqrt{3}} = \dfrac{3\sqrt{3}}{3} = \sqrt{3}$

97. $\dfrac{2}{\sqrt[3]{2}} = \dfrac{2}{\sqrt[3]{2}} \cdot \dfrac{\sqrt[3]{4}}{\sqrt[3]{4}} = \dfrac{2\sqrt[3]{4}}{\sqrt[3]{8}} = \dfrac{2\sqrt[3]{4}}{2} = \sqrt[3]{4}$

99. $\dfrac{2b}{\sqrt[4]{3a^2}} = \dfrac{2b}{\sqrt[4]{3a^2}} \cdot \dfrac{\sqrt[4]{27a^2}}{\sqrt[4]{27a^2}} = \dfrac{2b\sqrt[4]{27a^2}}{\sqrt[4]{81a^4}} = \dfrac{2b\sqrt[4]{27a^2}}{3a}$

101. $\sqrt[3]{\dfrac{2u^4}{9v}} = \dfrac{\sqrt[3]{2u^4}}{\sqrt[3]{9v}} = \dfrac{\sqrt[3]{u^3}\sqrt[3]{2u}}{\sqrt[3]{9v}} \cdot \dfrac{\sqrt[3]{3v^2}}{\sqrt[3]{3v^2}} = \dfrac{u\sqrt[3]{6uv^2}}{\sqrt[3]{27v^3}} = \dfrac{u\sqrt[3]{6uv^2}}{3v}$

103. $\dfrac{\sqrt{5}}{10} = \dfrac{\sqrt{5}}{10} \cdot \dfrac{\sqrt{5}}{\sqrt{5}} = \dfrac{5}{10\sqrt{5}} = \dfrac{1}{2\sqrt{5}}$

105. $\dfrac{\sqrt[3]{9}}{3} = \dfrac{\sqrt[3]{9}}{3} \cdot \dfrac{\sqrt[3]{3}}{\sqrt[3]{3}} = \dfrac{\sqrt[3]{27}}{3\sqrt[3]{3}} = \dfrac{3}{3\sqrt[3]{3}} = \dfrac{1}{\sqrt[3]{3}}$

107. $\dfrac{\sqrt[5]{16b^3}}{64a} = \dfrac{\sqrt[5]{16b^3}}{64a} \cdot \dfrac{\sqrt[5]{2b^2}}{\sqrt[5]{2b^2}} = \dfrac{\sqrt[5]{32b^5}}{64a\sqrt[5]{2b^2}} = \dfrac{2b}{64a\sqrt[5]{2b^2}} = \dfrac{b}{32a\sqrt[5]{2b^2}}$

109. $\sqrt{\dfrac{1}{3}} - \sqrt{\dfrac{1}{27}} = \dfrac{\sqrt{1}}{\sqrt{3}} - \dfrac{\sqrt{1}}{\sqrt{27}} = \dfrac{1}{\sqrt{3}} \cdot \dfrac{\sqrt{3}}{\sqrt{3}} - \dfrac{1}{\sqrt{27}} \cdot \dfrac{\sqrt{3}}{\sqrt{3}} = \dfrac{\sqrt{3}}{3} - \dfrac{\sqrt{3}}{\sqrt{81}} = \dfrac{\sqrt{3}}{3} - \dfrac{\sqrt{3}}{9}$
$$= \dfrac{3\sqrt{3}}{9} - \dfrac{\sqrt{3}}{9} = \dfrac{2\sqrt{3}}{9}$$

111. $\sqrt{\dfrac{x}{8}} - \sqrt{\dfrac{x}{2}} + \sqrt{\dfrac{x}{32}} = \dfrac{\sqrt{x}}{\sqrt{8}} - \dfrac{\sqrt{x}}{\sqrt{2}} + \dfrac{\sqrt{x}}{\sqrt{32}} = \dfrac{\sqrt{x}}{\sqrt{8}} \cdot \dfrac{\sqrt{2}}{\sqrt{2}} - \dfrac{\sqrt{x}}{\sqrt{2}} \cdot \dfrac{\sqrt{2}}{\sqrt{2}} + \dfrac{\sqrt{x}}{\sqrt{32}} \cdot \dfrac{\sqrt{2}}{\sqrt{2}}$
$$= \dfrac{\sqrt{2x}}{\sqrt{16}} - \dfrac{\sqrt{2x}}{\sqrt{4}} + \dfrac{\sqrt{2x}}{\sqrt{64}}$$
$$= \dfrac{\sqrt{2x}}{4} - \dfrac{\sqrt{2x}}{2} + \dfrac{\sqrt{2x}}{8}$$
$$= \dfrac{2\sqrt{2x}}{8} - \dfrac{4\sqrt{2x}}{8} + \dfrac{\sqrt{2x}}{8} = -\dfrac{\sqrt{2x}}{8}$$

113. $\sqrt[4]{9} = 9^{1/4} = (3^2)^{1/4} = 3^{2/4} = 3^{1/2} = \sqrt{3}$

115. $\sqrt[10]{16x^6} = (16x^6)^{1/10} = (2^4x^6)^{1/10} = 2^{4/10}x^{6/10} = 2^{2/5}x^{3/5} = (2^2x^3)^{1/5} = \sqrt[5]{4x^3}$

117. $\sqrt{2}\sqrt[3]{2} = 2^{1/2} \cdot 2^{1/3} = 2^{3/6} \cdot 2^{2/6} = \sqrt[6]{2^3}\sqrt[6]{2^2} = \sqrt[6]{8}\sqrt[6]{4} = \sqrt[6]{32}$

119. $\dfrac{\sqrt[4]{3}}{\sqrt{2}} = \dfrac{3^{1/4}}{2^{1/2}} = \dfrac{3^{1/4}}{2^{2/4}} = \dfrac{\sqrt[4]{3}}{\sqrt[4]{2^2}} = \dfrac{\sqrt[4]{3}}{\sqrt[4]{4}} = \dfrac{\sqrt[4]{3}}{\sqrt[4]{4}} \cdot \dfrac{\sqrt[4]{4}}{\sqrt[4]{4}} = \dfrac{\sqrt[4]{12}}{\sqrt[4]{16}} = \dfrac{\sqrt[4]{12}}{2}$

121. $\sqrt[4]{x^4} = |x|$. Since $|x| = x$ if $x \geq 0$, then $\sqrt[4]{x^4} = x$ if $x \geq 0$.

123. $\left(\dfrac{x}{y}\right)^{-m/n} = \dfrac{x^{-m/n}}{y^{-m/n}} = \dfrac{x^{-m/n}}{y^{-m/n}} \cdot \dfrac{x^{m/n}y^{m/n}}{x^{m/n}y^{m/n}} = \dfrac{y^{m/n}}{x^{m/n}} = \dfrac{(y^m)^{1/n}}{(x^m)^{1/n}} = \left(\dfrac{y^m}{x^m}\right)^{1/n} = \sqrt[n]{\dfrac{y^m}{x^m}}$

125. $-2 < x \leq 5 \Rightarrow (-2, 5]$ **127.** $x^2 - y^2 = (-2)^2 - 3^2 = 4 - 9 = -5$

129. $617{,}000{,}000 = 6.17 \times 10^8$

Exercise 0.4 (page 47)

1. monomial, variables   **3.** trinomial   **5.** one   **7.** like

9. coefficients, variables   **11.** yes, trinomial, 2nd degree   **13.** no

15. yes, binomial, 3rd degree   **17.** yes, monomial, 0th degree   **19.** yes, monomial, no degree

21. $(x^3 - 3x^2) + (5x^3 - 8x) = x^3 - 3x^2 + 5x^3 - 8x = x^3 + 5x^3 - 3x^2 - 8x = 6x^3 - 3x^2 - 8x$

23. $(y^5 + 2y^3 + 7) - (y^5 - 2y^3 - 7) = y^5 + 2y^3 + 7 - y^5 + 2y^3 + 7$
$$= y^5 - y^5 + 2y^3 + 2y^3 + 7 + 7 = 4y^3 + 14$$

25. $2(x^2 + 3x - 1) - 3(x^2 + 2x - 4) + 4 = 2(x^2) + 2(3x) + 2(-1) - 3(x^2) - 3(2x) - 3(-4) + 4$
$$= 2x^2 + 6x - 2 - 3x^2 - 6x + 12 + 4$$
$$= 2x^2 - 3x^2 + 6x - 6x - 2 + 12 + 4 = -x^2 + 14$$

27. $8(t^2 - 2t + 5) + 4(t^2 - 3t + 2) - 6(2t^2 - 8)$
$$= 8(t^2) + 8(-2t) + 8(5) + 4(t^2) + 4(-3t) + 4(2) - 6(2t^2) - 6(-8)$$
$$= 8t^2 - 16t + 40 + 4t^2 - 12t + 8 - 12t^2 + 48$$
$$= 8t^2 + 4t^2 - 12t^2 - 16t - 12t + 40 + 8 + 48 = -28t + 96$$

29. $y(y^2 - 1) - y^2(y + 2) - y(2y - 2) = y(y^2) + y(-1) - y^2(y) - y^2(2) - y(2y) - y(-2)$
$$= y^3 - y - y^3 - 2y^2 - 2y^2 + 2y$$
$$= y^3 - y^3 - 2y^2 - 2y^2 - y + 2y = -4y^2 + y$$

31. $xy(x - 4y) - y(x^2 + 3xy) + xy(2x + 3y)$
$$= xy(x) + xy(-4y) - y(x^2) - y(3xy) + xy(2x) + xy(3y)$$
$$= x^2y - 4xy^2 - x^2y - 3xy^2 + 2x^2y + 3xy^2$$
$$= x^2y - x^2y + 2x^2y - 4xy^2 - 3xy^2 + 3xy^2 = 2x^2y - 4xy^2$$

33. $2x^2y^3(4xy^4) = 2(4)x^2xy^3y^4 = 8x^3y^7$

35. $-3m^2n(2mn^2)\left(-\frac{mn}{12}\right) = (-3)(2)\left(-\frac{1}{12}\right)m^2mmnn^2n = \frac{6}{12}m^4n^4 = \frac{m^4n^4}{2}$

37. $-4rs(r^2 + s^2) = -4rs(r^2) - 4rs(s^2) = -4r^3s - 4rs^3$

39. $6ab^2c(2ac + 3bc^2 - 4ab^2c) = 6ab^2c(2ac) + 6ab^2c(3bc^2) + 6ab^2c(-4ab^2c)$
$$= 12a^2b^2c^2 + 18ab^3c^3 - 24a^2b^4c^2$$

41. $(a + 2)(a + 2) = a^2 + 2a + 2a + 4$
$$= a^2 + 4a + 4$$

43. $(a - 6)^2 = (a - 6)(a - 6)$
$$= a^2 - 6a - 6a + 36$$
$$= a^2 - 12a + 36$$

45. $(x + 4)(x - 4) = x^2 - 4x + 4x - 16$
$$= x^2 - 16$$

47. $(x - 3)(x + 5) = x^2 + 5x - 3x - 15$
$$= x^2 + 2x - 15$$

49. $(u + 2)(3u - 2) = 3u^2 - 2u + 6u - 4$
$$= 3u^2 + 4u - 4$$

51. $(5x - 1)(2x + 3) = 10x^2 + 15x - 2x - 3$
$$= 10x^2 + 13x - 3$$

53. $(3a - 2b)^2 = (3a - 2b)(3a - 2b) = 9a^2 - 6ab - 6ab + 4b^2 = 9a^2 - 12ab + 4b^2$

55. $(3m + 4n)(3m - 4n) = 9m^2 - 12mn + 12mn - 16n^2 = 9m^2 - 16n^2$

57. $(2y - 4x)(3y - 2x) = 6y^2 - 4xy - 12xy + 8x^2 = 6y^2 - 16xy + 8x^2$

59. $(9x - y)(x^2 - 3y) = 9x^3 - 27xy - x^2y + 3y^2$

61. $(5z + 2t)(z^2 - t) = 5z^3 - 5tz + 2tz^2 - 2t^2$

63. $(3x - 1)^3 = (3x - 1)(3x - 1)(3x - 1)$
$$= (9x^2 - 3x - 3x + 1)(3x - 1)$$
$$= (9x^2 - 6x + 1)(3x - 1)$$
$$= 9x^2(3x) + 9x^2(-1) - 6x(3x) - 6x(-1) + 1(3x) + 1(-1)$$
$$= 27x^3 - 9x^2 - 18x^2 + 6x + 3x - 1 = 27x^3 - 27x^2 + 9x - 1$$

65. $(3x+1)(2x^2+4x-3) = 3x(2x^2) + 3x(4x) + 3x(-3) + 1(2x^2) + 1(4x) + 1(-3)$
$$= 6x^3 + 12x^2 - 9x + 2x^2 + 4x - 3 = 6x^3 + 14x^2 - 5x - 3$$

67. $(3x+2y)(2x^2-3xy+4y^2)$
$$= 3x(2x^2) + 3x(-3xy) + 3x(4y^2) + 2y(2x^2) + 2y(-3xy) + 2y(4y^2)$$
$$= 6x^3 - 9x^2y + 12xy^2 + 4x^2y - 6xy^2 + 8y^3 = 6x^3 - 5x^2y + 6xy^2 + 8y^3$$

69. $2y^n(3y^n + y^{-n}) = 2y^n(3y^n) + 2y^n(y^{-n}) = 6y^{n+n} + 2y^{n+(-n)} = 6y^{2n} + 2y^0 = 6y^{2n} + 2$

71. $-5x^{2n}y^n(2x^{2n}y^{-n} + 3x^{-2n}y^n) = -5x^{2n}y^n(2x^{2n}y^{-n}) - 5x^{2n}y^n(3x^{-2n}y^n)$
$$= -10x^{2n+2n}y^{n+(-n)} - 15x^{2n+(-2n)}y^{n+n}$$
$$= -10x^{4n}y^0 - 15x^0y^{2n} = -10x^{4n} - 15y^{2n}$$

73. $(x^n+3)(x^n-4) = x^nx^n - 4x^n + 3x^n - 12 = x^{2n} - x^n - 12$

75. $(2r^n-7)(3r^n-2) = 2r^n(3r^n) - 2r^n(2) - 7(3r^n) + 14$
$$= 6r^{2n} - 4r^n - 21r^n + 14 = 6r^{2n} - 25r^n + 14$$

77. $x^{1/2}(x^{1/2}y + xy^{1/2}) = x^{1/2}x^{1/2}y + x^{1/2}xy^{1/2} = x^{2/2}y + x^{3/2}y^{1/2} = xy + x^{3/2}y^{1/2}$

79. $(a^{1/2}+b^{1/2})(a^{1/2}-b^{1/2}) = a^{1/2}a^{1/2} - a^{1/2}b^{1/2} + a^{1/2}b^{1/2} - b^{1/2}b^{1/2}$
$$= a^{2/2} - b^{2/2} = a - b$$

81. $\dfrac{2}{\sqrt{3}-1} = \dfrac{2}{\sqrt{3}-1} \cdot \dfrac{\sqrt{3}+1}{\sqrt{3}+1} = \dfrac{2(\sqrt{3}+1)}{(\sqrt{3})^2 - 1^2} = \dfrac{2(\sqrt{3}+1)}{3-1} = \dfrac{2(\sqrt{3}+1)}{2} = \sqrt{3}+1$

83. $\dfrac{3x}{\sqrt{7}+2} = \dfrac{3x}{\sqrt{7}+2} \cdot \dfrac{\sqrt{7}-2}{\sqrt{7}-2} = \dfrac{3x(\sqrt{7}-2)}{(\sqrt{7})^2 - 2^2} = \dfrac{3x(\sqrt{7}-2)}{7-4} = \dfrac{3x(\sqrt{7}-2)}{3} = x(\sqrt{7}-2)$

85. $\dfrac{x}{x-\sqrt{3}} = \dfrac{x}{x-\sqrt{3}} \cdot \dfrac{x+\sqrt{3}}{x+\sqrt{3}} = \dfrac{x(x+\sqrt{3})}{x^2 - (\sqrt{3})^2} = \dfrac{x(x+\sqrt{3})}{x^2 - 3}$

87. $\dfrac{y+\sqrt{2}}{y-\sqrt{2}} = \dfrac{y+\sqrt{2}}{y-\sqrt{2}} \cdot \dfrac{y+\sqrt{2}}{y+\sqrt{2}} = \dfrac{(y+\sqrt{2})(y+\sqrt{2})}{y^2 - (\sqrt{2})^2} = \dfrac{y^2 + 2y\sqrt{2} + 2}{y^2 - 2}$

89. $\dfrac{\sqrt{2}-\sqrt{3}}{1-\sqrt{3}} = \dfrac{\sqrt{2}-\sqrt{3}}{1-\sqrt{3}} \cdot \dfrac{1+\sqrt{3}}{1+\sqrt{3}} = \dfrac{\sqrt{2}+\sqrt{6}-\sqrt{3}-(\sqrt{3})^2}{1^2-(\sqrt{3})^2} = \dfrac{\sqrt{2}+\sqrt{6}-\sqrt{3}-3}{1-3}$

$$= \dfrac{\sqrt{2}+\sqrt{6}-\sqrt{3}-3}{-2}$$

$$= \dfrac{-(\sqrt{2}+\sqrt{6}-\sqrt{3}-3)}{2}$$

$$= \dfrac{\sqrt{3}+3-\sqrt{2}-\sqrt{6}}{2}$$

91. $\dfrac{\sqrt{x}-\sqrt{y}}{\sqrt{x}+\sqrt{y}} = \dfrac{\sqrt{x}-\sqrt{y}}{\sqrt{x}+\sqrt{y}} \cdot \dfrac{\sqrt{x}-\sqrt{y}}{\sqrt{x}-\sqrt{y}} = \dfrac{\sqrt{x^2}-\sqrt{xy}-\sqrt{xy}+\sqrt{y^2}}{\left(\sqrt{x}\right)^2-\left(\sqrt{y}\right)^2} = \dfrac{x-2\sqrt{xy}+y}{x-y}$

93. $\dfrac{\sqrt{2}+1}{2} = \dfrac{\sqrt{2}+1}{2} \cdot \dfrac{\sqrt{2}-1}{\sqrt{2}-1} = \dfrac{\left(\sqrt{2}\right)^2-1^2}{2\left(\sqrt{2}-1\right)} = \dfrac{2-1}{2\left(\sqrt{2}-1\right)} = \dfrac{1}{2\left(\sqrt{2}-1\right)}$

95. $\dfrac{y-\sqrt{3}}{y+\sqrt{3}} = \dfrac{y-\sqrt{3}}{y+\sqrt{3}} \cdot \dfrac{y+\sqrt{3}}{y+\sqrt{3}} = \dfrac{y^2-\left(\sqrt{3}\right)^2}{y^2+y\sqrt{3}+y\sqrt{3}+\sqrt{9}} = \dfrac{y^2-3}{y^2+2y\sqrt{3}+3}$

97. $\dfrac{\sqrt{x+3}-\sqrt{x}}{3} = \dfrac{\sqrt{x+3}-\sqrt{x}}{3} \cdot \dfrac{\sqrt{x+3}+\sqrt{x}}{\sqrt{x+3}+\sqrt{x}} = \dfrac{(\sqrt{x+3})^2-(\sqrt{x})^2}{3(\sqrt{x+3}+\sqrt{x})}$

$$= \dfrac{x+3-x}{3(\sqrt{x+3}+\sqrt{x})}$$

$$= \dfrac{3}{3(\sqrt{x+3}+\sqrt{x})} = \dfrac{1}{\sqrt{x+3}+\sqrt{x}}$$

99. $\dfrac{36a^2b^3}{18ab^6} = 2a^{2-1}b^{3-6} = 2a^1b^{-3} = \dfrac{2a}{b^3}$

101. $\dfrac{16x^6y^4z^9}{-24x^9y^6z^0} = -\dfrac{2}{3}x^{6-9}y^{4-6}z^{9-0} = -\dfrac{2}{3}x^{-3}y^{-2}z^9 = -\dfrac{2z^9}{3x^3y^2}$

103. $\dfrac{5x^3y^2+15x^3y^4}{10x^2y^3} = \dfrac{5x^3y^2}{10x^2y^3} + \dfrac{15x^3y^4}{10x^2y^3} = \dfrac{x}{2y} + \dfrac{3xy}{2}$

105. $\dfrac{24x^5y^7-36x^2y^5+12xy}{60x^5y^4} = \dfrac{24x^5y^7}{60x^5y^4} - \dfrac{36x^2y^5}{60x^5y^4} + \dfrac{12xy}{60x^5y^4} = \dfrac{2y^3}{5} - \dfrac{3y}{5x^3} + \dfrac{1}{5x^4y^3}$

107.
$$\begin{array}{r} 3x + 2 \\ x+3 \enclose{longdiv}{3x^2+11x+6} \\ \underline{3x^2+9x} \\ 2x+6 \\ \underline{2x+6} \\ 0 \end{array}$$

109.
$$\begin{array}{r} x - 7+\frac{2}{2x-5} \\ 2x-5 \enclose{longdiv}{2x^2-19x+37} \\ \underline{2x^2-5x} \\ -14x+37 \\ \underline{-14x+35} \\ 2 \end{array}$$

111.

$$x^2 + x - 1 \overline{\smash{\big)}\ x^3 - 2x^2 - 4x + 3} \quad \Big\vert\ x - 3$$

$$\underline{x^3 + x^2 - x}$$
$$-3x^2 - 3x + 3$$
$$\underline{-3x^2 - 3x + 3}$$
$$0$$

113.

$$x^3 - 2 \overline{\smash{\big)}\ x^5 + 0x^4 - 2x^3 - 3x^2 + 0x + 9} \quad \Big\vert\ x^2 - 2 + \frac{-x^2+5}{x^3-2}$$

$$\underline{x^5 \qquad\qquad - 2x^2}$$
$$-2x^3 - x^2 + 0x + 9$$
$$\underline{-2x^3 \qquad\qquad + 4}$$
$$-x^2 \qquad + 5$$

115.

$$x - 2 \overline{\smash{\big)}\ x^5 + 0x^4 + 0x^3 + 0x^2 + 0x - 32} \quad \Big\vert\ x^4 + 2x^3 + 4x^2 + 8x + 16$$

$$\underline{x^5 - 2x^4}$$
$$2x^4 + 0x^3$$
$$\underline{2x^4 - 4x^3}$$
$$4x^3 + 0x^2$$
$$\underline{4x^3 - 8x^2}$$
$$8x^2 + 0x$$
$$\underline{8x^2 - 16x}$$
$$16x - 32$$
$$\underline{16x - 32}$$
$$0$$

117.

$$6x^2 + 11x - 10 \overline{\smash{\big)}\ 36x^4 + 72x^3 - 121x^2 - 142x + 120} \quad \Big\vert\ 6x^2 + x - 12$$

$$\underline{36x^4 + 66x^3 - 60x^2}$$
$$6x^3 - 61x^2 - 142x$$
$$\underline{6x^3 + 11x^2 - 10x}$$
$$-72x^2 - 132x + 120$$
$$\underline{-72x^2 - 132x + 120}$$
$$0$$

119. Area $=$ length $\cdot$ width $= (x+5)(x-2)$ ft^2 $= (x^2 - 2x + 5x - 10)$ ft^2 $= (x^2 + 3x - 10)$ ft^2

121. $(a+b+c)^2 = (a+b+c)(a+b+c) = a(a+b+c) + b(a+b+c) + c(a+b+c)$
$$= a^2 + ab + ac + ab + b^2 + bc + ac + bc + c^2$$
$$= a^2 + b^2 + c^2 + 2ab + 2bc + 2ac$$

123. Answers may vary.

125. Check the formula with $a = 1$ and $b = 2$.

127. $9^{3/2} = \left(9^{1/2}\right)^3 = 3^3 = 27$

129. $\left(\dfrac{625x^4}{16y^8}\right)^{3/4} = \dfrac{625^{3/4}(x^4)^{3/4}}{16^{3/4}(y^8)^{3/4}} = \dfrac{125x^3}{8y^6}$

131. $\sqrt[3]{16ab^4} - b\sqrt[3]{54ab} = \sqrt[3]{8b^3}\sqrt[3]{2ab} - b\sqrt[3]{27}\sqrt[3]{2ab} = 2b\sqrt[3]{2ab} - b(3)\sqrt[3]{2ab} = 2b\sqrt[3]{2ab} - 3b\sqrt[3]{2ab}$
$$= -b\sqrt[3]{2ab}$$

Exercise 0.5 (page 56)

1. factor

3. $ax + bx = x(a + b)$

5. $x^2 + 2xy + y^2 = (x + y)(x + y)$

7. $x^3 + y^3 = (x + y)(x^2 - xy + y^2)$

9. $3x - 6 = 3(x - 2)$

11. $8x^2 + 4x^3 = 4x^2(2 + x)$

13. $7x^2y^2 + 14x^3y^2 = 7x^2y^2(1 + 2x)$

15. $3a^2bc + 6ab^2c + 9abc^2 = 3abc(a + 2b + 3c)$

17. $a(x + y) + b(x + y) = (x + y)(a + b)$

19. $4a + b - 12a^2 - 3ab = 4a + b - 3a(4a + b) = 1(4a + b) - 3a(4a + b) = (4a + b)(1 - 3a)$

21. $3x^3 + 3x^2 - x - 1 = 3x^2(x + 1) - 1(x + 1) = (x + 1)(3x^2 - 1)$

23. $2txy + 2ctx - 3ty - 3ct = t(2xy + 2cx - 3y - 3c) = t[2x(y + c) - 3(y + c)] = t(y + c)(2x - 3)$

25. $ax + bx + ay + by + az + bz = x(a + b) + y(a + b) + z(a + b) = (a + b)(x + y + z)$

27. $4x^2 - 9 = (2x)^2 - 3^2 = (2x + 3)(2x - 3)$

29. $4 - 9r^2 = 2^2 - (3r)^2 = (2 + 3r)(2 - 3r)$

31. $(x + z)^2 - 25 = (x + z)^2 - 5^2$
$= (x + z + 5)(x + z - 5)$

33. $25x^4 + 1 = (5x^2)^2 + 1^2$
prime (sum of squares)

35. $x^2 - (y - z)^2 = [x + (y - z)][x - (y - z)] = (x + y - z)(x - y + z)$

37. $(x - y)^2 - (x + y)^2 = [(x - y) + (x + y)][(x - y) - (x + y)] = (x - y + x + y)(x - y - x - y)$
$= (2x)(-2y) = -4xy$

39. $x^4 - y^4 = (x^2)^2 - (y^2)^2 = (x^2 + y^2)(x^2 - y^2) = (x^2 + y^2)(x + y)(x - y)$

41. $3x^2 - 12 = 3(x^2 - 4) = 3(x + 2)(x - 2)$

43. $18xy^2 - 8x = 2x(9y^2 - 4)$
$= 2x(3y + 2)(3y - 2)$

45. $x^2 + 8x + 16 = (x + 4)(x + 4) = (x + 4)^2$

47. $b^2 - 10b + 25 = (b - 5)(b - 5) = (b - 5)^2$

49. $m^2 + 4mn + 4n^2 = (m + 2n)(m + 2n)$
$= (m + 2n)^2$

51. $x^2 + 10x + 21 = (x + 3)(x + 7)$

53. $x^2 - 4x - 12 = (x - 6)(x + 2)$

55. $x^2 - 2x + 15 \Rightarrow$ prime

57. $12x^2 - xy - 6y^2 = (4x - 3y)(3x + 2y)$

59. $-15 + 2a + 24a^2 = 24a^2 + 2a - 15$
$= (6a + 5)(4a - 3)$

13

61. $6x^2 + 29xy + 35y^2 = (3x + 7y)(2x + 5y)$

63. $12p^2 - 58pq - 70q^2 = 2(6p^2 - 29pq - 35q^2) = 2(6p - 35q)(p + q)$

65. $-6m^2 + 47mn - 35n^2 = -(6m^2 - 47mn + 35n^2) = -(6m - 5n)(m - 7n)$

67. $-6x^3 + 23x^2 + 35x = -x(6x^2 - 23x - 35) = -x(6x + 7)(x - 5)$

69. $6x^4 - 11x^3 - 35x^2 = x^2(6x^2 - 11x - 35) = x^2(2x - 7)(3x + 5)$

71. $x^4 + 2x^2 - 15 = (x^2 + 5)(x^2 - 3)$ **73.** $a^{2n} - 2a^n - 3 = (a^n - 3)(a^n + 1)$

75. $6x^{2n} - 7x^n + 2 = (3x^n - 2)(2x^n - 1)$ **77.** $4x^{2n} - 9y^{2n} = (2x^n)^2 - (3y^n)^2$
$$= (2x^n + 3y^n)(2x^n - 3y^n)$$

79. $10y^{2n} - 11y^n - 6 = (5y^n + 2)(2y^n - 3)$

81. $8z^3 - 27 = (2z)^3 - 3^3 = (2z - 3)\left[(2z)^2 + (2z)(3) + 3^2\right] = (2z - 3)(4z^2 + 6z + 9)$

83. $2x^3 + 2000 = 2(x^3 + 1000) = 2(x^3 + 10^3) = 2(x + 10)(x^2 - 10x + 100)$

85. $(x + y)^3 - 64 = (x + y)^3 - 4^3 = [(x + y) - 4]\left[(x + y)^2 + 4(x + y) + 4^2\right]$
$$= (x + y - 4)\left(x^2 + 2xy + y^2 + 4x + 4y + 16\right)$$

87. $64a^6 - y^6 = \left(8a^3\right)^2 - \left(y^3\right)^2 = \left(8a^3 + y^3\right)\left(8a^3 - y^3\right)$
$$= (2a + y)\left(4a^2 - 2ay + y^2\right)(2a - y)\left(4a^2 + 2ay + y^2\right)$$

89. $a^3 - b^3 + a - b = (a - b)(a^2 + ab + b^2) + (a - b)1 = (a - b)(a^2 + ab + b^2 + 1)$

91. $64x^6 + y^6 = \left(4x^2\right)^3 + \left(y^2\right)^3 = \left(4x^2 + y^2\right)\left(\left(4x^2\right)^2 - 4x^2y^2 + \left(y^2\right)^2\right)$
$$= \left(4x^2 + y^2\right)\left(16x^4 - 4x^2y^2 + y^4\right)$$

93. $x^2 - 6x + 9 - 144y^2 = (x - 3)(x - 3) - 144y^2 = (x - 3)^2 - (12y)^2$
$$= (x - 3 + 12y)(x - 3 - 12y)$$

95. $(a + b)^2 - 3(a + b) - 10 = [(a + b) - 5][(a + b) + 2] = (a + b - 5)(a + b + 2)$

97. $x^6 + 7x^3 - 8 = (x^3 + 8)(x^3 - 1) = (x + 2)(x^2 - 2x + 4)(x - 1)(x^2 + x + 1)$

99. $x^4 + 3x^2 + 4 = x^4 + 4x^2 + 4 - x^2$ **101.** $x^4 + 7x^2 + 16 = x^4 + 8x^2 + 16 - x^2$
$$= \left(x^2 + 2\right)\left(x^2 + 2\right) - x^2 \qquad\qquad = \left(x^2 + 4\right)\left(x^2 + 4\right) - x^2$$
$$= \left(x^2 + 2\right)^2 - x^2 \qquad\qquad\qquad = \left(x^2 + 4\right)^2 - x^2$$
$$= \left(x^2 + 2 + x\right)\left(x^2 + 2 - x\right) \qquad = \left(x^2 + 4 + x\right)\left(x^2 + 4 - x\right)$$

103. $4a^4 + 1 + 3a^2 = 4a^4 + 4a^2 + 1 - a^2 = (2a^2 + 1)(2a^2 + 1) - a^2 = (2a^2 + 1)^2 - a^2$
$$= (2a^2 + 1 + a)(2a^2 + 1 - a)$$

105. Answers may vary.

107. Answers may vary.

109. $3x + 2 = 2\left(\frac{3}{2}x + 1\right)$

111. $x^2 + 2x + 4 = 2\left(\frac{1}{2}x^2 + x + 2\right)$

113. $a + b = a\left(1 + \frac{b}{a}\right)$

115. $x + x^{1/2} = x^{1/2}\left(x^{1/2} + 1\right)$

117. $2x + \sqrt{2}y = \sqrt{2}\left(\sqrt{2}x + y\right)$

119. $ab^{3/2} - a^{3/2}b = ab\left(b^{1/2} - a^{1/2}\right)$

121. $x^2 + x - 6 + xy - 2y = (x + 3)(x - 2) + y(x - 2) = (x - 2)(x + 3 + y)$

123. $a^4 + 2a^3 + a^2 + a + 1 = a^2\left(a^2 + 2a + 1\right) + a + 1 = a^2(a + 1)(a + 1) + 1(a + 1)$
$$= (a + 1)\left[a^2(a + 1) + 1\right]$$
$$= (a + 1)\left(a^3 + a^2 + 1\right)$$

125. The number 1 is neither prime nor composite.

127. $(x^3x^2)^4 = (x^5)^4 = x^{20}$

129. $\left(\dfrac{3x^4x^3}{6x^{-2}x^4}\right)^0 = 1$

131. $\sqrt{20x} - \sqrt{125x} = \sqrt{4}\sqrt{5x} - \sqrt{25}\sqrt{5x} = 2\sqrt{5x} - 5\sqrt{5x} = -3\sqrt{5x}$

Exercise 0.6 (page 66)

1. numerator

3. $ad = bc$

5. $\dfrac{ac}{bd}$

7. $\dfrac{a + c}{b}$

9.
$$\frac{8x}{3y} \overset{?}{=} \frac{16x}{6y}$$
$$8x \cdot 6y \overset{?}{=} 3y \cdot 16x$$
$$48xy = 48xy$$
EQUAL

11.
$$\frac{25xyz}{12ab^2c} \overset{?}{=} \frac{50a^2bc}{24xyz}$$
$$25xyz \cdot 24xyz \overset{?}{=} 12ab^2c \cdot 50a^2bc$$
$$600x^2y^2z^2 \neq 600a^3b^3c^2$$
NOT EQUAL

13. $\dfrac{7a^2b}{21ab^2} = \dfrac{a \cdot 7ab}{3b \cdot 7ab} = \dfrac{a}{3b} \cdot \dfrac{7ab}{7ab} = \dfrac{a}{3b}$

15. $\dfrac{4x}{7} \cdot \dfrac{2}{5a} = \dfrac{4x \cdot 2}{7 \cdot 5a} = \dfrac{8x}{35a}$

17. $\dfrac{8m}{5n} \div \dfrac{3m}{10n} = \dfrac{8m}{5n} \cdot \dfrac{10n}{3m} = \dfrac{80mn}{15mn} = \dfrac{16}{3}$

19. $\dfrac{3z}{5c} + \dfrac{2z}{5c} = \dfrac{3z + 2z}{5c} = \dfrac{5z}{5c} = \dfrac{z}{c}$

21. $\dfrac{15x^2y}{7a^2b^3} - \dfrac{x^2y}{7a^2b^3} = \dfrac{14x^2y}{7a^2b^3} = \dfrac{2x^2y}{a^2b^3}$

23. $\dfrac{2x - 4}{x^2 - 4} = \dfrac{2(x - 2)}{(x + 2)(x - 2)} = \dfrac{2}{x + 2}$

25. $\dfrac{25 - x^2}{x^2 + 10x + 25} = \dfrac{(5 + x)(5 - x)}{(x + 5)(x + 5)} = \dfrac{5 - x}{x + 5} = -\dfrac{x - 5}{x + 5}$

27. $\dfrac{6x^3 + x^2 - 12x}{4x^3 + 4x^2 - 3x} = \dfrac{x(6x^2 + x - 12)}{x(4x^2 + 4x - 3)} = \dfrac{x(2x + 3)(3x - 4)}{x(2x + 3)(2x - 1)} = \dfrac{3x - 4}{2x - 1}$

29. $\dfrac{x^3 - 8}{x^2 + ax - 2x - 2a} = \dfrac{x^3 - 2^3}{x(x + a) - 2(x + a)} = \dfrac{(x - 2)(x^2 + 2x + 4)}{(x + a)(x - 2)} = \dfrac{x^2 + 2x + 4}{x + a}$

31. $\dfrac{x^2 - 1}{x} \cdot \dfrac{x^2}{x^2 + 2x + 1} = \dfrac{(x + 1)(x - 1)}{x} \cdot \dfrac{x^2}{(x + 1)(x + 1)} = \dfrac{x(x - 1)}{x + 1}$

33. $\dfrac{3x^2 + 7x + 2}{x^2 + 2x} \cdot \dfrac{x^2 - x}{3x^2 + x} = \dfrac{(3x + 1)(x + 2)}{x(x + 2)} \cdot \dfrac{x(x - 1)}{x(3x + 1)} = \dfrac{x - 1}{x}$

35. $\dfrac{x^2 + x}{x - 1} \cdot \dfrac{x^2 - 1}{x + 2} = \dfrac{x(x + 1)}{x - 1} \cdot \dfrac{(x + 1)(x - 1)}{x + 2} = \dfrac{x(x + 1)^2}{x + 2}$

37. $\dfrac{2x^2 + 32}{8} \div \dfrac{x^2 + 16}{2} = \dfrac{2x^2 + 32}{8} \cdot \dfrac{2}{x^2 + 16} = \dfrac{2(x^2 + 16)}{8} \cdot \dfrac{2}{x^2 + 16} = \dfrac{1}{2}$

39. $\dfrac{z^2 + z - 20}{z^2 - 4} \div \dfrac{z^2 - 25}{z - 5} = \dfrac{z^2 + z - 20}{z^2 - 4} \cdot \dfrac{z - 5}{z^2 - 25} = \dfrac{(z + 5)(z - 4)}{(z + 2)(z - 2)} \cdot \dfrac{z - 5}{(z + 5)(z - 5)}$
$$= \dfrac{z - 4}{(z + 2)(z - 2)}$$

41. $\dfrac{3x^2 + 5x - 2}{x^3 + 2x^2} \div \dfrac{6x^2 + 13x - 5}{2x^3 + 5x^2} = \dfrac{3x^2 + 5x - 2}{x^3 + 2x^2} \cdot \dfrac{2x^3 + 5x^2}{6x^2 + 13x - 5}$
$$= \dfrac{(3x - 1)(x + 2)}{x^2(x + 2)} \cdot \dfrac{x^2(2x + 5)}{(3x - 1)(2x + 5)} = 1$$

43. $\dfrac{x^2 + 7x + 12}{x^3 - x^2 - 6x} \cdot \dfrac{x^2 - 3x - 10}{x^2 + 2x - 3} \cdot \dfrac{x^3 - 4x^2 + 3x}{x^2 - x - 20}$
$$= \dfrac{(x + 3)(x + 4)}{x(x - 3)(x + 2)} \cdot \dfrac{(x - 5)(x + 2)}{(x + 3)(x - 1)} \cdot \dfrac{x(x - 3)(x - 1)}{(x - 5)(x + 4)} = 1$$

45. $\dfrac{x^3+27}{x^2-4} \div \left(\dfrac{x^2+4x+3}{x^2+2x} \div \dfrac{x^2+x-6}{x^2-3x+9} \right) = \dfrac{x^3+27}{x^2-4} \div \left(\dfrac{x^2+4x+3}{x^2+2x} \cdot \dfrac{x^2-3x+9}{x^2+x-6} \right)$

$$= \dfrac{x^3+27}{x^2-4} \div \left(\dfrac{(x+3)(x+1)}{x(x+2)} \cdot \dfrac{x^2-3x+9}{(x+3)(x-2)} \right)$$

$$= \dfrac{x^3+27}{x^2-4} \div \dfrac{(x+1)(x^2-3x+9)}{x(x+2)(x-2)}$$

$$= \dfrac{(x+3)(x^2-3x+9)}{(x+2)(x-2)} \cdot \dfrac{x(x+2)(x-2)}{(x+1)(x^2-3x+9)}$$

$$= \dfrac{x(x+3)}{x+1}$$

47. $\dfrac{3}{x+3} + \dfrac{x+2}{x+3} = \dfrac{3+x+2}{x+3} = \dfrac{x+5}{x+3}$ **49.** $\dfrac{4x}{x-1} - \dfrac{4}{x-1} = \dfrac{4x-4}{x-1} = \dfrac{4(x-1)}{x-1} = 4$

51. $\dfrac{2}{5-x} + \dfrac{1}{x-5} = \dfrac{-2}{x-5} + \dfrac{1}{x-5} = \dfrac{-1}{x-5}$

53. $\dfrac{3}{x+1} + \dfrac{2}{x-1} = \dfrac{3(x-1)}{(x+1)(x-1)} + \dfrac{2(x+1)}{(x-1)(x+1)} = \dfrac{3x-3}{(x+1)(x-1)} + \dfrac{2x+2}{(x-1)(x+1)}$

$$= \dfrac{5x-1}{(x+1)(x-1)}$$

55. $\dfrac{a+3}{a^2+7a+12} + \dfrac{a}{a^2-16} = \dfrac{a+3}{(a+3)(a+4)} + \dfrac{a}{(a+4)(a-4)}$

$$= \dfrac{1}{a+4} + \dfrac{a}{(a+4)(a-4)}$$

$$= \dfrac{1(a-4)}{(a+4)(a-4)} + \dfrac{a}{(a+4)(a-4)}$$

$$= \dfrac{a-4}{(a+4)(a-4)} + \dfrac{a}{(a+4)(a-4)} = \dfrac{2a-4}{(a+4)(a-4)}$$

57. $\dfrac{x}{x^2-4} - \dfrac{1}{x+2} = \dfrac{x}{(x+2)(x-2)} - \dfrac{1}{x+2} = \dfrac{x}{(x+2)(x-2)} - \dfrac{1(x-2)}{(x+2)(x-2)}$

$$= \dfrac{x}{(x+2)(x-2)} - \dfrac{x-2}{(x+2)(x-2)}$$

$$= \dfrac{2}{(x+2)(x-2)}$$

59. $\dfrac{3x-2}{x^2+2x+1} - \dfrac{x}{x^2-1} = \dfrac{3x-2}{(x+1)(x+1)} - \dfrac{x}{(x+1)(x-1)}$

$$= \dfrac{(3x-2)(x-1)}{(x+1)(x+1)(x-1)} - \dfrac{x(x+1)}{(x+1)(x-1)(x+1)}$$

$$= \dfrac{3x^2-5x+2}{(x+1)(x+1)(x-1)} - \dfrac{x^2+x}{(x+1)(x+1)(x-1)}$$

$$= \dfrac{2x^2-6x+2}{(x+1)(x+1)(x-1)} = \dfrac{2(x^2-3x+1)}{(x+1)^2(x-1)}$$

61. $\dfrac{2}{y^2-1} + 3 + \dfrac{1}{y+1} = \dfrac{2}{(y+1)(y-1)} + \dfrac{3}{1} + \dfrac{1}{y+1}$

$$= \dfrac{2}{(y+1)(y-1)} + \dfrac{3(y+1)(y-1)}{1(y+1)(y-1)} + \dfrac{1(y-1)}{(y+1)(y-1)}$$

$$= \dfrac{2}{(y+1)(y-1)} + \dfrac{3y^2-3}{(y+1)(y-1)} + \dfrac{y-1}{(y+1)(y-1)}$$

$$= \dfrac{3y^2+y-2}{(y+1)(y-1)} = \dfrac{(3y-2)(y+1)}{(y+1)(y-1)} = \dfrac{3y-2}{y-1}$$

63. $\dfrac{1}{x-2} + \dfrac{3}{x+2} - \dfrac{3x-2}{x^2-4} = \dfrac{1}{x-2} + \dfrac{3}{x+2} - \dfrac{3x-2}{(x+2)(x-2)}$

$$= \dfrac{1(x+2)}{(x-2)(x+2)} + \dfrac{3(x-2)}{(x+2)(x-2)} - \dfrac{3x-2}{(x+2)(x-2)}$$

$$= \dfrac{x+2}{(x+2)(x-2)} + \dfrac{3x-6}{(x+2)(x-2)} - \dfrac{3x-2}{(x+2)(x-2)}$$

$$= \dfrac{x-2}{(x+2)(x-2)} = \dfrac{1}{x+2}$$

65. $\left(\dfrac{1}{x-2} + \dfrac{1}{x-3}\right) \cdot \dfrac{x-3}{2x} = \left(\dfrac{1(x-3)}{(x-2)(x-3)} + \dfrac{1(x-2)}{(x-3)(x-2)}\right) \cdot \dfrac{x-3}{2x}$

$$= \left(\dfrac{x-3}{(x-2)(x-3)} + \dfrac{x-2}{(x-2)(x-3)}\right) \cdot \dfrac{x-3}{2x}$$

$$= \dfrac{2x-5}{(x-2)(x-3)} \cdot \dfrac{x-3}{2x} = \dfrac{2x-5}{2x(x-2)}$$

67.
$$\frac{3x}{x-4} - \frac{x}{x+4} - \frac{3x+1}{16-x^2} = \frac{3x}{x-4} - \frac{x}{x+4} - \frac{3x+1}{(4+x)(4-x)}$$
$$= \frac{3x}{x-4} - \frac{x}{x+4} + \frac{3x+1}{(x+4)(x-4)}$$
$$= \frac{3x(x+4)}{(x-4)(x+4)} - \frac{x(x-4)}{(x+4)(x-4)} + \frac{3x+1}{(x+4)(x-4)}$$
$$= \frac{3x^2+12x}{(x+4)(x-4)} - \frac{x^2-4x}{(x+4)(x-4)} + \frac{3x+1}{(x+4)(x-4)}$$
$$= \frac{2x^2+19x+1}{(x+4)(x-4)}$$

69.
$$\frac{1}{x^2+3x+2} - \frac{2}{x^2+4x+3} + \frac{1}{x^2+5x+6} \cdots$$
$$= \frac{1}{(x+2)(x+1)} - \frac{2}{(x+3)(x+1)} + \frac{1}{(x+2)(x+3)}$$
$$= \frac{1(x+3)}{(x+2)(x+1)(x+3)} - \frac{2(x+2)}{(x+3)(x+1)(x+2)} + \frac{1(x+1)}{(x+2)(x+3)(x+1)}$$
$$= \frac{x+3}{(x+2)(x+1)(x+3)} - \frac{2x+4}{(x+2)(x+1)(x+3)} + \frac{x+1}{(x+2)(x+1)(x+3)}$$
$$= \frac{x+3-2x-4+x+1}{(x+2)(x+1)(x+3)} = \frac{0}{(x+2)(x+1)(x+3)} = 0$$

71.
$$\frac{3x-2}{x^2+x-20} - \frac{4x^2+2}{x^2-25} + \frac{3x^2-25}{x^2-16} \cdots$$
$$= \frac{3x-2}{(x+5)(x-4)} - \frac{4x^2+2}{(x+5)(x-5)} + \frac{3x^2-25}{(x+4)(x-4)}$$
$$= \frac{(3x-2)(x-5)(x+4)}{(x+5)(x-4)(x-5)(x+4)} - \frac{(4x^2+2)(x-4)(x+4)}{(x+5)(x-5)(x-4)(x+4)} \cdots$$
$$+ \frac{(3x^2-25)(x+5)(x-5)}{(x+4)(x-4)(x+5)(x-5)}$$
$$= \frac{3x^3-5x^2-58x+40}{(x+5)(x-4)(x-5)(x+4)} - \frac{4x^4-62x^2-32}{(x+5)(x-4)(x-5)(x+4)} \cdots$$
$$+ \frac{3x^4-100x^2+625}{(x+5)(x-4)(x-5)(x+4)}$$
$$= \frac{3x^3-5x^2-58x+40-4x^4+62x^2+32+3x^4-100x^2+625}{(x+5)(x-4)(x-5)(x+4)}$$
$$= \frac{-x^4+3x^3-43x^2-58x+697}{(x+5)(x-4)(x-5)(x+4)}$$

73.
$$\frac{\frac{3a}{b}}{\frac{6ac}{b^2}} = \frac{3a}{b} \div \frac{6ac}{b^2} = \frac{3a}{b} \cdot \frac{b^2}{6ac} = \frac{b}{2c}$$

75.
$$\frac{3a^2b}{\frac{ab}{27}} = \frac{3a^2b}{1} \div \frac{ab}{27} = \frac{3a^2b}{1} \cdot \frac{27}{ab} = 81a$$

77. $\dfrac{\frac{x-y}{ab}}{\frac{y-x}{ab}} = \dfrac{x-y}{ab} \div \dfrac{y-x}{ab} = \dfrac{x-y}{ab} \cdot \dfrac{ab}{y-x}$

$\qquad = -1$

79. $\dfrac{\frac{1}{x} + \frac{1}{y}}{xy} = \dfrac{xy\left(\frac{1}{x} + \frac{1}{y}\right)}{xy(xy)} = \dfrac{xy\left(\frac{1}{x}\right) + xy\left(\frac{1}{y}\right)}{x^2 y^2}$

$\qquad = \dfrac{y+x}{x^2 y^2}$

81. $\dfrac{\frac{1}{x} + \frac{1}{y}}{\frac{1}{x} - \frac{1}{y}} = \dfrac{xy\left(\frac{1}{x} + \frac{1}{y}\right)}{xy\left(\frac{1}{x} - \frac{1}{y}\right)} = \dfrac{xy\left(\frac{1}{x}\right) + xy\left(\frac{1}{y}\right)}{xy\left(\frac{1}{x}\right) - xy\left(\frac{1}{y}\right)} = \dfrac{y+x}{y-x}$

83. $\dfrac{\frac{3a}{b} - \frac{4a^2}{x}}{\frac{1}{b} + \frac{1}{ax}} = \dfrac{abx\left(\frac{3a}{b} - \frac{4a^2}{x}\right)}{abx\left(\frac{1}{b} + \frac{1}{ax}\right)} = \dfrac{abx\left(\frac{3a}{b}\right) - abx\left(\frac{4a^2}{x}\right)}{abx\left(\frac{1}{b}\right) + abx\left(\frac{1}{ax}\right)} = \dfrac{3a^2 x - 4a^3 b}{ax + b} = \dfrac{a^2(3x - 4ab)}{ax + b}$

85. $\dfrac{x + 1 - \frac{6}{x}}{x + 5 + \frac{6}{x}} = \dfrac{x\left(x + 1 - \frac{6}{x}\right)}{x\left(x + 5 + \frac{6}{x}\right)} = \dfrac{x(x) + x(1) - x\left(\frac{6}{x}\right)}{x(x) + x(5) + x\left(\frac{6}{x}\right)} = \dfrac{x^2 + x - 6}{x^2 + 5x + 6} = \dfrac{(x+3)(x-2)}{(x+2)(x+3)} = \dfrac{x-2}{x+2}$

87. $\dfrac{3xy}{1 - \frac{1}{xy}} = \dfrac{xy(3xy)}{xy\left(1 - \frac{1}{xy}\right)} = \dfrac{3x^2 y^2}{xy(1) - xy\left(\frac{1}{xy}\right)}$

$\qquad = \dfrac{3x^2 y^2}{xy - 1}$

89. $\dfrac{3x}{x + \frac{1}{x}} = \dfrac{x(3x)}{x\left(x + \frac{1}{x}\right)} = \dfrac{3x^2}{x^2 + 1}$

91. $\dfrac{\frac{x}{x+2} - \frac{2}{x-1}}{\frac{3}{x+2} + \frac{x}{x-1}} = \dfrac{(x+2)(x-1)\left(\frac{x}{x+2} - \frac{2}{x-1}\right)}{(x+2)(x-1)\left(\frac{3}{x+2} + \frac{x}{x-1}\right)} = \dfrac{(x+2)(x-1)\left(\frac{x}{x+2}\right) - (x+2)(x-1)\left(\frac{2}{x-1}\right)}{(x+2)(x-1)\left(\frac{3}{x+2}\right) + (x+2)(x-1)\left(\frac{x}{x-1}\right)}$

$\qquad = \dfrac{(x-1)(x) - (x+2)(2)}{(x-1)(3) + (x+2)(x)}$

$\qquad = \dfrac{x^2 - x - 2x - 4}{3x - 3 + x^2 + 2x} = \dfrac{x^2 - 3x - 4}{x^2 + 5x - 3}$

93. $\dfrac{1}{1 + x^{-1}} = \dfrac{1}{1 + \frac{1}{x}} = \dfrac{x(1)}{x\left(1 + \frac{1}{x}\right)} = \dfrac{x}{x+1}$

95. $\dfrac{3(x+2)^{-1} + 2(x-1)^{-1}}{(x+2)^{-1}} = \dfrac{\frac{3}{x+2} + \frac{2}{x-1}}{\frac{1}{x+2}} = \dfrac{(x+2)(x-1)\left(\frac{3}{x+2} + \frac{2}{x-1}\right)}{(x+2)(x-1)\left(\frac{1}{x+2}\right)}$

$\qquad = \dfrac{(x+2)(x-1)\left(\frac{3}{x+2}\right) + (x+2)(x-1)\left(\frac{2}{x-1}\right)}{x-1}$

$\qquad = \dfrac{(x-1)(3) + (x+2)(2)}{x-1}$

$\qquad = \dfrac{3x - 3 + 2x + 4}{x-1} = \dfrac{5x + 1}{x-1}$

97. $\dfrac{1}{\frac{1}{k_1}+\frac{1}{k_2}} = \dfrac{k_1 k_2 (1)}{k_1 k_2 \left(\frac{1}{k_1}+\frac{1}{k_2}\right)} = \dfrac{k_1 k_2}{k_1 k_2 \left(\frac{1}{k_1}\right)+k_1 k_2 \left(\frac{1}{k_2}\right)} = \dfrac{k_1 k_2}{k_2 + k_1}$

99. $\dfrac{x}{1+\frac{1}{3x^{-1}}} = \dfrac{x}{1+\frac{1}{\frac{3}{x}}} = \dfrac{x}{1+\frac{x(1)}{x\left(\frac{3}{x}\right)}} = \dfrac{x}{1+\frac{x}{3}} = \dfrac{3x}{3\left(1+\frac{x}{3}\right)} = \dfrac{3x}{3+x}$

101. $\dfrac{1}{1+\frac{1}{1+\frac{1}{x}}} = \dfrac{1}{1+\frac{x(1)}{x\left(1+\frac{1}{x}\right)}} = \dfrac{1}{1+\frac{x}{x+1}} = \dfrac{(x+1)1}{(x+1)\left(1+\frac{x}{x+1}\right)} = \dfrac{x+1}{x+1+x} = \dfrac{x+1}{2x+1}$

103. Answers may vary.   **105. Answers may vary.**   **107.** Since $-6 < 0$,
$|-6| = -(-6) = 6.$

109. $\left(\dfrac{x^3 y^{-2}}{x^{-1}y}\right)^{-3} = \left(\dfrac{x^{-1}y}{x^3 y^{-2}}\right)^3 = \left(\dfrac{y^2 y}{x^3 x}\right)^3 = \left(\dfrac{y^3}{x^4}\right)^3 = \dfrac{y^9}{x^{12}}$

111. $\sqrt{20} - \sqrt{45} = \sqrt{4}\sqrt{5} - \sqrt{9}\sqrt{5} = 2\sqrt{5} - 3\sqrt{5} = -\sqrt{5}$

Chapter 0 Summary   (page 71)

1. **a.** natural: $3, 6, 8$ **b.** whole: $0, 3, 6, 8$

c. integers: $-6, -3, 0, 3, 6, 8$ **d.** rational: $-6, -3, 0, \frac{1}{2}, 3, 6, 8$

e. irrational: $\pi, \sqrt{5}$ **f.** real: $-6, -3, 0, \frac{1}{2}, 3, \pi, \sqrt{5}, 6, 8$

2. **a.** prime: 3 **b.** composite: $6, 8$

c. even integers: $-6, 0, 6, 8$ **d.** odd integers: $-3, 3$

3. **a.** associative property of addition   **b.** commutative property of addition

c. associative property of multiplication   **d.** distributive property

e. commutative property of multiplication   **f.** commutative property of addition

g. double negative rule

4. **a.**

b.

5. **a.** $-3 < x \le 5$

b. $x \ge 0$ or $x < -1$

c. $(-2, 4]$

d. $(-\infty, 2) \cap (-5, \infty)$

$(-\infty, 2)$

$(-5, \infty)$

$(-\infty, 2) \cap (-5, \infty)$

e. $(-\infty, -4) \cup [6, \infty)$

6. **a.** Since $6 \geq 0$, $|6| = 6$. **b.** Since $-25 < 0$, $|-25| = -(-25) = 25$.

c. Since $1 - \sqrt{2} < 0$, **d.** Since $\sqrt{3} - 1 \geq 0$,
$$\left|1 - \sqrt{2}\right| = -\left(1 - \sqrt{2}\right) = \sqrt{2} - 1. \qquad \left|\sqrt{3} - 1\right| = \sqrt{3} - 1.$$

7. distance $= |-5 - 7| = |-12| = 12$

8. **a.** $-5a^3 = -5aaa$ **b.** $(-5a)^2 = (-5)(-5)aa = 25aa$

9. **a.** $3ttt = 3t^3$ **b.** $(-2b)(3b) = (-2)(3)bb = -6b^2$

10. **a.** $n^2 n^4 = n^{2+4} = n^6$ **b.** $\left(p^3\right)^2 = p^{3\cdot 2} = p^6$

c. $\left(x^3 y^2\right)^4 = \left(x^3\right)^4 \left(y^2\right)^4 = x^{12} y^8$ **d.** $\left(\dfrac{a^4}{b^2}\right)^3 = \dfrac{\left(a^4\right)^3}{\left(b^2\right)^3} = \dfrac{a^{12}}{b^6}$

e. $\left(m^{-3} n^0\right)^2 = \left(m^{-3}\right)^2 \left(n^0\right)^2 = m^{-6} n^0 = m^{-6} = \dfrac{1}{m^6}$

f. $\left(\dfrac{p^{-2} q^2}{2}\right)^3 = \left(\dfrac{q^2}{2p^2}\right)^3 = \dfrac{\left(q^2\right)^3}{\left(2p^2\right)^3} = \dfrac{q^6}{8p^6}$

g. $\dfrac{a^5}{a^8} = a^{5-8} = a^{-3} = \dfrac{1}{a^3}$ **h.** $\left(\dfrac{a^2}{b^3}\right)^{-2} = \left(\dfrac{b^3}{a^2}\right)^2 = \dfrac{b^6}{a^4}$

i. $\left(\dfrac{3x^2y^{-2}}{x^2y^2}\right)^{-2} = \left(\dfrac{x^2y^2}{3x^2y^{-2}}\right)^2 = \left(\dfrac{x^2y^2y^2}{3x^2}\right)^2 = \left(\dfrac{y^4}{3}\right)^2 = \dfrac{y^8}{9}$

j. $\left(\dfrac{a^{-3}b^2}{ab^{-3}}\right)^{-2} = \left(\dfrac{ab^{-3}}{a^{-3}b^2}\right)^2 = \left(\dfrac{aa^3}{b^2b^3}\right)^2 = \left(\dfrac{a^4}{b^5}\right)^2 = \dfrac{a^8}{b^{10}}$

k. $\left(\dfrac{-3x^3y}{xy^3}\right)^{-2} = \left(\dfrac{xy^3}{-3x^3y}\right)^2 = \left(\dfrac{y^2}{-3x^2}\right)^2 = \dfrac{y^4}{9x^4}$

l. $\left(-\dfrac{2m^{-2}n^0}{4m^2n^{-1}}\right)^{-3} = \left(-\dfrac{4m^2n^{-1}}{2m^{-2}n^0}\right)^3 = \left(-\dfrac{2m^2m^2}{n^1n^0}\right)^3 = \left(-\dfrac{2m^4}{n}\right)^3 = -\dfrac{8m^{12}}{n^3}$

11. $-x^2 - xy^2 = -(-3)^2 - (-3)(3)^2 = -(+9) - (-3)(9) = -9 - (-27) = -9 + 27 = 18$

12. a. $6750 = 6.750 \times 10^3$ **b.** $0.00023 = 2.3 \times 10^{-4}$

13. a. $4.8 \times 10^2 = 480$ **b.** $0.25 \times 10^{-3} = 0.00025$

14. $\dfrac{(45{,}000)(350{,}000)}{0.000105} = \dfrac{(4.5 \times 10^4)(3.5 \times 10^5)}{1.05 \times 10^{-4}} = \dfrac{4.5 \times 3.5 \times 10^4 \times 10^5}{1.05 \times 10^{-4}} = \dfrac{15.75 \times 10^9}{1.05 \times 10^{-4}}$

$$= 15 \times 10^{13}$$
$$= 1.5 \times 10^{14}$$

15. a. $121^{1/2} = (11^2)^{1/2} = 11$ **b.** $\left(\dfrac{27}{125}\right)^{1/3} = \left[\left(\dfrac{3}{5}\right)^3\right]^{1/3} = \dfrac{3}{5}$

c. $(32x^5)^{1/5} = 32^{1/5}(x^5)^{1/5} = 2x$ **d.** $(81a^4)^{1/4} = 81^{1/4}(a^4)^{1/4} = 3|a|$

e. $(-1000x^6)^{1/3} = (-1000)^{1/3}(x^6)^{1/3} = -10x^2$

f. $(-25x^2)^{1/2} = (-25)^{1/2}(x^2)^{1/2} \Rightarrow$ not a real number

g. $(x^{12}y^2)^{1/2} = (x^{12})^{1/2}(y^2)^{1/2} = x^6|y|$ **h.** $\left(\dfrac{x^{12}}{y^4}\right)^{-1/2} = \left(\dfrac{y^4}{x^{12}}\right)^{1/2} = \dfrac{y^2}{x^6}$

i. $\left(\dfrac{-c^{2/3}c^{5/3}}{c^{-2/3}}\right)^{1/3} = \left(\dfrac{-c^{7/3}}{c^{-2/3}}\right)^{1/3} = (-c^{9/3})^{1/3} = (-c^3)^{1/3} = -c$

j. $\left(\dfrac{a^{-1/4}a^{3/4}}{a^{9/2}}\right)^{-1/2} = \left(\dfrac{a^{9/2}}{a^{-1/4}a^{3/4}}\right)^{1/2} = \left(\dfrac{a^{9/2}}{a^{2/4}}\right)^{1/2} = \left(\dfrac{a^{9/2}}{a^{1/2}}\right)^{1/2} = \left(a^{8/2}\right)^{1/2}$

$$= \left(a^4\right)^{1/2} = a^2$$

16. **a.** $64^{2/3} = \left(64^{1/3}\right)^2 = 4^2 = 16$

b. $32^{-3/5} = \dfrac{1}{32^{3/5}} = \dfrac{1}{\left(32^{1/5}\right)^3} = \dfrac{1}{2^3} = \dfrac{1}{8}$

c. $\left(\dfrac{16}{81}\right)^{3/4} = \dfrac{16^{3/4}}{81^{3/4}} = \dfrac{\left(16^{1/4}\right)^3}{\left(81^{1/4}\right)^3} = \dfrac{2^3}{3^3} = \dfrac{8}{27}$

d. $\left(\dfrac{32}{243}\right)^{2/5} = \dfrac{32^{2/5}}{243^{2/5}} = \dfrac{\left(32^{1/5}\right)^2}{\left(243^{1/5}\right)^2} = \dfrac{2^2}{3^2} = \dfrac{4}{9}$

e. $\left(\dfrac{8}{27}\right)^{-2/3} = \left(\dfrac{27}{8}\right)^{2/3} = \dfrac{27^{2/3}}{8^{2/3}} = \dfrac{\left(27^{1/3}\right)^2}{\left(8^{1/3}\right)^2} = \dfrac{3^2}{2^2} = \dfrac{9}{4}$

f. $\left(\dfrac{16}{625}\right)^{-3/4} = \left(\dfrac{625}{16}\right)^{3/4} = \dfrac{625^{3/4}}{16^{3/4}} = \dfrac{\left(625^{1/4}\right)^3}{\left(16^{1/4}\right)^3} = \dfrac{5^3}{2^3} = \dfrac{125}{8}$

g. $\left(-216x^3\right)^{2/3} = (-216)^{2/3}\left(x^3\right)^{2/3} = 36x^2$

h. $\dfrac{p^{a/2}p^{a/3}}{p^{a/6}} = \dfrac{p^{3a/6}p^{2a/6}}{p^{a/6}} = \dfrac{p^{5a/6}}{p^{a/6}} = p^{4a/6} = p^{2a/3}$

17. **a.** $\sqrt{36} = 6$

b. $-\sqrt{49} = -7$

c. $\sqrt{\dfrac{9}{25}} = \dfrac{\sqrt{9}}{\sqrt{25}} = \dfrac{3}{5}$

d. $\sqrt[3]{\dfrac{27}{125}} = \dfrac{\sqrt[3]{27}}{\sqrt[3]{125}} = \dfrac{3}{5}$

e. $\sqrt{x^2y^4} = \sqrt{x^2}\sqrt{y^4} = |x|y^2$

f. $\sqrt[3]{x^3} = x$

g. $\sqrt[4]{\dfrac{m^8n^4}{p^{16}}} = \dfrac{\sqrt[4]{m^8}\sqrt[4]{n^4}}{\sqrt[4]{p^{16}}} = \dfrac{m^2|n|}{p^4}$

h. $\sqrt[5]{\dfrac{a^{15}b^{10}}{c^5}} = \dfrac{\sqrt[5]{a^{15}}\sqrt[5]{b^{10}}}{\sqrt[5]{c^5}} = \dfrac{a^3b^2}{c}$

18. **a.** $\sqrt{50} + \sqrt{8} = \sqrt{25}\sqrt{2} + \sqrt{4}\sqrt{2} = 5\sqrt{2} + 2\sqrt{2} = 7\sqrt{2}$

b. $\sqrt{12} + \sqrt{3} - \sqrt{27} = \sqrt{4}\sqrt{3} + \sqrt{3} - \sqrt{9}\sqrt{3} = 2\sqrt{3} + \sqrt{3} - 3\sqrt{3} = 3\sqrt{3} - 3\sqrt{3} = 0$

c. $\sqrt[3]{24x^4} - \sqrt[3]{3x^4} = \sqrt[3]{8x^3}\sqrt[3]{3x} - \sqrt[3]{x^3}\sqrt[3]{3x} = 2x\sqrt[3]{3x} - x\sqrt[3]{3x} = x\sqrt[3]{3x}$

19. **a.** $\dfrac{\sqrt{7}}{\sqrt{5}} = \dfrac{\sqrt{7}}{\sqrt{5}} \cdot \dfrac{\sqrt{5}}{\sqrt{5}} = \dfrac{\sqrt{35}}{5}$

b. $\dfrac{8}{\sqrt{8}} = \dfrac{8}{\sqrt{8}} \cdot \dfrac{\sqrt{2}}{\sqrt{2}} = \dfrac{8\sqrt{2}}{\sqrt{16}} = \dfrac{8\sqrt{2}}{4}$
$= 2\sqrt{2}$

c. $\dfrac{1}{\sqrt[3]{2}} = \dfrac{1}{\sqrt[3]{2}} \cdot \dfrac{\sqrt[3]{4}}{\sqrt[3]{4}} = \dfrac{\sqrt[3]{4}}{\sqrt[3]{8}} = \dfrac{\sqrt[3]{4}}{2}$

d. $\dfrac{2}{\sqrt[3]{25}} = \dfrac{2}{\sqrt[3]{25}} \cdot \dfrac{\sqrt[3]{5}}{\sqrt[3]{5}} = \dfrac{2\sqrt[3]{5}}{\sqrt[3]{125}} = \dfrac{2\sqrt[3]{5}}{5}$

20. **a.** $\dfrac{\sqrt{2}}{5} = \dfrac{\sqrt{2}}{5} \cdot \dfrac{\sqrt{2}}{\sqrt{2}} = \dfrac{2}{5\sqrt{2}}$

b. $\dfrac{\sqrt{5}}{5} = \dfrac{\sqrt{5}}{5} \cdot \dfrac{\sqrt{5}}{\sqrt{5}} = \dfrac{5}{5\sqrt{5}} = \dfrac{1}{\sqrt{5}}$

c. $\dfrac{\sqrt{2x}}{3} = \dfrac{\sqrt{2x}}{3} \cdot \dfrac{\sqrt{2x}}{\sqrt{2x}} = \dfrac{2x}{3\sqrt{2x}}$

d. $\dfrac{3\sqrt[3]{7x}}{2} = \dfrac{3\sqrt[3]{7x}}{2} \cdot \dfrac{\sqrt[3]{49x^2}}{\sqrt[3]{49x^2}} = \dfrac{3\sqrt[3]{343x^3}}{2\sqrt[3]{49x^2}}$

$$= \dfrac{21x}{2\sqrt[3]{49x^2}}$$

21. **a.** 3rd degree, binomial

b. 2nd degree, trinomial

c. 2nd degree, monomial

d. 4th degree, trinomial

22. **a.** $2(x+3) + 3(x-4) = 2x + 6 + 3x - 12 = 5x - 6$

b. $3x^2(x-1) - 2x(x+3) - x^2(x+2) = 3x^3 - 3x^2 - 2x^2 - 6x - x^3 - 2x^2 = 2x^3 - 7x^2 - 6x$

c. $(3x+2)(3x+2) = 9x^2 + 6x + 6x + 4 = 9x^2 + 12x + 4$

d. $(3x+y)(2x-3y) = 6x^2 - 9xy + 2xy - 3y^2 = 6x^2 - 7xy - 3y^2$

e. $(4a+2b)(2a-3b) = 8a^2 - 12ab + 4ab - 6b^2 = 8a^2 - 8ab - 6b^2$

f. $(z+3)(3z^2 + z - 1) = 3z^3 + z^2 - z + 9z^2 + 3z - 3 = 3z^3 + 10z^2 + 2z - 3$

g. $(a^n + 2)(a^n - 1) = a^{2n} - a^n + 2a^n - 2 = a^{2n} + a^n - 2$

h. $\left(\sqrt{2}+x\right)^2 = \left(\sqrt{2}+x\right)\left(\sqrt{2}+x\right) = \left(\sqrt{2}\right)^2 + x\sqrt{2} + x\sqrt{2} + x^2 = 2 + 2x\sqrt{2} + x^2$

i. $\left(\sqrt{2}+1\right)\left(\sqrt{3}+1\right) = \sqrt{6} + \sqrt{2} + \sqrt{3} + 1$

j. $\left(\sqrt[3]{3}-2\right)\left(\sqrt[3]{9} + 2\sqrt[3]{3} + 4\right) = \sqrt[3]{27} + 2\sqrt[3]{9} + 4\sqrt[3]{3} - 2\sqrt[3]{9} - 4\sqrt[3]{3} - 8 = 3 - 8 = -5$

23. **a.** $\dfrac{2}{\sqrt{3}-1} = \dfrac{2}{\sqrt{3}-1} \cdot \dfrac{\sqrt{3}+1}{\sqrt{3}+1} = \dfrac{2\left(\sqrt{3}+1\right)}{\left(\sqrt{3}\right)^2 - 1^2} = \dfrac{2\left(\sqrt{3}+1\right)}{3-1} = \dfrac{2\left(\sqrt{3}+1\right)}{2} = \sqrt{3}+1$

b. $\dfrac{-2}{\sqrt{3}-\sqrt{2}} = \dfrac{-2}{\sqrt{3}-\sqrt{2}} \cdot \dfrac{\sqrt{3}+\sqrt{2}}{\sqrt{3}+\sqrt{2}} = \dfrac{-2\left(\sqrt{3}+\sqrt{2}\right)}{\left(\sqrt{3}\right)^2 - \left(\sqrt{2}\right)^2} = \dfrac{-2\left(\sqrt{3}+\sqrt{2}\right)}{3-2}$

$$= \dfrac{-2\left(\sqrt{3}+\sqrt{2}\right)}{1}$$

$$= -2\left(\sqrt{3}+\sqrt{2}\right)$$

c. $\dfrac{2x}{\sqrt{x}-2} = \dfrac{2x}{\sqrt{x}-2} \cdot \dfrac{\sqrt{x}+2}{\sqrt{x}+2} = \dfrac{2x(\sqrt{x}+2)}{(\sqrt{x})^2 - 2^2} = \dfrac{2x(\sqrt{x}+2)}{x-4}$

d. $\dfrac{\sqrt{x}-\sqrt{y}}{\sqrt{x}+\sqrt{y}} = \dfrac{\sqrt{x}-\sqrt{y}}{\sqrt{x}+\sqrt{y}} \cdot \dfrac{\sqrt{x}-\sqrt{y}}{\sqrt{x}-\sqrt{y}} = \dfrac{\sqrt{x^2} - \sqrt{xy} - \sqrt{xy} + y}{(\sqrt{x})^2 - (\sqrt{y})^2} = \dfrac{x - 2\sqrt{xy} + y}{x-y}$

24. a. $\dfrac{\sqrt{x}+2}{5} = \dfrac{\sqrt{x}+2}{5} \cdot \dfrac{\sqrt{x}-2}{\sqrt{x}-2} = \dfrac{(\sqrt{x})^2 - 2^2}{5(\sqrt{x}-2)} = \dfrac{x-4}{5(\sqrt{x}-2)}$

b. $\dfrac{1-\sqrt{a}}{a} = \dfrac{1-\sqrt{a}}{a} \cdot \dfrac{1+\sqrt{a}}{1+\sqrt{a}} = \dfrac{1^2 - (\sqrt{a})^2}{a(1+\sqrt{a})} = \dfrac{1-a}{a(1+\sqrt{a})}$

25. a. $\dfrac{3x^2 y^2}{6x^3 y} = \dfrac{y}{2x}$

b. $\dfrac{4a^2 b^3 + 6ab^4}{2b^2} = \dfrac{4a^2 b^3}{2b^2} + \dfrac{6ab^4}{2b^2}$
$$= 2a^2 b + 3ab^2$$

c.
$$
\begin{array}{r}
x^2 + 2x + 1 \\
2x+3\overline{\smash{\big)}\,2x^3 + 7x^2 + 8x + 3} \\
\underline{2x^3 + 3x^2} \\
4x^2 + 8x \\
\underline{4x^2 + 6x} \\
2x + 3 \\
\underline{2x + 3} \\
0
\end{array}
$$

d.
$$
\begin{array}{r}
x^3 + 2x - 3 + \frac{-6}{x^2-1} \\
x^2-1\overline{\smash{\big)}\,x^5 + 0x^4 + x^3 - 3x^2 - 2x - 3} \\
\underline{x^5 - x^3} \\
2x^3 - 3x^2 - 2x \\
\underline{2x^3 - 2x} \\
-3x^2 - 3 \\
\underline{-3x^2 + 3} \\
-6
\end{array}
$$

26. a. $3t^3 - 3t = 3t(t^2 - 1) = 3t(t+1)(t-1)$

b. $5r^3 - 5 = 5(r^3 - 1) = 5(r^3 - 1^3) = 5(r-1)(r^2 + r + 1)$

c. $6x^2 + 7x - 24 = (3x + 8)(2x - 3)$

d. $3a^2 + ax - 3a - x = a(3a + x) - 1(3a + x) = (3a + x)(a - 1)$

e. $8x^3 - 125 = (2x)^3 - 5^3 = (2x - 5)\big[(2x)^2 + (2x)(5) + 5^2\big] = (2x - 5)(4x^2 + 10x + 25)$

f. $6x^2 - 20x - 16 = 2(3x^2 - 10x - 8) = 2(3x + 2)(x - 4)$

g. $x^2 + 6x + 9 - t^2 = (x+3)(x+3) - t^2 = (x+3)^2 - t^2 = (x + 3 + t)(x + 3 - t)$

h. $3x^2 - 1 + 5x = 3x^2 + 5x - 1 \Rightarrow$ prime

i. $8z^3 + 343 = (2z)^3 + 7^3 = (2z + 7)\big[(2z)^2 - (2z)(7) + 7^2\big] = (2z + 7)(4z^2 - 14z + 49)$

j. $1 + 14b + 49b^2 = 49b^2 + 14b + 1 = (7b + 1)(7b + 1) = (7b + 1)^2$

k. $121z^2 + 4 - 44z = 121z^2 - 44z + 4 = (11z - 2)(11z - 2) = (11z - 2)^2$

l. $64y^3 - 1000 = 8(8y^3 - 125) = 8[(2y)^3 - 5^3] = 8(2y - 5)(4y^2 + 10y + 25)$

m. $2xy - 4zx - wy + 2zw = 2x(y - 2z) - w(y - 2z) = (y - 2z)(2x - w)$

n. $x^8 + x^4 + 1 = x^8 + 2x^4 + 1 - x^4 = (x^4 + 1)(x^4 + 1) - x^4$
$$= (x^4 + 1)^2 - (x^2)^2$$
$$= (x^4 + 1 + x^2)(x^4 + 1 - x^2)$$
$$= (x^4 + 2x^2 + 1 - x^2)(x^4 - x^2 + 1)$$
$$= [(x^2 + 1)(x^2 + 1) - x^2](x^4 - x^2 + 1)$$
$$= (x^2 + 1 + x)(x^2 + 1 - x)(x^4 - x^2 + 1)$$

27. **a.** $\dfrac{x^2 - 4x + 4}{x + 2} \cdot \dfrac{x^2 + 5x + 6}{x - 2} = \dfrac{(x - 2)(x - 2)}{x + 2} \cdot \dfrac{(x + 2)(x + 3)}{x - 2} = (x - 2)(x + 3)$

b. $\dfrac{2y^2 - 11y + 15}{y^2 - 6y + 8} \cdot \dfrac{y^2 - 2y - 8}{y^2 - y - 6} = \dfrac{(2y - 5)(y - 3)}{(y - 4)(y - 2)} \cdot \dfrac{(y - 4)(y + 2)}{(y - 3)(y + 2)} = \dfrac{2y - 5}{y - 2}$

c. $\dfrac{2t^2 + t - 3}{3t^2 - 7t + 4} \div \dfrac{10t + 15}{3t^2 - t - 4} = \dfrac{2t^2 + t - 3}{3t^2 - 7t + 4} \cdot \dfrac{3t^2 - t - 4}{10t + 15}$
$$= \dfrac{(2t + 3)(t - 1)}{(3t - 4)(t - 1)} \cdot \dfrac{(3t - 4)(t + 1)}{5(2t + 3)} = \dfrac{t + 1}{5}$$

d. $\dfrac{p^2 + 7p + 12}{p^3 + 8p^2 + 4p} \div \dfrac{p^2 - 9}{p^2} = \dfrac{p^2 + 7p + 12}{p^3 + 8p^2 + 4p} \cdot \dfrac{p^2}{p^2 - 9} = \dfrac{(p + 3)(p + 4)}{p(p^2 + 8p + 4)} \cdot \dfrac{p^2}{(p + 3)(p - 3)}$
$$= \dfrac{p(p + 4)}{(p^2 + 8p + 4)(p - 3)}$$

e. $\dfrac{x^2 + x - 6}{x^2 - x - 6} \cdot \dfrac{x^2 - x - 6}{x^2 + x - 2} \div \dfrac{x^2 - 4}{x^2 - 5x + 6} \cdots$
$$= \dfrac{x^2 + x - 6}{x^2 - x - 6} \cdot \dfrac{x^2 - x - 6}{x^2 + x - 2} \cdot \dfrac{x^2 - 5x + 6}{x^2 - 4}$$
$$= \dfrac{(x + 3)(x - 2)}{(x - 3)(x + 2)} \cdot \dfrac{(x - 3)(x + 2)}{(x + 2)(x - 1)} \cdot \dfrac{(x - 2)(x - 3)}{(x + 2)(x - 2)} = \dfrac{(x + 3)(x - 2)(x - 3)}{(x + 2)^2(x - 1)}$$

f. $\left(\dfrac{2x + 6}{x + 5} \div \dfrac{2x^2 - 2x - 4}{x^2 - 25} \right) \dfrac{x^2 - x - 2}{x^2 - 2x - 15} = \dfrac{2x + 6}{x + 5} \cdot \dfrac{x^2 - 25}{2(x^2 - x - 2)} \cdot \dfrac{x^2 - x - 2}{x^2 - 2x - 15}$
$$= \dfrac{2(x + 3)}{x + 5} \cdot \dfrac{(x + 5)(x - 5)}{2(x - 2)(x + 1)} \cdot \dfrac{(x - 2)(x + 1)}{(x - 5)(x + 3)}$$
$$= 1$$

g. $\dfrac{2}{x-4} + \dfrac{3x}{x+5} = \dfrac{2(x+5)}{(x-4)(x+5)} + \dfrac{3x(x-4)}{(x+5)(x-4)} = \dfrac{2x+10}{(x-4)(x+5)} + \dfrac{3x^2-12x}{(x-4)(x+5)}$

$$= \dfrac{3x^2-10x+10}{(x-4)(x+5)}$$

h. $\dfrac{5x}{x-2} - \dfrac{3x+7}{x+2} + \dfrac{2x+1}{x+2} = \dfrac{5x}{x-2} + \dfrac{-x-6}{x+2}$

$$= \dfrac{5x(x+2)}{(x-2)(x+2)} + \dfrac{(-x-6)(x-2)}{(x+2)(x-2)}$$

$$= \dfrac{5x^2+10x}{(x-2)(x+2)} + \dfrac{-x^2-4x+12}{(x-2)(x+2)}$$

$$= \dfrac{4x^2+6x+12}{(x-2)(x+2)} = \dfrac{2(2x^2+3x+6)}{(x-2)(x+2)}$$

i. $\dfrac{x}{x-1} + \dfrac{x}{x-2} + \dfrac{x}{x-3}$

$$= \dfrac{x(x-2)(x-3)}{(x-1)(x-2)(x-3)} + \dfrac{x(x-1)(x-3)}{(x-2)(x-1)(x-3)} + \dfrac{x(x-1)(x-2)}{(x-3)(x-1)(x-2)}$$

$$= \dfrac{x^3-5x^2+6x}{(x-1)(x-2)(x-3)} + \dfrac{x^3-4x^2+3x}{(x-1)(x-2)(x-3)} + \dfrac{x^3-3x^2+2x}{(x-1)(x-2)(x-3)}$$

$$= \dfrac{3x^3-12x^2+11x}{(x-1)(x-2)(x-3)} = \dfrac{x(3x^2-12x+11)}{(x-1)(x-2)(x-3)}$$

j. $\dfrac{x}{x+1} - \dfrac{3x+7}{x+2} + \dfrac{2x+1}{x+2} = \dfrac{x}{x+1} + \dfrac{-3x-7}{x+2} + \dfrac{2x+1}{x+2}$

$$= \dfrac{x}{x+1} + \dfrac{-x-6}{x+2}$$

$$= \dfrac{x(x+2)}{(x+1)(x+2)} + \dfrac{(-x-6)(x+1)}{(x+2)(x+1)}$$

$$= \dfrac{x^2+2x}{(x+1)(x+2)} + \dfrac{-x^2-7x-6}{(x+1)(x+2)} = \dfrac{-5x-6}{(x+1)(x+2)}$$

k. $\dfrac{3(x+1)}{x} - \dfrac{5(x^2+3)}{x^2} + \dfrac{x}{x+1} = \dfrac{3x(x+1)(x+1)}{x^2(x+1)} - \dfrac{5(x^2+3)(x+1)}{x^2(x+1)} + \dfrac{x(x^2)}{x^2(x+1)}$

$$= \dfrac{3x^3+6x^2+3x}{x^2(x+1)} - \dfrac{5x^3+5x^2+15x+15}{x^2(x+1)} + \dfrac{x^3}{x^2(x+1)}$$

$$= \dfrac{-x^3+x^2-12x-15}{x^2(x+1)}$$

l. $\dfrac{3x}{x+1} + \dfrac{x^2+4x+3}{x^2+3x+2} - \dfrac{x^2+x-6}{x^2-4} = \dfrac{3x}{x+1} + \dfrac{(x+3)(x+1)}{(x+1)(x+2)} - \dfrac{(x+3)(x-2)}{(x+2)(x-2)}$

$$= \dfrac{3x}{x+1} + \dfrac{x+3}{x+2} - \dfrac{x+3}{x+2} = \dfrac{3x}{x+1}$$

28. **a.** $\dfrac{\frac{5x}{2}}{\frac{3x^2}{8}} = \dfrac{5x}{2} \div \dfrac{3x^2}{8} = \dfrac{5x}{2} \cdot \dfrac{8}{3x^2} = \dfrac{20}{3x}$ $\qquad$ **b.** $\dfrac{\frac{3x}{y}}{\frac{6x}{y^2}} = \dfrac{3x}{y} \div \dfrac{6x}{y^2} = \dfrac{3x}{y} \cdot \dfrac{y^2}{6x} = \dfrac{y}{2}$

c. $\dfrac{\frac{1}{x} + \frac{1}{y}}{x - y} = \dfrac{xy\left(\frac{1}{x} + \frac{1}{y}\right)}{xy(x - y)} = \dfrac{xy\left(\frac{1}{x}\right) + xy\left(\frac{1}{y}\right)}{xy(x - y)} = \dfrac{y + x}{xy(x - y)}$

d. $\dfrac{x^{-1} + y^{-1}}{y^{-1} - x^{-1}} = \dfrac{\frac{1}{x} + \frac{1}{y}}{\frac{1}{y} - \frac{1}{x}} = \dfrac{xy\left(\frac{1}{x} + \frac{1}{y}\right)}{xy\left(\frac{1}{y} - \frac{1}{x}\right)} = \dfrac{xy\left(\frac{1}{x}\right) + xy\left(\frac{1}{y}\right)}{xy\left(\frac{1}{y}\right) - xy\left(\frac{1}{x}\right)} = \dfrac{y + x}{x - y}$

Chapter 0 Test $\;$ (page 77)

1. odd integers: $-7, 1, 3$ $\qquad$ **2.** prime numbers: 3

3. commutative property of addition $\qquad$ **4.** distributive property

5. $-4 < x \le 2 \Rightarrow$
$\qquad$ **6.** $(-\infty, -3) \cup [6, \infty) \Rightarrow$

7. Since $-17 < 0$, $|-17| = -(-17) = 17$ $\qquad$ **8.** If $x < 0$, then $x - 7 < 0$. Then $|x - 7| = -(x - 7)$.

9. distance $= |-4 - 12| = |-16| = 16$ $\qquad$ **10.** distance $= |-20 - (-12)| = |-8| = 8$

11. $x^4 x^5 x^2 = x^{4+5+2} = x^{11}$ $\qquad$ **12.** $\dfrac{r^2 r^3 s}{r^4 s^2} = \dfrac{r^5 s}{r^4 s^2} = \dfrac{r}{s}$

13. $\dfrac{(a^{-1}a^2)^{-2}}{a^{-3}} = \dfrac{(a^1)^{-2}}{a^{-3}} = \dfrac{a^{-2}}{a^{-3}} = a$ $\qquad$ **14.** $\left(\dfrac{x^0 x^2}{x^{-2}}\right)^6 = \left(\dfrac{x^2}{x^{-2}}\right)^6 = \left(x^4\right)^6 = x^{24}$

15. $450{,}000 = 4.5 \times 10^5$ $\qquad$ **16.** $0.000345 = 3.45 \times 10^{-4}$

17. $3.7 \times 10^3 = 3{,}700$ $\qquad$ **18.** $1.2 \times 10^{-3} = 0.0012$

19. $(25a^4)^{1/2} = 25^{1/2}(a^4)^{1/2} = 5a^2$ $\qquad$ **20.** $\left(\dfrac{36}{81}\right)^{3/2} = \dfrac{36^{3/2}}{81^{3/2}} = \dfrac{\left(36^{1/2}\right)^3}{\left(81^{1/2}\right)^3} = \dfrac{216}{729}$

21. $\left(\dfrac{8t^6}{27s^9}\right)^{-2/3} = \left(\dfrac{27s^9}{8t^6}\right)^{2/3} = \dfrac{27^{2/3}(s^9)^{2/3}}{8^{2/3}(t^6)^{2/3}} = \dfrac{\left(27^{1/3}\right)^2 s^6}{\left(8^{1/3}\right)^2 t^4} = \dfrac{9s^6}{4t^4}$

22. $\sqrt[3]{27a^6} = \sqrt[3]{27}\sqrt[3]{a^6} = 3a^2$ $\qquad$ **23.** $\sqrt{12} + \sqrt{27} = \sqrt{4}\sqrt{3} + \sqrt{9}\sqrt{3}$
$= 2\sqrt{3} + 3\sqrt{3} = 5\sqrt{3}$

29

24. $2\sqrt[3]{3x^4} - 3x\sqrt[3]{24x} = 2\sqrt[3]{x^3}\sqrt[3]{3x} - 3x\sqrt[3]{8}\sqrt[3]{3x} = 2x\sqrt[3]{3x} - 3x(2)\sqrt[3]{3x} = 2x\sqrt[3]{3x} - 6x\sqrt[3]{3x}$

$$= -4x\sqrt[3]{3x}$$

25. $\dfrac{x}{\sqrt{x} - 2} = \dfrac{x}{\sqrt{x} - 2} \cdot \dfrac{\sqrt{x} + 2}{\sqrt{x} + 2} = \dfrac{x(\sqrt{x} + 2)}{(\sqrt{x})^2 - 2^2} = \dfrac{x(\sqrt{x} + 2)}{x - 4}$

26. $\dfrac{\sqrt{x} - \sqrt{y}}{\sqrt{x} + \sqrt{y}} = \dfrac{\sqrt{x} - \sqrt{y}}{\sqrt{x} + \sqrt{y}} \cdot \dfrac{\sqrt{x} + \sqrt{y}}{\sqrt{x} + \sqrt{y}} = \dfrac{(\sqrt{x})^2 - (\sqrt{y})^2}{\sqrt{x^2} + \sqrt{xy} + \sqrt{xy} + \sqrt{y^2}} = \dfrac{x - y}{x + 2\sqrt{xy} + y}$

27. $(a^2 + 3) - (2a^2 - 4) = a^2 + 3 - 2a^2 + 4$

$$= -a^2 + 7$$

28. $(3a^3b^2)(-2a^3b^4) = -6a^6b^6$

29. $(3x - 4)(2x + 7) = 6x^2 + 21x - 8x - 28$

$$= 6x^2 + 13x - 28$$

30. $(a^n + 2)(a^n - 3) = a^{2n} - 3a^n + 2a^n - 6$

$$= a^{2n} - a^n - 6$$

31. $(x^2 + 4)(x^2 - 4) = x^4 - 4x^2 + 4x^2 - 16 = x^4 - 16$

32. $(x^2 - x + 2)(2x - 3) = 2x^3 - 3x^2 - 2x^2 + 3x + 4x - 6 = 2x^3 - 5x^2 + 7x - 6$

33.
$$
\begin{array}{r}
6x + \ 19 + \frac{34}{x-3} \\
x - 3 \overline{)6x^2 + \quad x - 23} \\
\underline{6x^2 - 18x \quad\quad} \\
19x - \ 23 \\
\underline{19x - \ 57} \\
34
\end{array}
$$

34.
$$
\begin{array}{r}
x^2 + \ 2x + \ 1 \\
2x - 1 \overline{)2x^3 + 3x^2 + 0x - 1} \\
\underline{2x^3 - \ x^2 \quad\quad\quad} \\
4x^2 + 0x \\
\underline{4x^2 - 2x} \\
2x - 1 \\
\underline{2x - 1} \\
0
\end{array}
$$

35. $3x + 6y = 3(x + 2y)$

36. $x^2 - 100 = x^2 - 10^2 = (x + 10)(x - 10)$

37. $10t^2 - 19tw + 6w^2 = (5t - 2w)(2t - 3w)$

38. $3a^3 - 648 = 3(a^3 - 216)$

$$= 3(a - 6)(a^2 + 6a + 36)$$

39. $x^4 - x^2 - 12 = (x^2 - 4)(x^2 + 3)$

$$= (x + 2)(x - 2)(x^2 + 3)$$

40. $6x^4 + 11x^2 - 10 = (3x^2 - 2)(2x^2 + 5)$

41. $\dfrac{x}{x + 2} + \dfrac{2}{x + 2} = \dfrac{x + 2}{x + 2} = 1$

42. $\dfrac{x}{x + 1} - \dfrac{x}{x - 1} = \dfrac{x(x - 1)}{(x + 1)(x - 1)} - \dfrac{x(x + 1)}{(x + 1)(x - 1)} = \dfrac{x^2 - x - x^2 - x}{(x + 1)(x - 1)} = \dfrac{-2x}{(x + 1)(x - 1)}$

43. $\dfrac{x^2 + x - 20}{x^2 - 16} \cdot \dfrac{x^2 - 25}{x - 5} = \dfrac{(x + 5)(x - 4)}{(x + 4)(x - 4)} \cdot \dfrac{(x + 5)(x - 5)}{x - 5} = \dfrac{(x + 5)^2}{x + 4}$

44. $\dfrac{x+2}{x^2+2x+1} \div \dfrac{x^2-4}{x+1} = = \dfrac{x+2}{(x+1)(x+1)} \cdot \dfrac{x+1}{(x+2)(x-2)} = \dfrac{1}{(x+1)(x-2)}$

45. $\dfrac{\frac{1}{a}+\frac{1}{b}}{\frac{1}{b}} = \dfrac{ab\left(\frac{1}{a}+\frac{1}{b}\right)}{ab\left(\frac{1}{b}\right)} = \dfrac{ab\left(\frac{1}{a}\right)+ab\left(\frac{1}{b}\right)}{a} = \dfrac{b+a}{a}$

46. $\dfrac{x^{-1}}{x^{-1}+y^{-1}} = \dfrac{\frac{1}{x}}{\frac{1}{x}+\frac{1}{y}} = \dfrac{xy\left(\frac{1}{x}\right)}{xy\left(\frac{1}{x}+\frac{1}{y}\right)} = \dfrac{y}{xy\left(\frac{1}{x}\right)+xy\left(\frac{1}{y}\right)} = \dfrac{y}{y+x}$

Exercise 1.1 (page 86)

1. root, solution   **3.** no   **5.** linear   **7.** one

9. $x+3=1$ **11.** $\dfrac{1}{x}=12$ **13.** $\sqrt{x}=4$ **15.** $\dfrac{1}{x-3}=\dfrac{5}{x+2}$
no restrictions $x \neq 0$ $x \geq 0$ $x \neq 3, x \neq -2$

17.
$$2x+5=15$$
$$2x+5-5=15-5$$
$$2x=10$$
$$\frac{2x}{2}=\frac{10}{2}$$
$$x=5$$
conditional equation

19.
$$2(n+2)-5=2n$$
$$2n+4-5=2n$$
$$2n-1=2n$$
$$2n-2n-1=2n-2n$$
$$-1 \neq 0$$
no solution

21.
$$\frac{x+7}{2}=7$$
$$2 \cdot \frac{x+7}{2}=2(7)$$
$$x+7=14$$
$$x+7-7=14-7$$
$$x=7$$
conditional equation

23.
$$2(a+1)=3(a-2)-a$$
$$2a+2=3a-6-a$$
$$2a+2=2a-6$$
$$2a-2a+2=2a-2a-6$$
$$2 \neq -6$$
no solution

25.
$$3(x-3)=\frac{6x-18}{2}$$
$$3x-9=\frac{6x-18}{2}$$
$$2(3x-9)=2 \cdot \frac{6x-18}{2}$$
$$6x-18=6x-18$$
identity

27.
$$\frac{3}{b-3}=1$$
$$(b-3)\cdot\frac{3}{b-3}=(b-3)(1)$$
$$3=b-3$$
$$3+3=b-3+3$$
$$6=b$$
conditional equation

29.
$$2x^2+5x-3=(2x-1)(x+3)$$
$$2x^2+5x-3=2x^2+5x-3$$
identity

31. $2x + 7 = 10 - x$
$3x + 7 = 10$
$3x = 3$
$x = 1$

33. $\dfrac{5}{3}z - 8 = 7$
$\dfrac{5}{3}z = 15$
$3 \cdot \dfrac{5}{3}z = 3(15)$
$5z = 45$
$z = 9$

35. $\dfrac{z}{5} + 2 = 4$
$\dfrac{z}{5} = 2$
$5 \cdot \dfrac{z}{5} = 5(2)$
$z = 10$

37. $\dfrac{3x - 2}{3} = 2x + \dfrac{7}{3}$
$3 \cdot \dfrac{3x - 2}{3} = 3\left(2x + \dfrac{7}{3}\right)$
$3x - 2 = 6x + 7$
$-3x - 2 = 7$
$-3x = 9$
$x = -3$

39. $5(x - 2) = 2x + 8$
$5x - 10 = 2x + 8$
$3x - 10 = 8$
$3x = 18$
$x = 6$

41. $2(2x + 1) - \dfrac{3x}{2} = \dfrac{-3(4 + x)}{2}$
$2\left[2(2x + 1) - \dfrac{3x}{2}\right] = 2 \cdot \dfrac{-3(4 + x)}{2}$
$4(2x + 1) - 3x = -3(4 + x)$
$8x + 4 - 3x = -12 - 3x$
$5x + 4 = -12 - 3x$
$8x + 4 = -12$
$8x = -16$
$x = -2$

43. $7(2x + 5) - 6(x + 8) = 7$
$14x + 35 - 6x - 48 = 7$
$8x - 13 = 7$
$8x = 20$
$x = \dfrac{20}{8} = \dfrac{5}{2}$

45. $(x - 2)(x + 5) = (x - 3)(x + 2)$
$x^2 + 3x - 10 = x^2 - x - 6$
$3x - 10 = -x - 6$
$4x - 10 = -6$
$4x = 4$
$x = 1$

47. $\dfrac{3}{2}(3x - 2) - 10x - 4 = 0$
$2\left[\dfrac{3}{2}(3x - 2) - 10x - 4\right] = 2(0)$
$3(3x - 2) - 20x - 8 = 0$
$9x - 6 - 20x - 8 = 0$
$-11x - 14 = 0$
$-11x = 14$
$x = -\dfrac{14}{11}$

49. $x(x+2) = (x+1)^2 - 1$
$x^2 + 2x = (x+1)(x+1) - 1$
$x^2 + 2x = x^2 + 2x + 1 - 1$
$x^2 + 2x = x^2 + 2x$
$0 = 0 \Rightarrow$ identity

51. $\dfrac{(y+2)^2}{3} = y + 2 + \dfrac{y^2}{3}$
$3\left[\dfrac{(y+2)^2}{3}\right] = 3\left(y + 2 + \dfrac{y^2}{3}\right)$
$(y+2)^2 = 3y + 6 + y^2$
$y^2 + 4y + 4 = y^2 + 3y + 6$
$4y + 4 = 3y + 6$
$y + 4 = 6$
$y = 2$

53. $2(s+2) + (s+3)^2 = s(s+5) + 2\left(\dfrac{17}{2} + s\right)$
$2s + 4 + s^2 + 6s + 9 = s^2 + 5s + 17 + 2s$
$s^2 + 8s + 13 = s^2 + 7s + 17$
$8s + 13 = 7s + 17$
$s = 4$

55. $\dfrac{2}{x+1} + \dfrac{1}{3} = \dfrac{1}{x+1}$
$3(x+1)\left(\dfrac{2}{x+1} + \dfrac{1}{3}\right) = 3(x+1) \cdot \dfrac{1}{x+1}$
$6 + 1(x+1) = 3(1)$
$6 + x + 1 = 3$
$x + 7 = 3$
$x = -4$

57. $\dfrac{9t+6}{t(t+3)} = \dfrac{7}{t+3}$
$t(t+3)\left[\dfrac{9t+6}{t(t+3)}\right] = t(t+3) \cdot \dfrac{7}{t+3}$
$9t + 6 = 7t$
$2t + 6 = 0$
$2t = -6$
$t = -3$
The answer does not check. $\Rightarrow$ no solution

59. $\dfrac{2}{(a-7)(a+2)} = \dfrac{4}{(a+3)(a+2)}$
$(a-7)(a+2)(a+3) \cdot \dfrac{2}{(a-7)(a+2)} = (a-7)(a+2)(a+3) \cdot \dfrac{4}{(a+3)(a+2)}$
$2(a+3) = 4(a-7)$
$2a + 6 = 4a - 28$
$-2a = -34$
$a = 17$

61.
$$\frac{2x+3}{x^2+5x+6} + \frac{3x-2}{x^2+x-6} = \frac{5x-2}{x^2-4}$$

$$\frac{2x+3}{(x+3)(x+2)} + \frac{3x-2}{(x+3)(x-2)} = \frac{5x-2}{(x+2)(x-2)}$$

$$(x-2)(2x+3) + (x+2)(3x-2) = (x+3)(5x-2) \quad \text{\{multiply by common denominator\}}$$

$$2x^2 - x - 6 + 3x^2 + 4x - 4 = 5x^2 + 13x - 6$$

$$5x^2 + 3x - 10 = 5x^2 + 13x - 6$$

$$3x - 10 = 13x - 6$$

$$-10x = 4$$

$$x = -\frac{4}{10} = -\frac{2}{5}$$

63.
$$\frac{3x+5}{x^3+8} + \frac{3}{x^2-4} = \frac{2(3x-2)}{(x-2)(x^2-2x+4)}$$

$$\frac{3x+5}{(x+2)(x^2-2x+4)} + \frac{3}{(x+2)(x-2)} = \frac{2(3x-2)}{(x-2)(x^2-2x+4)}$$

$$(x-2)(3x+5) + (x^2-2x+4)(3) = 2(x+2)(3x-2) \quad \text{\{multiply by common denominator\}}$$

$$3x^2 - x - 10 + 3x^2 - 6x + 12 = 6x^2 + 8x - 8$$

$$6x^2 - 7x + 2 = 6x^2 + 8x - 8$$

$$-15x = -10$$

$$x = \frac{-10}{-15} = \frac{2}{3}$$

65.
$$\frac{1}{11-n} - \frac{2(3n-1)}{-7n^2+74n+33} = \frac{1}{7n+3}$$

$$\frac{-1}{n-11} + \frac{2(3n-1)}{7n^2-74n-33} = \frac{1}{7n+3}$$

$$\frac{-1}{n-11} + \frac{6n-2}{(7n+3)(n-11)} = \frac{1}{7n+3}$$

$$-(7n+3) + 6n - 2 = (n-11)1 \quad \text{\{multiply by common denominator\}}$$

$$-7n - 3 + 6n - 2 = n - 11$$

$$-n - 5 = n - 11$$

$$-2n = -6$$

$$n = 3$$

67.
$$\frac{5}{y+4} + \frac{2}{y+2} = \frac{6}{y+2} - \frac{1}{y^2+6y+8}$$

$$\frac{5}{y+4} = \frac{4}{y+2} - \frac{1}{(y+2)(y+4)}$$

$$5(y+2) = 4(y+4) - 1 \quad \text{\{multiply by common denominator\}}$$

$$5y + 10 = 4y + 16 - 1$$

$$5y + 10 = 4y + 15$$

$$y = 5$$

69.

$$\frac{3y}{6-3y} + \frac{2y}{2y+4} = \frac{8}{4-y^2}$$

$$\frac{3y}{3(2-y)} + \frac{2y}{2(y+2)} = \frac{8}{(2+y)(2-y)}$$

$$\frac{y}{2-y} + \frac{y}{2+y} = \frac{8}{(2+y)(2-y)}$$

$$y(2+y) + y(2-y) = 8 \quad \{\text{multiply by common denominator}\}$$

$$2y + y^2 + 2y - y^2 = 8$$

$$4y = 8$$

$$y = 2 \Rightarrow \text{The solution does not check, so the equation has no solution.}$$

71.

$$\frac{a}{a+2} - 1 = -\frac{3a+2}{a^2+4a+4}$$

$$\frac{a}{a+2} - \frac{1}{1} = -\frac{3a+2}{(a+2)(a+2)}$$

$$(a+2)(a+2)\left[\frac{a}{a+2} - 1\right] = (a+2)(a+2) \cdot \left[-\frac{3a+2}{(a+2)(a+2)}\right]$$

$$a(a+2) - (a+2)(a+2) = -(3a+2)$$

$$a^2 + 2a - \left(a^2+4a+4\right) = -3a-2$$

$$a^2 + 2a - a^2 - 4a - 4 = -3a-2$$

$$-2a - 4 = -3a - 2$$

$$a = 2$$

73.

$$k = 2.2p$$

$$\frac{k}{2.2} = \frac{2.2p}{2.2}$$

$$\frac{k}{2.2} = p$$

75.

$$p = 2l + 2w$$

$$p - 2l = 2w$$

$$\frac{p-2l}{2} = \frac{2w}{2}$$

$$\frac{p-2l}{2} = w$$

77.

$$V = \frac{1}{3}\pi r^2 h$$

$$3V = 3 \cdot \frac{1}{3}\pi r^2 h$$

$$3V = \pi r^2 h$$

$$\frac{3V}{\pi h} = \frac{\pi r^2 h}{\pi h}$$

$$\frac{3V}{\pi h} = r^2$$

79.

$$P_n = L + \frac{si}{f}$$

$$P_n - L = \frac{si}{f}$$

$$f(P_n - L) = f \cdot \frac{si}{f}$$

$$f(P_n - L) = si$$

$$\frac{f(P_n - L)}{i} = \frac{si}{i}$$

$$\frac{f(P_n - L)}{i} = s$$

81.

$$F = \frac{mMg}{r^2}$$

$$Fr^2 = \frac{mMg}{r^2} \cdot r^2$$

$$Fr^2 = mMg$$

$$\frac{Fr^2}{Mg} = \frac{mMg}{Mg}$$

$$\frac{Fr^2}{Mg} = m$$

83.

$$\frac{x}{a} + \frac{y}{b} = 1$$

$$\frac{y}{b} = 1 - \frac{x}{a}$$

$$b \cdot \frac{y}{b} = b\left(1 - \frac{x}{a}\right)$$

$$y = b\left(1 - \frac{x}{a}\right)$$

SECTION 1.1

85.
$$\frac{1}{r} = \frac{1}{r_1} + \frac{1}{r_2}$$
$$rr_1r_2 \cdot \frac{1}{r} = rr_1r_2\left(\frac{1}{r_1} + \frac{1}{r_2}\right)$$
$$r_1r_2 = rr_2 + rr_1$$
$$r_1r_2 = r(r_2 + r_1)$$
$$\frac{r_1r_2}{r_2 + r_1} = \frac{r(r_2 + r_1)}{r_2 + r_1}$$
$$\frac{r_1r_2}{r_2 + r_1} = r$$

87.
$$l = a + (n-1)d$$
$$l = a + nd - d$$
$$l - a + d = nd$$
$$\frac{l - a + d}{d} = \frac{nd}{d}$$
$$\frac{l - a + d}{d} = n$$

89.
$$a = (n-2)\frac{180}{n}$$
$$an = (n-2)\frac{180}{n} \cdot n$$
$$an = (n-2)180$$
$$an = 180n - 360$$
$$360 = 180n - an$$
$$360 = n(180 - a)$$
$$\frac{360}{180-a} = n$$

91.
$$R = \frac{1}{\frac{1}{r_1} + \frac{1}{r_2} + \frac{1}{r_3}}$$
$$R = \frac{r_1r_2r_3(1)}{r_1r_2r_3\left(\frac{1}{r_1} + \frac{1}{r_2} + \frac{1}{r_3}\right)}$$
$$R = \frac{r_1r_2r_3}{r_2r_3 + r_1r_3 + r_1r_2}$$
$$R(r_2r_3 + r_1r_3 + r_1r_2) = r_1r_2r_3$$
$$Rr_2r_3 + Rr_1r_3 + Rr_1r_2 = r_1r_2r_3$$
$$Rr_1r_3 + Rr_1r_2 - r_1r_2r_3 = -Rr_2r_3$$
$$r_1(Rr_3 + Rr_2 - r_2r_3) = -Rr_2r_3$$
$$r_1 = \frac{-Rr_2r_3}{Rr_3 + Rr_2 - r_2r_3}$$

93. Answers may vary.

95. $(25x^2)^{1/2} = \left[(5x)^2\right]^{1/2} = 5|x|$

97.
$$\left(\frac{125x^3}{8y^6}\right)^{-2/3} = \left(\frac{8y^6}{125x^3}\right)^{2/3}$$
$$= \frac{8^{2/3}(y^6)^{2/3}}{125^{2/3}(x^3)^{2/3}} = \frac{4y^4}{25x^2}$$

99. $\sqrt{25y^2} = \sqrt{(5y)^2} = 5|y|$

101. $\sqrt[4]{\frac{a^4b^{12}}{z^8}} = \sqrt[4]{\left(\frac{ab^3}{z^2}\right)^4} = \frac{|ab^3|}{z^2}$

Exercise 1.2 (page 94)

1. add

3. amount

5. rate, time

7. Let x = the score on the first exam.
Then $x + 5$ = the score on the midterm,
and $x + 13$ = the score on the final.

$$\frac{\boxed{\text{Sum of scores}}}{3} = 90$$

$$\frac{x + x + 5 + x + 13}{3} = 90$$

$$\frac{3x + 18}{3} = 90$$

$$3x + 18 = 270$$

$$3x = 252$$

$$x = 84$$

His score on the first exam was 84.

9. Let x = the program development score.

$$\frac{\boxed{\text{Sum of scores}}}{4} = 86$$

$$\frac{82 + 90 + x + 78}{4} = 86$$

$$\frac{x + 250}{4} = 86$$

$$x + 250 = 344$$

$$x = 94$$

The program development score was 94.

11. Let x = the number of locks replaced.

$$40 + 28 \cdot \boxed{\begin{array}{c}\text{Number}\\ \text{of locks}\end{array}} = 236$$

$$40 + 28x = 236$$

$$28x = 196$$

$$x = 7$$

7 locks can be changed for \$236.

13.

$$\boxed{\text{Perimeter}} = 14$$

$$x + (x + 2) + x + (x + 2) = 14$$

$$4x + 4 = 14$$

$$4x = 10$$

$$x = \frac{5}{2} = 2\frac{1}{2}$$

The width is $2\frac{1}{2}$ feet.

15.

$$\boxed{\text{Total Area}} = 2 \cdot \boxed{\text{Triangular Area}}$$

$$20x + \frac{1}{2}(16)(20) = 2 \cdot \frac{1}{2}(16)(20)$$

$$20x + 160 = 320$$

$$20x = 160$$

$$x = 8 \Rightarrow \text{The dimensions are 8 feet by 20 feet.}$$

17.

$$\boxed{\begin{array}{c}\text{New}\\ \text{Area}\end{array}} = \boxed{\begin{array}{c}\text{Old}\\ \text{Area}\end{array}} + 0.50 \cdot \boxed{\begin{array}{c}\text{Old}\\ \text{Area}\end{array}}$$

$$12(x + 10) + 12x = 12(x + 10) + 0.50 \cdot 12(x + 10)$$

$$12x + 120 + 12x = 12x + 120 + 6x + 60$$

$$24x + 120 = 18x + 180$$

$$6x = 60$$

$$x = 10 \Rightarrow \text{The length of the living room is } x + 10 = 20 \text{ feet.}$$

SECTION 1.2

19. Let x = the amount invested at 7%. Then $22000 - x$ = the amount invested at 6%.

$$\boxed{\text{Interest at 7\%}} + \boxed{\text{Interest at 6\%}} = \boxed{\text{Total interest}}$$

$$0.07x + 0.06(22000 - x) = 1420$$
$$0.07x + 1320 - 0.06x = 1420$$
$$0.01x = 100$$
$$x = 10000$$

$10,000 was invested at 7% and $12,000 was invested at 6%.

21. Let x = the number of full-price tickets sold. Then $585 - x$ = the number of student tickets sold.

$$2.50 \cdot \boxed{\begin{smallmatrix}\text{\# of}\\\text{full-price}\end{smallmatrix}} + 1.75 \cdot \boxed{\begin{smallmatrix}\text{\# of}\\\text{student}\end{smallmatrix}} = 1217.25$$

$$2.50x + 1.75(585 - x) = 1217.25$$
$$2.50x + 1023.75 - 1.75x = 1217.25$$
$$0.75x = 193.50$$
$$x = 258 \Rightarrow \text{There were 327 student tickets sold.}$$

23. Let x = the amount invested at 8%. Then $37,000 - x$ = the amount invested at $9\frac{1}{2}$%.

$$\boxed{\text{Interest at } 9\tfrac{1}{2}\%} = \boxed{\text{Interest at 8\%}} + 452.50$$

$$0.095(37,000 - x) = 0.08x + 452.50$$
$$3515 - 0.095x = 0.08x + 452.50$$
$$3062.50 = 0.175x$$
$$17500 = x$$

$17,500 is invested at 8% and $19,500 is invested at $9\frac{1}{2}$%.

25. Let p = the original price.

$$\boxed{\begin{smallmatrix}\text{Original}\\\text{price}\end{smallmatrix}} - \boxed{\text{Discount}} = \boxed{\begin{smallmatrix}\text{New}\\\text{price}\end{smallmatrix}}$$

$$p - 0.20p = 63.96$$
$$0.80p = 63.96$$
$$p = 79.95$$

The original price was $79.95.

27. Let x = # of plates for equal costs.

$$\boxed{\begin{smallmatrix}\text{Cost of 1st}\\\text{machine}\end{smallmatrix}} = \boxed{\begin{smallmatrix}\text{Cost of 2nd}\\\text{machine}\end{smallmatrix}}$$

$$600 + 3x = 800 + 2x$$
$$x = 200$$

The break point is 200 plates.

29. Let x = # of computers to break even.

$$\boxed{\text{Income}} = \boxed{\text{Expenses}}$$

$$1275x = 8925 + 850x$$
$$425x = 8925$$
$$x = 21$$

21 computers need to be sold to break even.

31. Let x = hours for both working together.

$$\boxed{\begin{smallmatrix}\text{Woman}\\\text{in 1 hour}\end{smallmatrix}} + \boxed{\begin{smallmatrix}\text{Man in}\\\text{1 hour}\end{smallmatrix}} = \boxed{\begin{smallmatrix}\text{Total in}\\\text{1 hour}\end{smallmatrix}}$$

$$\frac{1}{2} + \frac{1}{4} = \frac{1}{x}$$
$$4x\left(\frac{1}{2} + \frac{1}{4}\right) = 4x\left(\frac{1}{x}\right)$$
$$2x + x = 4$$
$$3x = 4$$
$$x = \frac{4}{3} = 1\frac{1}{3}$$

They can mow the lawn in $1\frac{1}{3}$ hours.

33. Let $x =$ hours for pool to fill with drain open.

$$\boxed{\begin{array}{c}\text{Pipe in}\\\text{1 hour}\end{array}} - \boxed{\begin{array}{c}\text{Drain in}\\\text{1 hour}\end{array}} = \boxed{\begin{array}{c}\text{Total in}\\\text{1 hour}\end{array}}$$

$$\frac{1}{10} - \frac{1}{19} = \frac{1}{x}$$

$$190x\left(\frac{1}{10} - \frac{1}{19}\right) = 190x\left(\frac{1}{x}\right)$$

$$19x - 10x = 190$$

$$9x = 190$$

$$x = \frac{190}{9} = 21\frac{1}{9}$$

The pool can be filled in $21\frac{1}{9}$ hours.

35. Let $x =$ the liters of liquid replaced with pure antifreeze.

$$\boxed{\begin{array}{c}\text{Liters of}\\\text{a.f. at start}\end{array}} - \boxed{\begin{array}{c}\text{Liters of}\\\text{a.f. removed}\end{array}} + \boxed{\begin{array}{c}\text{Liters of}\\\text{a.f. replaced}\end{array}} = \boxed{\begin{array}{c}\text{Liters of}\\\text{a.f. at end}\end{array}}$$

$$0.40(6) - 0.40x + x = 0.50(6)$$

$$2.4 + 0.6x = 3$$

$$0.6x = 0.6$$

$$x = 1 \Rightarrow 1 \text{ liter should be replaced with pure antifreeze.}$$

37. Let $x =$ the liters of pure alcohol added.

$$\boxed{\begin{array}{c}\text{Liters of}\\\text{alcohol at start}\end{array}} + \boxed{\begin{array}{c}\text{Liters of}\\\text{alcohol added}\end{array}} = \boxed{\begin{array}{c}\text{Liters of}\\\text{alcohol at end}\end{array}}$$

$$0.20(1) + x = 0.25(1 + x)$$

$$0.20 + x = 0.25 + 0.25x$$

$$0.75x = 0.05$$

$$x = \frac{0.05}{0.75} = \frac{1}{15} \Rightarrow \frac{1}{15} \text{ of a liter of pure alcohol should be added.}$$

39. Let $x =$ the gallons of pure chlorine added.

$$\boxed{\begin{array}{c}\text{Gallons of}\\\text{chlorine at start}\end{array}} + \boxed{\begin{array}{c}\text{Gallons of}\\\text{chlorine added}\end{array}} = \boxed{\begin{array}{c}\text{Gallons of}\\\text{chlorine at end}\end{array}}$$

$$0(15000) + x = 0.0003(15000 + x)$$

$$x = 4.5 + 0.0003x$$

$$0.9997x = 4.5$$

$$x \approx 4.5 \Rightarrow \text{About 4.5 gallons of pure chlorine should be added.}$$

41. Let $x =$ the liters of water evaporated.

$$\boxed{\begin{array}{c}\text{Liters of}\\\text{salt at start}\end{array}} - \boxed{\begin{array}{c}\text{Liters of}\\\text{salt evaporated}\end{array}} = \boxed{\begin{array}{c}\text{Liters of}\\\text{salt at end}\end{array}}$$

$$0.24(12) - 0(x) = 0.36(12 - x)$$

$$2.88 - 0 = 4.32 - 0.36x$$

$$0.36x = 1.44$$

$$x = 4 \Rightarrow 4 \text{ liters of water should be evaporated.}$$

43. Let $r =$ his first rate. Then $r + 26 =$ his return rate.

$$\boxed{\text{Distance to city}} = \boxed{\text{Return distance}}$$
$$5r = 3(r + 26)$$
$$5r = 3r + 78$$
$$2r = 78$$
$$r = 39 \Rightarrow \text{He drove 39 mph going and 65 mph returning.}$$

45. Let $t =$ the time the cars travel.

$$\boxed{\begin{array}{c}\text{Distance 1st}\\\text{car travels}\end{array}} + \boxed{\begin{array}{c}\text{Distance 2nd}\\\text{car travels}\end{array}} = \boxed{\text{Total distance}}$$
$$60t + 64t = 310$$
$$124t = 310$$
$$t = 2.5 \Rightarrow \text{They will be 310 miles apart after 2.5 hours.}$$

47. Let $t =$ the time the runners run.

$$\boxed{\begin{array}{c}\text{Distance}\\\text{1st runs}\end{array}} + \boxed{\begin{array}{c}\text{Distance}\\\text{2nd runs}\end{array}} = \boxed{\begin{array}{c}\text{Distance}\\\text{between them}\end{array}}$$
$$8t + 10t = \frac{440}{1760}$$
$$18t = \frac{1}{4}$$
$$t = \frac{1}{72} \text{ hour} = \frac{1}{72}(60) = \frac{5}{6} \text{ minute} = 50 \text{ seconds}$$

They will meet after 50 seconds.

49. Let $r =$ the speed of the boat in still water.

Then the speed of the boat is $r + 2$ downstream and $r - 2$ upstream.

$$\boxed{\text{Time upstream}} = \boxed{\text{Time downstream}} \qquad \{\text{Note: Time} = \text{Distance} \div \text{Rate}\}$$
$$\frac{5}{r - 2} = \frac{7}{r + 2}$$
$$(r + 2)(r - 2)\frac{5}{r - 2} = (r + 2)(r - 2)\frac{7}{r + 2}$$
$$5(r + 2) = 7(r - 2)$$
$$5r + 10 = 7r - 14$$
$$24 = 2r$$
$$12 = r \Rightarrow \text{The speed of the boat is 12 mph.}$$

51. Since the mixture is to be 25% barley, there will be $0.25(2400) = 600$ pounds of barley used. Thus, the other 1800 pounds will be either oats or soybean meal.

Let $x =$ the number of pounds of oats used. Then $1800 - x =$ the number of pounds of meal used.

$$\boxed{\begin{array}{c}\text{Pounds of protein} \\ \text{from barley}\end{array}} + \boxed{\begin{array}{c}\text{Pounds of protein} \\ \text{from oats}\end{array}} + \boxed{\begin{array}{c}\text{Pounds of protein} \\ \text{from soybean meal}\end{array}} = \boxed{\begin{array}{c}\text{Total pounds} \\ \text{of protein}\end{array}}$$

$$0.117(600) + 0.118x + 0.445(1800 - x) = 0.14(2400)$$
$$70.2 + 0.118x + 801 - 0.445x = 336$$
$$871.2 - 0.327x = 336$$
$$-0.327x = -535.2$$
$$x \approx 1637$$

The farmer should use 600 pounds of barley, 1637 pounds of oats and 163 pounds of soybean meal.

53.
$$V = \pi r^2 h$$
$$712.51 = \pi(4.5)^2 d$$
$$\frac{712.51}{\pi(4.5)^2} = d$$
$$11.2 \approx d$$
The hole is about 11.2 millimeters deep.

55. **Answers may vary.**

57. $x^2 - 2x - 63 = (x - 9)(x + 7)$

59. $9x^2 - 12x - 5 = (3x + 1)(3x - 5)$

61. $x^2 + 6x + 9 = (x + 3)(x + 3)$

63. $x^3 + 8 = (x + 2)(x^2 - 2x + 2^2)$
$$= (x + 2)(x^2 - 2x + 4)$$

Exercise 1.3 (page 106)

1. $ax^2 + bx + c = 0$

3. $\sqrt{c}, -\sqrt{c}$

5. equal real numbers

7.
$$x^2 - x - 6 = 0$$
$$(x + 2)(x - 3) = 0$$
$$x + 2 = 0 \quad \text{or} \quad x - 3 = 0$$
$$x = -2 \qquad\qquad x = 3$$

9.
$$x^2 - 144 = 0$$
$$(x + 12)(x - 12) = 0$$
$$x + 12 = 0 \quad \text{or} \quad x - 12 = 0$$
$$x = -12 \qquad\qquad x = 12$$

11.
$$2x^2 + x - 10 = 0$$
$$(2x + 5)(x - 2) = 0$$
$$2x + 5 = 0 \quad \text{or} \quad x - 2 = 0$$
$$2x = -5 \qquad\qquad x = 2$$
$$x = -\tfrac{5}{2} \qquad\qquad x = 2$$

13.
$$5x^2 - 13x + 6 = 0$$
$$(5x - 3)(x - 2) = 0$$
$$5x - 3 = 0 \quad \text{or} \quad x - 2 = 0$$
$$5x = 3 \qquad\qquad x = 2$$
$$x = \tfrac{3}{5} \qquad\qquad x = 2$$

15.
$$15x^2 + 16x = 15$$
$$15x^2 + 16x - 15 = 0$$
$$(3x + 5)(5x - 3) = 0$$
$$3x + 5 = 0 \quad \textbf{or} \quad 5x - 3 = 0$$
$$3x = -5 \qquad\qquad 5x = 3$$
$$x = -\tfrac{5}{3} \qquad\qquad x = \tfrac{3}{5}$$

17.
$$12x^2 + 9 = 24x$$
$$12x^2 - 24x + 9 = 0$$
$$3(4x^2 - 8x + 3) = 0$$
$$(2x - 1)(2x - 3) = 0$$
$$2x - 1 = 0 \quad \textbf{or} \quad 2x - 3 = 0$$
$$2x = 1 \qquad\qquad 2x = 3$$
$$x = \tfrac{1}{2} \qquad\qquad x = \tfrac{3}{2}$$

19.
$$x^2 = 9$$
$$x = \sqrt{9} \quad \textbf{or} \quad x = -\sqrt{9}$$
$$x = 3 \qquad\qquad x = -3$$

21.
$$y^2 - 50 = 0$$
$$y^2 = 50$$
$$y = \sqrt{50} \quad \textbf{or} \quad y = -\sqrt{50}$$
$$y = 5\sqrt{2} \qquad\qquad y = -5\sqrt{2}$$

23.
$$(x - 1)^2 = 4$$
$$x - 1 = \sqrt{4} \quad \textbf{or} \quad x - 1 = -\sqrt{4}$$
$$x - 1 = 2 \qquad\qquad x - 1 = -2$$
$$x = 3 \qquad\qquad x = -1$$

25.
$$a^2 + 2a + 1 = 9$$
$$(a + 1)^2 = 9$$
$$a + 1 = \sqrt{9} \quad \textbf{or} \quad a + 1 = -\sqrt{9}$$
$$a + 1 = 3 \qquad\qquad a + 1 = -3$$
$$a = 2 \qquad\qquad a = -4$$

27. $x^2 + 6x + \left[\tfrac{1}{2}(6)\right]^2 = x^2 + 6x + 3^2$
$$= x^2 + 6x + 9$$

29. $x^2 - 4x + \left[\tfrac{1}{2}(-4)\right]^2 = x^2 - 4x + (-2)^2$
$$= x^2 - 4x + 4$$

31. $a^2 + 5a + \left[\tfrac{1}{2}(5)\right]^2 = a^2 + 5a + \left(\dfrac{5}{2}\right)^2 = a^2 + 5a + \dfrac{25}{4}$

33. $r^2 - 11r + \left[\tfrac{1}{2}(-11)\right]^2 = r^2 - 11r + \left(\dfrac{-11}{2}\right)^2 = r^2 - 11r + \dfrac{121}{4}$

35. $y^2 + \dfrac{3}{4}y + \left[\dfrac{1}{2}\left(\dfrac{3}{4}\right)\right]^2 = y^2 + \dfrac{3}{4}y + \left(\dfrac{3}{8}\right)^2$
$$= y^2 + \dfrac{3}{4}y + \dfrac{9}{64}$$

37. $q^2 - \dfrac{1}{5}q + \left[\dfrac{1}{2}\left(-\dfrac{1}{5}\right)\right]^2 = q^2 - \dfrac{1}{5}q + \left(\dfrac{-1}{10}\right)^2$
$$= q^2 - \dfrac{1}{5}q + \dfrac{1}{100}$$

39.
$$x^2 - 8x + 15 = 0$$
$$x^2 - 8x = -15$$
$$x^2 - 8x + 16 = -15 + 16$$
$$(x - 4)^2 = 1$$
$$x - 4 = \sqrt{1} \quad \textbf{or} \quad x - 4 = -\sqrt{1}$$
$$x - 4 = 1 \qquad\qquad x - 4 = -1$$
$$x = 5 \qquad\qquad x = 3$$

41.
$$x^2 + x - 6 = 0$$
$$x^2 + x = 6$$
$$x^2 + x + \frac{1}{4} = 6 + \frac{1}{4}$$
$$\left(x + \frac{1}{2}\right)^2 = \frac{25}{4}$$
$$x + \frac{1}{2} = \sqrt{\frac{25}{4}} \quad \textbf{or} \quad x + \frac{1}{2} = -\sqrt{\frac{25}{4}}$$
$$x + \frac{1}{2} = \frac{5}{2} \qquad\qquad x + \frac{1}{2} = -\frac{5}{2}$$
$$x = \frac{4}{2} \qquad\qquad\quad x = -\frac{6}{2}$$
$$x = 2 \qquad\qquad\quad x = -3$$

43.
$$x^2 - 25x = 0$$
$$x^2 - 25x + \frac{625}{4} = 0 + \frac{625}{4}$$
$$\left(x - \frac{25}{2}\right)^2 = \frac{625}{4}$$
$$x - \frac{25}{2} = \sqrt{\frac{625}{4}} \quad \textbf{or} \quad x - \frac{25}{2} = -\sqrt{\frac{625}{4}}$$
$$x - \frac{25}{2} = \frac{25}{2} \qquad\qquad x - \frac{25}{2} = -\frac{25}{2}$$
$$x = \frac{50}{2} \qquad\qquad\qquad x = 0$$
$$x = 25 \qquad\qquad\qquad x = 0$$

45.
$$3x^2 + 4x = 4$$
$$x^2 + \frac{4}{3}x = \frac{4}{3}$$
$$x^2 + \frac{4}{3} + \frac{4}{9} = \frac{4}{3} + \frac{4}{9}$$
$$\left(x + \frac{2}{3}\right)^2 = \frac{16}{9}$$
$$x + \frac{2}{3} = \sqrt{\frac{16}{9}} \quad \textbf{or} \quad x + \frac{2}{3} = -\sqrt{\frac{16}{9}}$$
$$x + \frac{2}{3} = \frac{4}{3} \qquad\qquad x + \frac{2}{3} = -\frac{4}{3}$$
$$x = \frac{2}{3} \qquad\qquad\qquad x = -\frac{6}{3}$$
$$x = \frac{2}{3} \qquad\qquad\qquad x = -2$$

47.
$$x^2 + 5 = -5x$$
$$x^2 + 5x = -5$$
$$x^2 + 5x + \frac{25}{4} = -5 + \frac{25}{4}$$
$$\left(x + \frac{5}{2}\right)^2 = \frac{5}{4}$$

$$\left(x + \frac{5}{2}\right)^2 = \frac{5}{4}$$
$$x + \frac{5}{2} = \sqrt{\frac{5}{4}} \quad \textbf{or} \quad x + \frac{5}{2} = -\sqrt{\frac{5}{4}}$$
$$x + \frac{5}{2} = \frac{\sqrt{5}}{2} \qquad\qquad x + \frac{5}{2} = -\frac{\sqrt{5}}{2}$$
$$x = \frac{-5 + \sqrt{5}}{2} \qquad\qquad x = \frac{-5 - \sqrt{5}}{2}$$

49.
$$3x^2 = 1 - 4x$$
$$3x^2 + 4x = 1$$
$$x^2 + \frac{4}{3}x = \frac{1}{3}$$
$$x^2 + \frac{4}{3}x + \frac{4}{9} = \frac{1}{3} + \frac{4}{9}$$
$$\left(x + \frac{2}{3}\right)^2 = \frac{7}{9}$$

$$\left(x + \frac{2}{3}\right)^2 = \frac{7}{9}$$
$$x + \frac{2}{3} = \sqrt{\frac{7}{9}} \qquad \textbf{or} \qquad x + \frac{2}{3} = -\sqrt{\frac{7}{9}}$$
$$x + \frac{2}{3} = \frac{\sqrt{7}}{3} \qquad\qquad\qquad x + \frac{2}{3} = -\frac{\sqrt{7}}{3}$$
$$x = \frac{-2 + \sqrt{7}}{3} \qquad\qquad\qquad x = \frac{-2 - \sqrt{7}}{3}$$

51. $x^2 - 12 = 0 \Rightarrow a = 1, b = 0, c = -12$
$$x = \frac{-b \pm \sqrt{b^2 - 4ac}}{2a} = \frac{-(0) \pm \sqrt{(0)^2 - 4(1)(-12)}}{2(1)} = \frac{0 \pm \sqrt{0 + 48}}{2} = \frac{0 \pm \sqrt{48}}{2} = \frac{\pm \sqrt{48}}{2}$$
$$x = \frac{\sqrt{48}}{2} = \frac{4\sqrt{3}}{2} = 2\sqrt{3} \text{ or } x = \frac{-\sqrt{48}}{2} = \frac{-4\sqrt{3}}{2} = -2\sqrt{3}$$

53. $2x^2 - x - 15 = 0 \Rightarrow a = 2, b = -1, c = -15$
$$x = \frac{-b \pm \sqrt{b^2 - 4ac}}{2a} = \frac{-(-1) \pm \sqrt{(-1)^2 - 4(2)(-15)}}{2(2)} = \frac{1 \pm \sqrt{1 + 120}}{4} = \frac{1 \pm \sqrt{121}}{4}$$
$$= \frac{1 \pm 11}{4}$$
$$x = \frac{1 + 11}{4} = \frac{12}{4} = 3 \text{ or } x = \frac{1 - 11}{4} = \frac{-10}{4} = -\frac{5}{2}$$

55. $5x^2 - 9x - 2 = 0 \Rightarrow a = 5, b = -9, c = -2$
$$x = \frac{-b \pm \sqrt{b^2 - 4ac}}{2a} = \frac{-(-9) \pm \sqrt{(-9)^2 - 4(5)(-2)}}{2(5)} = \frac{9 \pm \sqrt{81 + 40}}{10} = \frac{9 \pm \sqrt{121}}{10}$$
$$= \frac{9 \pm 11}{10}$$
$$x = \frac{9 + 11}{10} = \frac{20}{10} = 2 \text{ or } x = \frac{9 - 11}{10} = \frac{-2}{10} = -\frac{1}{5}$$

57. $2x^2 + 2x - 4 = 0 \Rightarrow a = 2, b = 2, c = -4$
$$x = \frac{-b \pm \sqrt{b^2 - 4ac}}{2a} = \frac{-(2) \pm \sqrt{(2)^2 - 4(2)(-4)}}{2(2)} = \frac{-2 \pm \sqrt{4 + 32}}{4} = \frac{-2 \pm \sqrt{36}}{4} = \frac{-2 \pm 6}{4}$$
$$x = \frac{-2 + 6}{4} = \frac{4}{4} = 1 \text{ or } x = \frac{-2 - 6}{4} = \frac{-8}{4} = -2$$

59. $-3x^2 = 5x + 1 \Rightarrow -3x^2 - 5x - 1 = 0 \Rightarrow a = -3, b = -5, c = -1$

$$x = \frac{-b \pm \sqrt{b^2 - 4ac}}{2a} = \frac{-(-5) \pm \sqrt{(-5)^2 - 4(-3)(-1)}}{2(-3)} = \frac{5 \pm \sqrt{25 - 12}}{-6} = \frac{5 \pm \sqrt{13}}{-6}$$
$$= \frac{-5 \pm \sqrt{13}}{6}$$

61. $5x\left(x + \frac{1}{5}\right) = 3 \Rightarrow 5x^2 + x - 3 = 0 \Rightarrow a = 5, b = 1, c = -3$

$$x = \frac{-b \pm \sqrt{b^2 - 4ac}}{2a} = \frac{-(1) \pm \sqrt{(1)^2 - 4(5)(-3)}}{2(5)} = \frac{-1 \pm \sqrt{1 + 60}}{10} = \frac{-1 \pm \sqrt{61}}{10}$$

63.
$h = \frac{1}{2}gt^2$
$2h = gt^2$
$\frac{2h}{g} = t^2 \Rightarrow t = \pm\sqrt{\frac{2h}{g}}$

65.
$$h = 64t - 16t^2$$
$$16t^2 - 64t + h = 0; a = 16, b = -64, c = h$$
$$t = \frac{-b \pm \sqrt{b^2 - 4ac}}{2a}$$
$$= \frac{-(-64) \pm \sqrt{(-64)^2 - 4(16)h}}{2(16)}$$
$$= \frac{64 \pm \sqrt{4096 - 64h}}{32}$$
$$= \frac{64 \pm \sqrt{64(64 - h)}}{32}$$
$$= \frac{64 \pm 8\sqrt{64 - h}}{32} = \frac{8 \pm \sqrt{64 - h}}{4}$$

67.
$\dfrac{x^2}{a^2} + \dfrac{y^2}{b^2} = 1$

$\dfrac{y^2}{b^2} = 1 - \dfrac{x^2}{a^2}$

$\dfrac{y^2}{b^2} = \dfrac{a^2 - x^2}{a^2}$

$y^2 = \dfrac{b^2(a^2 - x^2)}{a^2}$

$y = \pm\sqrt{\dfrac{b^2(a^2 - x^2)}{a^2}}$

$y = \pm\dfrac{b\sqrt{a^2 - x^2}}{a}$

69.
$\dfrac{x^2}{a^2} - \dfrac{y^2}{b^2} = 1$

$a^2b^2\left(\dfrac{x^2}{a^2} - \dfrac{y^2}{b^2}\right) = a^2b^2(1)$

$b^2x^2 - a^2y^2 = a^2b^2$

$b^2x^2 = a^2b^2 + a^2y^2$

$b^2x^2 = a^2\left(b^2 + y^2\right)$

$\dfrac{b^2x^2}{b^2 + y^2} = a^2$

$\pm\sqrt{\dfrac{b^2x^2}{b^2 + y^2}} = a$

$\pm\dfrac{bx\sqrt{b^2 + y^2}}{b^2 + y^2} = a$

71. $x^2 + xy - y^2 = 0 \Rightarrow a = 1, b = y, c = -y^2$

$$x = \frac{-b \pm \sqrt{b^2 - 4ac}}{2a}$$

$$= \frac{-(y) \pm \sqrt{(y)^2 - 4(1)(-y^2)}}{2(1)}$$

$$= \frac{-y \pm \sqrt{y^2 + 4y^2}}{2}$$

$$= \frac{-y \pm \sqrt{5y^2}}{2} = \frac{-y \pm y\sqrt{5}}{2}$$

73. $x^2 + 6x + 9 = 0 \Rightarrow a = 1, b = 6, c = 9$
$b^2 - 4ac = 6^2 - 4(1)(9) = 36 - 36 = 0$
The solutions are rational and equal.

75. $3x^2 - 2x + 5 = 0 \Rightarrow a = 3, b = -2, c = 5$
$b^2 - 4ac = (-2)^2 - 4(3)(5)$
$\qquad = 4 - 60 = -56$
The solutions are not real numbers.

77. $10x^2 + 29x = 21 \Rightarrow 10x^2 + 29x - 21 = 0$
$a = 10, b = 29, c = -21$
$b^2 - 4ac = (29)^2 - 4(10)(-21)$
$\qquad = 841 + 840 = 1681$
The solutions are rational and unequal.

79. $-3x^2 + 2x = 21 \Rightarrow -3x^2 + 2x - 21 = 0$
$a = -3, b = 2, c = -21$
$b^2 - 4ac = (2)^2 - 4(-3)(-21)$
$\qquad = 4 - 252 = -248$
The solutions are not real numbers.

81. $1492x^2 + 1984x - 1776 = 0$
$a = 1492, b = 1984, c = -1776$
$b^2 - 4ac = (1984)^2 - 4(1492)(-1776)$
$\qquad = 3{,}936{,}256 + 10{,}599{,}168$
$\qquad = 14{,}535{,}424$
The solutions are real numbers.

83.
$$x^2 + kx + 3k - 5 = 0$$
$$a = 1, b = k, c = 3k - 5$$
Set the discriminant equal to 0:
$$b^2 - 4ac = 0$$
$$k^2 - 4(1)(3k - 5) = 0$$
$$k^2 - 4(3k - 5) = 0$$
$$k^2 - 12k + 20 = 0$$
$$(k - 2)(k - 10) = 0$$
$$k = 2 \text{ or } k = 10$$

85.
$$x + 1 = \frac{12}{x}$$
$$x(x + 1) = x\left(\frac{12}{x}\right)$$
$$x^2 + x = 12$$
$$x^2 + x - 12 = 0$$
$$(x + 4)(x - 3) = 0$$
$$x + 4 = 0 \quad \textbf{or} \quad x - 3 = 0$$
$$x = -4 \qquad\qquad x = 3$$

87.
$$8x - \frac{3}{x} = 10$$
$$x\left(8x - \frac{3}{x}\right) = x(10)$$
$$8x^2 - 3 = 10x$$
$$8x^2 - 10x - 3 = 0$$
$$(4x + 1)(2x - 3) = 0$$
$$4x + 1 = 0 \quad \textbf{or} \quad 2x - 3 = 0$$
$$4x = -1 \qquad\qquad 2x = 3$$
$$x = -\tfrac{1}{4} \qquad\qquad x = \tfrac{3}{2}$$

89.
$$\frac{5}{x} = \frac{4}{x^2} - 6$$
$$x^2\left(\frac{5}{x}\right) = x^2\left(\frac{4}{x^2} - 6\right)$$
$$5x = 4 - 6x^2$$
$$6x^2 + 5x - 4 = 0$$
$$(3x + 4)(2x - 1) = 0$$
$$3x + 4 = 0 \quad \textbf{or} \quad 2x - 1 = 0$$
$$3x = -4 \qquad\qquad 2x = 1$$
$$x = -\tfrac{4}{3} \qquad\qquad x = \tfrac{1}{2}$$

91.
$$x\left(30 - \frac{13}{x}\right) = \frac{10}{x}$$
$$30x - 13 = \frac{10}{x}$$
$$x(30x - 13) = x\left(\frac{10}{x}\right)$$
$$30x^2 - 13x = 10$$

$$30x^2 - 13x = 10$$
$$30x^2 - 13x - 10 = 0$$
$$(5x + 2)(6x - 5) = 0$$
$$5x + 2 = 0 \quad \textbf{or} \quad 6x - 5 = 0$$
$$5x = -2 \qquad\qquad 6x = 5$$
$$x = -\tfrac{2}{5} \qquad\qquad x = \tfrac{5}{6}$$

93.
$$(a - 2)(a + 4) = 2a(a - 3)$$
$$a^2 + 2a - 8 = 2a^2 - 6a$$
$$0 = a^2 - 8a + 8$$
$$a = 1, b = -8, c = 8$$
$$a = \frac{-b \pm \sqrt{b^2 - 4ac}}{2a}$$
$$= \frac{-(-8) \pm \sqrt{(-8)^2 - 4(1)(8)}}{2(1)}$$
$$= \frac{8 \pm \sqrt{32}}{2} = \frac{8 \pm 4\sqrt{2}}{2} = 4 \pm 2\sqrt{2}$$

95.
$$\frac{1}{x} + \frac{3}{x + 2} = 2$$
$$x(x + 2)\left(\frac{1}{x} + \frac{3}{x + 2}\right) = x(x + 2)(2)$$
$$1(x + 2) + 3x = 2x(x + 2)$$
$$x + 2 + 3x = 2x^2 + 4x$$
$$0 = 2x^2 - 2$$
$$0 = 2(x + 1)(x - 1)$$
$$x + 1 = 0 \quad \textbf{or} \quad x - 1 = 0$$
$$x = -1 \qquad\qquad x = 1$$

97.
$$\frac{1}{x + 1} + \frac{5}{2x - 4} = 1$$
$$(x + 1)(2x - 4)\left(\frac{1}{x + 1} + \frac{5}{2x - 4}\right) = (x + 1)(2x - 4)1$$
$$1(2x - 4) + 5(x + 1) = (x + 1)(2x - 4)$$
$$2x - 4 + 5x + 5 = 2x^2 - 2x - 4$$
$$0 = 2x^2 - 9x - 5$$
$$0 = (2x + 1)(x - 5)$$
$$2x + 1 = 0 \quad \textbf{or} \quad x - 5 = 0$$
$$2x = -1 \qquad\qquad x = 5$$
$$x = -\tfrac{1}{2} \qquad\qquad x = 5$$

99.
$$x + 1 + \frac{x+2}{x-1} = \frac{3}{x-1}$$
$$(x-1)\left(\frac{x+1}{1} + \frac{x+2}{x-1}\right) = (x-1)\frac{3}{x-1}$$
$$(x-1)(x+1) + x + 2 = 3$$
$$x^2 - 1 + x + 2 = 3$$
$$x^2 + x - 2 = 0$$
$$(x+2)(x-1) = 0$$
$$x + 2 = 0 \quad \text{or} \quad x - 1 = 0$$
$$x = -2 \qquad\qquad x = 1$$

Since $x = 1$ does not check, the only solution is $x = -2$.

101.
$$\frac{24}{a} - 11 = \frac{-12}{a+1}$$
$$a(a+1)\left(\frac{24}{a} - 11\right) = a(a+1)\frac{-12}{a+1}$$
$$24(a+1) - 11a(a+1) = a(-12)$$
$$24a + 24 - 11a^2 - 11a = -12a$$
$$0 = 11a^2 - 25a - 24$$
$$0 = (11a + 8)(a - 3)$$
$$11a + 8 = 0 \quad \text{or} \quad a - 3 = 0$$
$$11a = -8 \qquad\qquad a = 3$$
$$a = -\frac{8}{11} \qquad\qquad a = 3$$

103. If r_1 and r_2 are the roots of $ax^2 + bx + c = 0$, then their values are
$$r_1 = \frac{-b + \sqrt{b^2 - 4ac}}{2a} \text{ and } r_2 = \frac{-b - \sqrt{b^2 - 4ac}}{2a}.$$
$$r_1 + r_2 = \frac{-b + \sqrt{b^2 - 4ac}}{2a} + \frac{-b - \sqrt{b^2 - 4ac}}{2a} = \frac{-2b}{2a} = -\frac{b}{a}$$

105. Rewrite the equation as $16t^2 - v_0 t + h = 0$ and solve for t using the quadratic formula.
$a = 16, b = -v_0, c = h$
$$t = \frac{-b \pm \sqrt{b^2 - 4ac}}{2a} = \frac{-(-v_0) \pm \sqrt{(-v_0)^2 - 4(16)(h)}}{2(16)} = \frac{v_0 \pm \sqrt{v_0^2 - 64h}}{32}$$
Since t_1 and t_2 are the solutions to the equation, we have
$$t_1 = \frac{v_0 + \sqrt{v_0^2 - 64h}}{32} \text{ and } t_2 = \frac{v_0 - \sqrt{v_0^2 - 64h}}{32}. \text{ Calculate } 16t_1 t_2:$$
$$16t_1 t_2 = 16 \cdot \frac{v_0 + \sqrt{v_0^2 - 64h}}{32} \cdot \frac{v_0 - \sqrt{v_0^2 - 64h}}{32} = 16 \cdot \frac{v_0^2 - (v_0^2 - 64h)}{1024}$$
$$= \frac{16 \cdot 64h}{1024} = \frac{1024h}{1024} = h$$

Thus, $h = 16t_1 t_2$.

107. Answers may vary.

109. $5x(x-2) - x(3x-2) = 5x^2 - 10x - 3x^2 + 2x = 2x^2 - 8x$

111. $(m+3)^2 - (m-3)^2 = (m+3)(m+3) - (m-3)(m-3)$
$$= m^2 + 6m + 9 - (m^2 - 6m + 9)$$
$$= m^2 + 6m + 9 - m^2 + 6m - 9 = 12m$$

113. $\sqrt{50x^3} - x\sqrt{8x} = \sqrt{25x^2}\sqrt{2x} - x\sqrt{4}\sqrt{2x} = 5x\sqrt{2x} - 2x\sqrt{2x} = 3x\sqrt{2x}$

Exercise 1.4 (page 112)

1. $A = lw$

3. Let w = the width of the rectangle.
Then $w + 4$ = the length.

$$\boxed{\text{Width}} \cdot \boxed{\text{Length}} = \boxed{\text{Area}}$$

$$w(w + 4) = 32$$
$$w^2 + 4w = 32$$
$$w^2 + 4w - 32 = 0$$
$$(w + 8)(w - 4) = 0$$
$$w + 8 = 0 \quad \textbf{or} \quad w - 4 = 0$$
$$w = -8 \qquad\qquad w = 4$$

Since the width cannot be negative, the only reasonable solution is $w = 4$.
The dimensions are 4 feet by 8 feet.

5. Let s = the side of the second square.
Then $s - 4$ = the side of the first square.

$$\boxed{\begin{array}{c}\text{Area of}\\\text{first}\end{array}} + \boxed{\begin{array}{c}\text{Area of}\\\text{second}\end{array}} = 106$$

$$(s - 4)^2 + s^2 = 106$$
$$s^2 - 8s + 16 + s^2 = 106$$
$$2s^2 - 8s - 90 = 0$$
$$2(s^2 - 4s - 45) = 0$$
$$2(s + 5)(s - 9) = 0$$
$$s + 5 = 0 \quad \textbf{or} \quad s - 9 = 0$$
$$s = -5 \qquad\qquad s = 9$$

Since the side cannot be negative, the only reasonable solution is $s = 9$.
The larger square has a side of length 9 cm.

7. The floor area of the box is a square with a side of length $12 - 2x$.

$$\boxed{\text{Floor area}} = 64$$

$$(12 - 2x)^2 = 64$$
$$144 - 48x + 4x^2 = 64$$
$$4x^2 - 48x + 80 = 0$$
$$4(x^2 - 12x + 20) = 0$$
$$4(x - 2)(x - 10) = 0$$
$$x - 2 = 0 \quad \textbf{or} \quad x - 10 = 0$$
$$x = 2 \qquad\qquad x = 10$$

The solution $x = 10$ does not make sense in the problem, so the depth is 2 inches.

9. Let x = the height of the tube. Then $x + 6$ = the width. Use the Pythagorean Theorem:

$$\text{height}^2 + \text{width}^2 = \text{diagonal}^2$$
$$x^2 + (x + 6)^2 = 52^2$$
$$x^2 + x^2 + 12x + 36 = 2704$$
$$2x^2 + 12x - 2668 = 0$$
$$a = 2, b = 12, c = -2668$$
$$x = \frac{-b \pm \sqrt{b^2 - 4ac}}{2a}$$
$$= \frac{-12 \pm \sqrt{12^2 - 4(2)(-2668)}}{2(2)}$$
$$= \frac{-12 \pm \sqrt{21488}}{4} \approx \frac{-12 \pm 146.588}{4}$$

The only positive solution is 33.6. The dimensions are 33.6 inches by 39.6 inches.

11. Let r = the cyclist's rate from DeKalb to Rockford. Then his return rate is $r - 10$.

$$\boxed{\text{Return time}} = \boxed{\text{First time}} + 2$$

	Rate	Time	Dist.
First trip	r	$\frac{40}{r}$	40
Return trip	$r - 10$	$\frac{40}{r-10}$	40

$$\frac{40}{r - 10} = \frac{40}{r} + 2$$

$$r(r - 10)\frac{40}{r - 10} = r(r - 10)\left(\frac{40}{r} + 2\right)$$

$$40r = 40(r - 10) + 2r(r - 10)$$

$$40r = 40r - 400 + 2r^2 - 20r$$

$$0 = 2r^2 - 20r - 400$$

$$0 = 2(r - 20)(r + 10)$$

$r - 20 = 0$ **or** $r + 10 = 0$ Since $r = -10$ does not make sense, the solution is $r = 20$.

$\qquad r = 20 \qquad\qquad r = -10$ The cyclist rides 20 mph going and 10 mph returning.

13. Let r = the slower rate. Then the faster rate is $r + 10$.

$$\boxed{\text{Faster time}} = \boxed{\text{Slower time}} - 1$$

	Rate	Time	Dist.
Slower trip	r	$\frac{420}{r}$	420
Faster trip	$r + 10$	$\frac{420}{r+10}$	420

$$\frac{420}{r + 10} = \frac{420}{r} - 1$$

$$r(r + 10)\frac{420}{r + 10} = r(r + 10)\left(\frac{420}{r} - 1\right)$$

$$420r = 420(r + 10) - r(r + 10)$$

$$420r = 420r + 4200 - r^2 - 10r$$

$$r^2 + 10r - 4200 = 0$$

$$(r - 60)(r + 70) = 0$$

$r - 60 = 0$ **or** $r + 70 = 0$ Since $r = -70$ does not make sense, the solution is $r = 60$.

$\qquad r = 60 \qquad\qquad r = -70$ The slower speed results in a trip of length 7 hours.

15. Set $h = 0$:

$$h = -16t^2 + 400t$$

$$0 = -16t^2 + 400t$$

$$16t^2 - 400t = 0$$

$$16t(t - 25) = 0$$

$16t = 0$ **or** $t - 25 = 0$

$\quad t = 0 \qquad\qquad t = 25$

$t = 0$ represents when the projectile was fired, so it returns to earth after 25 seconds.

17. Set $s = 1454$:

$$s = 16t^2$$

$$1454 = 16t^2$$

$$\frac{1454}{16} = t^2$$

$$t = \sqrt{\frac{1454}{16}} \quad \text{or} \quad t = -\sqrt{\frac{1454}{16}}$$

$$t \approx 9.5 \qquad\qquad t \approx -9.5$$

$t = -9.5$ does not make sense, so it takes the penny about 9.5 seconds to hit the ground.

19. Let $x =$ the number of nickel increases.
The new fare $= 25 + 5x$ (in cents), while
the number of passengers $= 3000 - 80x$.

$$\boxed{\begin{array}{c}\text{Number of}\\\text{passengers}\end{array}} \cdot \boxed{\text{Fare}} = \boxed{\text{Revenue}}$$

$(3000 - 80x)(25 + 5x) = 99400$
$75000 + 13000x - 400x^2 = 99400$
$400x^2 - 13000x + 24400 = 0$
$200\left(2x^2 - 65x + 122\right) = 0$
$200(2x - 61)(x - 2) = 0$
$2x - 61 = 0 \quad \text{or} \quad x - 2 = 0$
$x = 30.5 \qquad\qquad x = 2$

Since you cannot have half of a nickel
increase, $x = 30.5$ does not make sense.
Thus, there should be 2 nickel increases,
for a fare increase of 10 cents.

21. Let $x =$ the number of $0.50 decreases.
The new price $= 15 - 0.5x$, while the
number attending $= 1200 + 40x$.

$$\boxed{\begin{array}{c}\text{Number}\\\text{attending}\end{array}} \cdot \boxed{\text{Price}} = \boxed{\text{Revenue}}$$

$(1200 + 40x)(15 - 0.5x) = 17280$
$18000 - 20x^2 = 17280$
$20x^2 = 720$
$x^2 = 36$
$x = \sqrt{36} \quad \text{or} \quad x = -\sqrt{36}$
$x = 6 \qquad\qquad x = -6$

$x = -6$ does not make sense. Thus, there
should be six 50-cent decreases, for a ticket
price of $12 and an attendance of 1440 people.

23. Let $h = 15$:

$$\frac{l}{h} = \frac{h}{l - h}$$
$$\frac{l}{15} = \frac{15}{l - 15}$$
$$15(l - 15) \cdot \frac{l}{15} = 15(l - 15) \cdot \frac{15}{l - 15}$$
$$l(l - 15) = 15^2$$
$$l^2 - 15l - 225 = 0$$

$a = 1, b = -15, c = -225$

$$l = \frac{-b \pm \sqrt{b^2 - 4ac}}{2a}$$
$$= \frac{-(-15) \pm \sqrt{(-15)^2 - 4(1)(-225)}}{2(1)}$$
$$= \frac{15 \pm \sqrt{1125}}{2} \approx \frac{15 \pm 33.541}{2}$$

The only positive solution is $l = 24.3$ ft.

25. Let $x =$ time for the second pipe to fill tank.

$$\boxed{\begin{array}{c}\text{First in}\\\text{1 hour}\end{array}} + \boxed{\begin{array}{c}\text{Second in}\\\text{1 hour}\end{array}} = \boxed{\begin{array}{c}\text{Total in}\\\text{1 hour}\end{array}}$$

$$\frac{1}{4} + \frac{1}{x} = \frac{1}{x - 2}$$
$$4x(x - 2)\left(\frac{1}{4} + \frac{1}{x}\right) = 4x(x - 2) \cdot \frac{1}{x - 2}$$
$$x(x - 2) + 4(x - 2) = 4x$$
$$x^2 - 2x + 4x - 8 = 4x$$
$$x^2 - 2x - 8 = 0$$

$x^2 - 2x - 8 = 0$
$(x - 4)(x + 2) = 0$
$x - 4 = 0 \quad \text{or} \quad x + 2 = 0$
$x = 4 \qquad\qquad x = -2$

Since $x = -2$ does not make sense, the
solution is $x = 4$. It takes the second pipe
4 hours to fill the tank alone.

27. Let x = time for the Steven to mow lawn.

Steven in 1 hour	+	Kristy in 1 hour	=	Total in 1 hour

$$\frac{1}{x} + \frac{1}{x-1} = \frac{1}{5}$$

$$5x(x-1)\left(\frac{1}{x} + \frac{1}{x-1}\right) = 5x(x-1)\cdot\frac{1}{5}$$

$$5(x-1) + 5x = x(x-1)$$

$$5x - 5 + 5x = x^2 - x$$

$$0 = x^2 - 11x + 5$$

$a = 1, b = -11, c = 5$

$$x = \frac{-b \pm \sqrt{b^2 - 4ac}}{2a}$$

$$= \frac{-(-11) \pm \sqrt{(-11)^2 - 4(1)(5)}}{2(1)}$$

$$= \frac{11 \pm \sqrt{101}}{2} \approx 10.5 \text{ or } 0.5$$

$x = 0.5$ does not make sense, so Kristy could mow the lawn in about 9.5 hours alone.

29. Let x = the length of the diagonal.

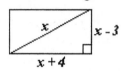

Using the Pythagorean Theorem:

$$(x+4)^2 + (x-3)^2 = x^2$$

$$x^2 + 8x + 16 + x^2 - 6x + 9 = x^2$$

$$x^2 + 2x + 25 = 0$$

$a = 1, b = 2, c = 25$

$$x = \frac{-b \pm \sqrt{b^2 - 4ac}}{2a}$$

$$= \frac{-2 \pm \sqrt{2^2 - 4(1)(25)}}{2(1)} = \frac{-2 \pm \sqrt{-96}}{2}$$

This does not equal a real number, so it is not possible.

31. Let x = Matilda's principal.

Maude's rate	=	Matilda's rate	− 0.01

$$\frac{280}{x+1000} = \frac{240}{x} - 0.01$$

	I	P	r
Matilda	240	x	$\frac{240}{x}$
Maude	280	$x + 1000$	$\frac{280}{x+1000}$

$$x(x+1000)\frac{280}{x+1000} = x(x+1000)\left(\frac{240}{x} - 0.01\right)$$

$$280x = 240(x+1000) - 0.01x(x+1000)$$

$$0.01x^2 + 50x - 240{,}000 = 0$$

$$x^2 + 5000x - 24{,}000{,}000 = 0$$

$$(x - 3000)(x + 8000) = 0$$

$x - 3000 = 0 \quad$ **or** $\quad x + 8000 = 0$

$x = 3000 \qquad\qquad x = -8000 \Rightarrow$

$x = -8000$ does not make sense. The principal amounts were $3000 and $4000. The interest rates were 8% for Matilda and 7% for Maude.

33. Let $x =$ the total number of professors.

$$\boxed{\begin{array}{c}\text{New share with}\\ \text{lower number}\end{array}} = \boxed{\begin{array}{c}\text{Original}\\ \text{share}\end{array}} + 10$$

$$\frac{150}{x-4} = \frac{150}{x} + 10$$

$$x(x-4)\frac{150}{x-4} = x(x-4)\left(\frac{150}{x} + 10\right)$$

$$150x = 150(x-4) + 10x(x-4)$$

$$150x = 150x - 600 + 10x^2 - 40x$$

$$0 = 10x^2 - 40x - 600$$

$$0 = 10x^2 - 40x - 600$$
$$0 = x^2 - 4x - 60$$
$$0 = (x-10)(x+6)$$
$$x - 10 = 0 \quad \text{or} \quad x + 6 = 0$$
$$x = 10 \qquad\qquad x = -6$$

$x = -6$ does not make sense, so there are 10 professors in the department.

35. Let $x =$ the actual number of spokes.

$$\boxed{\begin{array}{c}\text{Actual angles}\\ \text{between spokes}\end{array}} - 6 = \boxed{\begin{array}{c}\text{New angle}\\ \text{between spokes}\end{array}}$$

$$\frac{360}{x} - 6 = \frac{360}{x+10}$$

$$x(x+10)\left(\frac{360}{x} - 6\right) = x(x+10)\frac{360}{x+10}$$

$$360(x+10) - 6x(x+10) = 360x$$

$$6x^2 + 60x - 3600 = 0$$

$$6x^2 + 60x - 3600 = 0$$
$$x^2 + 10x - 600 = 0$$
$$(x-20)(x+30) = 0$$
$$x - 20 = 0 \quad \text{or} \quad x + 30 = 0$$
$$x = 20 \qquad\qquad x = -30$$

$x = -30$ does not make sense, so there are 20 spokes.

37. Let x and $x - 14$ represent the two legs.
Use the Pythagorean Theorem:

$$x^2 + (x-14)^2 = 26^2$$
$$x^2 + x^2 - 28x + 196 = 676$$
$$2x^2 - 28x - 480 = 0$$
$$x^2 - 14x - 240 = 0$$

$$x^2 - 14x - 240 = 0$$
$$(x-24)(x+10) = 0$$
$$x - 24 = 0 \quad \text{or} \quad x + 10 = 0$$
$$x = 24 \qquad\qquad x = -10$$

$x = -10$ does not make sense. The legs have lengths of 24 meters and 10 meters.

39. **Answers may vary.**

41. $\dfrac{2}{x} - \dfrac{1}{x-3} = \dfrac{2(x-3)}{x(x-3)} - \dfrac{1(x)}{(x-3)(x)} = \dfrac{2x-6}{x(x-3)} - \dfrac{x}{x(x-3)} = \dfrac{x-6}{x(x-3)}$

43. $\dfrac{x+3}{x^2-x-6} \div \dfrac{x^2+3x}{x^2-9} = \dfrac{x+3}{x^2-x-6} \cdot \dfrac{x^2-9}{x^2+3x} = \dfrac{x+3}{(x-3)(x+2)} \cdot \dfrac{(x+3)(x-3)}{x(x+3)} = \dfrac{x+3}{x(x+2)}$

45. $\dfrac{\frac{1}{x} + \frac{1}{y}}{\frac{1}{x} - \frac{1}{y}} = \dfrac{xy\left(\frac{1}{x} + \frac{1}{y}\right)}{xy\left(\frac{1}{x} - \frac{1}{y}\right)} = \dfrac{xy\left(\frac{1}{x}\right) + xy\left(\frac{1}{y}\right)}{xy\left(\frac{1}{x}\right) - xy\left(\frac{1}{y}\right)} = \dfrac{y+x}{y-x}$

Exercise 1.5 (page 123)

1. imaginary   **3.** imaginary   **5.** $2 - 5i$ **7.** real

9. Equate real parts: $\boxed{x = 3}$ Equate imaginary parts: $x + y = 8$

$$3 + y = 8$$

$$\boxed{y = 5}$$

11. Equate real parts: $3x = 2$ Equate imaginary parts: $-2y = x + y$

$$\boxed{x = \frac{2}{3}}$$

$$-3y = x$$

$$y = -\frac{1}{3}x$$

$$y = -\frac{1}{3} \cdot \frac{2}{3}$$

$$\boxed{y = -\frac{2}{9}}$$

13. $(2 - 7i) + (3 + i) = 2 - 7i + 3 + i$

$$= 5 - 6i$$

15. $(5 - 6i) - (7 + 4i) = 5 - 6i - 7 - 4i$

$$= -2 - 10i$$

17. $(14i + 2) + \left(2 - \sqrt{-16}\right) = (14i + 2) + (2 - 4i) = 14i + 2 + 2 - 4i = 4 + 10i$

19. $\left(3 + \sqrt{-4}\right) - \left(2 + \sqrt{-9}\right) = (3 + 2i) - (2 + 3i) = 3 + 2i - 2 - 3i = 1 - i$

21. $(2 + 3i)(3 + 5i) = 6 + 19i + 15(-1) = 6 + 19i - 15 = -9 + 19i$

23. $(2 + 3i)^2 = (2 + 3i)(2 + 3i) = 4 + 12i + 9i^2 = 4 + 12i + 9(-1) = 4 + 12i - 9 = -5 + 12i$

25. $\left(11 + \sqrt{-25}\right)\left(2 - \sqrt{-36}\right) = (11 + 5i)(2 - 6i) = 22 - 56i - 30i^2 = 22 - 56i - 30(-1)$

$$= 22 - 56i + 30 = 52 - 56i$$

27. $\left(\sqrt{-16} + 3\right)\left(2 + \sqrt{-9}\right) = (4i + 3)(2 + 3i) = 6 + 17i + 12i^2 = 6 + 17i + 12(-1)$

$$= 6 + 17i - 12 = -6 + 17i$$

29. $\dfrac{1}{i^3} = \dfrac{1 \cdot i}{i^3 \cdot i} = \dfrac{i}{i^4} = \dfrac{i}{1} = i = 0 + i$

31. $\dfrac{-4}{i^{10}} = \dfrac{-4 \cdot i^2}{i^{10} \cdot i^2} = \dfrac{-4i^2}{i^{12}} = \dfrac{-4i^2}{1} = -4(-1)$

$$= 4 + 0i$$

33. $\dfrac{1}{2 + i} = \dfrac{1(2 - i)}{(2 + i)(2 - i)} = \dfrac{2 - i}{2^2 - i^2} = \dfrac{2 - i}{4 - (-1)} = \dfrac{2 - i}{5} = \dfrac{2}{5} - \dfrac{1}{5}i$

35. $\dfrac{2i}{7 + i} = \dfrac{2i(7 - i)}{(7 + i)(7 - i)} = \dfrac{14i - 2i^2}{7^2 - i^2} = \dfrac{14i - 2(-1)}{49 - (-1)} = \dfrac{14i + 2}{50} = \dfrac{7i + 1}{25} = \dfrac{1}{25} + \dfrac{7}{25}i$

37. $\dfrac{2 + i}{3 - i} = \dfrac{(2 + i)(3 + i)}{(3 - i)(3 + i)} = \dfrac{6 + 5i + i^2}{9 - i^2} = \dfrac{5 + 5i}{10} = \dfrac{5}{10} + \dfrac{5}{10}i = \dfrac{1}{2} + \dfrac{1}{2}i$

39. $\dfrac{4 - 5i}{2 + 3i} = \dfrac{(4 - 5i)(2 - 3i)}{(2 + 3i)(2 - 3i)} = \dfrac{8 - 22i + 15i^2}{4 - 9i^2} = \dfrac{-7 - 22i}{13} = -\dfrac{7}{13} - \dfrac{22}{13}i$

41. $\dfrac{5 - \sqrt{-16}}{-8 + \sqrt{-4}} = \dfrac{5 - 4i}{-8 + 2i} = \dfrac{(5 - 4i)(-8 - 2i)}{(-8 + 2i)(-8 - 2i)} = \dfrac{-40 + 22i + 8i^2}{64 - 4i^2} = \dfrac{-48 + 22i}{68} = -\dfrac{48}{68} + \dfrac{22}{68}i$

$$= -\dfrac{12}{17} + \dfrac{11}{34}i$$

43. $\dfrac{2 + i\sqrt{3}}{3 + i} = \dfrac{\left(2 + i\sqrt{3}\right)(3 - i)}{(3 + i)(3 - i)} = \dfrac{6 - 2i + 3i\sqrt{3} - i^2\sqrt{3}}{9 - i^2} = \dfrac{6 + \sqrt{3} + \left(3\sqrt{3} - 2\right)i}{10}$

$$= \dfrac{6 + \sqrt{3}}{10} + \dfrac{3\sqrt{3} - 2}{10}i$$

45. $i^9 = i^8 i = (i^4)^2 i = 1^2 i = i$

47. $i^{38} = i^{36} i^2 = (i^4)^9 i^2 = 1^9 i^2 = i^2 = -1$

49. $i^{-6} = \dfrac{1}{i^6} = \dfrac{1 \cdot i^2}{i^6 \cdot i^2} = \dfrac{i^2}{i^8} = \dfrac{i^2}{1} = i^2 = -1$

51. $i^{-10} = \dfrac{1}{i^{10}} = \dfrac{1 \cdot i^2}{i^{10} \cdot i^2} = \dfrac{i^2}{i^{12}} = \dfrac{i^2}{1} = i^2$

$$= -1$$

53. $|3 + 4i| = \sqrt{3^2 + 4^2} = \sqrt{9 + 16}$

$$= \sqrt{25} = 5$$

55. $|2 + 3i| = \sqrt{2^2 + 3^2} = \sqrt{4 + 9} = \sqrt{13}$

57. $\left|-7 + \sqrt{-49}\right| = |-7 + 7i| = \sqrt{(-7)^2 + 7^2} = \sqrt{49 + 49} = \sqrt{98} = 7\sqrt{2}$

59. $\left|\dfrac{1}{2} + \dfrac{1}{2}i\right| = \sqrt{\left(\dfrac{1}{2}\right)^2 + \left(\dfrac{1}{2}\right)^2} = \sqrt{\dfrac{1}{4} + \dfrac{1}{4}} = \sqrt{\dfrac{1}{2}} = \dfrac{\sqrt{2}}{2}$

61. $|-6i| = |0 - 6i| = \sqrt{0^2 + (-6)^2} = \sqrt{0 + 36} = \sqrt{36} = 6$

63. $\left|\dfrac{2}{1 + i}\right| = \left|\dfrac{2(1 - i)}{(1 + i)(1 - i)}\right| = \left|\dfrac{2(1 - i)}{1 - i^2}\right| = \left|\dfrac{2(1 - i)}{2}\right| = |1 - i| = \sqrt{1^2 + (-1)^2} = \sqrt{2}$

65. $\left|\dfrac{-3i}{2 + i}\right| = \left|\dfrac{-3i(2 - i)}{(2 + i)(2 - i)}\right| = \left|\dfrac{-3i(2 - i)}{4 - i^2}\right| = \left|\dfrac{-3i(2 - i)}{5}\right| = \left|\dfrac{-6i + 3i^2}{5}\right|$

$$= \left|-\dfrac{3}{5} - \dfrac{6}{5}i\right|$$

$$= \sqrt{\left(-\dfrac{3}{5}\right)^2 + \left(-\dfrac{6}{5}\right)^2}$$

$$= \sqrt{\dfrac{9}{25} + \dfrac{36}{25}}$$

$$= \sqrt{\dfrac{45}{25}} = \dfrac{\sqrt{45}}{5} = \dfrac{3\sqrt{5}}{5}$$

67. $\left|\dfrac{i+2}{i-2}\right| = \left|\dfrac{(i+2)(i+2)}{(i-2)(i+2)}\right| = \left|\dfrac{i^2+4i+4}{i^2-4}\right| = \left|\dfrac{3+4i}{5}\right| = \left|\dfrac{3}{5}+\dfrac{4}{5}i\right| = \sqrt{\left(\dfrac{3}{5}\right)^2+\left(\dfrac{4}{5}\right)^2}$

$$= \sqrt{\tfrac{9}{25}+\tfrac{16}{25}}$$

$$= \sqrt{\tfrac{25}{25}} = \sqrt{1} = 1$$

69. $x^2+2x+2 = 0 \Rightarrow a=1, b=2, c=2$

$x = \dfrac{-b\pm\sqrt{b^2-4ac}}{2a} = \dfrac{-2\pm\sqrt{2^2-4(1)(2)}}{2(1)} = \dfrac{-2\pm\sqrt{4-8}}{2} = \dfrac{-2\pm\sqrt{-4}}{2} = \dfrac{-2\pm2i}{2}$

$$= -1\pm i$$

71. $y^2+4y+5 = 0 \Rightarrow a=1, b=4, c=5$

$y = \dfrac{-b\pm\sqrt{b^2-4ac}}{2a} = \dfrac{-4\pm\sqrt{4^2-4(1)(5)}}{2(1)} = \dfrac{-4\pm\sqrt{16-20}}{2} = \dfrac{-4\pm\sqrt{-4}}{2} = \dfrac{-4\pm2i}{2}$

$$= -2\pm i$$

73. $x^2-2x = -5 \Rightarrow x^2-2x+5 = 0 \Rightarrow a=1, b=-2, c=5$

$x = \dfrac{-b\pm\sqrt{b^2-4ac}}{2a} = \dfrac{-(-2)\pm\sqrt{(-2)^2-4(1)(5)}}{2(1)} = \dfrac{2\pm\sqrt{4-20}}{2} = \dfrac{2\pm\sqrt{-16}}{2} = \dfrac{2\pm4i}{2}$

$$= 1\pm2i$$

75. $x^2-\dfrac{2}{3}x = -\dfrac{2}{9} \Rightarrow x^2-\dfrac{2}{3}x+\dfrac{2}{9} = 0 \Rightarrow 9x^2-6x+2 = 0 \Rightarrow a=9, b=-6, c=2$

$x = \dfrac{-b\pm\sqrt{b^2-4ac}}{2a} = \dfrac{-(-6)\pm\sqrt{(-6)^2-4(9)(2)}}{2(9)} = \dfrac{6\pm\sqrt{36-72}}{18} = \dfrac{6\pm\sqrt{-36}}{18}$

$$= \dfrac{6\pm6i}{18} = \dfrac{1}{3}\pm\dfrac{1}{3}i$$

77. $x^2+4 = x^2-(-4) = x^2-(2i)^2 = (x+2i)(x-2i)$

79. $25p^2+36q^2 = (5p)^2-(-36q^2) = (5p)^2-(6qi)^2 = (5p+6qi)(5p-6qi)$

81. $2y^2+8z^2 = 2(y^2+4z^2) = 2[y^2-(-4z^2)] = 2\left[y^2-(2zi)^2\right] = 2(y+2zi)(y-2zi)$

83. $50m^2+2n^2 = 2(25m^2+n^2) = 2\left[(5m)^2-(-n^2)\right] = 2\left[(5m)^2-(ni)^2\right] = 2(5m+ni)(5m-ni)$

85. $V = IR = (3-2i)(3+6i) = 9+18i-6i-12i^2 = 9+12i+12 = 21+12i$

87. $(a+bi)+(c+di) = a+bi+c+di \qquad (c+di)+(a+bi) = c+di+a+bi$

$\qquad\qquad\qquad\qquad = a+c+bi+di \qquad\qquad\qquad\qquad\quad = c+a+di+bi$

$\qquad\qquad\qquad\qquad = (a+c)+(b+d)i \qquad\qquad\qquad\quad = (c+a)+(d+b)i$

$\qquad\qquad\qquad\qquad\qquad\qquad\qquad\qquad\qquad\qquad\qquad\qquad\quad = (a+c)+(b+d)i$

89.
$$[(a+bi)+(c+di)]+(e+fi) = a+bi+c+di+e+fi$$
$$= a+c+e+bi+di+fi$$
$$= (a+c+e)+(b+d+f)i$$
$$(a+bi)+[(c+di)+(e+fi)] = a+bi+c+di+e+fi$$
$$= a+c+e+bi+di+fi$$
$$= (a+c+e)+(b+d+f)i$$

91. **Answers may vary.**

93. $\sqrt{8x^3}\sqrt{4x} = \sqrt{32x^4} = \sqrt{16x^4}\sqrt{2} = 4x^2\sqrt{2}$

95. $\left(\sqrt{x+1}-2\right)^2 = \left(\sqrt{x+1}-2\right)\left(\sqrt{x+1}-2\right) = x+1-4\sqrt{x+1}+4 = x-4\sqrt{x+1}+5$

97. $\dfrac{4}{\sqrt{5}-1} = \dfrac{4}{\sqrt{5}-1}\cdot\dfrac{\sqrt{5}+1}{\sqrt{5}+1} = \dfrac{4\left(\sqrt{5}+1\right)}{\left(\sqrt{5}\right)^2-1^2} = \dfrac{4\left(\sqrt{5}+1\right)}{5-1} = \dfrac{4\left(\sqrt{5}+1\right)}{4} = \sqrt{5}+1$

Exercise 1.6 (page 130)

1. equal

3. extraneous

5.
$$x^3 + 9x^2 + 20x = 0$$
$$x\left(x^2 + 9x + 20\right) = 0$$
$$x(x+5)(x+4) = 0$$
$$x=0 \quad \text{or} \quad x+5=0 \quad \text{or} \quad x+4=0$$
$$x=0 \qquad\qquad x=-5 \qquad\qquad x=-4$$

7.
$$6a^3 - 5a^2 - 4a = 0$$
$$a\left(6a^2 - 5a - 4\right) = 0$$
$$a(2a+1)(3a-4) = 0$$
$$a=0 \quad \text{or} \quad 2a+1=0 \quad \text{or} \quad 3a-4=0$$
$$a=0 \qquad\qquad 2a=-1 \qquad\qquad 3a=4$$
$$a=0 \qquad\qquad a=-\tfrac{1}{2} \qquad\qquad a=\tfrac{4}{3}$$

9.
$$y^4 - 26y^2 + 25 = 0$$
$$\left(y^2 - 25\right)\left(y^2 - 1\right) = 0$$
$$y^2 - 25 = 0 \quad \text{or} \quad y^2 - 1 = 0$$
$$y^2 = 25 \qquad\qquad y^2 = 1$$
$$y = \pm 5 \qquad\qquad y = \pm 1$$

11.
$$x^4 - 37x^2 + 36 = 0$$
$$\left(x^2 - 36\right)\left(x^2 - 1\right) = 0$$
$$x^2 - 36 = 0 \quad \text{or} \quad x^2 - 1 = 0$$
$$x^2 = 36 \qquad\qquad x^2 = 1$$
$$x = \pm 6 \qquad\qquad x = \pm 1$$

13.
$$2y^4 - 46y^2 = -180$$
$$2\left(y^4 - 23y^2 + 90\right) = 0$$
$$2\left(y^2 - 18\right)\left(y^2 - 5\right) = 0$$
$$y^2 - 18 = 0 \quad \text{or} \quad y^2 - 5 = 0$$
$$y^2 = 18 \qquad\qquad y^2 = 5$$
$$y = \pm\sqrt{18} \qquad\qquad y = \pm\sqrt{5}$$
$$y = \pm 3\sqrt{2} \qquad\qquad y = \pm\sqrt{5}$$

15.
$$z^{3/2} - z^{1/2} = 0$$
$$z^{1/2}\left(z^{2/2} - 1\right) = 0$$
$$z^{1/2}(z - 1) = 0$$
$$z^{1/2} = 0 \quad \text{or} \quad z - 1 = 0$$
$$\left(z^{1/2}\right)^2 = 0^2 \qquad\qquad z = 1$$
$$z = 0 \qquad\qquad z = 1$$
Both answers check.

SECTION 1.6

17.

$$2m^{2/3} + 3m^{1/3} - 2 = 0$$
$$\left(2m^{1/3} - 1\right)\left(m^{1/3} + 2\right) = 0$$

$2m^{1/3} - 1 = 0$	**or**	$m^{1/3} + 2 = 0$

$$m^{1/3} = \tfrac{1}{2} \qquad\qquad m^{1/3} = -2$$
$$\left(m^{1/3}\right)^3 = \left(\tfrac{1}{2}\right)^3 \qquad\qquad \left(m^{1/3}\right)^3 = (-2)^3$$
$$m = \tfrac{1}{8} \qquad\qquad m = -8$$

Both answers check.

19.

$$x - 13x^{1/2} + 12 = 0$$
$$\left(x^{1/2} - 12\right)\left(x^{1/2} - 1\right) = 0$$

$x^{1/2} - 12 = 0$	**or**	$x^{1/2} - 1 = 0$

$$x^{1/2} = 12 \qquad\qquad x^{1/2} = 1$$
$$\left(x^{1/2}\right)^2 = (12)^2 \qquad\qquad \left(x^{1/2}\right)^2 = (1)^2$$
$$x = 144 \qquad\qquad x = 1$$

Both answers check.

21.

$$2t^{1/3} + 3t^{1/6} - 2 = 0$$
$$\left(2t^{1/6} - 1\right)\left(t^{1/6} + 2\right) = 0$$

$2t^{1/6} - 1 = 0$	**or**	$t^{1/6} + 2 = 0$

$$t^{1/6} = \tfrac{1}{2} \qquad\qquad t^{1/6} = -2$$
$$\left(t^{1/6}\right)^6 = \left(\tfrac{1}{2}\right)^6 \qquad\qquad \left(t^{1/6}\right)^6 = (-2)^6$$
$$t = \tfrac{1}{64} \qquad\qquad t = 64$$

$t = 64$ does not check and is extraneous.

23.

$$6p + p^{1/2} - 1 = 0$$
$$\left(2p^{1/2} + 1\right)\left(3p^{1/2} - 1\right) = 0$$

$2p^{1/2} + 1 = 0$	**or**	$3p^{1/2} - 1 = 0$

$$p^{1/2} = -\tfrac{1}{2} \qquad\qquad p^{1/2} = \tfrac{1}{3}$$
$$\left(p^{1/2}\right)^2 = \left(-\tfrac{1}{2}\right)^2 \qquad\qquad \left(p^{1/2}\right)^2 = \left(\tfrac{1}{3}\right)^2$$
$$p = \tfrac{1}{4} \qquad\qquad p = \tfrac{1}{9}$$

$p = \tfrac{1}{4}$ does not check and is extraneous.

25.

$$\sqrt{x - 2} = 5$$
$$\left(\sqrt{x - 2}\right)^2 = 5^2$$
$$x - 2 = 25$$
$$x = 27$$

The solution checks.

27.

$$3\sqrt{x + 1} = \sqrt{6}$$
$$\left(3\sqrt{x + 1}\right)^2 = \left(\sqrt{6}\right)^2$$
$$9(x + 1) = 6$$
$$9x + 9 = 6$$
$$9x = -3$$
$$x = -\tfrac{1}{3}$$

The solution checks.

29.

$$\sqrt{5a - 2} = \sqrt{a + 6}$$
$$\left(\sqrt{5a - 2}\right)^2 = \left(\sqrt{a + 6}\right)^2$$
$$5a - 2 = a + 6$$
$$4a = 8$$
$$a = 2$$

The solution checks.

55. $\sqrt{y+8} - \sqrt{y-4} = -2$

$\sqrt{y+8} = \sqrt{y-4} - 2$

$\left(\sqrt{y+8}\right)^2 = \left(\sqrt{y-4} - 2\right)^2$

$y+8 = y-4 - 4\sqrt{y-4} + 4$

$4\sqrt{y-4} = -8$

$\left(4\sqrt{y-4}\right)^2 = (-8)^2$

$16(y-4) = 64$

$16y - 64 = 64$

$16y = 128$

$y = 8$

The solution does not check.

57.
$$\sqrt{2b+3} - \sqrt{b+1} = \sqrt{b-2}$$
$$\left(\sqrt{2b+3} - \sqrt{b+1}\right)^2 = \left(\sqrt{b-2}\right)^2$$
$$2b+3 - 2\sqrt{(2b+3)(b+1)} + b+1 = b-2$$
$$3b+4 - 2\sqrt{2b^2+5b+3} = b-2$$
$$2b+6 = 2\sqrt{2b^2+5b+3}$$
$$(2b+6)^2 = \left(2\sqrt{2b^2+5b+3}\right)^2$$
$$4b^2 + 24b + 36 = 4\left(2b^2+5b+3\right)$$
$$4b^2 + 24b + 36 = 8b^2 + 20b + 12$$
$$0 = 4b^2 - 4b - 24$$
$$0 = 4(b-3)(b+2)$$

$b-3=0$ **or** $b+2=0$ $b=-2$ does not check, so it

$b=3$ $b=-2$ is an extraneous solution.

59. $\sqrt{\sqrt{b} + \sqrt{b+8}} = 2$

$\left(\sqrt{\sqrt{b} + \sqrt{b+8}}\right)^2 = 2^2$

$\sqrt{b} + \sqrt{b+8} = 4$

$\sqrt{b+8} = 4 - \sqrt{b}$

$\left(\sqrt{b+8}\right)^2 = \left(4 - \sqrt{b}\right)^2$

$b+8 = 16 - 8\sqrt{b} + b$

$8\sqrt{b} = 8$

$\sqrt{b} = 1$

$\left(\sqrt{b}\right)^2 = 1^2$

$b = 1 \Rightarrow$ The solution checks.

61.
$$t = \sqrt{\frac{d}{16}}$$
$$5 = \sqrt{\frac{d}{16}}$$
$$5^2 = \left(\sqrt{\frac{d}{16}}\right)^2$$
$$25 = \frac{d}{16}$$
$$400 = d \Rightarrow \text{The bridge is 400 feet high.}$$

63.
$$l = \sqrt{f^2 + h^2}$$
$$10 = \sqrt{f^2 + 6^2}$$
$$10^2 = \left(\sqrt{f^2 + 36}\right)^2$$
$$100 = f^2 + 36$$
$$64 = f^2$$
$$\pm 8 = f$$
He should nail the brace to the floor 8 feet from the wall.

65. **Answers may vary.**

67. Natural numbers between -4 and 4:

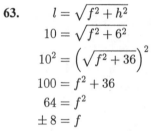

69. $x \geq 3 \Rightarrow [3, \infty)$

71. $-2 \leq x < 1 \Rightarrow [-2, 1)$

73. $x < 1$ or $x \geq 2$
$(-\infty, 1) \cup [2, \infty)$

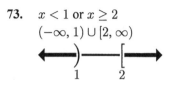

Exercise 1.7 (page 140)

1. right

3. $a < c$

5. $b - c$

7. $>$

9. linear

11. equivalent

13. $3x + 2 < 5$
$\quad\; 3x < 3$
$\qquad x < 1 \Rightarrow (-\infty, 1)$

15. $3x + 2 \geq 5$
$\quad\; 3x \geq 3$
$\qquad x \geq 1 \Rightarrow [1, \infty)$

17. $-5x + 3 > -2$
$\quad\;\; -5x > -5$
$\qquad x < 1 \Rightarrow (-\infty, 1)$

19. $-5x + 3 \leq -2$
$\quad\;\; -5x \leq -5$
$\qquad x \geq 1 \Rightarrow [1, \infty)$

21. $2(x-3) \le -2(x-3)$
$2x - 6 \le -2x + 6$
$4x \le 12$
$x \le 3 \Rightarrow (-\infty, 3]$

23. $\frac{3}{5}x + 4 > 2$

$5\left(\frac{3}{5}x + 4\right) > 5(2)$

$3x + 20 > 10$

$3x > -10$

$x > -\frac{10}{3} \Rightarrow \left(-\frac{10}{3}, \infty\right)$

25. $\frac{x+3}{4} < \frac{2x-4}{3}$

$12 \cdot \frac{x+3}{4} < 12 \cdot \frac{2x-4}{3}$

$3(x+3) < 4(2x-4)$

$3x + 9 < 8x - 16$

$-5x < -25$

$x > 5 \Rightarrow (5, \infty)$

27. $\frac{6(x-4)}{5} \ge \frac{3(x+2)}{4}$

$20 \cdot \frac{6x-24}{5} \ge 20 \cdot \frac{3x+6}{4}$

$4(6x - 24) \ge 5(3x + 6)$

$24x - 96 \ge 15x + 30$

$9x \ge 126$

$x \ge 14 \Rightarrow [14, \infty)$

29. $\frac{5}{9}(a+3) - a \ge \frac{4}{3}(a-3) - 1$

$9\left[\frac{5}{9}(a+3) - a\right] \ge 9\left[\frac{4}{3}(a-3) - 1\right]$

$5(a+3) - 9a \ge 12(a-3) - 9$

$5a + 15 - 9a \ge 12a - 36 - 9$

$-16a \ge -60$

$a \le \frac{-60}{-16}$

$a \le \frac{15}{4} \Rightarrow \left(-\infty, \frac{15}{4}\right]$

31. $\frac{2}{3}a - \frac{3}{4}a < \frac{3}{5}\left(a + \frac{2}{3}\right) + \frac{1}{3}$

$60\left(\frac{2}{3}a - \frac{3}{4}a\right) < 60\left[\frac{3}{5}\left(a + \frac{2}{3}\right) + \frac{1}{3}\right]$

$40a - 45a < 36\left(a + \frac{2}{3}\right) + 20$

$-5a < 36a + 24 + 20$

$-41a < 44$

$a > -\frac{44}{41} \Rightarrow \left(-\frac{44}{41}, \infty\right)$

33. $4 < 2x - 8 \le 10$
$12 < \quad 2x \quad \le 18$
$6 < \quad x \quad \le 9 \Rightarrow (6, 9]$

35. $9 \ge \frac{x-4}{2} > 2$
$18 \ge x - 4 > 4$
$22 \ge \quad x \quad > 8$
$8 < \quad x \quad \le 22 \Rightarrow (8, 22]$

37.
$$0 \leq \frac{4-x}{3} \leq 5$$
$$0 \leq 4-x \leq 15$$
$$-4 \leq -x \leq 11$$
$$4 \geq x \geq -11$$
$$-11 \leq x \leq 4 \Rightarrow [-11, 4]$$

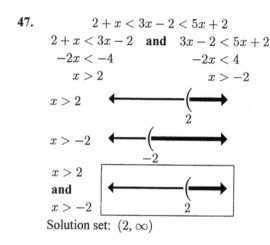

39.
$$-2 \geq \frac{1-x}{2} \geq -10$$
$$-4 \geq 1-x \geq -20$$
$$-5 \geq -x \geq -21$$
$$5 \leq x \leq 21 \Rightarrow [5, 21]$$

41.
$$-3x > -2x > -x$$
$$-3x > -2x \quad \textbf{and} \quad -2x > -x$$
$$-x > 0 \qquad\qquad -x > 0$$
$$x < 0 \qquad\qquad x < 0$$
$$x < 0 \Rightarrow (-\infty, 0)$$

43.
$$x < 2x < 3x$$
$$x < 2x \quad \textbf{and} \quad 2x < 3x$$
$$-x < 0 \qquad\qquad -x < 0$$
$$x > 0 \qquad\qquad x > 0$$
$$x > 0 \Rightarrow (0, \infty)$$

45.
$$2x+1 < 3x-2 < 12$$
$$2x+1 < 3x-2 \quad \textbf{and} \quad 3x-2 < 12$$
$$-x < -3 \qquad\qquad 3x < 14$$
$$x > 3 \qquad\qquad x < \frac{14}{3}$$

$x > 3$

$x < \frac{14}{3}$

$x > 3$
and
$x < \frac{14}{3}$

Solution set: $\left(3, \frac{14}{3}\right)$

47.
$$2+x < 3x-2 < 5x+2$$
$$2+x < 3x-2 \quad \textbf{and} \quad 3x-2 < 5x+2$$
$$-2x < -4 \qquad\qquad -2x < 4$$
$$x > 2 \qquad\qquad x > -2$$

$x > 2$

$x > -2$

$x > 2$
and
$x > -2$

Solution set: $(2, \infty)$

49.
$$3 + x > 7x - 2 > 5x - 10$$
$$3 + x > 7x - 2 \quad \text{and} \quad 7x - 2 > 5x - 10$$
$$-6x > -5 \qquad\qquad 2x > -8$$
$$x < \tfrac{5}{6} \qquad\qquad\quad x > -4$$

$x < \tfrac{5}{6}$

$x > -4$

$x < \tfrac{5}{6}$
and
$x > -4$

Solution set: $\left(-4, \tfrac{5}{6}\right)$

51.
$$x \leq x + 1 \leq 2x + 3$$
$$x \leq x + 1 \quad \text{and} \quad x + 1 \leq 2x + 3$$
$$0 \leq 1 \qquad\qquad -x \leq 2$$
true for all real $\qquad\qquad x \geq -2$
numbers x

$0 \leq 1$

$x \geq -2$

$0 \leq 1$
and
$x \geq -2$

Solution set: $[-2, \infty)$

53.
$$x^2 + 7x + 12 < 0$$
$$(x + 3)(x + 4) < 0$$
factors $= 0$: $x = -3$, $x = -4$
intervals: $(-\infty, -4)$, $(-4, -3)$, $(-3, \infty)$

interval	test number	value of $x^2+7x+12$
$(-\infty, -4)$	-5	$+2$
$(-4, -3)$	-3.5	-0.25
$(-3, \infty)$	0	$+12$

Solution: $(-4, -3)$

55.
$$x^2 - 5x + 6 \geq 0$$
$$(x - 3)(x - 2) \geq 0$$
factors $= 0$: $x = 3$, $x = 2$
intervals: $(-\infty, 2)$, $(2, 3)$, $(3, \infty)$

interval	test number	value of x^2-5x+6
$(-\infty, 2)$	0	$+6$
$(2, 3)$	2.5	-0.25
$(3, \infty)$	4	$+2$

Solution: $(-\infty, 2] \cup [3, \infty)$

57.
$$x^2 + 5x + 6 < 0$$
$$(x + 3)(x + 2) < 0$$
factors $= 0$: $x = -3$, $x = -2$
intervals: $(-\infty, -3)$, $(-3, -2)$, $(-2, \infty)$

interval	test number	value of x^2+5x+6
$(-\infty, -3)$	-4	$+2$
$(-3, -2)$	-2.5	-0.25
$(-2, \infty)$	0	$+6$

Solution: $(-3, -2)$

59.
$$6x^2 + 5x + 1 \geq 0$$
$$(2x + 1)(3x + 1) \geq 0$$
factors $= 0$: $x = -\tfrac{1}{2}$, $x = -\tfrac{1}{3}$
intervals: $\left(-\infty, -\tfrac{1}{2}\right)$, $\left(-\tfrac{1}{2}, -\tfrac{1}{3}\right)$, $\left(-\tfrac{1}{3}, \infty\right)$

interval	test number	value of $6x^2+5x+1$
$\left(-\infty, -\tfrac{1}{2}\right)$	-1	$+2$
$\left(-\tfrac{1}{2}, -\tfrac{1}{3}\right)$	-0.4	-0.04
$\left(-\tfrac{1}{3}, \infty\right)$	0	$+1$

Solution: $\left(-\infty, -\tfrac{1}{2}\right] \cup \left[-\tfrac{1}{3}, \infty\right)$

SECTION 1.7

61.
$$6x^2 - 5x < -1$$
$$6x^2 - 5x + 1 < 0$$
$$(2x - 1)(3x - 1) < 0$$
factors = 0: $x = \frac{1}{2}, x = \frac{1}{3}$
intervals: $\left(-\infty, \frac{1}{3}\right), \left(\frac{1}{3}, \frac{1}{2}\right), \left(\frac{1}{2}, \infty\right)$

interval	test number	value of $6x^2-5x+1$
$\left(-\infty, \frac{1}{3}\right)$	0	$+1$
$\left(\frac{1}{3}, \frac{1}{2}\right)$	0.4	-0.04
$\left(\frac{1}{2}, \infty\right)$	1	$+2$

Solution: $\left(\frac{1}{3}, \frac{1}{2}\right)$

63.
$$2x^2 \geq 3 - x$$
$$2x^2 + x - 3 \geq 0$$
$$(2x + 3)(x - 1) \geq 0$$
factors = 0: $x = -\frac{3}{2}, x = 1$
intervals: $\left(-\infty, -\frac{3}{2}\right), \left(-\frac{3}{2}, 1\right), (1, \infty)$

interval	test number	value of $2x^2+x-3$
$\left(-\infty, -\frac{3}{2}\right)$	-2	$+3$
$\left(-\frac{3}{2}, 1\right)$	0	-3
$(1, \infty)$	2	$+7$

Solution: $\left(-\infty, -\frac{3}{2}\right] \cup [1, \infty)$

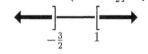

65. $\dfrac{x + 3}{x - 2} < 0$
factors = 0: $x = -3, x = 2$
intervals: $(-\infty, -3), (-3, 2), (2, \infty)$

interval	test number	sign of $\frac{x+3}{x-2}$
$(-\infty, -3)$	-4	$+$
$(-3, 2)$	0	$-$
$(2, \infty)$	3	$+$

Solution: $(-3, 2)$

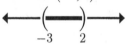

67. $\dfrac{x^2 + x}{x^2 - 1} > 0$

$$\dfrac{x(x + 1)}{(x + 1)(x - 1)} > 0$$

factors = 0: $x = 0, x = -1, x = 1$
intervals: $(-\infty, -1), (-1, 0), (0, 1), (1, \infty)$

interval	test number	sign of $\frac{x^2+x}{x^2-1}$
$(-\infty, -1)$	-2	$+$
$(-1, 0)$	$-\frac{1}{2}$	$+$
$(0, 1)$	$\frac{1}{2}$	$-$
$(1, \infty)$	2	$+$

Solution: $(-\infty, -1) \cup (-1, 0) \cup (1, \infty)$

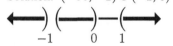

69. $\dfrac{x^2 + 5x + 6}{x^2 + x - 6} \geq 0 \Rightarrow \dfrac{(x + 3)(x + 2)}{(x + 3)(x - 2)} \geq 0$
factors = 0: $x = -3, x = \pm 2$
intervals: $(-\infty, -3), (-3, -2), (-2, 2),$
$\qquad\qquad (2, \infty)$

interval	test number	sign of $\frac{x^2+5x+6}{x^2+x-6}$
$(-\infty, -3)$	-4	$+$
$(-3, -2)$	-2.5	$+$
$(-2, 2)$	0	$-$
$(2, \infty)$	3	$+$

Include endpoints which make the numerator equal to 0. Do not include endpoints which make the denominator equal to 0.

Solution: $(-\infty, -3) \cup (-3, -2] \cup (2, \infty)$

71. $\dfrac{6x^2 - x - 1}{x^2 + 4x + 4} > 0 \Rightarrow \dfrac{(2x-1)(3x+1)}{(x+2)(x+2)} > 0$

factors $= 0$: $x = \frac{1}{2},\ x = -\frac{1}{3},\ x = -2$

intervals: $(-\infty, -2),\ \left(-2, -\frac{1}{3}\right),\ \left(-\frac{1}{3}, \frac{1}{2}\right),$
$\qquad \left(\frac{1}{2}, \infty\right)$

interval	test number	sign of $\frac{6x^2-x-1}{x^2+4x+4}$
$(-\infty, -2)$	-3	$+$
$\left(-2, -\frac{1}{3}\right)$	-1	$+$
$\left(-\frac{1}{3}, \frac{1}{2}\right)$	0	$-$
$\left(\frac{1}{2}, \infty\right)$	1	$+$

Solution: $(-\infty, -2) \cup \left(-2, -\frac{1}{3}\right) \cup \left(\frac{1}{2}, \infty\right)$

$\xleftarrow{\hspace{1cm}})\ (\!\!\relbar\!\!\relbar\!\!)\relbar(\xrightarrow{\hspace{1cm}}$
$\qquad\qquad -2 \qquad -\frac{1}{3}\ \ \frac{1}{2}$

73. $\dfrac{3}{x} > 2$

$\dfrac{3}{x} - 2 > 0$

$\dfrac{3 - 2x}{x} > 0$

factors $= 0$: $x = \frac{3}{2},\ x = 0$

intervals: $(-\infty, 0),\ \left(0, \frac{3}{2}\right),\ \left(\frac{3}{2}, \infty\right)$

interval	test number	sign of $\frac{3-2x}{x}$
$(-\infty, 0)$	-1	$-$
$\left(0, \frac{3}{2}\right)$	1	$+$
$\left(\frac{3}{2}, \infty\right)$	2	$-$

Solution: $\left(0, \frac{3}{2}\right)$

$\xleftarrow{\hspace{1cm}} (\!\!\relbar\!\!\relbar\!\!) \xrightarrow{\hspace{1cm}}$
$\qquad\quad 0 \qquad \frac{3}{2}$

75. $\dfrac{6}{x} < 4$

$\dfrac{6}{x} - 4 < 0$

$\dfrac{6 - 4x}{x} < 0$

factors $= 0$: $x = \frac{3}{2},\ x = 0$

intervals: $(-\infty, 0),\ \left(0, \frac{3}{2}\right),\ \left(\frac{3}{2}, \infty\right)$

interval	test number	sign of $\frac{6-4x}{x}$
$(-\infty, 0)$	-1	$-$
$\left(0, \frac{3}{2}\right)$	1	$+$
$\left(\frac{3}{2}, \infty\right)$	2	$-$

Solution: $(-\infty, 0) \cup \left(\frac{3}{2}, \infty\right)$

$\xleftarrow{\hspace{1cm}})\relbar\relbar(\xrightarrow{\hspace{1cm}}$
$\qquad\quad 0 \qquad \frac{3}{2}$

77. $\dfrac{3}{x-2} \le 5$

$\dfrac{3}{x-2} - 5 \le 0$

$\dfrac{3}{x-2} - \dfrac{5(x-2)}{x-2} \le 0$

$\dfrac{3 - 5x + 10}{x-2} \le 0$

$\dfrac{13 - 5x}{x-2} \le 0$

factors $= 0$: $x = \frac{13}{5},\ x = 2$

intervals: $(-\infty, 2),\ \left(2, \frac{13}{5}\right),\ \left(\frac{13}{5}, \infty\right)$

interval	test number	sign of $\frac{13-5x}{x-2}$
$(-\infty, 2)$	0	$-$
$\left(2, \frac{13}{5}\right)$	$\frac{11}{5}$	$+$
$\left(\frac{13}{5}, \infty\right)$	3	$-$

Solution: $(-\infty, 2) \cup \left[\frac{13}{5}, \infty\right)$

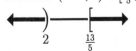
$\quad 2 \qquad \frac{13}{5}$

Include endpoints which make the numerator equal to 0. Do not include endpoints which make the denominator equal to 0.

79.

$$\frac{6}{x^2 - 1} < 1$$

$$\frac{6}{x^2 - 1} - 1 < 0$$

$$\frac{6}{x^2 - 1} - \frac{x^2 - 1}{x^2 - 1} < 0$$

$$\frac{7 - x^2}{x^2 - 1} < 0$$

$$\frac{7 - x^2}{(x + 1)(x - 1)} < 0$$

factors $= 0$: $x = \pm\sqrt{7}, x = \pm 1$

intervals: $\left(-\infty, -\sqrt{7}\right), \left(-\sqrt{7}, -1\right), (-1, 1),$

$\left(1, \sqrt{7}\right), \left(\sqrt{7}, \infty\right)$

interval	test number	sign of $\frac{7-x^2}{x^2-1}$
$\left(-\infty, -\sqrt{7}\right)$	-3	$-$
$\left(-\sqrt{7}, -1\right)$	-2	$+$
$(-1, 1)$	0	$-$
$\left(1, \sqrt{7}\right)$	2	$+$
$\left(\sqrt{7}, \infty\right)$	3	$-$

Sol'n: $\left(-\infty, -\sqrt{7}\right) \cup (-1, 1) \cup \left(\sqrt{7}, \infty\right)$

81. Let $x =$ the number of minutes after 3 minutes. The total cost $= 36 + 11x$ cents.

$\boxed{\text{Total cost}} < 200$

$$36 + 11x < 200$$
$$11x < 164$$
$$x < 14.9$$

For \$2, a person can talk for up to an additional 14 min, for a total of up to 17 min.

83. Let $x =$ the number of CDs. Then the total cost $= 150 + 9.75x$.

$\boxed{\text{Total cost}} < 275$

$$150 + 9.75x < 275$$
$$9.75x < 125$$
$$x < 12.8$$

He can buy up to 12 disks.

85. Let $p =$ the price of the refrigerator. Then the total cost $= p + 0.065p + 0.0025p$.

$\boxed{\text{Total cost}} < 1200$

$$p + 0.065p + 0.0025p < 1200$$
$$1.0675p < 1200$$
$$p < 1124.122$$
$$p \leq \$1124.12$$

87. Let $a =$ the assessed value. Find when Method $1 <$ Method 2:

$$2200 + 0.04a < 1200 + 0.06a$$
$$1000 < 0.02a$$
$$50000 < a$$

The first method will benefit the taxpayer when $a > \$50,000$.

89. Let $P =$ the perimeter. Then the length of one side is equal to $\frac{P}{3}$.

$$50 < P < 60$$
$$\frac{50}{3} < \frac{P}{3} < \frac{60}{3}$$
$$16\tfrac{2}{3} < \text{length} < 20$$

The length of a side is between $16\tfrac{2}{3}$ and 20 cm.

91.

$$20 < l < 30$$
$$40 < 2l < 60$$
$$40 + 2w < 2l + 2w < 60 + 2w$$
$$40 + 2w < P < 60 + 2w$$

93. **Answers may vary.**

95. even integers: $-2, 2$

97. prime numbers: 2

99. real numbers: all in the list

Exercise 1.8 (page 148)

1. x

3. $x = k$ or $x = -k$

5. $-k < x < k$

7. $x \leq -k$ or $x \geq k$

9. $|7| = 7$

11. $|0| = 0$

13. $|5| - |-3| = 5 - 3 = 2$

15. $|\pi - 2| = \pi - 2$

17. $x > 5 \Rightarrow |x - 5| = x - 5$

19. $|x^3| = \begin{cases} x^3 & \text{if } x \geq 0 \\ -x^3 & \text{if } x < 0 \end{cases}$

21.
$$|x + 2| = 2$$
$x + 2 = 2 \quad \text{or} \quad x + 2 = -2$
$x = 0 \qquad\qquad x = -4$

23.
$$|3x - 1| = 5$$
$3x - 1 = 5 \quad \text{or} \quad 3x - 1 = -5$
$3x = 6 \qquad\qquad 3x = -4$
$x = 2 \qquad\qquad x = -\frac{4}{3}$

25.
$$\left|\frac{3x - 4}{2}\right| = 5$$
$\dfrac{3x - 4}{2} = 5 \quad \text{or} \quad \dfrac{3x - 4}{2} = -5$
$3x - 4 = 10 \qquad\qquad 3x - 4 = -10$
$3x = 14 \qquad\qquad 3x = -6$
$x = \frac{14}{3} \qquad\qquad x = -2$

27.
$$\left|\frac{2x - 4}{5}\right| = 2$$
$\dfrac{2x - 4}{5} = 2 \quad \text{or} \quad \dfrac{2x - 4}{5} = -2$
$2x - 4 = 10 \qquad\qquad 2x - 4 = -10$
$2x = 14 \qquad\qquad 2x = -6$
$x = 7 \qquad\qquad x = -3$

29.
$$\left|\frac{x - 3}{4}\right| = -2$$
An absolute value can never
equal a negative number.
no solution

31.
$$\left|\frac{4x - 2}{x}\right| = 3$$
$\dfrac{4x - 2}{x} = 3 \quad \text{or} \quad \dfrac{4x - 2}{x} = -3$
$4x - 2 = 3x \qquad\qquad 4x - 2 = -3x$
$x = 2 \qquad\qquad 7x = 2$
$x = 2 \qquad\qquad x = \frac{2}{7}$

33. $|x| = x$
True for all $x \geq 0$.

35.
$$|x + 3| = |x|$$
$x + 3 = x \quad \text{or} \quad x + 3 = -x$
$0 = 3 \qquad\qquad 2x = -3$
not true $\qquad\qquad x = -\frac{3}{2}$

37.
$$|x - 3| = |2x + 3|$$
$x - 3 = 2x + 3 \quad \text{or} \quad x - 3 = -(2x + 3)$
$-x = 6 \qquad\qquad x - 3 = -2x - 3$
$x = -6 \qquad\qquad 3x = 0$
$\qquad\qquad\qquad x = 0$

39.
$$|x + 2| = |x - 2|$$
$x + 2 = x - 2 \quad \text{or} \quad x + 2 = -(x - 2)$
$\qquad 0 = -4 \qquad\qquad x + 2 = -x + 2$
$\qquad \text{not true} \qquad\qquad\quad 2x = 0$
$\qquad\qquad\qquad\qquad\qquad\quad\; x = 0$

41.
$$\left|\frac{x + 3}{2}\right| = |2x - 3|$$
$\dfrac{x + 3}{2} = 2x - 3 \quad \text{or} \quad \dfrac{x + 3}{2} = -(2x - 3)$
$x + 3 = 4x - 6 \qquad\qquad \dfrac{x + 3}{2} = -2x + 3$
$\quad -3x = -9 \qquad\qquad\quad x + 3 = -4x + 6$
$\qquad\; x = 3 \qquad\qquad\qquad\quad 5x = 3$
$\qquad\qquad\qquad\qquad\qquad\qquad\quad x = \frac{3}{5}$

43.
$$\left|\frac{3x - 1}{2}\right| = \left|\frac{2x + 3}{3}\right|$$
$\dfrac{3x - 1}{2} = \dfrac{2x + 3}{3} \quad \text{or} \quad \dfrac{3x - 1}{2} = -\dfrac{2x + 3}{3}$
$6\left(\dfrac{3x - 1}{2} = \dfrac{2x + 3}{3}\right) \qquad 6\left(\dfrac{3x - 1}{2} = -\dfrac{2x + 3}{3}\right)$
$3(3x - 1) = 2(2x + 3) \qquad 3(3x - 1) = -2(2x + 3)$
$9x - 3 = 4x + 6 \qquad\qquad 9x - 3 = -4x - 6$
$\quad 5x = 9 \qquad\qquad\qquad\quad 13x = -3$
$\qquad x = \frac{9}{5} \qquad\qquad\qquad\quad x = -\frac{3}{13}$

45.
$$|x - 3| < 6$$
$-6 < x - 3 < 6$
$-3 < \quad x \quad < 9$
$(-3, 9)$

47.
$$|x + 3| > 6$$
$x + 3 > 6 \quad \text{or} \quad x + 3 < -6$
$\quad x > 3 \qquad\qquad\quad x < -9$
$(-\infty, -9) \cup (3, \infty)$

49.
$$|2x + 4| \geq 10$$
$2x + 4 \geq 10 \quad \text{or} \quad 2x + 4 \leq -10$
$\quad 2x \geq 6 \qquad\qquad\quad 2x \leq -14$
$\quad\; x \geq 3 \qquad\qquad\qquad x \leq -7$
$(-\infty, -7] \cup [3, \infty)$

51.
$$|3x + 5| + 1 \leq 9$$
$|3x + 5| \leq 8$
$-8 \leq 3x + 5 \leq 8$
$-13 \leq \quad 3x \quad \leq 3$
$-\frac{13}{3} \leq \quad x \quad \leq 1$
$\left[-\frac{13}{3}, 1\right]$

53.
$$|x + 3| > 0$$
$$x + 3 > 0 \quad \textbf{or} \quad x + 3 < -0$$
$$x > -3 \qquad\qquad x < -3$$
$$(-\infty, -3) \cup (-3, \infty)$$

-3

55.
$$\left|\frac{5x+2}{3}\right| < 1$$
$$-1 < \frac{5x+2}{3} < 1$$
$$-3 < 5x + 2 < 3$$
$$-5 < \quad 5x \quad < 1$$
$$-1 < \quad x \quad < \frac{1}{5}$$
$$\left(-1, \tfrac{1}{5}\right)$$

$-1 \qquad \frac{1}{5}$

57.
$$3\left|\frac{3x-1}{2}\right| > 5$$
$$\left|\frac{3x-1}{2}\right| > \frac{5}{3}$$
$$\frac{3x-1}{2} > \frac{5}{3} \quad \textbf{or} \quad \frac{3x-1}{2} < -\frac{5}{3}$$
$$6 \cdot \frac{3x-1}{2} > 6 \cdot \frac{5}{3} \qquad 6 \cdot \frac{3x-1}{2} < 6\left(-\frac{5}{3}\right)$$
$$9x - 3 > 10 \qquad\qquad 9x - 3 < -10$$
$$9x > 13 \qquad\qquad 9x < -7$$
$$x > \frac{13}{9} \qquad\qquad x < -\frac{7}{9}$$
$$\left(-\infty, -\tfrac{7}{9}\right) \cup \left(\tfrac{13}{9}, \infty\right)$$

$-\frac{7}{9} \qquad \frac{13}{9}$

59.
$$\frac{|x-1|}{-2} > -3$$
$$|x - 1| < 6$$
$$-6 < x - 1 < 6$$
$$-5 < \quad x \quad < 7$$
$$(-5, 7)$$

$-5 \qquad 7$

61.
$$0 < |2x + 1| < 3$$
$$0 < |2x + 1| \qquad \textbf{and} \qquad |2x + 1| < 3$$
$$\textbf{(1)} \quad |2x + 1| > 0 \qquad\qquad \textbf{(2)} \quad |2x + 1| < 3$$
$$2x + 1 > 0 \quad \textbf{or} \quad 2x + 1 < -0 \qquad -3 < 2x + 1 < 3$$
$$2x > -1 \qquad\qquad 2x < -1 \qquad\qquad -4 < \quad 2x \quad < 2$$
$$x > -\frac{1}{2} \qquad\qquad x < -\frac{1}{2} \qquad\qquad -2 < \quad x \quad < 1$$

(1)

$-\frac{1}{2}$

(2)

$-2 \qquad 1$

(1)

$-\frac{1}{2}$

(2)

$-2 \qquad\qquad 1$

(1) and (2) $\Rightarrow \left(-2, -\tfrac{1}{2}\right) \cup \left(-\tfrac{1}{2}, 1\right)$

$-2 \quad -\frac{1}{2} \qquad 1$

63.

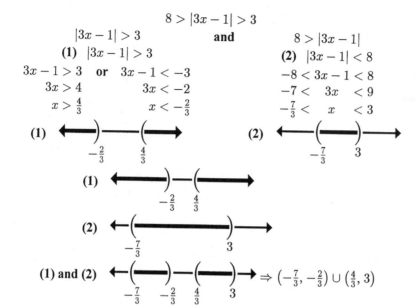

$$8 > |3x - 1| > 3$$

$|3x - 1| > 3$ **and** $8 > |3x - 1|$

(1) $|3x - 1| > 3$ **(2)** $|3x - 1| < 8$

$3x - 1 > 3$ **or** $3x - 1 < -3$ $\qquad -8 < 3x - 1 < 8$

$\qquad 3x > 4 \qquad\qquad 3x < -2 \qquad\qquad -7 < 3x < 9$

$\qquad x > \frac{4}{3} \qquad\qquad x < -\frac{2}{3} \qquad\qquad -\frac{7}{3} < x < 3$

(1) and **(2)** $\Rightarrow \left(-\frac{7}{3}, -\frac{2}{3}\right) \cup \left(\frac{4}{3}, 3\right)$

65.

$$2 < \left|\frac{x - 5}{3}\right| < 4$$

$2 < \left|\frac{x - 5}{3}\right|$ **and** $\left|\frac{x - 5}{3}\right| < 4$

(1) $\left|\frac{x - 5}{3}\right| > 2$ **(2)** $\left|\frac{x - 5}{3}\right| < 4$

$\dfrac{x - 5}{3} > 2$ **or** $\dfrac{x - 5}{3} < -2 \qquad\qquad -4 < \dfrac{x - 5}{3} < 4$

$x - 5 > 6 \qquad\qquad x - 5 < -6 \qquad\qquad -12 < x - 5 < 12$

$x > 11 \qquad\qquad x < -1 \qquad\qquad -7 < x < 17$

(1) and **(2)** $\Rightarrow (-7, -1) \cup (11, 17)$

67.
$$10 > \left|\frac{x-2}{2}\right| > 4$$

$$\left|\frac{x-2}{2}\right| > 4 \qquad \textbf{and} \qquad 10 > \left|\frac{x-2}{2}\right|$$

(1) $\left|\dfrac{x-2}{2}\right| > 4$ ⟶ **(2)** $\left|\dfrac{x-2}{2}\right| < 10$

$\dfrac{x-2}{2} > 4$ **or** $\dfrac{x-2}{2} < -4$ ⟶ $-10 < \dfrac{x-2}{2} < 10$

$x - 2 > 8 \qquad x - 2 < -8$ ⟶ $-20 < x - 2 < 20$

$x > 10 \qquad x < -6$ ⟶ $-18 < \quad x \quad < 22$

(1) ←——)——(——→
  $-6 \qquad 10$

(2) ←——(———)——→
  $-18 \qquad 22$

(1) ←——)—(——→
  $-6 \ 10$

(2) ←——(————————)——→
  $-18 \qquad\qquad 22$

(1) and (2) ←——(———)—(———)——→ $\Rightarrow (-18, -6) \cup (10, 22)$
  $-18 \ -6 \ 10 \ 22$

69.
$$2 \le \left|\frac{x+1}{3}\right| < 3$$

$$2 \le \left|\frac{x+1}{3}\right| \qquad \textbf{and} \qquad \left|\frac{x+1}{3}\right| < 3$$

(1) $\left|\dfrac{x+1}{3}\right| \ge 2$ ⟶ **(2)** $\left|\dfrac{x+1}{3}\right| < 3$

$\dfrac{x+1}{3} \ge 2$ **or** $\dfrac{x+1}{3} \le -2$ ⟶ $-3 < \dfrac{x+1}{3} < 3$

$x + 1 \ge 6 \qquad x + 1 \le -6$ ⟶ $-9 < x + 1 < 9$

$x \ge 5 \qquad x \le -7$ ⟶ $-10 < \quad x \quad < 8$

(1) ←——]——[——→
  $-7 \quad 5$

(2) ←——(————)——→
  $-10 \qquad 8$

(1) ←——]—[——→
  $-7 \ 5$

(2) ←——(————————)——→
  $-10 \qquad\qquad 8$

(1) and (2) ←——(———]—[———)——→ $\Rightarrow (-10, -7] \cup [5, 8)$
  $-10 \ -7 \ 5 \ 8$

71.
$$|x+1| \geq |x|$$
$$\sqrt{(x+1)^2} \geq \sqrt{x^2}$$
$$(x+1)^2 \geq x^2$$
$$x^2 + 2x + 1 \geq x^2$$
$$2x \geq -1$$
$$x \geq -\tfrac{1}{2}$$
Solution: $\left[-\tfrac{1}{2}, \infty\right)$

73.
$$|2x+1| < |2x-1|$$
$$\sqrt{(2x+1)^2} < \sqrt{(2x-1)^2}$$
$$(2x+1)^2 < (2x-1)^2$$
$$4x^2 + 4x + 1 < 4x^2 - 4x + 1$$
$$8x < 0$$
$$x < 0$$
Solution: $(-\infty, 0)$

75.
$$|x+1| < |x|$$
$$\sqrt{(x+1)^2} < \sqrt{x^2}$$
$$(x+1)^2 < x^2$$
$$x^2 + 2x + 1 < x^2$$
$$2x < -1$$
$$x < -\tfrac{1}{2}$$
Solution: $\left(-\infty, -\tfrac{1}{2}\right)$

77.
$$|2x+1| \geq |2x-1|$$
$$\sqrt{(2x+1)^2} \geq \sqrt{(2x-1)^2}$$
$$(2x+1)^2 \geq (2x-1)^2$$
$$4x^2 + 4x + 1 \geq 4x^2 - 4x + 1$$
$$8x \geq 0$$
$$x \geq 0$$
Solution: $[0, \infty)$

79.
$$|t - 78°| \leq 8°$$
$$-8° \leq t - 78° \leq 8°$$
$$70° \leq \quad t \quad \leq 86°$$

81.
$$0.6° + 0.5° = 1.1°$$
$$0.6° - 0.5° = 0.1°$$
$$0.1° \leq \quad c \quad \leq 1.1°$$
$$0.6° - 0.5° \leq \quad c \quad \leq 0.6° + 0.5°$$
$$-0.5° \leq c - 0.6° \leq 0.5°$$
$$|c - 0.6°| \leq 0.5°$$

83.
$$\frac{38 + 72}{2} = \frac{110}{2} = 55$$
$$38 = 55 - 17$$
$$72 = 55 + 17$$
$$38 < \quad h \quad < 72$$
$$55 - 17 < \quad h \quad < 55 + 17$$
$$-17 < h - 55 < 17$$
$$|h - 55| < 17$$

85. **Answers may vary.**

87. **Answers may vary.**

89. $37{,}250 = 3.725 \times 10^4$

91. $5.23 \times 10^5 = 523{,}000$

93. $(x-y)^2 - (x+y)^2 = [(x-y)(x-y)] - [(x+y)(x+y)] = \left[x^2 - 2xy + y^2\right] - \left[x^2 + 2xy + y^2\right]$
$$= x^2 - 2xy + y^2 - x^2 - 2xy - y^2$$
$$= -4xy$$

Chapter 1 Summary (page 152)

1. **a.** $3x + 7 = 4$
no restrictions on x

b. $x + \dfrac{1}{x} = 2$
restriction: $x \neq 0$

c. $\sqrt{x} = 4$
restriction: $x \geq 0$

d. $\dfrac{1}{x-2} = \dfrac{2}{x-3}$
restriction: $x \neq 2$, $x \neq 3$

2. **a.** $3(9x + 4) = 28$
$27x + 12 = 28$
$27x = 16$
$x = \frac{16}{27}$
conditional equation

b. $\frac{3}{2}a = 7(a + 11)$
$2 \cdot \frac{3}{2}a = 2 \cdot 7(a + 11)$
$3a = 14a + 154$
$-11a = 154$
$a = -\frac{154}{11} = -14$
conditional equation

c. $8(3x - 5) - 4(x + 3) = 12$
$24x - 40 - 4x - 12 = 12$
$20x - 52 = 12$
$20x = 64$
$x = \frac{64}{20} = \frac{16}{5} \Rightarrow$ conditional equation

d.
$$\frac{x+3}{x+4} + \frac{x+3}{x+2} = 2$$
$$(x+4)(x+2)\left(\frac{x+3}{x+4} + \frac{x+3}{x+2}\right) = (x+4)(x+2) \cdot 2$$
$$(x+2)(x+3) + (x+4)(x+3) = (x^2 + 6x + 8) \cdot 2$$
$$x^2 + 5x + 6 + x^2 + 7x + 12 = 2x^2 + 12x + 16$$
$$2x^2 + 12x + 18 = 2x^2 + 12x + 16$$
$$18 \neq 16 \Rightarrow \text{no solution}$$

e.
$$\frac{3}{x-1} = \frac{1}{2}$$
$$2(x-1) \cdot \frac{3}{x-1} = 2(x-1) \cdot \frac{1}{2}$$
$$6 = x - 1$$
$$7 = x$$
conditional equation

f.
$$\frac{8x^2 + 72x}{9 + x} = 8x$$
$$(9 + x) \cdot \frac{8x^2 + 72x}{9 + x} = (9 + x) \cdot 8x$$
$$8x^2 + 72x = 72x + 8x^2$$
identity

g.

$$\frac{3x}{x-1} - \frac{5}{x+3} = 3$$

$$(x-1)(x+3)\left(\frac{3x}{x-1} - \frac{5}{x+3}\right) = (x-1)(x+3) \cdot 3$$

$$3x(x+3) - 5(x-1) = \left(x^2 + 2x - 3\right) \cdot 3$$

$$3x^2 + 9x - 5x + 5 = 3x^2 + 6x - 9$$

$$4x + 5 = 6x - 9$$

$$-2x = -14$$

$$x = 7 \Rightarrow \text{conditional equation}$$

h.

$$x + \frac{1}{2x-3} = \frac{2x^2}{2x-3}$$

$$(2x-3)\left(x + \frac{1}{2x-3}\right) = (2x-3) \cdot \frac{2x^2}{2x-3}$$

$$(2x-3)x + 1 = 2x^2$$

$$2x^2 - 3x + 1 = 2x^2$$

$$-3x = -1$$

$$x = \tfrac{1}{3} \Rightarrow \text{conditional equation}$$

3. **a.**

$$C = \frac{5}{9}(F - 32)$$

$$\frac{9}{5}C = \frac{9}{5} \cdot \frac{5}{9}(F - 32)$$

$$\frac{9}{5}C = F - 32$$

$$\frac{9}{5}C + 32 = F$$

b.

$$P_n = l + \frac{si}{f}$$

$$P_n - l = \frac{si}{f}$$

$$f(P_n - l) = f \cdot \frac{si}{f}$$

$$f(P_n - l) = si$$

$$\frac{f(P_n - l)}{P_n - l} = \frac{si}{P_n - l}$$

$$f = \frac{si}{P_n - l}$$

c.

$$\frac{1}{f} = \frac{1}{f_1} + \frac{1}{f_2}$$

$$ff_1 f_2 \cdot \frac{1}{f} = ff_1 f_2 \left(\frac{1}{f_1} + \frac{1}{f_2}\right)$$

$$f_1 f_2 = ff_2 + ff_1$$

$$f_1 f_2 - ff_1 = ff_2$$

$$f_1(f_2 - f) = ff_2$$

$$\frac{f_1(f_2 - f)}{f_2 - f} = \frac{ff_2}{f_2 - f}$$

$$f_1 = \frac{ff_2}{f_2 - f}$$

d.

$$S = \frac{a - lr}{1 - r}$$

$$S(1 - r) = \frac{a - lr}{1 - r}(1 - r)$$

$$S(1 - r) = a - lr$$

$$S - Sr = a - lr$$

$$lr = a - S + Sr$$

$$\frac{lr}{r} = \frac{a - S + Sr}{r}$$

$$l = \frac{a - S + Sr}{r}$$

4. Let x = the liters of water added.

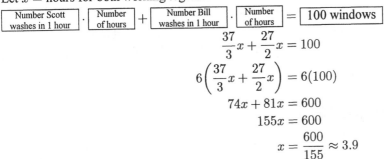

$$0.50(1) + 0 = 0.20(1 + x)$$
$$0.50 = 0.20 + 0.20x$$
$$0.30 = 0.20x$$
$$1.5 = x \Rightarrow 1.5 \text{ liters of water should be added.}$$

5. Let x = hours for both working together.

| Number Scott washes in 1 hour | · | Number of hours | + | Number Bill washes in 1 hour | · | Number of hours | = | 100 windows |

$$\frac{37}{3}x + \frac{27}{2}x = 100$$
$$6\left(\frac{37}{3}x + \frac{27}{2}x\right) = 6(100)$$
$$74x + 81x = 600$$
$$155x = 600$$
$$x = \frac{600}{155} \approx 3.9$$

They can wash 100 windows together in about 3.9 hours.

6. Let x = hours for both pipes to fill the tank.

| 1st pipe in 1 hour | + | 2nd pipe in 1 hour | = | Total in 1 hour |

$$\frac{1}{9} + \frac{1}{12} = \frac{1}{x}$$
$$36x\left(\frac{1}{9} + \frac{1}{12}\right) = 36x\left(\frac{1}{x}\right)$$
$$4x + 3x = 36$$
$$7x = 36$$
$$x = \frac{36}{7} = 5\frac{1}{7}$$

The tank can be filled in $5\frac{1}{7}$ hours.

7. Let x = the ounces of pure zinc added.

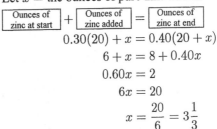

$$0.30(20) + x = 0.40(20 + x)$$
$$6 + x = 8 + 0.40x$$
$$0.60x = 2$$
$$6x = 20$$
$$x = \frac{20}{6} = 3\frac{1}{3}$$

$3\frac{1}{3}$ ounces of zinc should be added.

8. Let x = the amount invested at 11%. Then $10,000 - x$ = the amount invested at 14%.

| Interest at 11% | + | Interest at 14% | = | Total interest |

$$0.11x + 0.14(10,000 - x) = 1,265$$
$$0.11x + 1,400 - 0.14x = 1,265$$
$$-0.03x = -135$$
$$x = 4,500$$

$4,500 was invested at 11% and $5,500 was invested at 14%.

9. Let x = # of rugs for equal costs.

| Cost of 1st loom | = | Cost of 2nd loom |

$$750 + 115x = 950 + 95x$$
$$20x = 200$$
$$x = 10$$

The costs are the same on either loom for 10 rugs.

10. a.
$$2x^2 - x - 6 = 0$$
$$(2x + 3)(x - 2) = 0$$
$$2x + 3 = 0 \quad \text{or} \quad x - 2 = 0$$
$$2x = -3 \qquad\qquad x = 2$$
$$x = -\tfrac{3}{2} \qquad\qquad x = 2$$

b.
$$12x^2 + 13x = 4$$
$$12x^2 + 13x - 4 = 0$$
$$(4x - 1)(3x + 4) = 0$$
$$4x - 1 = 0 \quad \text{or} \quad 3x + 4 = 0$$
$$4x = 1 \qquad\qquad 3x = -4$$
$$x = \tfrac{1}{4} \qquad\qquad x = -\tfrac{4}{3}$$

c.
$$5x^2 - 8x = 0$$
$$x(5x - 8) = 0$$
$$x = 0 \quad \text{or} \quad 5x - 8 = 0$$
$$x = 0 \qquad\qquad 5x = 8$$
$$x = 0 \qquad\qquad x = \tfrac{8}{5}$$

d.
$$27x^2 = 30x - 8$$
$$27x^2 - 30x + 8 = 0$$
$$(9x - 4)(3x - 2) = 0$$
$$9x - 4 = 0 \quad \text{or} \quad 3x - 2 = 0$$
$$9x = 4 \qquad\qquad 3x = 2$$
$$x = \tfrac{4}{9} \qquad\qquad x = \tfrac{2}{3}$$

11. a.
$$x^2 - 8x + 15 = 0$$
$$x^2 - 8x = -15$$
$$x^2 - 8x + 16 = -15 + 16$$
$$(x - 4)^2 = 1$$
$$x - 4 = \sqrt{1} \quad \text{or} \quad x - 4 = -\sqrt{1}$$
$$x - 4 = 1 \qquad\qquad x - 4 = -1$$
$$x = 5 \qquad\qquad x = 3$$

b.
$$3x^2 + 18x = -24$$
$$\frac{3x^2 + 18x}{3} = \frac{-24}{3}$$
$$x^2 + 6x = -8$$
$$x^2 + 6x + 9 = -8 + 9$$
$$(x + 3)^2 = 1$$
$$x + 3 = \sqrt{1} \quad \text{or} \quad x + 3 = -\sqrt{1}$$
$$x + 3 = 1 \qquad\qquad x + 3 = -1$$
$$x = -2 \qquad\qquad x = -4$$

c.
$$5x^2 - x - 1 = 0$$
$$5x^2 - x = 1$$
$$x^2 - \frac{1}{5}x = \frac{1}{5}$$
$$x^2 - \frac{1}{5}x + \frac{1}{100} = \frac{1}{5} + \frac{1}{100}$$
$$\left(x - \frac{1}{10}\right)^2 = \frac{21}{100}$$
$$x - \frac{1}{10} = \sqrt{\frac{21}{100}} \quad \text{or} \quad x - \frac{1}{10} = -\sqrt{\frac{21}{100}}$$
$$x - \frac{1}{10} = \frac{\sqrt{21}}{10} \qquad\qquad x - \frac{1}{10} = -\frac{\sqrt{21}}{10}$$
$$x = \frac{1 + \sqrt{21}}{10} \qquad\qquad x = \frac{1 - \sqrt{21}}{10}$$

d.
$$5x^2 - x = 0$$
$$x^2 - \frac{1}{5}x = 0$$
$$x^2 - \frac{1}{5}x + \frac{1}{100} = 0 + \frac{1}{100}$$
$$\left(x - \frac{1}{10}\right)^2 = \frac{1}{100}$$
$$x - \frac{1}{10} = \sqrt{\frac{1}{100}} \quad \textbf{or} \quad x - \frac{1}{10} = -\sqrt{\frac{1}{100}}$$
$$x - \frac{1}{10} = \frac{1}{10} \qquad\qquad x - \frac{1}{10} = -\frac{1}{10}$$
$$x = \frac{2}{10} = \frac{1}{5} \qquad\qquad x = \frac{0}{10} = 0$$

12. a. $x^2 + 5x - 14 = 0 \Rightarrow a = 1, b = 5, c = -14$
$$x = \frac{-b \pm \sqrt{b^2 - 4ac}}{2a} = \frac{-(5) \pm \sqrt{(5)^2 - 4(1)(-14)}}{2(1)} = \frac{-5 \pm \sqrt{25 + 56}}{2} = \frac{-5 \pm \sqrt{81}}{2}$$
$$= \frac{-5 \pm 9}{2}$$
$$x = \frac{-5 + 9}{2} = \frac{4}{2} = 2 \text{ or } x = \frac{-5 - 9}{2} = \frac{-14}{2} = -7$$

b. $3x^2 - 25x = 18 \Rightarrow 3x^2 - 25x - 18 = 0 \Rightarrow a = 3, b = -25, c = -18$
$$x = \frac{-b \pm \sqrt{b^2 - 4ac}}{2a} = \frac{-(-25) \pm \sqrt{(-25)^2 - 4(3)(-18)}}{2(3)} = \frac{25 \pm \sqrt{625 + 216}}{6}$$
$$= \frac{25 \pm \sqrt{841}}{6} = \frac{25 \pm 29}{6}$$
$$x = \frac{25 + 29}{6} = \frac{54}{6} = 9 \text{ or } x = \frac{25 - 29}{6} = \frac{-4}{6} = -\frac{2}{3}$$

c. $5x^2 = 1 - x \Rightarrow 5x^2 + x - 1 = 0 \Rightarrow a = 5, b = 1, c = -1$
$$x = \frac{-b \pm \sqrt{b^2 - 4ac}}{2a} = \frac{-(1) \pm \sqrt{(1)^2 - 4(5)(-1)}}{2(5)} = \frac{-1 \pm \sqrt{1 + 20}}{10} = \frac{-1 \pm \sqrt{21}}{10}$$

d. $-5 = a^2 + 2a \Rightarrow a^2 + 2a + 5 = 0 \Rightarrow a = 1, b = 2, c = 5$
$$a = \frac{-b \pm \sqrt{b^2 - 4ac}}{2a} = \frac{-(2) \pm \sqrt{(2)^2 - 4(1)(5)}}{2(1)} = \frac{-2 \pm \sqrt{4 - 20}}{2} = \frac{-2 \pm \sqrt{-16}}{2}$$
$$= \frac{-2 \pm 4i}{2}$$
$$= -1 \pm 2i$$

13. $kx^2 + 4x + 12 = 0$

$a = k, b = 4, c = 12$

Set the discriminant equal to 0:

$$b^2 - 4ac = 0$$
$$4^2 - 4(k)(12) = 0$$
$$16 - 48k = 0$$
$$-48k = -16$$
$$k = \frac{1}{3}$$

14.

$$4y^2 + (k+2)y = 1 - k$$
$$4y^2 + (k+2)y - 1 + k = 0$$
$$a = 4, b = k+2, c = -1 + k$$

Set the discriminant equal to 0:

$$b^2 - 4ac = 0$$
$$(k+2)^2 - 4(4)(-1+k) = 0$$
$$k^2 + 4k + 4 + 16 - 16k = 0$$
$$k^2 - 12k + 20 = 0$$
$$(k-10)(k-2) = 0$$
$$k - 10 = 0 \quad \textbf{or} \quad k - 2 = 0$$
$$k = 10 \qquad\qquad k = 2$$

15.

$$\frac{4}{a-4} + \frac{4}{a-1} = 5$$
$$(a-4)(a-1)\left(\frac{4}{a-4} + \frac{4}{a-1}\right) = (a-4)(a-1)5$$
$$4(a-1) + 4(a-4) = 5(a^2 - 5a + 4)$$
$$4a - 4 + 4a - 16 = 5a^2 - 25a + 20$$
$$0 = 5a^2 - 33a + 40$$
$$0 = (5a-8)(a-5)$$
$$5a - 8 = 0 \quad \textbf{or} \quad a - 5 = 0$$
$$5a = 8 \qquad\qquad a = 5$$
$$a = \tfrac{8}{5} \qquad\qquad a = 5$$

16. Let $x =$ one side of the garden.

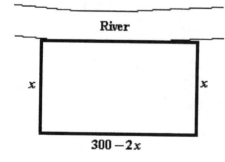

River

x x

$\mathbf{300 - 2x}$

Area $= 10450$
$$x(300 - 2x) = 10450$$
$$-2x^2 + 300x = 10450$$
$$0 = 2x^2 - 300x + 10450$$
$$0 = 2(x^2 - 150x + 5225)$$
$$0 = 2(x-95)(x-55)$$
$$x - 95 = 0 \quad \textbf{or} \quad x - 55 = 0$$
$$x = 95 \qquad\qquad x = 55$$

The dimensions are 95 yards by 110 yards or 55 yards by 190 yards.

17. Let $r =$ the rate of the propeller-driven plane. Then the rate of the jet plane is $r + 120$.

$$\boxed{\text{Jet time}} = \boxed{\text{Propeller time}} - 3$$

$$\frac{3520}{r + 120} = \frac{3520}{r} - 3$$

	Rate	Time	Dist.
Propeller	r	$\frac{3520}{r}$	3520
Jet	$r + 120$	$\frac{3520}{r+120}$	3520

$$r(r + 120)\frac{3520}{r + 120} = r(r + 120)\left(\frac{3520}{r} - 3\right)$$

$$3520r = 3520(r + 120) - 3r(r + 120)$$

$$3520r = 3520r + 422400 - 3r^2 - 360r$$

$$3r^2 + 360r + 422{,}400 = 0$$

$$3(r - 320)(r + 440) = 0$$

$r - 320 = 0$ **or** $r + 440 = 0$

$\qquad r = 320 \qquad\qquad r = -440$

Since $r = -440$ does not make sense, the solution is $r = 320$. The prop. plane's rate is 320 mph, while the jet plane's rate is 440 mph.

18. Set $h = 48$:

$$h = -16t^2 + 64t$$
$$48 = -16t^2 + 64t$$
$$16t^2 - 64t + 48 = 0$$
$$16(t - 1)(t - 3) = 0$$

$t - 1 = 0$ **or** $t - 3 = 0$

$\quad t = 1 \qquad\qquad t = 3$

The shortest time required for the ball to reach a height of 48 feet is 1 second.

19. Let $x =$ the width of the walk. Then the total dimensions are $16 + 2x$ by $20 + 2x$.

$$\boxed{\begin{array}{c}\text{Total}\\\text{area}\end{array}} - \boxed{\begin{array}{c}\text{Area of}\\\text{pool}\end{array}} = \boxed{\begin{array}{c}\text{Area}\\\text{of walk}\end{array}}$$

$$(16 + 2x)(20 + 2x) - (16)(20) = 117$$
$$320 + 72x + 4x^2 - 320 = 117$$
$$4x^2 + 72x - 117 = 0$$
$$(2x + 39)(2x - 3) = 0$$

$2x + 39 = 0$ **or** $2x - 3 = 0$

$\quad x = -\frac{39}{2} \qquad\qquad x = \frac{3}{2}$

Since $x = -\frac{39}{2}$ does not make sense, the only solution is $x = \frac{3}{2}$. The walk is $1\frac{1}{2}$ feet wide.

20

a. $(2 - 3i) + (-4 + 2i) = 2 - 3i - 4 + 2i$
$$= -2 - i$$

b. $(2 - 3i) - (4 + 2i) = 2 - 3i - 4 - 2i$
$$= -2 - 5i$$

c. $\left(3 - \sqrt{-36}\right) + \left(\sqrt{-16} + 2\right) = (3 - 6i) + (4i + 2) = 3 - 6i + 4i + 2 = 5 - 2i$

d. $\left(3 + \sqrt{-9}\right)\left(2 - \sqrt{-25}\right) = (3 + 3i)(2 - 5i) = 6 - 9i - 15i^2 = 6 - 9i + 15 = 21 - 9i$

e. $\dfrac{3}{i} = \dfrac{3i}{ii} = \dfrac{3i}{i^2} = \dfrac{3i}{-1} = 0 - 3i$

f. $-\dfrac{2}{i^3} = -\dfrac{2i}{i^3 i} = -\dfrac{2i}{i^4} = -\dfrac{2i}{1} = 0 - 2i$

g. $\dfrac{3}{1 + i} = \dfrac{3(1 - i)}{(1 + i)(1 - i)} = \dfrac{3(1 - i)}{1^2 - i^2} = \dfrac{3 - 3i}{2} = \dfrac{3}{2} - \dfrac{3}{2}i$

h. $\dfrac{2i}{2 - i} = \dfrac{2i(2 + i)}{(2 - i)(2 + i)} = \dfrac{4i + 2i^2}{2^2 - i^2} = \dfrac{-2 + 4i}{5} = -\dfrac{2}{5} + \dfrac{4}{5}i$

CHAPTER 1 SUMMARY

i. $\dfrac{3+i}{3-i} = \dfrac{(3+i)(3+i)}{(3-i)(3+i)} = \dfrac{9+6i+i^2}{3^2-i^2} = \dfrac{8+6i}{10} = \dfrac{8}{10} + \dfrac{6}{10}i = \dfrac{4}{5} + \dfrac{3}{5}i$

j. $\dfrac{3-2i}{1+i} = \dfrac{(3-2i)(1-i)}{(1+i)(1-i)} = \dfrac{3-5i+2i^2}{1^2-i^2} = \dfrac{1-5i}{2} = \dfrac{1}{2} - \dfrac{5}{2}i$

k. $i^{53} = i^{52}i = (i^4)^{13}i = 1^{13}i = 0 + i$ **l.** $i^{103} = i^{100}i^3 = (i^4)^{25}i^3 = 1^{25}i^3 = 0 - i$

m. $|3 - i| = \sqrt{3^2 + (-1)^2} = \sqrt{9+1} = \sqrt{10}$

n. $\left|\dfrac{1+i}{1-i}\right| = \left|\dfrac{(1+i)(1+i)}{(1-i)(1+i)}\right| = \left|\dfrac{1+2i+i^2}{1^2-i^2}\right| = \left|\dfrac{2i}{2}\right| = |0+i| = \sqrt{0^2+1^2} = 1$

21. a.
$$\dfrac{3x}{2} - \dfrac{2x}{x-1} = x - 3$$
$$2(x-1)\left(\dfrac{3x}{2} - \dfrac{2x}{x-1}\right) = 2(x-1)(x-3)$$
$$3x(x-1) - 2(2x) = 2(x^2 - 4x + 3)$$
$$3x^2 - 3x - 4x = 2x^2 - 8x + 6$$
$$x^2 + x - 6 = 0$$
$$(x+3)(x-2) = 0$$
$$x + 3 = 0 \quad \textbf{or} \quad x - 2 = 0$$
$$x = -3 \qquad\qquad x = 2$$

b.
$$\dfrac{12}{x} - \dfrac{x}{2} = x - 3$$
$$2x\left(\dfrac{12}{x} - \dfrac{x}{2}\right) = 2x(x-3)$$
$$2(12) - x(x) = 2x^2 - 6x$$
$$24 - x^2 = 2x^2 - 6x$$
$$0 = 3x^2 - 6x - 24$$
$$0 = 3(x+2)(x-4)$$
$$x + 2 = 0 \quad \textbf{or} \quad x - 4 = 0$$
$$x = -2 \qquad\qquad x = 4$$

c.
$$x^4 - 2x^2 + 1 = 0$$
$$(x^2 - 1)(x^2 - 1) = 0$$
$$x^2 - 1 = 0 \quad \textbf{or} \quad x^2 - 1 = 0$$
$$x^2 = 1 \qquad\qquad x^2 = 1$$
$$x = \pm 1 \qquad\qquad x = \pm 1$$

d.
$$x^4 + 36 = 37x^2$$
$$x^4 - 37x^2 + 36 = 0$$
$$(x^2 - 36)(x^2 - 1) = 0$$
$$x^2 - 36 = 0 \quad \textbf{or} \quad x^2 - 1 = 0$$
$$x^2 = 36 \qquad\qquad x^2 = 1$$
$$x = \pm 6 \qquad\qquad x = \pm 1$$

e.
$$a - a^{1/2} - 6 = 0$$
$$(a^{1/2} + 2)(a^{1/2} - 3) = 0$$

$$a^{1/2} + 2 = 0 \quad \textbf{or} \quad a^{1/2} - 3 = 0$$
$$a^{1/2} = -2 \qquad\qquad a^{1/2} = 3$$
$$(a^{1/2})^2 = (-2)^2 \qquad (a^{1/2})^2 = (3)^2$$
$$a = 4 \qquad\qquad a = 9$$

$a = 4$ does not check and is extraneous.

f.
$$x^{2/3} + x^{1/3} - 6 = 0$$
$$(x^{1/3} - 2)(x^{1/3} + 3) = 0$$

$$x^{1/3} - 2 = 0 \quad \textbf{or} \quad x^{1/3} + 3 = 0$$
$$x^{1/3} = 2 \qquad\qquad x^{1/3} = -3$$
$$(x^{1/3})^3 = (2)^3 \qquad (x^{1/3})^3 = (-3)^3$$
$$x = 8 \qquad\qquad x = -27$$

Both answers check.

g.
$$\sqrt{x-1}+x=7$$
$$\sqrt{x-1}=7-x$$
$$\left(\sqrt{x-1}\right)^2=(7-x)^2$$
$$x-1=49-14x+x^2$$
$$0=x^2-15x+50$$
$$0=(x-5)(x-10)$$
$$x-5=0 \quad\textbf{or}\quad x-10=0$$
$$x=5 \qquad\qquad x=10$$
$x=10$ does not check and is extraneous.

h.
$$\sqrt{a+9}-\sqrt{a}=3$$
$$\sqrt{a+9}=3+\sqrt{a}$$
$$\left(\sqrt{a+9}\right)^2=(3+\sqrt{a})^2$$
$$a+9=9+6\sqrt{a}+a$$
$$0=6\sqrt{a}$$
$$0^2=(6\sqrt{a})^2$$
$$0=36a \Rightarrow a=0$$
The solution checks.

i.
$$\sqrt{5-x}+\sqrt{5+x}=4$$
$$\sqrt{5+x}=4-\sqrt{5-x}$$
$$\left(\sqrt{5+x}\right)^2=\left(4-\sqrt{5-x}\right)^2$$
$$5+x=16-8\sqrt{5-x}+5-x$$
$$8\sqrt{5-x}=16-2x$$
$$\left(8\sqrt{5-x}\right)^2=(16-2x)^2$$
$$64(5-x)=256-64x+4x^2$$
$$320-64x=4x^2-64x+256$$
$$0=4x^2-64$$
$$0=4(x+4)(x-4)$$
$$x+4=0 \quad\textbf{or}\quad x-4=0$$
$$x=-4 \qquad\qquad x=4$$
Both solutions check.

j.
$$\sqrt{y+5}+\sqrt{y}=1$$
$$\sqrt{y+5}=1-\sqrt{y}$$
$$\left(\sqrt{y+5}\right)^2=(1-\sqrt{y})^2$$
$$y+5=1-2\sqrt{y}+y$$
$$2\sqrt{y}=-4$$
$$(2\sqrt{y})^2=(-4)^2$$
$$4y=16$$
$$y=4$$
The solution does not check.

22. **a.**
$$2x-9<5$$
$$2x<14$$
$$x<7 \Rightarrow (-\infty,7)$$

$$7$$

b.
$$5x+3\geq 2$$
$$5x\geq -1$$
$$x\geq -\tfrac{1}{5} \Rightarrow \left[-\tfrac{1}{5},\infty\right)$$

$$-\tfrac{1}{5}$$

c.
$$\frac{5(x-1)}{2}<x$$
$$5(x-1)<2x$$
$$5x-5<2x$$
$$3x<5$$
$$x<\tfrac{5}{3} \Rightarrow \left(-\infty,\tfrac{5}{3}\right)$$

$$\tfrac{5}{3}$$

d.
$$0 \leq \frac{3+x}{2} < 4$$
$$0 \leq 3+x < 8$$
$$-3 \leq \quad x \quad < 5 \Rightarrow [-3,5)$$

$$-3 \qquad 5$$

e. $(x+2)(x-4) > 0$

factors $= 0$: $x = -2$, $x = 4$

intervals: $(-\infty, -2), (-2, 4), (4, \infty)$

interval	test number	value of $(x+2)(x-4)$
$(-\infty, -2)$	-3	$+7$
$(-2, 4)$	0	-8
$(4, \infty)$	5	$+10$

Solution: $(-\infty, -2) \cup (4, \infty)$

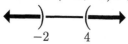

$-2 \qquad 4$

f. $(x-1)(x+4) < 0$

factors $= 0$: $x = 1$, $x = -4$

intervals: $(-\infty, -4), (-4, 1), (1, \infty)$

interval	test number	value of $(x-1)(x+4)$
$(-\infty, -4)$	-5	$+6$
$(-4, 1)$	0	-4
$(1, \infty)$	2	$+8$

Solution: $(-4, 1)$

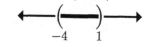

$-4 \qquad 1$

g. $x^2 - 2x - 3 < 0$

$(x-3)(x+1) < 0$

factors $= 0$: $x = 3$, $x = -1$

intervals: $(-\infty, -1), (-1, 3), (3, \infty)$

interval	test number	value of x^2-2x-3
$(-\infty, -1)$	-2	$+5$
$(-1, 3)$	0	-3
$(3, \infty)$	4	$+5$

Solution: $(-1, 3)$

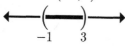

$-1 \qquad 3$

h. $2x^2 + x - 3 > 0$

$(2x+3)(x-1) > 0$

factors $= 0$: $x = -\frac{3}{2}$, $x = 1$

intervals: $\left(-\infty, -\frac{3}{2}\right), \left(-\frac{3}{2}, 1\right), (1, \infty)$

interval	test number	value of $2x^2+x-3$
$\left(-\infty, -\frac{3}{2}\right)$	-2	$+3$
$\left(-\frac{3}{2}, 1\right)$	0	-3
$(1, \infty)$	2	$+7$

Solution: $\left(-\infty, -\frac{3}{2}\right) \cup (1, \infty)$

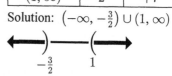

$-\frac{3}{2} \qquad 1$

i. $\dfrac{x+2}{x-3} \geq 0$

factors $= 0$: $x = -2$, $x = 3$

intervals: $(-\infty, -2), (-2, 3), (3, \infty)$

interval	test number	sign of $\frac{x+2}{x-3}$
$(-\infty, -2)$	-3	$+$
$(-2, 3)$	0	$-$
$(3, \infty)$	4	$+$

Include endpoints which make the numerator equal to 0. Do not include endpoints which make the denominator equal to 0.

Solution: $(-\infty, -2] \cup (3, \infty)$

$-2 \qquad 3$

j. $\dfrac{x-1}{x+4} \leq 0$

factors $= 0$: $x = 1$, $x = -4$

intervals: $(-\infty, -4), (-4, 1), (1, \infty)$

interval	test number	sign of $\frac{x-1}{x+4}$
$(-\infty, -4)$	-5	$+$
$(-4, 1)$	0	$-$
$(1, \infty)$	2	$+$

Include endpoints which make the numerator equal to 0. Do not include endpoints which make the denominator equal to 0.

Solution: $(-4, 1]$

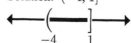

$-4 \qquad 1$

k.

$$\frac{x^2 + x - 2}{x - 3} \geq 0$$

$$\frac{(x + 2)(x - 1)}{x - 3} \geq 0$$

factors $= 0$: $x = -2$, $x = 1$, $x = 3$

int.: $(-\infty, -2)$, $(-2, 1)$, $(1, 3)$, $(3, \infty)$

interval	test number	sign of $\frac{x^2+x-2}{x-3}$
$(-\infty, -2)$	-3	$-$
$(-2, 1)$	0	$+$
$(1, 3)$	2	$-$
$(3, \infty)$	4	$+$

Include endpoints which make the numerator equal to 0. Do not include endpoints which make the denominator equal to 0.

Solution: $[-2, 1] \cup (3, \infty)$

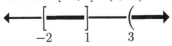

l.

$$\frac{5}{x} < 2$$

$$\frac{5}{x} - 2 < 0$$

$$\frac{5 - 2x}{x} < 0$$

factors $= 0$: $x = \frac{5}{2}$, $x = 0$

intervals: $(-\infty, 0)$, $\left(0, \frac{5}{2}\right)$, $\left(\frac{5}{2}, \infty\right)$

interval	test number	value of $\frac{5-2x}{x}$
$(-\infty, 0)$	-1	-7
$\left(0, \frac{5}{2}\right)$	1	$+3$
$\left(\frac{5}{2}, \infty\right)$	3	$-\frac{1}{3}$

Solution: $(-\infty, 0) \cup \left(\frac{5}{2}, \infty\right)$

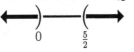

23. a.

$$|x + 1| = 6$$

$x + 1 = 6$ or $x + 1 = -6$

$\quad x = 5 \qquad\qquad x = -7$

b.

$$|2x - 1| = |2x + 1|$$

$2x - 1 = 2x + 1$ or $2x - 1 = -(2x + 1)$

$\quad 0 = 2 \qquad\qquad 2x - 1 = -2x - 1$

never true $\qquad\qquad\quad 4x = 0$

$\qquad\qquad\qquad\qquad\quad x = 0$

c.

$$|x + 3| < 3$$

$$-3 < x + 3 < 3$$

$$-6 < \quad x \quad < 0$$

$$(-6, 0)$$

d.

$$|3x - 7| \geq 1$$

$3x - 7 \geq 1$ or $3x - 7 \leq -1$

$\quad 3x \geq 8 \qquad\qquad 3x \leq 6$

$\quad x \geq \frac{8}{3} \qquad\qquad x \leq 2$

$$(-\infty, 2] \cup \left[\frac{8}{3}, \infty\right)$$

e.

$$\left|\frac{x+2}{3}\right| < 1$$
$$-1 < \frac{x+2}{3} < 1$$
$$-3 < x+2 < 3$$
$$-5 < \quad x \quad < 1$$
$$(-5, 1)$$

f.

$$\left|\frac{x-3}{4}\right| > 8$$
$$\frac{x-3}{4} > 8 \quad \textbf{or} \quad \frac{x-3}{4} < -8$$
$$x-3 > 32 \qquad x-3 < -32$$
$$x > 35 \qquad\quad x < -29$$
$$(-\infty, -29) \cup (35, \infty)$$

g.

$$1 < |2x+3| < 4$$

$$1 < |2x+3| \qquad\qquad \textbf{and} \qquad\qquad |2x+3| < 4$$

(1) $|2x+3| > 1$ | **(2)** $|2x+3| < 4$

$$2x+3 > 1 \quad \textbf{or} \quad 2x+3 < -1 \qquad\qquad -4 < 2x+3 < 4$$
$$2x > -2 \qquad\qquad 2x < -4 \qquad\qquad -7 < \quad 2x \quad < 1$$
$$x > -1 \qquad\qquad x < -2 \qquad\qquad -\frac{7}{2} < \quad x \quad < \frac{1}{2}$$

(1)

(2)

(1)

(2)

(1) and (2) $\Rightarrow \left(-\frac{7}{2}, -2\right) \cup \left(-1, \frac{1}{2}\right)$

h.

$$0 < |3x-4| < 7$$

$$0 < |3x-4| \qquad\qquad \textbf{and} \qquad\qquad |3x-4| < 7$$

(1) $|3x-4| > 0$ | **(2)** $|3x-4| < 7$

$$3x-4 > 0 \quad \textbf{or} \quad 3x-4 < -0 \qquad\qquad -7 < 3x-4 < 7$$
$$3x > 4 \qquad\qquad 3x < 4 \qquad\qquad -3 < \quad 3x \quad < 11$$
$$x > \frac{4}{3} \qquad\qquad x < \frac{4}{3} \qquad\qquad -1 < \quad x \quad < \frac{11}{3}$$

(1)

(2)

(1)

(2)

(1) and (2) $\Rightarrow \left(-1, \frac{4}{3}\right) \cup \left(\frac{4}{3}, \frac{11}{3}\right)$

Chapter 1 Test (page 157)

1. $\dfrac{x}{x(x-1)}$

restrictions: $x \neq 0, \ x \neq 1$

2. $\sqrt{x}$

restrictions: $x \geq 0$

3.
$$7(2a+5) - 7 = 6(a+8)$$
$$14a + 35 - 7 = 6a + 48$$
$$8a = 20$$
$$a = \frac{20}{8} = \frac{5}{2}$$

4.
$$\frac{3}{x^2 - 5x - 14} = \frac{4}{x^2 + 5x + 6}$$
$$\frac{3}{(x-7)(x+2)} = \frac{4}{(x+2)(x+3)}$$
$$(x-7)(x+2)(x+3)\frac{3}{(x-7)(x+2)} = (x-7)(x+2)(x+3)\frac{4}{(x+2)(x+3)}$$
$$3(x+3) = 4(x-7)$$
$$3x + 9 = 4x - 28$$
$$37 = x$$

5.
$$z = \frac{x - \mu}{\sigma}$$
$$\sigma z = \sigma \cdot \frac{x - \mu}{\sigma}$$
$$z\sigma = x - \mu$$
$$z\sigma + \mu = x$$

6.
$$\frac{1}{a} = \frac{1}{b} + \frac{1}{c}$$
$$abc \cdot \frac{1}{a} = abc \left(\frac{1}{b} + \frac{1}{c} \right)$$
$$bc = ac + ab$$
$$bc = a(c + b)$$
$$\frac{bc}{c+b} = \frac{a(c+b)}{c+b}$$
$$\frac{bc}{c+b} = a$$

7. Let $x =$ the score on the final.

Note: This score is counted twice.

$$\frac{\boxed{\text{Sum of scores}}}{5} = 80$$
$$\frac{75 + 75 + 75 + x + x}{5} = 80$$
$$\frac{2x + 225}{5} = 80$$
$$2x + 225 = 400$$
$$2x = 175$$
$$x = 87.5$$

The student needs to score 87.5.

8. Let $x =$ the amount invested at 6%. Then $20{,}000 - x =$ the amount invested at 7%.

$$\boxed{\begin{array}{c}\text{Interest} \\ \text{at 6\%}\end{array}} + \boxed{\begin{array}{c}\text{Interest} \\ \text{at 7\%}\end{array}} = \boxed{\begin{array}{c}\text{Total} \\ \text{interest}\end{array}}$$
$$0.06x + 0.07(20{,}000 - x) = 1{,}260$$
$$0.06x + 1{,}400 - 0.07x = 1{,}260$$
$$-0.01x = -140$$
$$x = 14{,}000$$

$14,000 was invested at 6%.

9.
$$4x^2 - 8x + 3 = 0$$
$$(2x - 3)(2x - 1) = 0$$
$$2x - 3 = 0 \quad \textbf{or} \quad 2x - 1 = 0$$
$$2x = 3 \qquad\qquad 2x = 1$$
$$x = \tfrac{3}{2} \qquad\qquad x = \tfrac{1}{2}$$

10.
$$2b^2 - 12 = -5b$$
$$2b^2 + 5b - 12 = 0$$
$$(2b - 3)(b + 4) = 0$$
$$2b - 3 = 0 \quad \textbf{or} \quad b + 4 = 0$$
$$2b = 3 \qquad\qquad b = -4$$
$$b = \tfrac{3}{2} \qquad\qquad b = -4$$

11. $x = \dfrac{-b \pm \sqrt{b^2 - 4ac}}{2a}$

12. $3x^2 - 5x - 9 = 0 \Rightarrow a = 3, b = -5, c = -9$

$$x = \frac{-b \pm \sqrt{b^2 - 4ac}}{2a} = \frac{-(-5) \pm \sqrt{(-5)^2 - 4(3)(-9)}}{2(3)} = \frac{5 \pm \sqrt{25 + 108}}{6} = \frac{5 \pm \sqrt{133}}{6}$$

13. $x^2 + (k + 1)x + k + 4 = 0$
$a = 1, b = k + 1, c = k + 4$
Set the discriminant equal to 0:
$$b^2 - 4ac = 0$$
$$(k + 1)^2 - 4(1)(k + 4) = 0$$
$$k^2 + 2k + 1 - 4k - 16 = 0$$
$$k^2 - 2k - 15 = 0$$
$$(k - 5)(k + 3) = 0$$
$$k - 5 = 0 \quad \textbf{or} \quad k + 3 = 0$$
$$k = 5 \qquad\qquad k = -3$$

14. Set $h = 0$:
$$h = -16t^2 + 128t$$
$$0 = -16t^2 + 128t$$
$$0 = -16t(t - 8)$$
$$-16t = 0 \quad \textbf{or} \quad t - 8 = 0$$
$$t = 0 \qquad\qquad t = 8$$
The ball will return after 8 seconds.

15. $(4 - 5i) - (-3 + 7i) = 4 - 5i + 3 - 7i$
$$= 7 - 12i$$

16. $(4 - 5i)(3 - 7i) = 12 - 43i + 35i^2$
$$= 12 - 43i - 35$$
$$= -23 - 43i$$

17. $\dfrac{2}{2 - i} = \dfrac{2(2 + i)}{(2 - i)(2 + i)} = \dfrac{4 + 2i}{2^2 - i^2} = \dfrac{4 + 2i}{5} = \dfrac{4}{5} + \dfrac{2}{5}i$

18. $\dfrac{1 + i}{1 - i} = \dfrac{(1 + i)(1 + i)}{(1 - i)(1 + i)} = \dfrac{1 + 2i + i^2}{1^2 - i^2} = \dfrac{2i}{2} = 0 + i$

19. $i^{13} = i^{12}i = (i^4)^3 i = 1^3 i = i$

20. $i^0 = 1$

21. $|5 - 12i| = \sqrt{5^2 + (-12)^2} = \sqrt{25 + 144} = \sqrt{169} = 13$

22. $\left|\dfrac{1}{3+i}\right| = \left|\dfrac{1(3-i)}{(3+i)(3-i)}\right| = \left|\dfrac{3-i}{3^2-i^2}\right| = \left|\dfrac{3-i}{10}\right| = \left|\dfrac{3}{10}-\dfrac{1}{10}i\right| = \sqrt{\left(\dfrac{3}{10}\right)^2 + \left(-\dfrac{1}{10}\right)^2}$

$$= \sqrt{\dfrac{9}{100} + \dfrac{1}{100}}$$

$$= \sqrt{\dfrac{10}{100}} = \dfrac{\sqrt{10}}{10}$$

23. $\quad z^4 - 13z^2 + 36 = 0 \qquad z^2 - 4 = 0 \qquad \textbf{or} \quad z^2 - 9 = 0$

$\quad\;\; \left(z^2 - 4\right)\left(z^2 - 9\right) = 0 \qquad\;\; z^2 = 4 \qquad\qquad\quad z^2 = 9$

$$z = \pm 2 \qquad\qquad\quad z = \pm 3$$

24. $\qquad 2p^{2/5} - p^{1/5} - 1 = 0 \qquad 2p^{1/5} + 1 = 0 \qquad \textbf{or} \quad p^{1/5} - 1 = 0$

$\qquad \left(2p^{1/5}+1\right)\left(p^{1/5}-1\right) = 0 \qquad\quad p^{1/5} = -\tfrac{1}{2} \qquad\qquad\quad p^{1/5} = 1$

$$\left(p^{1/5}\right)^5 = \left(-\tfrac{1}{2}\right)^5 \qquad\quad \left(p^{1/5}\right)^5 = (1)^5$$

$$p = -\tfrac{1}{32} \qquad\qquad\qquad p = 1$$

Both answers check.

25. $\qquad \sqrt{x+5} = 12$

$\qquad \left(\sqrt{x+5}\right)^2 = 12^2$

$\qquad\qquad x + 5 = 144$

$\qquad\qquad\quad\; x = 139$

The answer checks.

26. $\qquad \sqrt{2z+3} = 1 - \sqrt{z+1}$

$\qquad \left(\sqrt{2z+3}\right)^2 = \left(1 - \sqrt{z+1}\right)^2$

$\qquad\qquad 2z + 3 = 1 - 2\sqrt{z+1} + z + 1$

$\qquad\qquad 2\sqrt{z+1} = -z - 1$

$\qquad \left(2\sqrt{z+1}\right)^2 = (-z-1)^2$

$\qquad\qquad 4(z+1) = z^2 + 2z + 1$

$\qquad\qquad 4z + 4 = z^2 + 2z + 1$

$\qquad\qquad\quad\; 0 = z^2 - 2z - 3$

$\qquad\qquad\quad\; 0 = (z+1)(z-3)$

$\qquad z + 1 = 0 \quad \textbf{or} \quad z - 3 = 0$

$\qquad\quad z = -1 \qquad\qquad z = 3$

The answer $z = 3$ is extraneous.

27. $\;\; 5x - 3 \le 7$

$\qquad\; 5x \le 10$

$\qquad\;\; x \le 2 \Rightarrow (-\infty, 2]$

$\qquad\qquad 2$

28. $\qquad \dfrac{x+3}{4} > \dfrac{2x-4}{3}$

$\qquad 12 \cdot \tfrac{x+3}{4} > 12 \cdot \tfrac{2x-4}{3}$

$\qquad 3(x+3) > 4(2x-4)$

$\qquad\;\; 3x + 9 > 8x - 16$

$\qquad\qquad -5x > -25$

$\qquad\qquad\quad\; x < 5 \Rightarrow (-\infty, 5)$

$\qquad\qquad 5$

29.

$$1 + x < 3x - 3 < 4x - 2$$

$$1 + x < 3x - 3 \quad \text{and} \quad 3x - 3 < 4x - 2$$
$$-2x < -4 \qquad\qquad -x < 1$$
$$x > 2 \qquad\qquad\qquad x > -1$$

$x > 2$

$x > -1$

$x > 2$
and
$x > -1$
Solution set: $(2, \infty)$

30. $\dfrac{x+2}{x-1} \le 0$

factors $= 0$: $x = -2, x = 1$

intervals: $(-\infty, -2), (-2, 1), (1, \infty)$

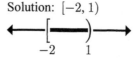

interval	test number	sign of $\frac{x+2}{x-1}$
$(-\infty, -2)$	-3	$+$
$(-2, 1)$	0	$-$
$(1, \infty)$	2	$+$

Include endpoints which make the numerator equal to 0. Do not include endpoints which make the denominator equal to 0.

Solution: $[-2, 1)$

31.

$$\left| \dfrac{3x+2}{2} \right| = 4$$

$$\dfrac{3x+2}{2} = 4 \quad \text{or} \quad \dfrac{3x+2}{2} = -4$$
$$3x + 2 = 8 \qquad\qquad 3x + 2 = -8$$
$$3x = 6 \qquad\qquad\quad 3x = -10$$
$$x = 2 \qquad\qquad\quad x = -\tfrac{10}{3}$$

32.

$$|x + 3| = |x - 3|$$

$$x + 3 = x - 3 \quad \text{or} \quad x + 3 = -(x - 3)$$
$$0 = -6 \qquad\qquad\quad x + 3 = -x + 3$$
$$\text{not true} \qquad\qquad\qquad 2x = 0$$
$$\qquad\qquad\qquad\qquad\quad x = 0$$

33.

$$|2x - 5| > 2$$

$$2x - 5 > 2 \quad \text{or} \quad 2x - 5 < -2$$
$$2x > 7 \qquad\qquad\quad 2x < 3$$
$$x > \tfrac{7}{2} \qquad\qquad\quad x < \tfrac{3}{2}$$

$$\left(-\infty, \tfrac{3}{2}\right) \cup \left(\tfrac{7}{2}, \infty\right)$$

$\tfrac{3}{2} \qquad \tfrac{7}{2}$

34.

$$\left| \dfrac{2x+3}{3} \right| \le 5$$

$$-5 \le \dfrac{2x+3}{3} \le 5$$
$$-15 \le 2x + 3 \le 15$$
$$-18 \le 2x \le 12$$
$$-9 \le x \le 6$$
$$[-9, 6]$$

$-9 \qquad 6$

Cumulative Review Exercises (page 158)

1. even integers: $-2, 0, 2, 6$

2. prime numbers: $2, 11$

3. $-4 \le x < 7 \Rightarrow [-4, 7)$

$-4 \qquad 7$

4. $x \ge 2$ or $x < 0 \Rightarrow (-\infty, 0) \cup [2, \infty)$

$0 \qquad 2$

5. commutative property of addition

6. transitive property

7. $(81a^4)^{1/2} = \left[(9a^2)^2\right]^{1/2} = 9a^2$

8. $81(a^4)^{1/2} = 81\left[(a^2)^2\right]^{1/2} = 81a^2$

9. $(a^{-3}b^{-2})^{-2} = (a^{-3})^{-2}(b^{-2})^{-2} = a^6b^4$

10. $\left(\dfrac{4x^4}{12x^2y}\right)^{-2} = \left(\dfrac{12x^2y}{4x^4}\right)^2 = \left(\dfrac{3y}{x^2}\right)^2 = \dfrac{9y^2}{x^4}$

11. $\left(\dfrac{4x^0y^2}{x^2y}\right)^{-2} = \left(\dfrac{x^2y}{4x^0y^2}\right)^2 = \left(\dfrac{x^2}{4y}\right)^2$
$= \dfrac{x^4}{16y^2}$

12. $\left(\dfrac{4x^{-5}y^2}{6x^{-2}y^{-3}}\right)^2 = \left(\dfrac{2y^5}{3x^3}\right)^2 = \dfrac{4y^{10}}{9x^6}$

13. $\left(a^{1/2}b\right)^2\left(ab^{1/2}\right)^2 = (ab^2)(a^2b) = a^3b^3$

14. $\left(a^{1/2}b^{1/2}c\right)^2 = abc^2$

15. $\dfrac{3}{\sqrt{3}} = \dfrac{3\sqrt{3}}{\sqrt{3}\sqrt{3}} = \dfrac{3\sqrt{3}}{3} = \sqrt{3}$

16. $\dfrac{2}{\sqrt[3]{4x}} = \dfrac{2\sqrt[3]{2x^2}}{\sqrt[3]{4x}\sqrt[3]{2x^2}} = \dfrac{2\sqrt[3]{2x^2}}{\sqrt[3]{8x^3}} = \dfrac{2\sqrt[3]{2x^2}}{2x}$
$= \dfrac{\sqrt[3]{2x^2}}{x}$

17. $\dfrac{3}{y-\sqrt{3}} = \dfrac{3\left(y+\sqrt{3}\right)}{\left(y-\sqrt{3}\right)\left(y+\sqrt{3}\right)} = \dfrac{3\left(y+\sqrt{3}\right)}{y^2 - \left(\sqrt{3}\right)^2} = \dfrac{3\left(y+\sqrt{3}\right)}{y^2 - 3}$

18. $\dfrac{3x}{\sqrt{x}-1} = \dfrac{3x\left(\sqrt{x}+1\right)}{\left(\sqrt{x}-1\right)\left(\sqrt{x}+1\right)} = \dfrac{3x\left(\sqrt{x}+1\right)}{\left(\sqrt{x}\right)^2 - 1^2} = \dfrac{3x\left(\sqrt{x}+1\right)}{x-1}$

19. $\sqrt{75} - 3\sqrt{5} = \sqrt{25}\sqrt{3} - 3\sqrt{5} = 5\sqrt{3} - 3\sqrt{5}$

20. $\sqrt{18} + \sqrt{8} - 2\sqrt{2} = \sqrt{9}\sqrt{2} + \sqrt{4}\sqrt{2} - 2\sqrt{2} = 3\sqrt{2} + 2\sqrt{2} - 2\sqrt{2} = 3\sqrt{2}$

21. $\left(\sqrt{2}-\sqrt{3}\right)^2 = \left(\sqrt{2}-\sqrt{3}\right)\left(\sqrt{2}-\sqrt{3}\right) = \sqrt{4} - 2\sqrt{6} + \sqrt{9} = 5 - 2\sqrt{6}$

22. $\left(3-\sqrt{5}\right)\left(3+\sqrt{5}\right) = 9 - \sqrt{25} = 9 - 5 = 4$

23. $(3x^2 - 2x + 5) - 3(x^2 + 2x - 1) = 3x^2 - 2x + 5 - 3x^2 - 6x + 3 = -8x + 8$

24. $5x^2(2x^2 - x) + x(x^2 - x^3) = 10x^4 - 5x^3 + x^3 - x^4 = 9x^4 - 4x^3$

25. $(3x - 5)(2x + 7) = 6x^2 + 21x - 10x - 35 = 6x^2 + 11x - 35$

26. $(z + 2)(z^2 - z + 2) = z^3 - z^2 + 2z + 2z^2 - 2z + 4 = z^3 + z^2 + 4$

27.

$$\begin{array}{r} 2x^2 - x + 1 \\ 3x+2\overline{\smash{\big)}\,6x^3 + x^2 + x + 2} \\ \underline{6x^3 + 4x^2} \\ -3x^2 + x \\ \underline{-3x^2 - 2x} \\ 3x + 2 \\ \underline{3x + 2} \\ 0 \end{array}$$

28.

$$\begin{array}{r} 3x^2 + 1 + \frac{-x}{x^2+2} \\ x^2+2\overline{\smash{\big)}\,3x^4 + 0x^3 + 7x^2 - x + 2} \\ \underline{3x^4 + 6x^2} \\ x^2 - x + 2 \\ \underline{x^2 + 2} \\ -x \end{array}$$

29. $3t^2 - 6t = 3t(t-2)$

30. $3x^2 - 10x - 8 = (3x+2)(x-4)$

31. $x^8 - 2x^4 + 1 = \left(x^4 - 1\right)\left(x^4 - 1\right) = \left(x^2 + 1\right)\left(x^2 - 1\right)\left(x^2 + 1\right)\left(x^2 - 1\right)$
$$= \left(x^2 + 1\right)^2(x+1)(x-1)(x+1)(x-1)$$
$$= \left(x^2 + 1\right)^2(x+1)^2(x-1)^2$$

32. $x^6 - 1 = (x^3)^2 - 1^2 = (x^3 + 1)(x^3 - 1) = (x+1)(x^2 - x + 1)(x-1)(x^2 + x + 1)$

33. $\dfrac{x^2 - 4}{x^2 + 5x + 6} \cdot \dfrac{x^2 - 2x - 15}{x^2 + 3x - 10} = \dfrac{(x+2)(x-2)}{(x+2)(x+3)} \cdot \dfrac{(x-5)(x+3)}{(x+5)(x-2)} = \dfrac{x-5}{x+5}$

34. $\dfrac{6x^3 + x^2 - x}{x + 2} \div \dfrac{3x^2 - x}{x^2 + 4x + 4} = \dfrac{x(6x^2 + x - 1)}{x + 2} \cdot \dfrac{x^2 + 4x + 4}{3x^2 - x}$
$$= \dfrac{x(2x+1)(3x-1)}{x+2} \cdot \dfrac{(x+2)(x+2)}{x(3x-1)} = (2x+1)(x+2)$$

35. $\dfrac{2}{x+3} + \dfrac{5x}{x-3} = \dfrac{2(x-3)}{(x+3)(x-3)} + \dfrac{5x(x+3)}{(x-3)(x+3)} = \dfrac{2x-6}{(x+3)(x-3)} + \dfrac{5x^2 + 15x}{(x+3)(x-3)}$
$$= \dfrac{5x^2 + 17x - 6}{(x+3)(x-3)}$$

36. $\dfrac{x-2}{x+3}\left(\dfrac{x+3}{x^2-4} - 1\right) = \dfrac{x-2}{x+3}\left(\dfrac{x+3}{x^2-4} - \dfrac{x^2-4}{x^2-4}\right) = \dfrac{x-2}{x+3}\left(\dfrac{-x^2+x+7}{(x+2)(x-2)}\right) = \dfrac{-x^2+x+7}{(x+3)(x+2)}$

37. $\dfrac{\frac{1}{a} + \frac{1}{b}}{\frac{1}{ab}} = \dfrac{ab\left(\frac{1}{a} + \frac{1}{b}\right)}{ab\left(\frac{1}{ab}\right)} = \dfrac{b+a}{1} = b+a$

38. $\dfrac{x^{-1} - y^{-1}}{x - y} = \dfrac{\frac{1}{x} - \frac{1}{y}}{x - y} = \dfrac{xy\left(\frac{1}{x} - \frac{1}{y}\right)}{xy(x - y)}$
$$= \dfrac{y - x}{xy(x - y)} = -\dfrac{1}{xy}$$

39.
$$\frac{3x}{x+5} = \frac{x}{x-5}$$
$$3x(x-5) = x(x+5)$$
$$3x^2 - 15x = x^2 + 5x$$
$$2x^2 - 20x = 0$$
$$2x(x-10) = 0$$
$$2x = 0 \quad \text{or} \quad x - 10 = 0$$
$$x = 0 \qquad\qquad x = 10$$

40.
$$8(2x-3) - 3(5x+2) = 4$$
$$16x - 24 - 15x - 6 = 4$$
$$x = 34$$

41.
$$\frac{1}{R} = \frac{1}{R_1} + \frac{1}{R_2}$$
$$RR_1R_2 \cdot \frac{1}{R} = RR_1R_2\left(\frac{1}{R_1} + \frac{1}{R_2}\right)$$
$$R_1R_2 = RR_2 + RR_1$$
$$R_1R_2 = R(R_2 + R_1)$$
$$\frac{R_1R_2}{R_2 + R_1} = \frac{R(R_2 + R_1)}{R_2 + R_1}$$
$$\frac{R_1R_2}{R_2 + R_1} = R$$

42.
$$S = \frac{a - lr}{1 - r}$$
$$S(1-r) = \frac{a-lr}{1-r}(1-r)$$
$$S(1-r) = a - lr$$
$$S - Sr = a - lr$$
$$S - a = Sr - lr$$
$$S - a = r(S - l)$$
$$\frac{S-a}{S-l} = r$$

43.

$$\text{Area} = 192$$
$$x(40 - 2x) = 192$$
$$40x - 2x^2 = 192$$
$$0 = 2x^2 - 40x + 192$$
$$0 = 2(x-8)(x-12)$$
$$x - 8 = 0 \quad \text{or} \quad x - 12 = 0$$
$$x = 8 \qquad\qquad x = 12$$

If $x = 8$, then the dimensions are 8 feet by 24 feet.
If $x = 12$, then the dimensions are 12 feet by 16 feet.

44. Let $x =$ the amount invested at 6%. Then $25,000 - x =$ the amount invested at 7%.

$$\boxed{\text{Interest at 6\%}} + \boxed{\text{Interest at 7\%}} = \boxed{\text{Total interest}}$$
$$0.06x + 0.07(25,000 - x) = 1,670$$
$$0.06x + 1,750 - 0.07x = 1,670$$
$$-0.01x = -80$$
$$x = 8,000 \Rightarrow \$8,000 \text{ was invested at 6\%.}$$

45.
$$\frac{2+i}{2-i} = \frac{(2+i)(2+i)}{(2-i)(2+i)} = \frac{4 + 4i + i^2}{4 - i^2} = \frac{3 + 4i}{5} = \frac{3}{5} + \frac{4}{5}i$$

46.
$$\frac{i(3-i)}{(1+i)(1+i)} = \frac{i(3-i)(1-i)(1-i)}{(1+i)(1-i)(1+i)(1-i)} = \frac{(3i-i^2)(1-2i+i^2)}{(1-i^2)(1-i^2)}$$
$$= \frac{(1+3i)(-2i)}{1-2i^2+i^4}$$
$$= \frac{-2i-6i^2}{4}$$
$$= \frac{6-2i}{4} = \frac{3}{2} - \frac{1}{2}i$$

47. $|3+4i| = \sqrt{3^2+4^2} = \sqrt{9+16} = \sqrt{25} = 5$

48. $\dfrac{5}{i^7} + 5i = \dfrac{5i}{i^7 i} + 5i = \dfrac{5i}{i^8} + 5i = \dfrac{5i}{(i^4)^2} + 5i = \dfrac{5i}{1^2} + 5i = 5i + 5i = 10i = 0 + 10i$

49.
$$\frac{x+3}{x-1} - \frac{6}{x} = 1$$
$$x(x-1)\left(\frac{x+3}{x-1} - \frac{6}{x}\right) = x(x-1)(1)$$
$$x(x+3) - 6(x-1) = x^2 - x$$
$$x^2 + 3x - 6x + 6 = x^2 - x$$
$$-2x = -6$$
$$x = 3$$

50.
$$x^4 + 36 = 13x^2$$
$$x^4 - 13x^2 + 36 = 0$$
$$(x^2-4)(x^2-9) = 0$$
$$x^2 - 4 = 0 \quad \textbf{or} \quad x^2 - 9 = 0$$
$$x^2 = 4 \qquad\qquad x^2 = 9$$
$$x = \pm 2 \qquad\qquad x = \pm 3$$

51.
$$\sqrt{y+2} + \sqrt{11-y} = 5$$
$$\sqrt{y+2} - 5 = -\sqrt{11-y}$$
$$\left(\sqrt{y+2} - 5\right)^2 = \left(-\sqrt{11-y}\right)^2$$
$$y+2 - 10\sqrt{y+2} + 25 = 11 - y$$
$$-10\sqrt{y+2} = -2y - 16$$
$$\left(-10\sqrt{y+2}\right)^2 = (-2y+16)^2$$
$$100(y+2) = 4y^2 - 64y + 256$$
$$100y + 200 = 4y^2 - 64y + 256$$
$$0 = 4y^2 - 36y + 56$$
$$0 = 4(y-2)(y-7)$$
$$y - 2 = 0 \quad \textbf{or} \quad y - 7 = 0$$
$$y = 2 \qquad\qquad y = 7$$
Both solutions check.

52.
$$z^{2/3} - 13z^{1/3} + 36 = 0$$
$$\left(z^{1/3} - 4\right)\left(z^{1/3} - 9\right) = 0$$
$$z^{1/3} - 4 = 0 \quad \textbf{or} \quad z^{1/3} - 9 = 0$$
$$z^{1/3} = 4 \qquad\qquad z^{1/3} = 9$$
$$\left(z^{1/3}\right)^3 = 4^3 \qquad \left(z^{1/3}\right)^3 = 9^3$$
$$z = 64 \qquad\qquad z = 729$$
Both solutions check.

53. $5x - 7 \le 4$

$\qquad 5x \le 11$

$\qquad x \le \frac{11}{5} \Rightarrow \left(-\infty, \frac{11}{5}\right]$

54. $x^2 - 8x + 15 > 0$

$(x - 3)(x - 5) > 0$

factors $= 0$: $x = 3$, $x = 5$

intervals: $(-\infty, 3), (3, 5), (5, \infty)$

interval	test number	value of $x^2-8x+15$
$(-\infty, 3)$	0	$+15$
$(3, 5)$	4	-1
$(5, \infty)$	6	$+3$

Solution: $(-\infty, 3) \cup (5, \infty)$

55. $\dfrac{x^2 + 4x + 3}{x - 2} \ge 0$

$\dfrac{(x + 3)(x + 1)}{x - 2} \ge 0$

factors $= 0$: $x = -3$, $x = -1$, $x = 2$

int.: $(-\infty, -3), (-3, -1), (-1, 2), (2, \infty)$

interval	test number	sign of $\frac{x^2+4x+3}{x-2}$
$(-\infty, -3)$	-4	$-$
$(-3, -1)$	-2	$+$
$(-1, 2)$	0	$-$
$(2, \infty)$	3	$+$

Include endpoints which make the numerator equal to 0. Do not include endpoints which make the denominator equal to 0.

Solution: $[-3, -1] \cup (2, \infty)$

56. $\dfrac{9}{x} > x$

$\dfrac{9}{x} - x > 0$

$\dfrac{9 - x^2}{x} > 0$

$\dfrac{(3 + x)(3 - x)}{x} > 0$

factors $= 0$: $x = -3$, $x = 3$, $x = 0$

int.: $(-\infty, -3), (-3, 0), (0, 3), (3, \infty)$

interval	test number	sign of $\frac{9-x^2}{x}$
$(-\infty, -3)$	-4	$+$
$(-3, 0)$	-1	$-$
$(0, 3)$	1	$+$
$(3, \infty)$	4	$-$

Solution: $(-\infty, -3) \cup (0, 3)$

57. $|2x - 3| \ge 5$

$2x - 3 \ge 5 \quad \textbf{or} \quad 2x - 3 \le -5$

$\quad 2x \ge 8 \qquad\qquad 2x \le -2$

$\quad\; x \ge 4 \qquad\qquad\; x \le -1$

$(-\infty, -1] \cup [4, \infty)$

58. $\left|\dfrac{3x - 5}{2}\right| < 2$

$-2 < \dfrac{3x - 5}{2} < 2$

$-4 < 3x - 5 < 4$

$1 < 3x < 9$

$\frac{1}{3} < x < 3$

$\left(\frac{1}{3}, 3\right)$

Exercise 2.1 (page 173)

1. quadrants **3.** to the right **5.** first **7.** linear

9. x-intercept **11.** horizontal **13.** midpoint **15.** $A(2, 3)$

17. $C(-2, -3)$ **19.** $E(0, 0)$ **21.** $G(-5, -5)$

23, 25, 27, 29.

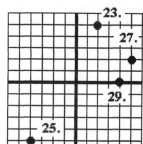

23. QI

25. QIII

27. QI

29. $+ x$-axis

31.
$$\begin{array}{ll} x + y = 5 & x + y = 5 \\ x + 0 = 5 & 0 + y = 5 \\ x = 5 & y = 5 \\ (5, 0) & (0, 5) \end{array}$$

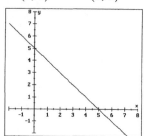

33.
$$\begin{array}{ll} 2x - y = 4 & 2x - y = 4 \\ 2x - 0 = 4 & 2(0) - y = 4 \\ 2x = 4 & -y = 4 \\ x = 2 & y = -4 \\ (2, 0) & (0, -4) \end{array}$$

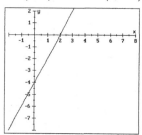

35.
$$\begin{array}{ll} 3x + 2y = 6 & 3x + 2y = 6 \\ 3x + 2(0) = 6 & 3(0) + 2y = 6 \\ 3x = 6 & 2y = 6 \\ x = 2 & y = 3 \\ (2, 0) & (0, 3) \end{array}$$

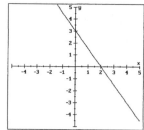

37.
$$4x - 5y = 20 \qquad 4x - 5y = 20$$
$$4x - 5(0) = 20 \qquad 4(0) - 5y = 20$$
$$4x = 20 \qquad -5y = 20$$
$$x = 5 \qquad y = -4$$
$$(5, 0) \qquad (0, -4)$$

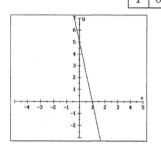

39.
$$y - 2x = 7$$
$$y = 2x + 7$$

x	y
0	7
-2	3

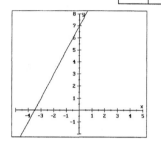

41.
$$y + 5x = 5$$
$$y = -5x + 5$$

x	y
0	5
1	0

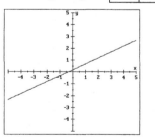

43.
$$6x - 3y = 10$$
$$-3y = -6x + 10$$
$$y = 2x - \frac{10}{3}$$

x	y
0	$-\frac{10}{3}$
2	$\frac{2}{3}$

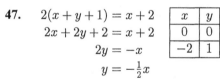
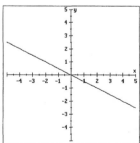

45.
$$3x = 6y - 1$$
$$-6y = -3x - 1$$
$$y = \frac{1}{2}x + \frac{1}{6}$$

x	y
0	$\frac{1}{6}$
-2	$-\frac{5}{6}$

47.
$$2(x + y + 1) = x + 2$$
$$2x + 2y + 2 = x + 2$$
$$2y = -x$$
$$y = -\frac{1}{2}x$$

x	y
0	0
-2	1

49. $y = 3$

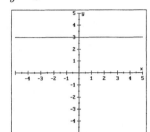

51. $3x + 5 = -1$
$$3x = -6 \Rightarrow x = -2$$

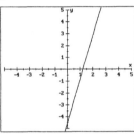

53. $3(y + 2) = y$
$$3y + 6 = y$$
$$2y = -6 \Rightarrow y = -3$$

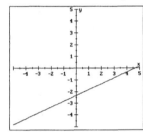

55. $3(y + 2x) = 6x + y$
$$3y + 6x = 6x + y$$
$$2y = 0 \Rightarrow y = 0$$

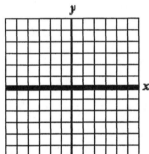

57. $y = 3.7x - 4.5$

x-int: $x = 1.22$

59. $1.5x - 3y = 7$
$$-3y = -1.5x + 7$$
$$y = 0.5x - \frac{7}{3}$$

x-int: $x = 4.67$

61. $d = \sqrt{(x_2 - x_1)^2 + (y_2 - y_1)^2}$
$$= \sqrt{(4 - 0)^2 + (-3 - 0)^2}$$
$$= \sqrt{4^2 + (-3)^2}$$
$$= \sqrt{16 + 9} = \sqrt{25} = 5$$

63. $d = \sqrt{(x_2 - x_1)^2 + (y_2 - y_1)^2}$
$$= \sqrt{(-3 - 0)^2 + (2 - 0)^2}$$
$$= \sqrt{(-3)^2 + (2)^2}$$
$$= \sqrt{9 + 4} = \sqrt{13}$$

65. $d = \sqrt{(x_2 - x_1)^2 + (y_2 - y_1)^2}$
$$= \sqrt{(1 - 0)^2 + (1 - 0)^2}$$
$$= \sqrt{(1)^2 + (1)^2}$$
$$= \sqrt{1 + 1} = \sqrt{2}$$

67. $d = \sqrt{(x_2 - x_1)^2 + (y_2 - y_1)^2}$
$$= \sqrt{\left(\sqrt{3} - 0\right)^2 + (1 - 0)^2}$$
$$= \sqrt{\left(\sqrt{3}\right)^2 + (1)^2}$$
$$= \sqrt{3 + 1} = \sqrt{4} = 2$$

69. $d = \sqrt{(x_2 - x_1)^2 + (y_2 - y_1)^2}$

$= \sqrt{(3 - 6)^2 + (7 - 3)^2}$

$= \sqrt{(-3)^2 + (4)^2}$

$= \sqrt{9 + 16} = \sqrt{25} = 5$

71. $d = \sqrt{(x_2 - x_1)^2 + (y_2 - y_1)^2}$

$= \sqrt{[4 - (-1)]^2 + [-6 - 6]^2}$

$= \sqrt{(5)^2 + (-12)^2}$

$= \sqrt{25 + 144} = \sqrt{169} = 13$

73. $d = \sqrt{(x_2 - x_1)^2 + (y_2 - y_1)^2}$

$= \sqrt{[-2 - (-9)]^2 + [-15 - (-39)]^2}$

$= \sqrt{(7)^2 + (24)^2}$

$= \sqrt{49 + 576} = \sqrt{625} = 25$

75. $d = \sqrt{(x_2 - x_1)^2 + (y_2 - y_1)^2}$

$= \sqrt{[3 - (-5)]^2 + [-3 - 5]^2}$

$= \sqrt{(8)^2 + (-8)^2}$

$= \sqrt{64 + 64} = \sqrt{128} = 8\sqrt{2}$

77. $d = \sqrt{(x_2 - x_1)^2 + (y_2 - y_1)^2} = \sqrt{[\pi - \pi]^2 + [-2 - 5]^2} = \sqrt{(0)^2 + (-7)^2}$

$= \sqrt{0 + 49} = \sqrt{49} = 7$

79. $M\left(\dfrac{x_1 + x_2}{2}, \dfrac{y_1 + y_2}{2}\right) = M\left(\dfrac{2 + 6}{2}, \dfrac{4 + 8}{2}\right) = M\left(\dfrac{8}{2}, \dfrac{12}{2}\right) = M(4, 6)$

81. $M\left(\dfrac{x_1 + x_2}{2}, \dfrac{y_1 + y_2}{2}\right) = M\left(\dfrac{2 + (-2)}{2}, \dfrac{-5 + 7}{2}\right) = M\left(\dfrac{0}{2}, \dfrac{2}{2}\right) = M(0, 1)$

83. $M\left(\dfrac{x_1 + x_2}{2}, \dfrac{y_1 + y_2}{2}\right) = M\left(\dfrac{-8 + 8}{2}, \dfrac{5 + (-5)}{2}\right) = M\left(\dfrac{0}{2}, \dfrac{0}{2}\right) = M(0, 0)$

85. $M\left(\dfrac{x_1 + x_2}{2}, \dfrac{y_1 + y_2}{2}\right) = M\left(\dfrac{0 + \sqrt{5}}{2}, \dfrac{0 + \sqrt{5}}{2}\right) = M\left(\dfrac{\sqrt{5}}{2}, \dfrac{\sqrt{5}}{2}\right)$

87. Let Q have coordinates (x, y):

$M\left(\dfrac{x_1+x_2}{2}, \dfrac{y_1+y_2}{2}\right) = (3, 5)$

$\dfrac{x_1 + x_2}{2} = 3 \qquad \dfrac{y_1 + y_2}{2} = 5$

$\dfrac{1 + x}{2} = 3 \qquad \dfrac{4 + y}{2} = 5$

$1 + x = 6 \qquad 4 + y = 10$

$x = 5 \qquad y = 6$

$Q(5, 6)$

89. Let Q have coordinates (x, y):

$M\left(\dfrac{x_1+x_2}{2}, \dfrac{y_1+y_2}{2}\right) = (5, 5)$

$\dfrac{x_1 + x_2}{2} = 5 \qquad \dfrac{y_1 + y_2}{2} = 5$

$\dfrac{5 + x}{2} = 5 \qquad \dfrac{-5 + y}{2} = 5$

$5 + x = 10 \qquad -5 + y = 10$

$x = 5 \qquad y = 15$

$Q(5, 15)$

91. Let the points be identified as $A(13, -2)$, $B(9, -8)$ and $C(5, -2)$.

$AB = \sqrt{(x_2 - x_1)^2 + (y_2 - y_1)^2} = \sqrt{(13 - 9)^2 + (-2 - (-8))^2} = \sqrt{16 + 36} = \sqrt{52} = 2\sqrt{13}$

$BC = \sqrt{(x_2 - x_1)^2 + (y_2 - y_1)^2} = \sqrt{(9 - 5)^2 + (-8 - (-2))^2} = \sqrt{16 + 36} = \sqrt{52} = 2\sqrt{13}$

Since AB and BC have the same length, the triangle is isosceles.

93. $M = \left(\dfrac{2 + 6}{2}, \dfrac{4 + 10}{2}\right) = \left(\dfrac{8}{2}, \dfrac{14}{2}\right) = (4, 7);$ $N = \left(\dfrac{4 + 6}{2}, \dfrac{6 + 10}{2}\right) = \left(\dfrac{10}{2}, \dfrac{16}{2}\right) = (5, 8)$

$MN = \sqrt{(x_2 - x_1)^2 + (y_2 - y_1)^2} = \sqrt{(4 - 5)^2 + (7 - 8)^2} = \sqrt{1 + 1} = \sqrt{2}$

95. $M = \left(\dfrac{0 + a}{2}, \dfrac{b + 0}{2}\right) = \left(\dfrac{a}{2}, \dfrac{b}{2}\right);$ $L = \left(\dfrac{a}{2}, 0\right);$ $N = \left(0, \dfrac{b}{2}\right)$

Area of $AOB = \frac{1}{2} \cdot$ base $\cdot$ height $= \frac{1}{2}(OA)(OB) = \frac{1}{2}(a)(b) = \frac{1}{2}ab$

Area of $OLMN =$ length $\cdot$ width $= (OL)(ON) = \frac{a}{2} \cdot \frac{b}{2} = \frac{1}{4}ab = \frac{1}{2}$(Area of AOB)

97. $y = 7500x + 125{,}000$
$y = 7500(5) + 125{,}000$
$y = 37{,}500 + 125{,}000$
$y = 162{,}500$
The value will be \$162,500.

99. $p = -\dfrac{1}{10}q + 170$

$150 = -\dfrac{1}{10}q + 170$

$\dfrac{1}{10}q = 20$

$q = 200$

200 TVs will be sold.

101. $V = \dfrac{nv}{N}$

$60 = \dfrac{12v}{20}$

$1200 = 12v$

$100 = v$

The smaller gear is spinning at 100 rpm.

103.

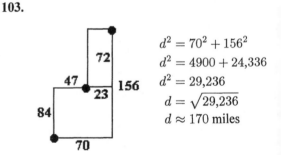

$d^2 = 70^2 + 156^2$
$d^2 = 4900 + 24{,}336$
$d^2 = 29{,}236$
$d = \sqrt{29{,}236}$
$d \approx 170$ miles

105. Answers may vary.

107. Answers may vary.

109. $[-3, 2)$

$(-2, 3]$

$[-3, 2) \cup (-2, 3]$

111. $[-3, -2)$

$(2, 3]$

$[-3, 2) \cap (-2, 3]$ no intersection

113.
$$\frac{3}{y+6} = \frac{4}{y+4}$$
$$(y+6)(y+4) \cdot \frac{3}{y+6} = (y+6)(y+4) \cdot \frac{4}{y+4}$$
$$3(y+4) = 4(y+6)$$
$$3y + 12 = 4y + 24$$
$$-12 = y$$

Exercise 2.2 (page 185)

1. divided   **3.** run   **5.** the change in   **7.** vertical

9. perpendicular

11. $m = \dfrac{y_2 - y_1}{x_2 - x_1} = \dfrac{10 - 5}{3 - 2} = \dfrac{5}{1} = 5$

13. $m = \dfrac{y_2 - y_1}{x_2 - x_1} = \dfrac{5 - (-2)}{-1 - 3} = \dfrac{7}{-4} = -\dfrac{7}{4}$

15. $m = \dfrac{y_2 - y_1}{x_2 - x_1} = \dfrac{1 - (-7)}{4 - 8} = \dfrac{8}{-4} = -2$

17. $m = \dfrac{y_2 - y_1}{x_2 - x_1} = \dfrac{-3 - 3}{-4 - (-4)} = \dfrac{-6}{0} \Rightarrow$ und.

19. $m = \dfrac{y_2 - y_1}{x_2 - x_1} = \dfrac{\frac{7}{3} - \frac{2}{3}}{\frac{5}{2} - \frac{3}{2}} = \dfrac{\frac{5}{3}}{\frac{2}{2}} = \dfrac{\frac{5}{3}}{1} = \dfrac{5}{3}$

21. $m = \dfrac{y_2 - y_1}{x_2 - x_1} = \dfrac{a - c}{(b+c) - (a+b)}$
$$= \dfrac{a - c}{c - a} = -1$$

23. $y = 3x + 2$ $m = \dfrac{y_2 - y_1}{x_2 - x_1} = \dfrac{5 - 2}{1 - 0}$

x	y
0	2
1	5

$$= \dfrac{3}{1} = 3$$

25. $5x - 10y = 3$ $m = \dfrac{y_2 - y_1}{x_2 - x_1}$

x	y
0	$-\frac{3}{10}$
1	$\frac{1}{5}$

$$= \dfrac{\frac{1}{5} - \left(-\frac{3}{10}\right)}{1 - 0}$$
$$= \dfrac{\frac{5}{10}}{1} = \dfrac{1}{2}$$

27. $3(y + 2) = 2x - 3$ $m = \dfrac{y_2 - y_1}{x_2 - x_1}$
$$3y - 2x = -9$$

x	y
0	-3
3	-1

$$= \dfrac{-1 - (-3)}{3 - 0}$$
$$= \dfrac{2}{3}$$

29. $3(y + x) = 3(x - 1)$ $m = \dfrac{y_2 - y_1}{x_2 - x_1}$
$$3y = -3$$
$$y = -1$$

x	y
0	-1
1	-1

$$= \dfrac{-1 - (-1)}{1 - 0}$$
$$= \dfrac{0}{1} = 0$$

31. The slope is negative.   **33.** The slope is positive.

35. The slope is undefined.

37. $m_1 m_2 = 3\left(-\dfrac{1}{3}\right) = -1$
perpendicular

39. $m_1 = \sqrt{8} = 2\sqrt{2} = m_2$
parallel

41. $m_1 m_2 = -\sqrt{2}\left(\dfrac{\sqrt{2}}{2}\right) = -1$
perpendicular

43. $m_1 m_2 = -0.125(8) = -1$
perpendicular

45. $m_1 m_2 = ab^{-1}\left(-a^{-1}b\right) = -a^0 b^0 = -1$
perpendicular

For Exercises 47-51 use the slope of line through R and S calculated below:

$$m_{RS} = \frac{y_2 - y_1}{x_2 - x_1} = \frac{7-5}{2-(-3)} = \frac{2}{5}$$

47. $m_{PQ} = \dfrac{y_2 - y_1}{x_2 - x_1} = \dfrac{6-4}{7-2} = \dfrac{2}{5} = m_{RS} \Rightarrow$ parallel

49. $m_{PQ} = \dfrac{y_2 - y_1}{x_2 - x_1} = \dfrac{1-6}{-2-(-4)} = \dfrac{-5}{2} = -\dfrac{5}{2} \Rightarrow$ perpendicular

51. $m_{PQ} = \dfrac{y_2 - y_1}{x_2 - x_1} = \dfrac{6a-a}{3a-a} = \dfrac{5a}{2a} = \dfrac{5}{2} \Rightarrow$ neither

53. $m_{PQ} = \dfrac{y_2 - y_1}{x_2 - x_1} = \dfrac{9-8}{-6-(-2)} = \dfrac{1}{-4} = -\dfrac{1}{4}$

$m_{PR} = \dfrac{y_2 - y_1}{x_2 - x_1} = \dfrac{5-8}{2-(-2)} = \dfrac{-3}{4} = -\dfrac{3}{4} \Rightarrow$ not on same line

55. $m_{PQ} = \dfrac{y_2 - y_1}{x_2 - x_1} = \dfrac{0-a}{0-(-a)} = \dfrac{-a}{a} = -1$

$m_{PR} = \dfrac{y_2 - y_1}{x_2 - x_1} = \dfrac{-a-a}{a-(-a)} = \dfrac{-2a}{2a} = -1 \Rightarrow$ on same line

57. $m_{PQ} = \dfrac{y_2 - y_1}{x_2 - x_1} = \dfrac{-5-4}{2-5} = \dfrac{-9}{-3} = 3$

$m_{PR} = \dfrac{y_2 - y_1}{x_2 - x_1} = \dfrac{-3-4}{8-5} = \dfrac{-7}{3} = -\dfrac{7}{3}$

$m_{QR} = \dfrac{y_2 - y_1}{x_2 - x_1} = \dfrac{-3-(-5)}{8-2} = \dfrac{2}{6} = \dfrac{1}{3} \Rightarrow$ None are perpendicular.

59. $m_{PQ} = \dfrac{y_2 - y_1}{x_2 - x_1} = \dfrac{9-3}{1-1} = \dfrac{6}{0} \Rightarrow$ undefined $\Rightarrow$ vertical

$m_{PR} = \dfrac{y_2 - y_1}{x_2 - x_1} = \dfrac{3-3}{7-1} = \dfrac{0}{6} = 0 \Rightarrow$ horizontal

$m_{QR} = \dfrac{y_2 - y_1}{x_2 - x_1} = \dfrac{3-9}{7-1} = \dfrac{-6}{6} = -1 \Rightarrow PQ$ and PR are perpendicular.

61. $m_{PQ} = \dfrac{y_2 - y_1}{x_2 - x_1} = \dfrac{b - 0}{a - 0} = \dfrac{b}{a}; m_{PR} = \dfrac{y_2 - y_1}{x_2 - x_1} = \dfrac{a - 0}{-b - 0} = \dfrac{a}{-b} = -\dfrac{a}{b}$

$m_{QR} = \dfrac{y_2 - y_1}{x_2 - x_1} = \dfrac{a - b}{-b - a} = \dfrac{a - b}{-b - a} \Rightarrow PQ$ and PR are perpendicular.

63. $m_{AB} = \dfrac{y_2 - y_1}{x_2 - x_1} = \dfrac{4 - (-1)}{-3 - (-1)} = \dfrac{5}{-2} = -\dfrac{5}{2}; m_{AC} = \dfrac{y_2 - y_1}{x_2 - x_1} = \dfrac{1 - (-1)}{4 - (-1)} = \dfrac{2}{5} = \dfrac{2}{5}$

AB and AC are perpendicular. $\Rightarrow$ right triangle

65. $m_{AB} = \dfrac{y_2 - y_1}{x_2 - x_1} = \dfrac{0 - (-1)}{3 - 1} = \dfrac{1}{2}; \ d(A, B) = \sqrt{(1 - 3)^2 + (-1 - 0)^2} = \sqrt{5}$

$m_{BC} = \dfrac{y_2 - y_1}{x_2 - x_1} = \dfrac{2 - 0}{2 - 3} = \dfrac{2}{-1} = -2; \ d(B, C) = \sqrt{(3 - 2)^2 + (0 - 2)^2} = \sqrt{5}$

$m_{CD} = \dfrac{y_2 - y_1}{x_2 - x_1} = \dfrac{1 - 2}{0 - 2} = \dfrac{-1}{-2} = \dfrac{1}{2}; \ d(C, D) = \sqrt{(2 - 0)^2 + (2 - 1)^2} = \sqrt{5}$

$m_{DA} = \dfrac{y_2 - y_1}{x_2 - x_1} = \dfrac{1 - (-1)}{0 - 1} = \dfrac{2}{-1} = -2; \ d(D, A) = \sqrt{(-1 - 1)^2 + (1 - 0)^2} = \sqrt{5}$

Adjacent sides are perpendicular and congruent, so the figure is a square.

67. $m_{AB} = \dfrac{y_2 - y_1}{x_2 - x_1} = \dfrac{3 - (-2)}{3 - (-2)} = \dfrac{5}{5} = 1; m_{BC} = \dfrac{y_2 - y_1}{x_2 - x_1} = \dfrac{6 - 3}{2 - 3} = \dfrac{3}{-1} = -3$

$m_{CD} = \dfrac{y_2 - y_1}{x_2 - x_1} = \dfrac{1 - 6}{-3 - 2} = \dfrac{-5}{-5} = 1; m_{DA} = \dfrac{y_2 - y_1}{x_2 - x_1} = \dfrac{1 - (-2)}{-3 - (-2)} = \dfrac{3}{-1} = -3$

Opposite sides are parallel, so the figure is a parallelogram.

69. $M\left(\dfrac{5+7}{2}, \dfrac{9+5}{2}\right) = M\left(\dfrac{12}{2}, \dfrac{14}{2}\right) = M(6, 7); N\left(\dfrac{1+7}{2}, \dfrac{3+5}{2}\right) = N\left(\dfrac{8}{2}, \dfrac{8}{2}\right) = N(4, 4)$

$m_{MN} = \dfrac{y_2 - y_1}{x_2 - x_1} = \dfrac{4 - 7}{4 - 6} = \dfrac{-3}{-2} = \dfrac{3}{2}; m_{AC} = \dfrac{y_2 - y_1}{x_2 - x_1} = \dfrac{9 - 3}{5 - 1} = \dfrac{6}{4} = \dfrac{3}{2} \Rightarrow MN \parallel AC$

71. $m = \dfrac{y_2 - y_1}{x_2 - x_1} = \dfrac{26 - 12}{5 - 1} = \dfrac{14}{4} = 3.5$

The rate of growth was 3.5 students per year.

73. $m = \dfrac{y_2 - y_1}{x_2 - x_1} = \dfrac{6700 - 2200}{10 - 3}$

$= \dfrac{4500}{7} \approx 642.86$

The cost decreased about \$642.86 per year.

75. $\dfrac{\Delta T}{\Delta t} =$ the hourly rate of change of temperature.

(Let $t = x$ and $T = y$.)

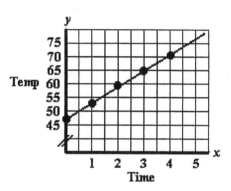

77. $D = 590t$; The slope is the speed of the plane.

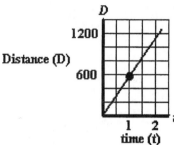

79. **Answers may vary.**

81. $3x + 7y = 21$
$$7y = -3x + 21$$
$$\frac{7y}{7} = \frac{-3x}{7} + \frac{21}{7}$$
$$y = -\frac{3}{7}x + 3$$

83. $\dfrac{x}{5} + \dfrac{y}{2} = 1$
$$2\left(\frac{x}{5} + \frac{y}{2}\right) = 2(1)$$
$$\frac{2}{5}x + y = 2$$
$$y = -\frac{2}{5}x + 2$$

85. $6p^2 + p - 12 = (2p + 3)(3p - 4)$

87. $mp + mq + np + nq = m(p+q) + n(p+q)$
$$= (p+q)(m+n)$$

Exercise 2.3 (page 197)

1. $y - y_1 = m(x - x_1)$

3. slope-intercept

5. $-\dfrac{A}{B}$

7. $y - y_1 = m(x - x_1)$
$$y - 4 = 2(x - 2)$$
$$y - 4 = 2x - 4$$
$$-2x + y = 0$$
$$2x - y = 0$$

9. $y - y_1 = m(x - x_1)$
$$y - \frac{1}{2} = 2\left(x + \frac{3}{2}\right)$$
$$y - \frac{1}{2} = 2x + 3$$
$$2y - 1 = 4x + 6$$
$$-4x + 2y = 7$$
$$4x - 2y = -7$$

11. $y - y_1 = m(x - x_1)$
$$y - 0 = \pi(x - \pi)$$
$$y = \pi x - \pi^2$$
$$-\pi x + y = -\pi^2$$
$$\pi x - y = \pi^2$$

13. From the graph, $m = \dfrac{2}{3}$ and the line passes through $(2, 5)$.

$$y - y_1 = m(x - x_1)$$
$$y - 5 = \frac{2}{3}(x - 2)$$
$$3(y - 5) = 3 \cdot \frac{2}{3}(x - 2)$$
$$3y - 15 = 2(x - 2)$$
$$3y - 15 = 2x - 4$$
$$-2x + 3y = 11$$
$$2x - 3y = -11$$

15. $m = \dfrac{y_2 - y_1}{x_2 - x_1} = \dfrac{4 - 0}{4 - 0} = \dfrac{4}{4} = 1$

$$y - y_1 = m(x - x_1)$$
$$y - 0 = 1(x - 0)$$
$$y = x$$

17. $m = \dfrac{y_2 - y_1}{x_2 - x_1} = \dfrac{-3 - 4}{0 - 3} = \dfrac{-7}{-3} = \dfrac{7}{3}$

$$y - y_1 = m(x - x_1)$$
$$y + 3 = \frac{7}{3}(x - 0)$$
$$y = \frac{7}{3}x - 3$$

19. From the graph, $m = -\dfrac{9}{5}$ and the line passes through $(-2, 4)$

$$y - y_1 = m(x - x_1)$$
$$y - 4 = -\frac{9}{5}(x + 2)$$
$$y - 4 = -\frac{9}{5}x - \frac{18}{5}$$
$$y = -\frac{9}{5}x - \frac{18}{5} + 4$$
$$y = -\frac{9}{5}x + \frac{2}{5}$$

21. $y = mx + b$
$y = 3x - 2$

23. $y = mx + b$
$$y = 5x - \frac{1}{5}$$

25. $y = mx + b$
$$y = ax + \frac{1}{a}$$

27. $y = mx + b$
$y = ax + a$

29. $y = mx + b$
$$0 = \frac{3}{2}(0) + b$$
$$0 = b$$

$y = mx + b$
$$y = \frac{3}{2}x + 0$$
$$2y = 3x$$
$$-3x + 2y = 0$$
$$3x - 2y = 0$$

31. $y = mx + b$
$5 = -3(-3) + b$
$5 = 9 + b$
$-4 = b$

$y = mx + b$
$y = -3x - 4$
$3x + y = -4$

33. $y = mx + b$
$\sqrt{2} = \sqrt{2}(0) + b$
$\sqrt{2} = b$

$y = mx + b$
$y = \sqrt{2}x + \sqrt{2}$
$-\sqrt{2}x + y = \sqrt{2}$
$\sqrt{2}x - y = -\sqrt{2}$

35. $y + 1 = x$

$\quad y = x - 1 \Rightarrow m = 1, (0, -1)$

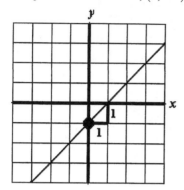

37. $x = \dfrac{3}{2}y - 3$

$\quad 2x = 3y - 6$

$\quad -3y = -2x - 6$

$\quad\quad y = \dfrac{2}{3}x + 2 \Rightarrow m = \dfrac{2}{3}, (0, 2)$

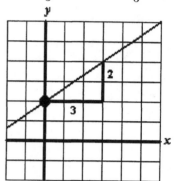

39. $3(y - 4) = -2(x - 3)$

$\quad 3y - 12 = -2x + 6$

$\quad\quad 3y = -2x + 18$

$\quad\quad\quad y = -\dfrac{2}{3}x + 6 \Rightarrow m = -\dfrac{2}{3}, (0, 6)$

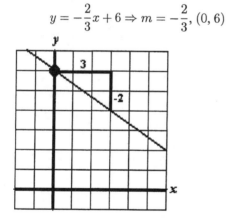

41. $3x - 2y = 8$

$\quad -2y = -3x + 8$

$\quad\quad y = \dfrac{3}{2}x - 4$

$\quad m = \dfrac{3}{2}, (0, -4)$

43. $-2(x + 3y) = 5$

$\quad -2x - 6y = 5$

$\quad\quad -6y = 2x + 5$

$\quad\quad\quad y = -\dfrac{1}{3}x - \dfrac{5}{6}$

$\quad m = -\dfrac{1}{3}, \left(0, -\dfrac{5}{6}\right)$

45. $x = \dfrac{2y - 4}{7}$

$\quad 7x = 2y - 4$

$\quad -2y = -7x - 4$

$\quad\quad y = \dfrac{7}{2}x + 2$

$\quad m = \dfrac{7}{2}, (0, 2)$

47. $y = 3x + 4 \quad y = 3x - 7$
$m = 3 \qquad m = 3$
The lines are parallel.

49. $x + y = 2 \qquad y = x + 5$
$\qquad y = -x + 2 \quad m = 1$
$m = -1$
The lines are perpendicular.

51. $y = 3x + 7 \quad 2y = 6x - 9$
$m = 3 \qquad\qquad y = 3x - \dfrac{9}{2}$
$\qquad\qquad\qquad m = 3$
The lines are parallel.

53. $\qquad x = 3y + 4 \quad y = -3x + 7$
$-3y = -x + 4 \quad m = -3$
$\qquad y = \dfrac{1}{3}x - \dfrac{4}{3}$
$\qquad m = \dfrac{1}{3}$
The lines are perpendicular.

55. $y = 3 \qquad x = 4$
horizontal $\quad$ vertical
The lines are perpendicular.

57. $\qquad x = \dfrac{y - 2}{3} \qquad 3(y - 3) + x = 0$
$\quad 3x = y - 2 \qquad\quad 3y - 9 + x = 0$
$\quad -y = -3x - 2 \qquad\qquad 3y = -x + 9$
$\qquad y = 3x + 2$
$\qquad m = 3 \qquad\qquad\qquad y = -\dfrac{1}{3}x + 3$
$\qquad\qquad\qquad\qquad m = -\dfrac{1}{3}$
The lines are perpendicular.

59. $y = 4x - 7 \quad y - y_1 = m(x - x_1)$
$m = 4 \qquad\quad y - 0 = 4(x - 0)$
Use $m = 4$. $\qquad \boxed{y = 4x}$

61. $4x - y = 7 \qquad y - y_1 = m(x - x_1)$
$\quad -y = -4x + 7 \quad y - 5 = 4(x - 2)$
$\quad y = 4x - 7 \qquad y - 5 = 4x - 8$
$m = 4 \qquad\qquad\qquad \boxed{y = 4x - 3}$
Use $m = 4$.

63. $\qquad x = \dfrac{5}{4}y - 2 \quad y - y_1 = m(x - x_1)$
$\quad 4x = 5y - 8 \qquad y + 2 = \dfrac{4}{5}(x - 4)$
$-5y = -4x - 8$
$\quad y = \dfrac{4}{5}x + \dfrac{8}{5} \qquad y + 2 = \dfrac{4}{5}x - \dfrac{16}{5}$
$m = \dfrac{4}{5} \qquad\qquad\quad \boxed{y = \dfrac{4}{5}x - \dfrac{26}{5}}$
Use $m = \dfrac{4}{5}$.

65. $y = 4x - 7 \qquad y - y_1 = m(x - x_1)$
$m = 4 \qquad\qquad\quad y - 0 = -\dfrac{1}{4}(x - 0)$
Use $m = -\dfrac{1}{4}$. $\qquad \boxed{y = -\dfrac{1}{4}x}$

67. $4x - y = 7 \qquad\quad y - y_1 = m(x - x_1)$
$\quad -y = -4x + 7 \qquad y - 5 = -\dfrac{1}{4}(x - 2)$
$\quad y = 4x - 7$
$m = 4 \qquad\qquad\quad y - 5 = -\dfrac{1}{4}x + \dfrac{1}{2}$
Use $m = -\dfrac{1}{4}$. $\qquad \boxed{y = -\dfrac{1}{4}x + \dfrac{11}{2}}$

69.
$$x = \frac{5}{4}y - 2$$
$$4x = 5y - 8$$
$$-5y = -4x - 8$$
$$y = \frac{4}{5}x + \frac{8}{5}$$
$$m = \frac{4}{5}$$
Use $m = -\frac{5}{4}$.

$$y - y_1 = m(x - x_1)$$
$$y + 2 = -\frac{5}{4}(x - 4)$$
$$y + 2 = -\frac{5}{4}x + 5$$
$$\boxed{y = -\frac{5}{4}x + 3}$$

71. $4x + 5y = 20 \Rightarrow A = 4, B = 5, C = 20$
$$m = -\frac{A}{B} = -\frac{4}{5}$$
$$b = \frac{C}{B} = \frac{20}{5} = 4 \Rightarrow (0, 4)$$

73. $2x + 3y = 12 \Rightarrow A = 2, B = 3, C = 12$
$$m = -\frac{A}{B} = -\frac{2}{3}$$
$$b = \frac{C}{B} = \frac{12}{3} = 4 \Rightarrow (0, 4)$$

75. Since $y = 3$ is the equation of a horizontal line, any perpendicular line will be vertical. Find the midpoint:
$$x = \frac{2 + (-6)}{2} = -2; y = \frac{4 + 10}{2} = 7$$
The vertical line through $(-2, 7)$ is $x = -2$.

77. Since $x = 3$ is the equation of a vertical line, any parallel line will be vertical. Find the midpoint:
$$x = \frac{2 + 8}{2} = 5; y = \frac{-4 + 12}{2} = 4$$
The vertical line through $(5, 4)$ is $x = 5$.

79. Let $x =$ the number of years the cab has been owned and let $y =$ the value of the cab. Then two points on the line are given: $(0, 24300)$ and $(7, 1900)$.
$$m = \frac{24300 - 1900}{0 - 7} = \frac{22400}{-7} = -3200$$
$$y - y_1 = m(x - x_1)$$
$$y - 24300 = -3200(x - 0)$$
$$y - 24300 = -3200x$$
$$y = -3200x + 24300$$

81. Let $x =$ the number of years the building has been owned and let $y =$ the value of the building. Then two points on the line are given: $(0, 475000)$ and $(10, 950000)$.
$$m = \frac{950000 - 475000}{10 - 0} = \frac{475000}{10}$$
$$= 47500$$
$$y - y_1 = m(x - x_1)$$
$$y - 475000 = 47500(x - 0)$$
$$y - 475000 = 47500x$$
$$y = 47500x + 475000$$

83. Let $x =$ the number of years the TV has been owned and let $y =$ the value of the TV. Then two points on the line are given: $(0, 1900)$ and $(3, 1190)$.
$$m = \frac{1900 - 1190}{0 - 3} = \frac{710}{-3} = -\frac{710}{3}$$
$$y - y_1 = m(x - x_1)$$
$$y - 1900 = -\frac{710}{3}(x - 0)$$
$$y - 1900 = -\frac{710}{3}x$$
$$y = -\frac{710}{3}x + 1900$$

85. Let $x =$ the number of years the copier has been owned and let $y =$ the value of the copier. Then one point on the line is given: $(0, 1050)$. Since the copier depreciates by \$120 per year, $m = -120$.
$$y - y_1 = m(x - x_1)$$
$$y - 1050 = -120(x - 0)$$
$$y - 1050 = -120x$$
$$y = -120x + 1050$$
Let $x = 8$ and find the value of y:
$$y = -120x + 1050$$
$$= -120(8) + 1050 = 90$$
The salvage value will be \$90.

87. Let $x =$ the number of years the table has been owned and let $y =$ the value of the table. Then one point on the line is given: $(2, 450)$. Since the table appreciates by \$40 per year, $m = 40$.
$$y - y_1 = m(x - x_1)$$
$$y - 450 = 40(x - 2)$$
$$y - 450 = 40x - 80$$
$$y = 40x + 370$$
Let $x = 13$ and find the value of y:
$$y = 40x + 370$$
$$= 40(13) + 370 = 890$$
The value will be \$890.

89. Let $x =$ the number of years the cottage has been owned and let $y =$ the value of the cottage. Then one point on the line is given: $(3, 47700)$. Since the cottage appreciates by \$3500 per year, $m = 3500$.
$$y - y_1 = m(x - x_1)$$
$$y - 47700 = 3500(x - 3)$$
$$y - 47700 = 3500x - 10500$$
$$y = 3500x + 37200$$
Let $x = 0$ and find the value of y:
$$y = 3500x + 37200$$
$$= 3500(0) + 37200 = 37200$$
The purchase price was \$37,200.

91. Let $x =$ the hours of labor and let $y =$ the labor charge. Then $m =$ the hourly charge.
$$y = mx \qquad y = 46x$$
$$69 = m(1.5) \qquad y = 46(5) = 230$$
$$46 = m \qquad \text{The charge will be \$230.}$$

93. Let $x =$ the number of fires and let $y =$ the population. Then two points on the line are given: $(300, 57000)$ and $(325, 59000)$.
$$m = \frac{59000 - 57000}{325 - 300} = \frac{2000}{25} = 80$$
$$y - y_1 = m(x - x_1)$$
$$y - 57000 = 80(x - 300)$$
$$y - 57000 = 80x - 24000$$
$$y = 80x + 33000$$
Let $y = 100000$ and find the value of x:
$$y = 80x + 33000$$
$$100000 = 80x + 33000$$
$$67000 = 80x$$
$$837.5 = x \Rightarrow \text{There will be about 838}$$
fires when the population is 100,000.

95. Let F replace x and C replace y. Then two points on the line are given: $(32, 0)$ and $(212, 100)$.
$$m = \frac{100 - 0}{212 - 32} = \frac{100}{180} = \frac{5}{9}$$
$$C - C_1 = m(F - F_1)$$
$$C - 0 = \frac{5}{9}(F - 32)$$
$$C = \frac{5}{9}(F - 32)$$

97. Let $y =$ the percent who smoke and let $x =$ the # of years since 1974. Two points are given: $(0, 47)$ and $(20, 29)$.

$m = \frac{29-47}{20-0} = \frac{-18}{20} = -\frac{9}{10}$

$y - y_1 = m(x - x_1)$

$y - 47 = -\frac{9}{10}(x - 0)$

$\boxed{y = -\frac{9}{10}x + 47}$

Let $x = 30$:

$y = -\frac{9}{10}(30) + 47 = -27 + 47 = 20$

20% will smoke in 2004.

99. Two points on the line are given: $(0, 37.5)$ and $(2, 45)$.

$m = \frac{45-37.5}{2-0} = \frac{7.5}{2} = 3.75$

$y - y_1 = m(x - x_1)$

$y - 37.5 = 3.75(x - 0)$

$y = 3.75x + 37.5$

Let $x = 4$ and find the value of y:

$y = 3.75x + 37.5$

$= 3.75(4) + 37.5$

$= 15 + 37.5 = 52.5$

The price will be \$52.50 in the year 2000.

101. The equation describing the production is $y = -70x + 1900$, where x represents the number of years and y is the level of production. Let $x = 3\frac{1}{2} = \frac{7}{2}$.

$y = -70x + 1900 = -70\left(\frac{7}{2}\right) + 1900 = 1655 \Rightarrow$ The production will be 1655 barrels per day.

103. a.

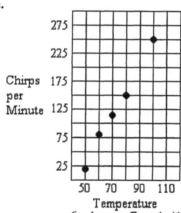

Chirps per Minute / Temperature (in degrees Farenheit)

b. Use $(50, 20)$ and $(100, 250)$ for the regression line.

$m = \frac{250 - 20}{100 - 50} = \frac{230}{50} = \frac{23}{5}$

$y - y_1 = m(x - x_1)$

$y - 20 = \frac{23}{5}(x - 50)$

$y - 20 = \frac{23}{5}x - 230$

$y = \frac{23}{5}x - 210$

c. $y = \frac{23}{5}(90) - 210 = 204$

The rate will be about 204 chirps per minute.

105. Answers may vary.

107. $m = \frac{b-0}{0-a} = -\frac{b}{a}$

$y - y_1 = m(x - x_1)$

$y - b = -\frac{b}{a}(x - 0)$

$y - b = -\frac{b}{a}x$

$ay - ab = -bx$

$bx + ay = ab$

$\frac{bx+ay}{ab} = \frac{ab}{ab}$

$\frac{x}{a} + \frac{y}{b} = 1$

109. Answers may vary.　　　**111.** Answers may vary.　　　**113.** Answers may vary.

115. $x^7 x^3 x^{-5} = x^{7+3-5} = x^5$

117. $\left(\dfrac{81}{25}\right)^{-3/2} = \left(\dfrac{25}{81}\right)^{3/2} = \left[\dfrac{5}{9}\right]^3 = \dfrac{125}{729}$

119. $\sqrt{27} - 2\sqrt{12} = \sqrt{9}\sqrt{3} - 2\sqrt{4}\sqrt{3} = 3\sqrt{3} - 2(2)\sqrt{3} = 3\sqrt{3} - 4\sqrt{3} = -\sqrt{3}$

121. $\dfrac{5}{\sqrt{x}+2} = \dfrac{5}{\sqrt{x}+2} \cdot \dfrac{\sqrt{x}-2}{\sqrt{x}-2} = \dfrac{5(\sqrt{x}-2)}{(\sqrt{x})^2 - 2^2} = \dfrac{5(\sqrt{x}-2)}{x-4}$

Exercise 2.4 (page 213)

1. x-intercept   **3.** axis of symmetry   **5.** x-axis   **7.** circle, center

9. $x^2 + y^2 = r^2$

11.
$y = x^2 - 4$ $y = x^2 - 4$
$0 = (x+2)(x-2)$ $y = 0^2 - 4$
$x = -2, x = 2$ $y = -4$
x-int: $(-2, 0), (2, 0)$ y-int: $(0, -4)$

13.
$y = 4x^2 - 2x$ $y = 4x^2 - 2x$
$0 = 2x(2x - 1)$ $y = 4(0)^2 - 2(0)$
$x = 0, x = \frac{1}{2}$ $y = 0$
x-int: $(0, 0), \left(\frac{1}{2}, 0\right)$ y-int: $(0, 0)$

15.
$y = x^2 - 4x - 5$ $y = x^2 - 4x - 5$
$0 = (x+1)(x-5)$ $y = 0^2 - 4(0) - 5$
$x = -1, x = 5$ $y = -5$
x-int: $(-1, 0), (5, 0)$ y-int: $(0, -5)$

17.
$y = x^2 + x - 2$ $y = x^2 + x - 2$
$0 = (x+2)(x-1)$ $y = 0^2 + 0 - 2$
$x = -2, x = 1$ $y = -2$
x-int: $(-2, 0), (1, 0)$ y-int: $(0, -2)$

19.
$y = x^3 - 9x$ $y = x^3 - 9x$
$0 = x(x^2 - 9)$ $y = 0^3 - 9(0)$
$0 = x(x+3)(x-3)$ $y = 0$
$x = 0, x = -3, x = 3$ y-int: $(0, 0)$
x-int: $(0, 0), (-3, 0), (3, 0)$

21.
$y = x^4 - 1$ $y = x^4 - 1$
$0 = (x^2 + 1)(x^2 - 1)$ $y = 0^4 - 1$
$0 = (x^2 + 1)(x+1)(x-1)$ $y = -1$
$x = -1, x = 1$ y-int: $(0, -1)$
x-int: $(-1, 0), (1, 0)$

23.
$y = x^2$
x-int: $(0, 0)$
y-int: $(0, 0)$

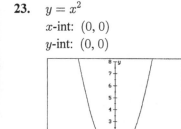

25.
$y = -x^2 + 2$
x-int: $\left(\sqrt{2}, 0\right), \left(-\sqrt{2}, 0\right)$
y-int: $(0, 2)$

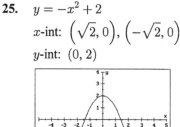

27. $y = x^2 - 4x$

x-int: $(0, 0), (4, 0)$

y-int: $(0, 0)$

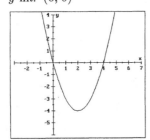

29. $y = \dfrac{1}{2}x^2 - 2x$

x-int: $(0, 0), (4, 0)$

y-int: $(0, 0)$

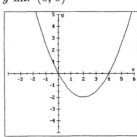

31.

$$y = x^2 + 2$$

x-axis	y-axis	origin
$-y = x^2 + 2$	$y = (-x)^2 + 2$	$-y = (-x)^2 + 2$
not equivalent: no symmetry	$y = x^2 + 2$	$-y = x^2 + 2$
	equivalent: $\boxed{\text{symmetry}}$	not equivalent: no symmetry

33.

$$y^2 + 1 = x$$

x-axis	y-axis	origin
$(-y)^2 + 1 = x$	$y^2 + 1 = -x$	$(-y)^2 + 1 = -x$
$y^2 + 1 = x$	not equivalent: no symmetry	$y^2 + 1 = -x$
equivalent: $\boxed{\text{symmetry}}$		not equivalent: no symmetry

35.

$$y^2 = x^2$$

x-axis	y-axis	origin
$(-y)^2 = x^2$	$y^2 = (-x)^2$	$(-y)^2 = (-x)^2$
$y^2 = x^2$	$y^2 = x^2$	$y^2 = x^2$
equivalent: $\boxed{\text{symmetry}}$	equivalent: $\boxed{\text{symmetry}}$	equivalent: $\boxed{\text{symmetry}}$

37.

$$y = 3x^2 + 7$$

x-axis	y-axis	origin
$-y = 3x^2 + 7$	$y = 3(-x)^2 + 7$	$-y = 3(-x)^2 + 7$
not equivalent: no symmetry	$y = 3x^2 + 7$	$-y = 3x^2 + 7$
	equivalent: $\boxed{\text{symmetry}}$	not equivalent: no symmetry

39.

$$y = 3x^3 + 7$$

x-axis	y-axis	origin
$-y = 3x^3 + 7$	$y = 3(-x)^3 + 7$	$-y = 3(-x)^3 + 7$
not equivalent: no symmetry	$y = -3x^3 + 7$	$-y = -3x^3 + 7$
	not equivalent: no symmetry	$y = 3x^3 - 7$
		not equivalent: no symmetry

41.

$$y^2 = 3x$$

x-axis	y-axis	origin
$(-y)^2 = 3x$	$y^2 + 1 = 3(-x)$	$(-y)^2 = 3(-x)$
$y^2 = 3x$	$y^2 + 1 = -3x$	$y^2 = -3x$
equivalent: $\boxed{\text{symmetry}}$	not equivalent: no symmetry	not equivalent: no symmetry

43.

$$y = |x|$$

x-axis	y-axis	origin
$-y = |x|$	$y = |-x|$	$-y = |-x|$
not equivalent: no symmetry	$y = |-1||x|$	$-y = |-1||x|$
	$y = |x|$	$-y = |x|$
	equivalent: $\boxed{\text{symmetry}}$	not equivalent: no symmetry

45.

$$|y| = x$$

x-axis	y-axis	origin
$|-y| = x$	$|y| = -x$	$|-y| = -x$
$|-1||y| = x$	not equivalent: no symmetry	$|-1||y| = -x$
$|y| = x$		$|y| = -x$
equivalent: $\boxed{\text{symmetry}}$		not equivalent: no symmetry

47. $y = x^2 + 4x$
x-int: $(0, 0)$, $(-4, 0)$
y-int: $(0, 0)$
symmetry: none

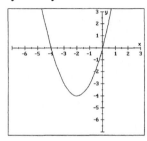

49. $y = x^3$
x-int: $(0, 0)$
y-int: $(0, 0)$
symmetry: origin

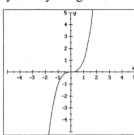

51. $y = |x - 2|$
x-int: $(2, 0)$
y-int: $(0, 2)$
symmetry: none

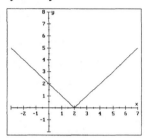

53. $y = 3 - |x|$
x-int: $(-3, 0), (3, 0)$
y-int: $(0, 3)$
symmetry: y-axis

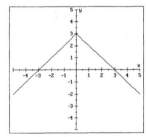

55. $y^2 = -x$
x-int: $(0, 0)$
y-int: $(0, 0)$
symmetry: x-axis

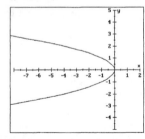

57. $y^2 = 9x$
x-int: $(0, 0)$
y-int: $(0, 0)$
symmetry: x-axis

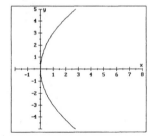

59. $y = \sqrt{x} - 1$
x-int: $(1, 0)$
y-int: $(0, -1)$
symmetry: none

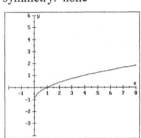

61. $xy = 4$
x-int: none
y-int: none
symmetry: origin

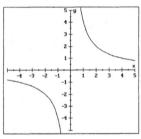

63. $x^2 + y^2 = 1^2$
$x^2 + y^2 - 1 = 0$

65.
$$(x - 6)^2 + (y - 8)^2 = 4^2$$
$$x^2 - 12x + 36 + y^2 - 16y + 64 = 16$$
$$x^2 + y^2 - 12x - 16y + 84 = 0$$

67.
$$(x - 3)^2 + (y + 4)^2 = \left(\sqrt{2}\right)^2$$
$$x^2 - 6x + 9 + y^2 + 8y + 16 = 2$$
$$x^2 + y^2 - 6x + 8y + 23 = 0$$

69. Center: $x = \dfrac{3 + 3}{2} = 3, y = \dfrac{-2 + 8}{2} = 3$
r = distance from center to endpoint
$$= \sqrt{(3 - 3)^2 + (3 - 8)^2} = 5$$
$$(x - 3)^2 + (y - 3)^2 = 5^2$$
$$x^2 - 6x + 9 + y^2 - 6y + 9 = 25$$
$$x^2 + y^2 - 6x - 6y - 7 = 0$$

71. r = distance from center to origin
$$= \sqrt{(0 - (-3))^2 + (0 - 4)^2} = 5$$
$$(x + 3)^2 + (y - 4)^2 = 5^2$$
$$x^2 + 6x + 9 + y^2 - 8y + 16 = 25$$
$$x^2 + y^2 + 6x - 8y = 0$$

73. $x^2 + y^2 - 25 = 0$
$$x^2 + y^2 = 25$$
$$C(0, 0), r = 5$$

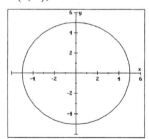

75. $(x - 1)^2 + (y + 2)^2 = 4$
$$C(1, -2), r = 2$$

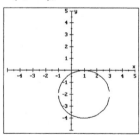

77. $x^2 + y^2 + 2x - 24 = 0$
$$x^2 + 2x + y^2 = 24$$
$$x^2 + 2x + 1 + y^2 = 24 + 1$$
$$(x + 1)^2 + y^2 = 25$$
$$C(-1, 0), r = 5$$

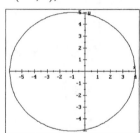

79. $9x^2 + 9y^2 - 12y = 5$
$$x^2 + y^2 - \frac{4}{3}y = \frac{5}{9}$$
$$x^2 + y^2 - \frac{4}{3}y + \frac{4}{9} = \frac{5}{9} + \frac{4}{9}$$
$$x^2 + \left(y - \frac{2}{3}\right)^2 = 1$$
$$C\left(0, \frac{2}{3}\right), r = 1$$

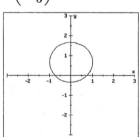

81. $4x^2 + 4y^2 - 4x + 8y + 1 = 0$
$$x^2 + y^2 - x + 2y = -\frac{1}{4}$$
$$x^2 - x + \frac{1}{4} + y^2 + 2y + 1 = -\frac{1}{4} + \frac{1}{4} + 1$$
$$\left(x - \frac{1}{2}\right)^2 + (y + 1)^2 = 1$$
$$C\left(\frac{1}{2}, -1\right), r = 1$$

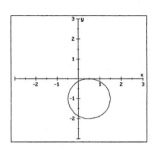

83. $y = 2x^2 - x + 1$
Vertex: $(0.25, 0.88)$

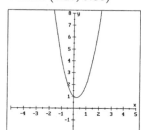

85. $y = 7 + x - x^2$
Vertex: $(0.50, 7.25)$

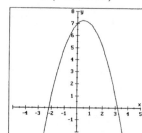

87. Graph $y = x^2 - 7$.
Find the x-intercepts.
$x = -2.65, \ x = 2.65$

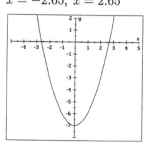

89. Graph $y = x^3 - 3$.
Find the x-intercepts.
$x = 1.44$

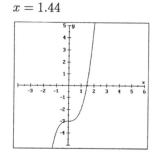

91. Let $y = 0$:
$y = 64t - 16t^2$
$0 = 16t(4 - t)$
$t = 0$ or $t = 4$
It strikes the ground after
4 seconds.

93. $y = 0.08x^2 + 0.9x$

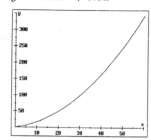

95. $r = \sqrt{(10 - 7)^2 + (0 - 4)^2} = 5$
$$(x - 7)^2 + (y - 4)^2 = 5^2$$
$$x^2 - 14x + 49 + y^2 - 8y + 16 = 25$$
$$x^2 + y^2 - 14x - 8y + 40 = 0$$

97. Graph $y = x^2 + x - 6$. Find the x-coordinates where the graph is below the x-axis.
Solution: $(-3, 2)$

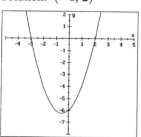

99. $3(x + 2) + x = 5x$
$3x + 6 + x = 5x$
$4x + 6 = 5x$
$6 = x$

101.

$$\frac{5(2-x)}{3} - 1 = x + 5$$

$$3\left[\frac{10-5x}{3} - 1\right] = 3(x+5)$$

$$10 - 5x - 3 = 3x + 15$$

$$-5x + 7 = 3x + 15$$

$$-8x = 8$$

$$x = -1$$

103. Let x = the ounces of copper added. Since 1 ounce out of 4 ounces are to be gold, the final result should be 25% gold.

Ounces of gold at start		Ounces of gold added		Ounces of gold at end

$$20 + 0 = 0.25(60 + x)$$

$$20 = 15 + 0.25x$$

$$5 = 0.25x$$

$$20 = x$$

20 ounces of copper should be added.

Exercise 2.5 (page 223)

1. quotient

3. means

5. extremes, means

7. inverse

9. joint

11.

$$\frac{4}{x} = \frac{2}{7}$$

$$4 \cdot 7 = 2 \cdot x$$

$$28 = 2x$$

$$14 = x$$

13.

$$\frac{x}{2} = \frac{3}{x+1}$$

$$x(x+1) = 3 \cdot 2$$

$$x^2 + x = 6$$

$$x^2 + x - 6 = 0$$

$$(x+3)(x-2) = 0$$

$$x = -3 \text{ or } x = 2$$

15. Let x = the number of women.

$$\frac{3}{5} = \frac{x}{30}$$

$$3 \cdot 30 = 5 \cdot x$$

$$90 = 5x$$

$$18 = x \Rightarrow \text{There are 18 women.}$$

17.

$$y = kx$$

$$15 = k(30)$$

$$\frac{1}{2} = k$$

19.

$$I = \frac{k}{R}$$

$$50 = \frac{k}{20}$$

$$1000 = k$$

21.

$$E = kIR$$

$$125 = k(5)(25)$$

$$125 = 125k$$

$$1 = k$$

23.

$$y = kx \qquad y = \frac{15}{4}x$$

$$15 = k(4) \qquad y = \frac{15}{4} \cdot \frac{7}{5}$$

$$\frac{15}{4} = k \qquad y = \frac{21}{4}$$

25.

$$P = krs \qquad P = -\frac{2}{5}rs$$

$$16 = k(5)(-8) \qquad P = -\frac{2}{5}(2)(10)$$

$$16 = -40k \qquad P = -8$$

$$-\frac{16}{40} = k$$

$$-\frac{2}{5} = k$$

27. direct

29. neither

SECTION 2.5

31.

$$V = \frac{kT}{P} \qquad V = \frac{\frac{80}{33}T}{P}$$

$$20 = \frac{k(330)}{40} \qquad V = \frac{\frac{80}{33}(300)}{50}$$

$$800 = 330k$$

$$\frac{800}{330} = k \qquad V = \frac{\frac{8000}{11}}{50}$$

$$\frac{80}{33} = k \qquad V = \frac{160}{11} = 14\frac{6}{11} \text{ ft}^3$$

33.

$$d = kt^2 \qquad d = 16t^2$$

$$16 = k(1)^2 \qquad 144 = 16t^2$$

$$16 = k \qquad 9 = t^2$$

$$3 = t \Rightarrow 3 \text{ seconds}$$

35.

$$P = \frac{kV^2}{R} \qquad P = \frac{V^2}{R}$$

$$20 = \frac{k(20)^2}{20} \qquad 40 = \frac{V^2}{10}$$

$$400 = 400k \qquad 400 = V^2$$

$$1 = k \qquad 20 = V \Rightarrow 20 \text{ volts}$$

37.

$$f = k\sqrt{T} \qquad f = \frac{144}{\sqrt{2}}\sqrt{T}$$

$$144 = k\sqrt{2}$$

$$\frac{144}{\sqrt{2}} = k \qquad f = \frac{144}{\sqrt{2}}\sqrt{18}$$

$$f = 144\sqrt{9}$$

$$f = 144(3) = 432 \text{ hertz}$$

39. Consider this figure:

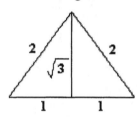

$h = \sqrt{3}$ can be computed using the Pythagorean Theorem.

$$A = \frac{1}{2}bh = \frac{1}{2}(2)\sqrt{3} = \sqrt{3}$$

$$A = ks^2$$

$$\sqrt{3} = k(2)^2$$

$$\sqrt{3} = 4k$$

$$\frac{\sqrt{3}}{4} = k$$

41. Answers may vary.　　　**43.** Answers may vary.　　　**45.** Answers may vary.

47.
$$\frac{1}{x+2} + \frac{2}{x+1} = \frac{1(x+1)}{(x+2)(x+1)} + \frac{2(x+2)}{(x+1)(x+2)} = \frac{x+1+2x+4}{(x+1)(x+2)} = \frac{3x+5}{(x+1)(x+2)}$$

49.
$$\frac{x^2+3x-4}{x^2-5x+4} \div \frac{x-1}{x^2-3x-4} = \frac{x^2+3x-4}{x^2-5x+4} \cdot \frac{x^2-3x-4}{x-1} = \frac{(x+4)(x-1)}{(x-4)(x-1)} \cdot \frac{(x-4)(x+1)}{x-1}$$

$$= \frac{(x+4)(x+1)}{x-1}$$

51.
$$\frac{x^2+4-(x+2)^2}{4x^2} = \frac{x^2+4-(x^2+4x+4)}{4x^2} = \frac{x^2+4-x^2-4x-4}{4x^2} = \frac{-4x}{4x^2} = -\frac{1}{x}$$

Chapter 2 Summary (page 226)

1. **a.** $A(2, 0)$ **b.** $B(-2, 1)$ **c.** $C(0, -1)$ **d.** $D(3, -1)$

2. **a.** $A(-3, 5)$: QII

b. $B(5, -3)$: QIV

c. $C(0, -7)$: negative y-axis

d. $D\left(-\dfrac{1}{2}, 0\right)$: negative x-axis

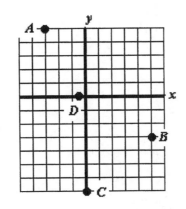

3. **a.**

$$3x - 5y = 15 \qquad 3x - 5y = 15$$
$$3x - 5(0) = 15 \qquad 3(0) - 5y = 15$$
$$3x = 15 \qquad -5y = 15$$
$$x = 5 \qquad y = -3$$
$$(5, 0) \qquad (0, -3)$$

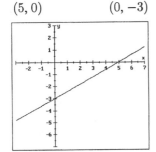

b.

$$x + y = 7 \qquad x + y = 7$$
$$x + 0 = 7 \qquad 0 + y = 7$$
$$x = 7 \qquad y = 7$$
$$(7, 0) \qquad (0, 7)$$

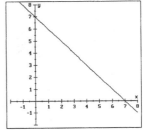

c.

$$x + y = -7 \qquad x + y = -7$$
$$x + 0 = -7 \qquad 0 + y = -7$$
$$x = -7 \qquad y = -7$$
$$(-7, 0) \qquad (0, -7)$$

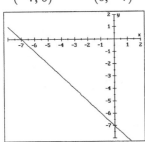

d.

$$x - 5y = 5 \qquad x - 5y = 5$$
$$x - 5(0) = 5 \qquad 0 - 5y = 5$$
$$x = 5 \qquad -5y = 5$$
$$(5, 0) \qquad y = -1$$
$$(0, -1)$$

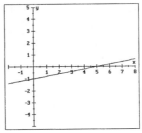

4. **a.** $y = 4 \Rightarrow$ horizontal

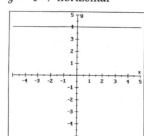

b. $x = -2 \Rightarrow$ vertical

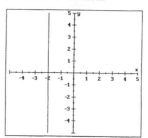

5. Let $x = 3$: $y = -2200x + 18{,}750 = -2200(3) + 18{,}750 = -6600 + 18{,}750 = \$12{,}150$

6. **a.** $d = \sqrt{(x_2 - x_1)^2 + (y_2 - y_1)^2}$

$ = \sqrt{(-3 - 3)^2 + (7 - (-1))^2}$

$ = \sqrt{(-6)^2 + (8)^2}$

$ = \sqrt{36 + 64} = \sqrt{100} = 10$

b. $d = \sqrt{(x_2 - x_1)^2 + (y_2 - y_1)^2}$

$ = \sqrt{(0 - (-12))^2 + (5 - 10)^2}$

$ = \sqrt{12^2 + (-5)^2}$

$ = \sqrt{144 + 25} = \sqrt{169} = 13$

c. $d = \sqrt{(x_2 - x_1)^2 + (y_2 - y_1)^2}$

$ = \sqrt{\left(\sqrt{3} - \sqrt{3}\right)^2 + (9 - 7)^2}$

$ = \sqrt{0^2 + (2)^2}$

$ = \sqrt{0 + 4} = \sqrt{4} = 2$

d. $d = \sqrt{(x_2 - x_1)^2 + (y_2 - y_1)^2}$

$ = \sqrt{(a - (-a))^2 + (-a - a)^2}$

$ = \sqrt{(2a)^2 + (-2a)^2}$

$ = \sqrt{4a^2 + 4a^2} = \sqrt{8a^2} = 2\sqrt{2}|a|$

7. **a.** $M\left(\dfrac{x_1 + x_2}{2}, \dfrac{y_1 + y_2}{2}\right) = M\left(\dfrac{-3 + 3}{2}, \dfrac{7 + (-1)}{2}\right) = M\left(\dfrac{0}{2}, \dfrac{6}{2}\right) = M(0, 3)$

b. $M\left(\dfrac{x_1 + x_2}{2}, \dfrac{y_1 + y_2}{2}\right) = M\left(\dfrac{0 + (-12)}{2}, \dfrac{5 + 10}{2}\right) = M\left(\dfrac{-12}{2}, \dfrac{15}{2}\right) = M\left(-6, \dfrac{15}{2}\right)$

c. $M\left(\dfrac{x_1 + x_2}{2}, \dfrac{y_1 + y_2}{2}\right) = M\left(\dfrac{\sqrt{3} + \sqrt{3}}{2}, \dfrac{9 + 7}{2}\right) = M\left(\dfrac{2\sqrt{3}}{2}, \dfrac{16}{2}\right) = M\left(\sqrt{3}, 8\right)$

d. $M\left(\dfrac{x_1 + x_2}{2}, \dfrac{y_1 + y_2}{2}\right) = M\left(\dfrac{a + (-a)}{2}, \dfrac{-a + a}{2}\right) = M\left(\dfrac{0}{2}, \dfrac{0}{2}\right) = M(0, 0)$

8. **a.** $m = \dfrac{y_2 - y_1}{x_2 - x_1} = \dfrac{7 - (-5)}{1 - 3} = \dfrac{12}{-2} = -6$

b. $m = \dfrac{y_2 - y_1}{x_2 - x_1} = \dfrac{-7 - 7}{-5 - 2} = \dfrac{-14}{-7} = 2$

c. $m = \dfrac{y_2 - y_1}{x_2 - x_1} = \dfrac{b - a}{a - b} = -1$

d. $m = \dfrac{y_2 - y_1}{x_2 - x_1} = \dfrac{(b - a) - b}{b - (a + b)} = \dfrac{-a}{-a} = 1$

9. $m = \dfrac{\Delta y}{\Delta x} = \dfrac{3000}{15} = 200$ ft. per minute

10. **a.** The slope is zero.   **b.** The slope is undefined.

11. **a.** The slope is negative.   **b.** The slope is positive.

12. $m = \dfrac{y_2 - y_1}{x_2 - x_1} = \dfrac{10 - 5}{6 - (-2)} = \dfrac{5}{8}$

$m = \dfrac{y_2 - y_1}{x_2 - x_1} = \dfrac{y - 2}{10 - 2} = \dfrac{5}{8}$

$$8(y - 2) = 5(8)$$
$$8y - 16 = 40$$
$$8y = 56$$
$$y = 7$$

13. $m = \dfrac{y_2 - y_1}{x_2 - x_1} = \dfrac{10 - 5}{6 - (-2)} = \dfrac{5}{8}$

$m = \dfrac{y_2 - y_1}{x_2 - x_1} = \dfrac{-3 - 5}{x - (-2)} = \dfrac{-8}{5}$

$$5(-8) = -8(x + 2)$$
$$-40 = -8x - 16$$
$$8x = 24$$
$$x = 3$$

14. **a.** $m = \dfrac{y_2 - y_1}{x_2 - x_1} = \dfrac{7 - 0}{-5 - 0} = -\dfrac{7}{5}$

$y - y_1 = m(x - x_1)$

$y - 0 = -\dfrac{7}{5}(x - 0)$

$y = -\dfrac{7}{5}x$

     b. $y - y_1 = m(x - x_1)$

$y - 1 = -4(x + 2)$

     c. $m = \dfrac{y_2 - y_1}{x_2 - x_1} = \dfrac{1 - (-5)}{4 - 7} = \dfrac{6}{-3} = -2$

$y - y_1 = m(x - x_1)$ **or** $y - y_1 = m(x - x_1)$

$y + 5 = -2(x - 7)$ $y - 1 = -2(x - 4)$

15. **a.** $y = mx + b$

$y = \dfrac{2}{3}x + 3$

     b. $y = mx + b$

$y = -\dfrac{3}{2}x - 5$

16. **a.** $y = \dfrac{3}{5}x - 2$

$m = \dfrac{3}{5},\ b = -2$

     b. $y = -\dfrac{4}{3}x + 3$

$m = -\dfrac{4}{3},\ b = 3$

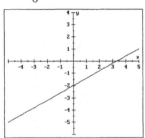

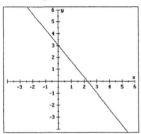

17. a. $m = \dfrac{y_2 - y_1}{x_2 - x_1} = \dfrac{-10 - 4}{4 - 2} = -7$

$y - y_1 = m(x - x_1)$

$y + 2 = -7(x - 7)$

$y + 2 = -7x + 49$

$y = -7x + 47$

b. $m = \dfrac{y_2 - y_1}{x_2 - x_1} = \dfrac{-10 - 4}{4 - 2} = -7$

$y - y_1 = m(x - x_1)$

$y + 2 = \frac{1}{7}(x - 7)$

$y + 2 = \frac{1}{7}x - 1$

$y = \frac{1}{7}x - 3$

c. $3x - 4y = 7$

$-4y = -3x + 7$

$y = \dfrac{3}{4}x - \dfrac{7}{4}$

$m = \dfrac{3}{4}$

Use $m = \dfrac{3}{4}$.

$y - y_1 = m(x - x_1)$

$y - 0 = \dfrac{3}{4}(x - 2)$

$\boxed{y = \frac{3}{4}x - \frac{3}{2}}$

d. $3y + x - 4 = 0$

$3y = -x + 4$

$y = -\dfrac{1}{3}x + \dfrac{4}{3}$

$m = -\dfrac{1}{3}$

Use $m = 3$.

$y - y_1 = m(x - x_1)$

$y - 5 = 3(x - 0)$

$y - 5 = 3x$

$\boxed{y = 3x + 5}$

18. a. $5x + 2y = 7$

$2y = -5x + 7$

$y = -\dfrac{5}{2}x + \dfrac{7}{2}$

$m = -\dfrac{5}{2}, \left(0, \dfrac{7}{2}\right)$

b. $3x - 4y = 14$

$-4y = -3x + 14$

$y = \dfrac{3}{4}x - \dfrac{7}{2}$

$m = \dfrac{3}{4}, \left(0, -\dfrac{7}{2}\right)$

19 a. If the slope is 0, then the line is horizontal: $y = 17$

b. If the slope is undefined, then the line is vertical: $x = -5$

20. a. $y = x^2 + 2$

x-int: none, y-int: $(0, 2)$

symmetry: y-axis

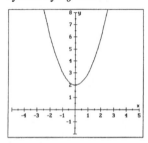

b. $y = x^3 - 2$

x-int: $\left(\sqrt[3]{2}, 0\right)$, y-int: $(0, -2)$

symmetry: none

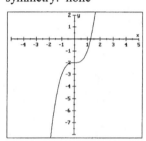

c. $y = \frac{1}{2}|x|$

x-int: $(0,0)$, y-int: $(0,0)$

symmetry: y-axis

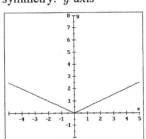

d. $y = -\sqrt{x-4}$

x-int: $(4,0)$, y-int: none

symmetry: none

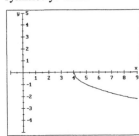

e. $y = \sqrt{x} + 2$

x-int: none, y-int: $(0,2)$

symmetry: none

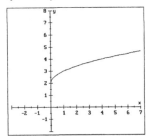

f. $y = |x+1| + 2$

x-int: none, y-int: $(0,3)$

symmetry: none

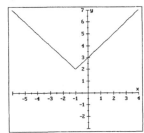

21. a. $y = |x-4| + 2$

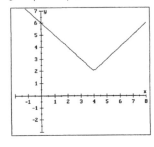

b. $y = -\sqrt{x+2} + 3$

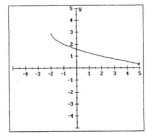

c. $y = x + 2|x|$

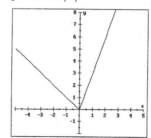

d. $y^2 = x - 3$

Graph $y = \pm \sqrt{x - 3}$.

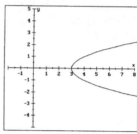

22 a. $C(-3, 4); r = 12$

$(x - h)^2 + (y - k)^2 = r^2$

$(x + 3)^2 + (y - 4)^2 = 144$

or $x^2 + y^2 + 6x - 8y - 119 = 0$

b. Center: $x = \dfrac{-6 + 5}{2} = -\dfrac{1}{2}$

$y = \dfrac{-3 + 8}{2} = \dfrac{5}{2}$

$r =$ distance from center to endpoint

$= \sqrt{\left(-\frac{1}{2} - 5\right)^2 + \left(\frac{5}{2} - 8\right)^2} = \sqrt{\frac{121}{2}}$

$\left(x + \frac{1}{2}\right)^2 + \left(y - \frac{5}{2}\right)^2 = \frac{121}{2}$, or

$x^2 + y^2 + x - 5y - 54 = 0$

23. a. $x^2 + y^2 - 2y = 15$

$x^2 + y^2 - 2y + 1 = 15 + 1$

$x^2 + (y - 1)^2 = 16$

$C(0, 1), r = 4$

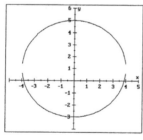

b. $x^2 + y^2 - 4x + 2y = 4$

$x^2 - 4x + 4 + y^2 + 2y + 1 = 4 + 4 + 1$

$(x - 2)^2 + (y + 1)^2 = 9$

$C(2, -1), r = 3$

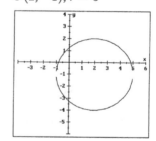

24. **a.** Graph $y = x^2 - 11$.
Find the x-intercepts.
$x = -3.32,\ x = 3.32$

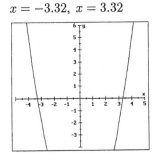

b. Graph $y = x^3 - x$.
Find the x-intercepts.
$x = -1,\ x = 0,\ x = 1$

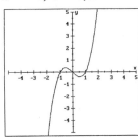

c. Graph $y = |x^2 - 2| - 1$.
Find the x-intercepts.
$x = -1.73,\ x = -1,\ x = 1,\ x = 1.73$

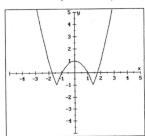

d. Graph $y = x^2 - 3x - 5$.
Find the x-intercepts.
$x = -1.19,\ x = 4.19$

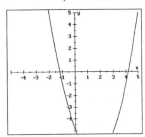

25. **a.**
$$\frac{x+3}{10} = \frac{x-1}{x}$$
$$x(x+3) = 10(x-1)$$
$$x^2 + 3x = 10x - 10$$
$$x^2 - 7x + 10 = 0$$
$$(x-5)(x-2) = 0$$
$$x = 5 \text{ or } x = 2$$

b.
$$\frac{x-1}{2} = \frac{12}{x+1}$$
$$(x+1)(x-1) = 2(12)$$
$$x^2 - 1 = 24$$
$$x^2 = 25$$
$$x = \pm 5$$

26. $f = ks \qquad f = \dfrac{3}{5}s$
$3 = k(5)$
$\dfrac{3}{5} = k \qquad f = \dfrac{3}{5}(3)$
$f = \dfrac{9}{5}\text{pounds}$

27. $$E = kv^2$$

30 mph	50 mph
$E = k(30)^2$	$E = k(50)^2$
$E = 900k$	$E = 2500k$

Factor of increase $= \dfrac{2500k}{900k} = \dfrac{25}{9}$

28.
$$V = \frac{kT}{P} \qquad V = \frac{\frac{100}{3}T}{P}$$
$$400 = \frac{k(300)}{25} \qquad V = \frac{\frac{100}{3}(200)}{20}$$
$$10000 = 300k \qquad V = \frac{\frac{1000}{3}}{3}$$
$$\frac{100}{3} = k \qquad V = 333\frac{1}{3} \text{ cm}^3$$

29.
$$A = klw$$
$$A = 1lw \Rightarrow k = 1$$

30.
$$R = \frac{kL}{D^2} \qquad R = \frac{0.0005L}{D^2}$$
$$200 = \frac{k(1000)}{(.05)^2} \qquad V = \frac{0.0005(1500)}{(0.08)^2}$$
$$200 = \frac{1000k}{.0025} \qquad V \approx 117 \text{ ohms}$$
$$0.0005 = k$$

31. Let $x = \#$ rolls and $y = $ total charge. Then the equation is $y = mx + b$, where m is the charge per roll and b is the fixed amount.
Given points: $(11, 177)$ and $(20, 294)$
$$m = \frac{y_2 - y_1}{x_2 - x_1} = \frac{294 - 177}{20 - 11} = \frac{117}{9} = 13$$
$$y - y_1 = m(x - x_1) \qquad \text{Let } x = 27:$$
$$y - 177 = 13(x - 11) \qquad y = 13x + 34$$
$$y = 13x + 34 \qquad y = 13(27) + 34$$
$$y = \$385$$

32. $14x + 18y = 5040$
Let $x = 180$:
$$14(180) + 18y = 5040$$
$$2520 + 18y = 5040$$
$$18y = 2520$$
$$y = 140 \text{ hours of tutoring Spanish.}$$

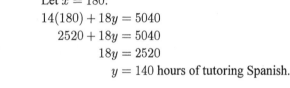

Chapter 2 Test (page 232)

1. $(-3, \pi) \Rightarrow$ QII

2. $(0, -8) \Rightarrow$ negative y-axis

3.
$$x + 3y = 6 \qquad x + 3y = 6$$
$$x + 3(0) = 6 \qquad 0 + 3y = 6$$
$$x = 6 \qquad y = 2$$
$$(6, 0) \qquad (0, 2)$$

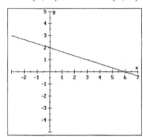

4.
$$2x - 5y = 10 \qquad 2x - 5y = 10$$
$$2x - 5(0) = 10 \qquad 2(0) - 5y = 10$$
$$x = 5 \qquad y = -2$$
$$(5, 0) \qquad (0, -2)$$

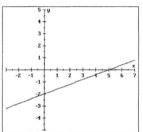

5. $2(x + y) = 3x + 5$

$2x + 2y = 3x + 5$

$2y = x + 5$

$y = \frac{1}{2}x + \frac{5}{2}$

x	y
0	$\frac{5}{2}$
1	3

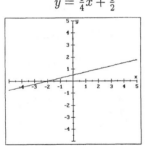

6. $3x - 5y = 3(x - 5)$

$3x - 5y = 3x - 15$

$-5y = -15$

$y = 3$

x	y
0	3
-2	3

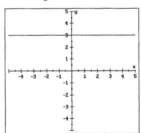

7. $\frac{1}{2}(x - 2y) = y - 1$

$\frac{1}{2}x - y = y - 1$

$x - 2y = 2y - 2$

$-4y = -x - 2$

$y = \frac{1}{4}x + \frac{1}{2}$

x	y
0	$\frac{1}{2}$
2	1

8. $\frac{x + y - 5}{7} = 3x$

$x + y - 5 = 21x$

$y = 20x + 5$

x	y
0	5
$-\frac{1}{4}$	0

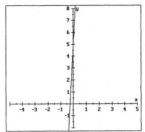

9. $d = \sqrt{(x_2 - x_1)^2 + (y_2 - y_1)^2}$

$= \sqrt{(1 - (-3))^2 + (-1 - 4)^2}$

$= \sqrt{(4)^2 + (-5)^2}$

$= \sqrt{16 + 25} = \sqrt{41}$

10. $d = \sqrt{(x_2 - x_1)^2 + (y_2 - y_1)^2}$

$= \sqrt{(0 - (-\pi))^2 + (\pi - 0)^2}$

$= \sqrt{\pi^2 + \pi^2}$

$= \sqrt{2\pi^2} = \pi\sqrt{2} \approx 4.44$

11. $M\left(\frac{x_1 + x_2}{2}, \frac{y_1 + y_2}{2}\right) = M\left(\frac{3 + (-3)}{2}, \frac{-7 + 7}{2}\right) = M\left(\frac{0}{2}, \frac{0}{2}\right) = M(0, 0)$

12. $M\left(\frac{x_1 + x_2}{2}, \frac{y_1 + y_2}{2}\right) = M\left(\frac{0 + \sqrt{8}}{2}, \frac{\sqrt{2} + \sqrt{18}}{2}\right) = M\left(\frac{2\sqrt{2}}{2}, \frac{4\sqrt{2}}{2}\right) = M\left(\sqrt{2}, 2\sqrt{2}\right)$

13. $m = \frac{y_2 - y_1}{x_2 - x_1} = \frac{1 - (-9)}{-5 - 3} = \frac{10}{-8} = -\frac{5}{4}$

14. $m = \dfrac{y_2 - y_1}{x_2 - x_1} = \dfrac{0 - 3}{-\sqrt{12} - \sqrt{3}} = \dfrac{-3}{-3\sqrt{3}} = \dfrac{1}{\sqrt{3}} = \dfrac{\sqrt{3}}{3}$

15. $y = 3x - 2 \qquad y = 2x - 3$
  $m = 3 \qquad\quad m = 2$
  neither

16. $2x - 3y = 5 \qquad\qquad 3x + 2y = 7$
  $-3y = -2x + 5 \qquad 2y = -3x + 7$
  $y = \frac{2}{3}x - \frac{5}{3} \qquad\quad y = -\frac{3}{2}x + \frac{7}{2}$
  $m = \frac{2}{3} \qquad\qquad\quad m = -\frac{3}{2}$
  perpendicular

17. $y - y_1 = m(x - x_1)$
  $y + 5 = 2(x - 3)$
  $y + 5 = 2x - 6$
  $y = 2x - 11$

18. $y = mx + b$
  $y = 3x + \dfrac{1}{2}$

19. $2x - y = 3$
  $-y = -2x + 3$
  $y = 2x - 3$
  $m = 2 \qquad y = 2x + 5$

20. $2x - y = 3$
  $-y = -2x + 3$
  $y = 2x - 3$
  $m = 2 \qquad y = -\frac{1}{2}x + 5$

21. $m = \dfrac{y_2 - y_1}{x_2 - x_1} = \dfrac{\frac{1}{2} - \left(-\frac{3}{2}\right)}{3 - 2} = \dfrac{\frac{4}{2}}{1} = 2$
  $y - y_1 = m(x - x_1)$
  $y - \dfrac{1}{2} = 2(x - 3)$
  $y - \dfrac{1}{2} = 2x - 6$
  $y = 2x - \dfrac{11}{2}$

22. If the line is parallel to the y-axis, then it is a vertical line: $x = 3$

23. $y = x^3 - 16x \qquad\qquad y = x^3 - 16x$
  $0 = x\left(x^2 - 16\right) \qquad\quad y = 0^3 - 16(0)$
  $0 = x(x + 4)(x - 4) \qquad y = 0$
  $x = 0,\ x = -4,\ x = 4 \qquad y\text{-int: } (0, 0)$
  $x\text{-int: } (0, 0),\ (-4, 0),\ (4, 0)$

24. $y = |x - 4| \qquad y = |x - 4|$
  $0 = |x - 4| \qquad y = |0 - 4|$
  $0 = x - 4 \qquad\ y = |-4|$
  $4 = x \qquad\qquad y = 4$
  $x\text{-int: } (4, 0) \qquad y\text{-int: } (0, 4)$

25.

$$y^2 = x - 1$$

x-axis	y-axis	origin
$(-y)^2 = x - 1$	$y^2 = -x - 1$	$(-y)^2 = -x - 1$
$y^2 = x - 1$	not equivalent: no symmetry	$y^2 = -x - 1$
equivalent: $\boxed{\text{symmetry}}$		not equivalent: no symmetry

26.

$$y = x^4 + 1$$

x-axis	y-axis	origin
$-y = x^4 + 1$	$y = (-x)^4 + 1$	$-y = (-x)^4 + 1$
not equivalent: no symmetry	$y = x^4 + 1$	$-y = x^4 + 1$
	equivalent: $\boxed{\text{symmetry}}$	not equivalent: no symmetry

27. $y = x^2 - 9$
x-int: $(3, 0)$, $(-3, 0)$
y-int: $(0, -9)$
symmetry: y-axis

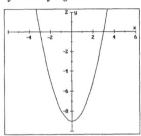

28. $x = |y|$
x-int: $(0, 0)$
y-int: $(0, 0)$
symmetry: x-axis

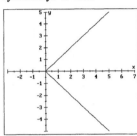

29. $y = 2\sqrt{x}$
x-int: $(0, 0)$
y-int: $(0, 0)$
symmetry: none

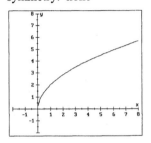

30. $x = y^3$
x-int: $(0, 0)$
y-int: $(0, 0)$
symmetry: origin

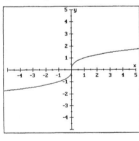

31. $C(5, 7); r = 8$
$(x - h)^2 + (y - k)^2 = r^2$
$(x - 5)^2 + (y - 7)^2 = 64$

32. $r = \sqrt{(2 - 6)^2 + (4 - 8)^2}$
$\quad = \sqrt{32}$
$(x - h)^2 + (y - k)^2 = r^2$
$(x - 2)^2 + (y - 4)^2 = 32$

33. $x^2 + y^2 = 9$
$C(0, 0), r = 3$

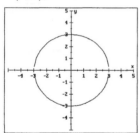

34. $x^2 - 4x + y^2 + 3 = 0$
$$x^2 - 4x + y^2 = -3$$
$$x^2 - 4x + 4 + y^2 = -3 + 4$$
$$(x - 2)^2 + y^2 = 1$$
$$C(2, 0), r = 1$$

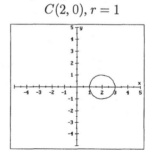

35. $y = kz^2$

36. $w = krs^2$

37. $P = kQ$ $P = \dfrac{7}{2}Q$
$7 = k(2)$
$\dfrac{7}{2} = k$ $P = \dfrac{7}{2}(5)$
$$P = \dfrac{35}{2}$$

38. $y = \dfrac{kx}{z^2}$ $y = \dfrac{\frac{64}{3}x}{z^2}$

$16 = \dfrac{k(3)}{2^2}$ $2 = \dfrac{\frac{64}{3}x}{3^2}$

$16 = \dfrac{3k}{4}$ $2 = \dfrac{\frac{64}{3}x}{9}$

$64 = 3k$ $18 = \dfrac{64}{3}x$

$\dfrac{64}{3} = k$ $54 = 64x$

$\dfrac{54}{64} = x$

$\dfrac{27}{32} = x$

39. Graph $y = x^2 - 7$.
Find any positive x-intercepts.
$x = 2.65$

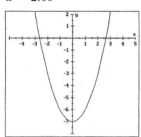

40. Graph $y = x^2 - 5x - 5$.
Find any positive x-intercepts.
$x = 5.85$

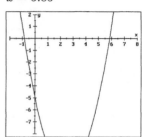

Exercise 3.1 (page 245)

1. function

3. domain

5. $y = f(x)$

7. x

9. vertical, once

11. $y = x$
Each value of x is paired
with only one value of y.
function

13. $y^2 = x$
$$y = \pm \sqrt{x}$$
At least one value of x is
paired with more than one
value of y. **not a function**

15. $y = x^2$
Each value of x is paired
with only one value of y.
function

17. $y^2 - 4x = 1$
$$y^2 = 4x + 1$$
$$y = \pm \sqrt{4x + 1}$$
At least one value of x is
paired with more than one
value of y. **not a function**

19. $|x| = |y|$
$$|y| = |x|$$
$$y = \pm |x|$$
At least one value of x is
paired with more than one
value of y. **not a function**

21. $y = 7$
Each value of x is paired
with only one value of y.
function

23. $f(x) = y = 3x + 5 \Rightarrow \text{domain} = (-\infty, \infty)$
$x = \dfrac{y - 5}{3} \Rightarrow \text{range} = (-\infty, \infty)$

25. $f(x) = y = x^2 \Rightarrow \text{domain} = (-\infty, \infty)$
$x = \pm \sqrt{y} \Rightarrow \text{range} = [0, \infty)$

27. $f(x) = y = \dfrac{3}{x + 1}$
$x \neq -1$
$\text{domain} = (-\infty, -1) \cup (-1, \infty)$

$y = \dfrac{3}{x + 1}$
$(x + 1)y = 3$
$x + 1 = \dfrac{3}{y}$
$x = \dfrac{3}{y} - 1 \Rightarrow \text{range} = (-\infty, 0) \cup (0, \infty)$

29. $f(x) = y = \sqrt{x} \Rightarrow x \geq 0 \Rightarrow \text{domain} = [0, \infty)$
note: $\sqrt{x} \geq 0$, so $y \geq 0 \Rightarrow \text{range} = [0, \infty)$

31. $f(x) = y = \dfrac{x}{x + 3}$
$x \neq -3$
$\text{domain} = (-\infty, -3) \cup (-3, \infty)$

$y = \dfrac{x}{x + 3}$
$(x + 3)y = x$
$xy + 3y = x$
$xy - x = -3y$
$x(y - 1) = -3y$
$x = \dfrac{-3y}{y - 1} \Rightarrow y \neq 1$
$\text{range} = (-\infty, 1) \cup (1, \infty)$

33. $f(x) = y = \dfrac{x-2}{x+3}$

$x \neq -3$

domain $= (-\infty, -3) \cup (-3, \infty)$

$y = \dfrac{x-2}{x+3}$

$(x+3)y = x - 2$

$xy + 3y = x - 2$

$xy - x = -3y - 2$

$x(y-1) = -3y - 2$

$x = \dfrac{-3y-2}{y-1} \Rightarrow y \neq 1$

range $= (-\infty, 1) \cup (1, \infty)$

35.
$$f(x) = 3x - 2$$

$f(2) = 3(2) - 2$	$f(-3) = 3(-3) - 2$	$f(k) = 3k - 2$	$f(k^2 - 1) = 3(k^2 - 1) - 2$
$= 6 - 2$	$= -9 - 2$		$= 3k^2 - 3 - 2$
$= 4$	$= -11$		$= 3k^2 - 5$

37.
$$f(x) = \tfrac{1}{2}x + 3$$

$f(2) = \tfrac{1}{2}(2) + 3$	$f(-3) = \tfrac{1}{2}(-3) + 3$	$f(k) = \tfrac{1}{2}k + 3$	$f(k^2 - 1) = \tfrac{1}{2}(k^2 - 1) + 3$
$= 1 + 3$	$= -\tfrac{3}{2} + 3$		$= \tfrac{1}{2}k^2 - \tfrac{1}{2} + 3$
$= 4$	$= \tfrac{3}{2}$		$= \tfrac{1}{2}k^2 + \tfrac{5}{2}$

39.
$$f(x) = x^2$$

$f(2) = 2^2$	$f(-3) = (-3)^2$	$f(k) = k^2$	$f(k^2 - 1) = (k^2 - 1)^2$
$= 4$	$= 9$		$= (k^2 - 1)(k^2 - 1)$
			$= k^4 - 2k^2 + 1$

41.
$$f(x) = |x^2 + 1|$$

$f(2) = |2^2 + 1|$	$f(-3) = |(-3)^2 + 1|$	$f(k) = |k^2 + 1|$	$f(k^2 - 1) = |(k^2 - 1)^2 + 1|$
$= |5|$	$= |10|$	$= k^2 + 1$	$= (k^2 - 1)^2 + 1$
$= 5$	$= 10$	$[k^2 + 1 \geq 0]$	$= k^4 - 2k^2 + 1 + 1$
			$= k^4 - 2k^2 + 2$
			$\left[(k^2 - 1)^2 + 1 \geq 0\right]$

43.
$$f(x) = \dfrac{2}{x+4}$$

$f(2) = \dfrac{2}{2+4}$	$f(-3) = \dfrac{2}{-3+4}$	$f(k) = \dfrac{2}{k+4}$	$f(k^2 - 1) = \dfrac{2}{k^2 - 1 + 4}$
$= \dfrac{2}{6} = \dfrac{1}{3}$	$= \dfrac{2}{1} = 2$		$= \dfrac{2}{k^2 + 3}$

45.
$$f(x) = \frac{1}{x^2 - 1}$$

$f(2) = \dfrac{1}{2^2 - 1}$	$f(-3) = \dfrac{1}{(-3)^2 - 1}$	$f(k) = \dfrac{1}{k^2 - 1}$	$f(k^2 - 1) = \dfrac{1}{(k^2 - 1)^2 - 1}$
$= \dfrac{1}{4 - 1}$	$= \dfrac{1}{9 - 1}$		$= \dfrac{1}{k^4 - 2k^2 + 1 - 1}$
$= \dfrac{1}{3}$	$= \dfrac{1}{8}$		$= \dfrac{1}{k^4 - 2k^2}$

47.
$$f(x) = \sqrt{x^2 + 1}$$

$f(2) = \sqrt{2^2 + 1}$	$f(-3) = \sqrt{(-3)^2 + 1}$	$f(k) = \sqrt{k^2 + 1}$	$f(k^2 - 1) = \sqrt{(k^2 - 1)^2 + 1}$
$= \sqrt{4 + 1}$	$= \sqrt{9 + 1}$		$= \sqrt{k^4 - 2k^2 + 1 + 1}$
$= \sqrt{5}$	$= \sqrt{10}$		$= \sqrt{k^4 - 2k^2 + 2}$

49.
$$\frac{f(x+h) - f(x)}{h} = \frac{[3(x+h) + 1] - [3x + 1]}{h} = \frac{[3x + 3h + 1] - [3x + 1]}{h}$$
$$= \frac{3x + 3h + 1 - 3x - 1}{h} = \frac{3h}{h} = 3$$

51.
$$\frac{f(x+h) - f(x)}{h} = \frac{[(x+h)^2 + 1] - [x^2 + 1]}{h} = \frac{[x^2 + 2xh + h^2 + 1] - [x^2 + 1]}{h}$$
$$= \frac{x^2 + 2xh + h^2 + 1 - x^2 - 1}{h}$$
$$= \frac{2xh + h^2}{h} = \frac{h(2x + h)}{h} = 2x + h$$

53.

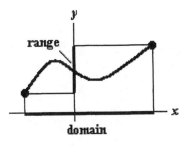

55. function

57. function

59. function

61. $f(x) = 2x + 3$

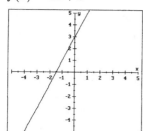

63. $f(x) = \frac{1}{2}x - 3$

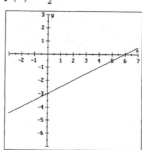

65. $2x = 3y - 3$
$-3y = -2x - 3$
$y = \frac{2}{3}x + 1$

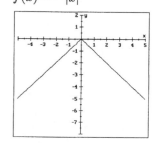

67. $f(x) = -\sqrt{x}$

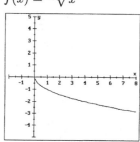

69. $f(x) = -\sqrt{x+1}$

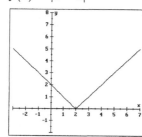

71. $f(x) = -|x|$

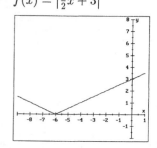

73. $f(x) = |x - 2|$

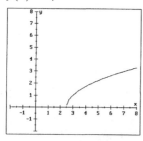

75. $f(x) = \sqrt{2x - 4}$

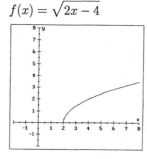

77. $f(x) = \left|\frac{1}{2}x + 3\right|$

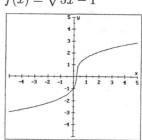

79. $f(x) = \sqrt{2x - 5}$

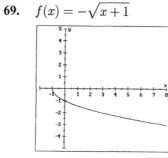

domain: $\left[\frac{5}{2}, \infty\right)$; range: $[0, \infty)$

81. $f(x) = \sqrt[3]{5x - 1}$

domain: $(-\infty, \infty)$; range: $(-\infty, \infty)$

83. $F = mC + b$ $F = mC + 32$ $F = \frac{9}{5}C + 32$
$32 = m(0) + b$ $-40 = m(-40) + 32$
$32 = b$ $-72 = -40m$
           $\frac{9}{5} = m$

85. $c = mn + b$ $c = 0.0012n + b$ $c = 0.0012n + 3.50$
$m = \dfrac{14.30 - 4.70}{9000 - 1000} = \dfrac{9.60}{8000} = 0.0012$ $4.70 = 0.0012(1000) + b$
                                         $3.50 = b$

87. **a.** The \$14,000 cost is fixed regardless of the number of square feet. Thus, the cost is \$14,000 when the house has an area of 0 square feet ($c = 14,000$ when $f = 0$). If the house has an area of 1 square foot, the cost is $14,000 + 95 = 14,095$. Thus,
     $c = mf + b$ $c = mf + 14000$ $\boxed{c = 95f + 14000}$
   $14000 = m(0) + b$ $14095 = m(1) + 14000$
   $14000 = b$ $95 = m$

     b. $c = 95f + 14000 = 95(2600) + 14000 = 261000 \Rightarrow$ It will cost \$261,000.

89. **Answers may vary.**   **91.** **Answers may vary.**   **93.** natural: $1, 7, 8$

95. prime: 7 **97.** $(-4, 7]$ **99.** $(-3, 5) \cup [6, \infty)$

Exercise 3.2 (page 254)

1. $y = ax^2 + bx + c$ **3.** $-\dfrac{b}{2a}$

5. $y = x^2 - x;\ a = 1, b = -1, c = 0$
vertex: $x = -\dfrac{b}{2a} = -\dfrac{-1}{2(1)} = \dfrac{1}{2}$
$y = x^2 - x = \left(\dfrac{1}{2}\right)^2 - \dfrac{1}{2} = -\dfrac{1}{4}$

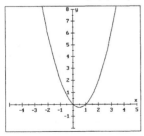

7. $y = -3x^2 + 2;\ a = -3, b = 0, c = 2$
vertex: $x = -\dfrac{b}{2a} = -\dfrac{0}{2(-3)} = 0$
$y = -3x^2 + 2 = -3(0)^2 + 2 = 2$

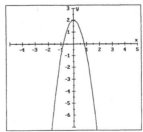

9. $y = -\dfrac{1}{2}x^2 + 3$; $a = -\dfrac{1}{2}$, $b = 0$, $c = 3$

vertex: $x = -\dfrac{b}{2a} = -\dfrac{0}{2\left(-\frac{1}{2}\right)} = 0$

$y = -\dfrac{1}{2}x^2 + 3 = -\dfrac{1}{2}(0)^2 + 3 = 3$

11. $y = x^2 - 4x + 1$; $a = 1$, $b = -4$, $c = 1$

vertex: $x = -\dfrac{b}{2a} = -\dfrac{-4}{2(1)} = 2$

$y = x^2 - 4x + 1 = 2^2 - 4(2) + 1 = -3$

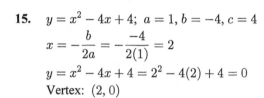

13. $y = x^2 - 1$; $a = 1$, $b = 0$, $c = -1$

$x = -\dfrac{b}{2a} = -\dfrac{0}{2(1)} = 0$

$y = x^2 - 1 = 0^2 - 1 = -1$

Vertex: $(0, -1)$

15. $y = x^2 - 4x + 4$; $a = 1$, $b = -4$, $c = 4$

$x = -\dfrac{b}{2a} = -\dfrac{-4}{2(1)} = 2$

$y = x^2 - 4x + 4 = 2^2 - 4(2) + 4 = 0$

Vertex: $(2, 0)$

17. $y = x^2 + 6x - 3$; $a = 1$, $b = 6$, $c = -3$

$x = -\dfrac{b}{2a} = -\dfrac{6}{2(1)} = -3$

$y = x^2 + 6x - 3 = (-3)^2 + 6(-3) - 3$
$\qquad = -12$

Vertex: $(-3, -12)$

19. $y = -2x^2 + 12x - 17$;

$a = -2$, $b = 12$, $c = -17$

$x = -\dfrac{b}{2a} = -\dfrac{12}{2(-2)} = 3$

$y = -2x^2 + 12x - 17$
$\qquad = -2(3)^2 + 12(3) - 17 = 1$

Vertex: $(3, 1)$

21. $y = 3x^2 - 4x + 5$;

$a = 3$, $b = -4$, $c = 5$

$x = -\dfrac{b}{2a} = -\dfrac{-4}{2(3)} = \dfrac{4}{6} = \dfrac{2}{3}$

$y = 3x^2 - 4x + 5 = 3\left(\dfrac{2}{3}\right)^2 - 4\left(\dfrac{2}{3}\right) + 5$
$\qquad\qquad = \dfrac{11}{3}$

Vertex: $\left(\dfrac{2}{3}, \dfrac{11}{3}\right)$

23. $y = \dfrac{1}{2}x^2 + 4x - 3$;

$a = \dfrac{1}{2}$, $b = 4$, $c = -3$

$x = -\dfrac{b}{2a} = -\dfrac{4}{2\left(\frac{1}{2}\right)} = -4$

$y = \dfrac{1}{2}x^2 + 4x - 3$
$\qquad = \dfrac{1}{2}(-4)^2 + 4(-4) - 3 = -11$

Vertex: $(-4, -11)$

25. $x^2 + 20y - 400 = 0$

$20y = -x^2 + 400$

$y = -\dfrac{1}{20}x^2 + 20$

$a = -\dfrac{1}{20}, b = 0, c = 20$

$x = -\dfrac{b}{2a} = -\dfrac{0}{2\left(-\frac{1}{20}\right)} = 0$

$y = -\dfrac{1}{20}x^2 + 20 = -\dfrac{1}{20}(0)^2 + 20 = 20$

The maximum height is 20 feet.

27. Since the triangle is a 45°-45°-90° triangle, we get the figure below:

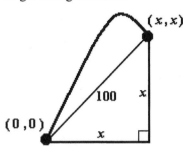

Use the Pythagorean Theorem to find x:

$x^2 + x^2 = 100^2$

$2x^2 = 100^2$

$x^2 = \dfrac{100^2}{2}$

$x = \pm\sqrt{\dfrac{100^2}{2}} = \pm\dfrac{100}{\sqrt{2}}$

Use the positive value for x.

The point $\left(\dfrac{100}{\sqrt{2}}, \dfrac{100}{\sqrt{2}}\right)$ must be on the graph: $y = -x^2 + ax$

$$\dfrac{100}{\sqrt{2}} = -\left(\dfrac{100}{\sqrt{2}}\right)^2 + a\left(\dfrac{100}{\sqrt{2}}\right)$$

$$\dfrac{100\sqrt{2}}{2} = -\dfrac{100^2}{2} + a\left(\dfrac{100\sqrt{2}}{2}\right)$$

$$100\sqrt{2} = -100^2 + \left(100\sqrt{2}\right)a$$

$$100\sqrt{2} + 100^2 = \left(100\sqrt{2}\right)a$$

$$\dfrac{100\sqrt{2} + 100^2}{100\sqrt{2}} = a \Rightarrow a = 1 + \dfrac{100}{\sqrt{2}} \Rightarrow a = 1 + 50\sqrt{2}$$

29. Let $x =$ the width.

Then $50 - x =$ the length.

Area $= lw$

$y = (50 - x)x$

$y = 50x - x^2$

$y = -x^2 + 50x$

$a = -1, b = 50, c = 0$

Find the vertex:

$x = -\dfrac{b}{2a} = -\dfrac{50}{2(-1)} = 25$

$y = -x^2 + 50x = -\left(25^2\right) + 50(25) = 625$

The maximum area occurs when the dimensions are 25 ft by 25 ft.

31. Set up the variables as indicated in the figure:

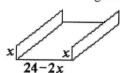

24−2x

$\text{Area} = lw$
$y = x(24 - 2x)$
$y = 24x - 2x^2$
$y = -2x^2 + 24x$
$a = -2, b = 24, c = 0$

Find the vertex:

$$x = -\frac{b}{2a} = -\frac{24}{2(-2)} = 6$$
$$y = -2x^2 + 24x = -2(6^2) + 24(6) = 72$$

The maximum area occurs when the depth is 6 inches and the width is 12 inches.

33. $\text{Revenue} = \text{Price} \cdot \text{\# Sold}$
$$y = p(1200 - p)$$
$$y = 1200p - p^2$$
$$y = -p^2 + 1200p$$
$$a = -1, b = 1200, c = 0$$

Find the vertex:

$$p = -\frac{b}{2a} = -\frac{1200}{2(-1)} = 600$$

The maximum revenue occurs when the price is $600.

35. Let $x = $ # of $5 increases.
Then Rate $= 90 + 5x$
Rooms $= 200 - 10x$
Revenue $= \text{Rate} \cdot \text{\# Rooms}$
$$y = (90 + 5x)(200 - 10x)$$
$$y = 18{,}000 + 100x - 50x^2$$
$$a = -50, b = 100, c = 18{,}000$$

Find the vertex:

$$x = -\frac{b}{2a} = -\frac{100}{2(-50)} = 1$$

The maximum revenue occurs when the room rate increases by 1 five-dollar increment, or when the rate is $95.

37. $s = 80t - 16t^2$
$a = -16, b = 80, c = 0$
Find the x-coord. of the vertex:
$$x = t = -\frac{b}{2a} = -\frac{80}{2(-16)} = \frac{5}{2} = 2.5$$
The max. height occurs after 2.5 seconds.

39. $s = 80t - 16t^2$
$a = -16, b = 80, c = 0$
Find the y-coord. of the vertex.
Note: The x-coord. was found in **#37**.
$$y = s = 80t - 16t^2 = 80(2.5) - 16(2.5)^2$$
$$= 100$$
The max. height is 100 ft.

41. $y = 2x^2 + 9x - 56$
Vertex: $(-2.25, 66.13)$

43. $y = (x - 7)(5x + 2)$
Vertex: $(3.3, -68.5)$

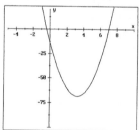

45. The equation of the line is $y = -\frac{3}{4}x + 9$.

Thus the point $(x, y) = \left(x, -\frac{3}{4}x + 9\right)$.

Area $= x\left(-\frac{3}{4}x + 9\right)$

$y = -\frac{3}{4}x^2 + 9x$

$a = -\frac{3}{4}, b = 9, c = 0$

Find the x-coord. of the vertex:

$$x = -\frac{b}{2a} = -\frac{9}{2\left(-\frac{3}{4}\right)} = 6$$

Thus, the dimensions are 6 by $4\frac{1}{2}$ units.

47. Let $x =$ one number.

Then $6 - x =$ the other number.

Sum of squares $= x^2 + (6 - x)^2$

$y = x^2 + 36 - 12x + x^2$

$y = 2x^2 - 12x + 36$

$a = 2, b = -12, c = 36$

Find the x-coord. of the vertex:

$$x = -\frac{b}{2a} = -\frac{-12}{2(2)} = 3$$

Thus, the numbers are both 3.

49. $f(a) = a^2 - 3a$

$f(-a) = (-a)^2 - 3(-a)$

$\quad = a^2 + 3a$

51. $f(a) = (5 - a)^2$

$f(-a) = [5 - (-a)]^2$

$\quad = (5 + a)^2$

53. $f(a) = 7$

$f(-a) = 7$

Exercise 3.3 (page 265)

1. 4

3. $n - 1$

5. odd

7. piecewise-defined

9. 3

11. $f(x) = x^3 - x$

$f(-x) = (-x)^3 - (-x)$

$\quad = -x^3 + x$

$\quad = -f(x) \Rightarrow$ odd

x-int.	y-int.
$x^3 - x = 0$	$f(0) = 0^3 - 0$
$x(x^2 - 1) = 0$	$y = 0$
$x(x + 1)(x - 1) = 0$	$(0, 0)$
$x = 0, x = -1, x = 1$	
$(0, 0), (-1, 0), (1, 0)$	

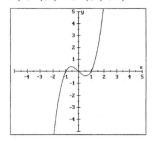

13. $f(x) = -x^3$

$f(-x) = -(-x)^3$

$\quad = -(-x^3)$

$\quad = x^3 = -f(x) \Rightarrow$ odd

x-int.	y-int.
$-x^3 = 0$	$f(0) = -0^3$
$x^3 = 0$	$y = 0$
$x = 0$	$(0, 0)$
$(0, 0)$	

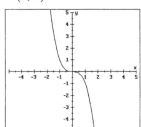

15. $f(x) = x^4 - 2x^2 + 1$

$f(-x) = (-x)^4 - 2(-x)^2 + 1$

$= x^4 + 2x^2 + 1 \Rightarrow$ even

x-int.	y-int.
$x^4 - 2x^2 + 1 = 0$	$f(0) = 1$
$(x^2 - 1)(x^2 - 1) = 0$	$y = 1$
$x^2 = 1$	$(0, 1)$
$x = 1, x = -1$	
$(1, 0), (-1, 0)$	

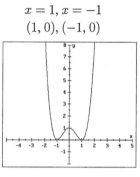

17. $f(x) = x^3 - x^2 - 4x + 4$

$f(-x) = (-x)^3 - (-x)^2 - 4(-x) + 4$

$= -x^3 - x^2 + 4x + 4$

$\Rightarrow$ neither even nor odd

x-int.	y-int.
$x^3 - x^2 - 4x + 4 = 0$	$f(0) = 4$
$x^2(x - 1) - 4(x - 1) = 0$	$y = 4$
$(x - 1)(x^2 - 4) = 0$	$(0, 4)$
$x = 1$ or $x^2 = 4$	
$x = 1, x = 2, x = -2$	
$(1, 0), (2, 0), (-2, 0)$	

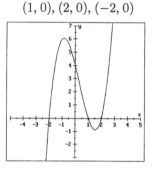

19. $f(x) = -x^4 + 5x^2 - 4$

$f(-x) = -(-x)^4 + 5(-x)^2 - 4$

$= -x^4 + 5x^2 - 4 \Rightarrow$ even

x-int.	y-int.
$-x^4 + 5x^2 - 4 = 0$	$f(0) = -4$
$-(x^4 - 5x^2 + 4) = 0$	$y = -4$
$(x^2 - 1)(x^2 - 4) = 0$	$(0, -4)$
$x^2 = 1$ or $x^2 = 4$	
$x = \pm 1, x = \pm 2$	
$(\pm 1, 0), (\pm 2, 0)$	

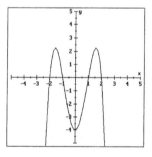

21. $y = f(x) = x^4 + x^2$

$f(-x) = (-x)^4 + (-x)^2$

$= x^4 + x^2 = f(x) \Rightarrow$ even

23. $y = f(x) = x^3 + x^2$

$f(-x) = (-x)^3 + (-x)^2$

$= -x^3 + x^2 \Rightarrow$ neither

25. $y = f(x) = x^5 + x^3$

$f(-x) = (-x)^5 + (-x)^3$

$= -x^5 - x^3 = -f(x) \Rightarrow$ odd

27. $y = f(x) = 2x^3 - 3x$

$f(-x) = 2(-x)^3 - 3(-x)$

$= -2x^3 + 3x = -f(x) \Rightarrow$ odd

29. increasing: $x > 0$

decreasing: $x < 0$

31. increasing: $x < 0$

decreasing: $x > 0$

33. $y = f(x) = x^2 - 4x + 4$

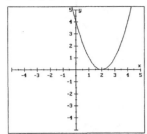

increasing: $x > 2$
decreasing: $x < 2$

35. $y = f(x) = \begin{cases} x + 2 & \text{if } x < 0 \\ 2 & \text{if } x \geq 0 \end{cases}$

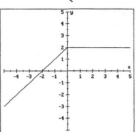

37. $y = f(x) = \begin{cases} -x & \text{if } x < 0 \\ x^2 & \text{if } x \geq 0 \end{cases}$

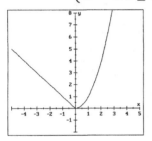

39. $y = f(x) = \begin{cases} 0 & \text{if } x < 0 \\ x^2 & \text{if } 0 \leq x \leq 2 \\ 4 - 2x & \text{if } x > 2 \end{cases}$

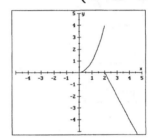

41. $y = [[2x]]$

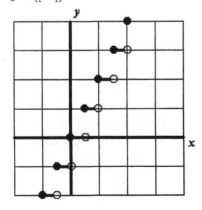

43. $y = [[x]] - 1$

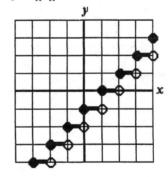

45.

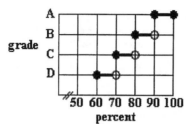

$$\frac{67 + 73 + 84 + 87 + 93}{5} = \frac{404}{5} = 80.8$$

The student's grade is B.

47. $26 for 275 miles

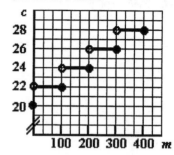

49. $12 per hour $\Rightarrow$ $0.20 per minute

$1.60 for $7\frac{1}{2}$ minutes

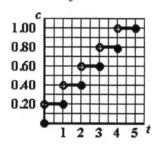

51.

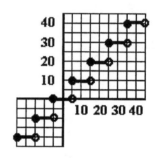

53. $y = \dfrac{|x|}{x}$; NO $\Rightarrow$ not defined at $x = 0$

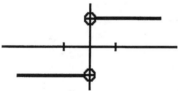

55. Answers may vary.

57. Answers may vary.

59. $f(x + 1) = 3(x + 1) + 2 = 3x + 5$
$f(x) + 1 = (3x + 2) + 1 = 3x + 3$

61. $f(x - 3) = \dfrac{3(x - 3) + 1}{5} = \dfrac{3x - 8}{5}$

$f(x) - 3 = \dfrac{3x + 1}{5} - 3 = \dfrac{3x - 14}{5}$

63.
$$2x^2 - 3 = x$$
$$2x^2 - x - 3 = 0$$
$$(2x - 3)(x + 1) = 0$$
$$2x = 3 \text{ or } x = -1$$
$$x = \tfrac{3}{2}, x = -1$$

Exercise 3.4 (page 276)

1. up

3. to the right

5. 2, down

7. y-axis

9. horizontally

11. $y = x^2 - 2$
Shift $y = x^2$ D 2

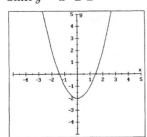

13. $y = (x + 3)^2$
Shift $y = x^2$ L 3

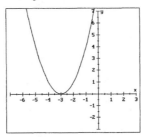

15. $y = (x + 1)^2 + 2$
Shift $y = x^2$ U 2, L 1

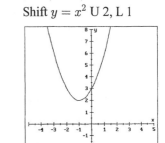

17. $y = \left(x + \dfrac{1}{2}\right)^2 - \dfrac{1}{2}$
Shift $y = x^2$ D $\frac{1}{2}$, L $\frac{1}{2}$

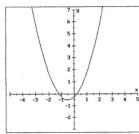

19. $y = x^3 + 1$
Shift $y = x^3$ U 1

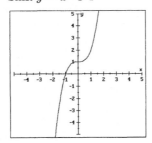

21. $y = (x - 2)^3$
Shift $y = x^3$ R 2

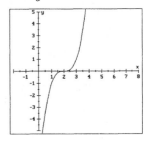

23. $y = (x - 2)^3 - 3$
Shift $y = x^3$ D 3, R 2

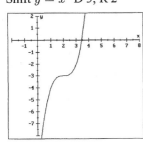

25. $y + 2 = x^3$
$y = x^3 - 2$
Shift $y = x^3$ D 2

27. $y = -x^2$
Reflect $y = x^2$ about x

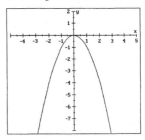

SECTION 3.4

29. $y = -x^3$
Reflect $y = x^3$ about x

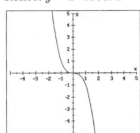

31. $y = 2x^2$: Stretch
$y = x^2$ vert. by a factor of 2

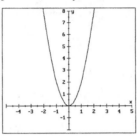

33. $y = -3x^2$: Stretch
$y = x^2$ vert. by a factor of 3
Reflect about x

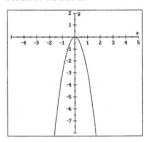

35. $y = \left(\frac{1}{2}x\right)^3$: Stretch
$y = x^3$ hor. by a factor of 2

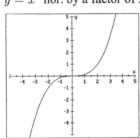

37. $y = -8x^3$: Stretch
$y = x^3$ vert. by a factor of 8
Reflect about x

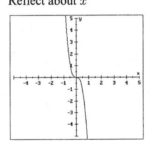

39. $y = |x - 2| + 1$
Shift $y = |x|$ U 1, R 2

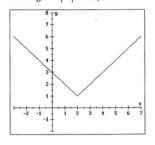

41. $y = |3x|$: Stretch
$y = |x|$ hor. by a factor of $\frac{1}{3}$

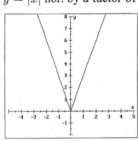

43. $y = \sqrt{x - 2} + 1$
Shift $y = \sqrt{x}$ U 1, R 2

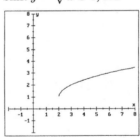

45. $y = 2\sqrt{x} + 3$: Stretch
$y = \sqrt{x}$ vert. by a factor
of 2; Shift U 3

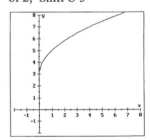

47. $y = -2|x| + 3$: Stretch $y = |x|$ vert. by a factor of 2; Reflect x; Shift U 3

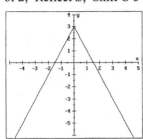

49. Shift $y = f(x)$ U 1

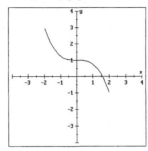

51. Stretch $y = f(x)$ vert. by a factor of 2

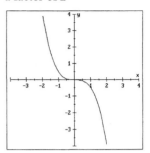

53. Shift $y = f(x)$ U 1, 2 R

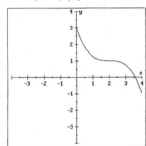

55. Stretch $y = f(x)$ vert. by a factor of 2, reflect about y

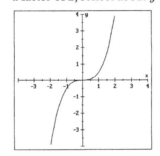

57. Answers may vary.

59. Answers may vary.

61. Answers may vary.

63. $\dfrac{x^2 + x - 6}{x^2 + 5x + 6} = \dfrac{(x+3)(x-2)}{(x+2)(x+3)} = \dfrac{x-2}{x+2}$

65. $f(x) = \dfrac{x+7}{x-3}$; domain $= (-\infty, 3) \cup (3, \infty)$

67.
$$
\begin{array}{r}
x + 2 + \frac{-2}{x+1} \\
x+1 \overline{\smash{\big)}\, x^2 + 3x + 0} \\
\underline{x^2 + x } \\
2x + 0 \\
\underline{2x + 2} \\
-2
\end{array}
$$

Exercise 3.5 (page 290)

1. asymptote

3. vertical

5. x-intercept

7. the same

9. horizontal

11. $t = f(30) = \dfrac{600}{30} = 20$ hr

13. $t = f(50) = \dfrac{600}{50} = 12$ hr

15. $c = f(10) = \dfrac{50,000(10)}{100 - 10} \approx \$5555.56$

17. $c = f(50) = \dfrac{50,000(50)}{100 - 50} = \$50,000.00$

19. $f(x) = \dfrac{x^2}{x-2}$; domain $= (-\infty, 2) \cup (2, \infty)$ **21.** $f(x) = \dfrac{2x^2 + 7x - 2}{x^2 - 25} = \dfrac{2x^2 + 7x - 2}{(x+5)(x-5)}$

domain $= (-\infty, -5) \cup (-5, 5) \cup (5, \infty)$

23. $f(x) = \dfrac{x-1}{x^3 - x} = \dfrac{x-1}{x(x+1)(x-1)}$; domain $= (-\infty, -1) \cup (-1, 0) \cup (0, 1) \cup (1, \infty)$

25. $\dfrac{3x^2 + 5}{x^2 + 1}$; domain $= (-\infty, \infty)$

27. $y = \dfrac{1}{x-2}$

Vertical: $x = 2$; Horizontal: $y = 0$
Slant: none; x-intercepts: none
y-intercepts: $\left(0, -\frac{1}{2}\right)$; Symmetry: none

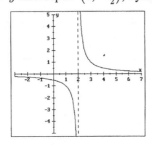

29. $y = \dfrac{x}{x-1} = 1 + \dfrac{1}{x-1}$

Vertical: $x = 1$; Horizontal: $y = 1$
Slant: none; x-intercepts: $(0, 0)$
y-intercepts: $(0, 0)$; Symmetry: none

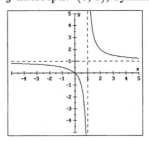

31. $y = \dfrac{x+1}{x+2} = 1 + \dfrac{-1}{x+2}$

Vertical: $x = -2$; Horizontal: $y = 1$
Slant: none; x-intercepts: $(-1, 0)$
y-intercepts: $\left(0, \frac{1}{2}\right)$; Symmetry: none

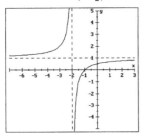

33. $y = \dfrac{2x-1}{x-1} = 2 + \dfrac{1}{x-1}$

Vertical: $x = 1$; Horizontal: $y = 2$
Slant: none; x-intercepts: $\left(\frac{1}{2}, 0\right)$
y-intercepts: $(0, 1)$; Symmetry: none

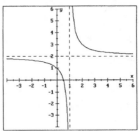

35. $y = \dfrac{x^2 - 9}{x^2 - 4} = \dfrac{(x+3)(x-3)}{(x+2)(x-2)} = 1 + \frac{-5}{x^2-4}$

Vertical: $x = -2$, $x = 2$; Horizontal: $y = 1$
Slant: none; x-intercepts: $(-3, 0)$, $(3, 0)$
y-intercepts: $\left(0, \frac{9}{4}\right)$; Symmetry: y-axis

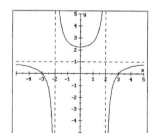

37. $y = \dfrac{x^2 - x - 2}{x^2 - 4x + 3} = \dfrac{(x+1)(x-2)}{(x-3)(x-1)}$

$$= 1 + \dfrac{3x - 5}{x^2 - 4x + 3}$$

Vertical: $x = 3$, $x = 1$; Horizontal: $y = 1$
Slant: none; x-intercepts: $(-1, 0)$, $(2, 0)$
y-intercepts: $\left(0, -\frac{2}{3}\right)$; Symmetry: none

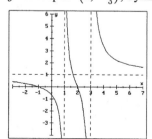

39. $y = \dfrac{x^2 + 2x - 3}{x^3 - 4x} = \dfrac{(x-1)(x+3)}{x(x+2)(x-2)}$

Vert: $x = 0$, $x = -2$, $x = 2$; Horiz: $y = 0$
Slant: none; x-intercepts: $(1, 0)$, $(-3, 0)$
y-intercepts: none; Symmetry: none

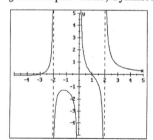

41. $y = \dfrac{x^2 - 9}{x^2} = \dfrac{(x+3)(x-3)}{x^2} = 1 + \frac{-9}{x^2}$

Vert: $x = 0$; Horiz: $y = 1$
Slant: none; x-intercepts: $(3, 0)$, $(-3, 0)$
y-intercepts: none; Symmetry: y-axis

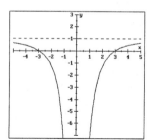

43. $y = \dfrac{x}{(x+3)^2}$

Vert: $x = -3$; Horiz: $y = 0$
Slant: none; x-intercepts: $(0, 0)$
y-intercepts: $(0, 0)$; Symmetry: none

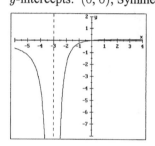

45. $y = \dfrac{x + 1}{x^2(x-2)}$

Vert: $x = 0$, $x = 2$; Horiz: $y = 0$
Slant: none; x-intercepts: $(-1, 0)$
y-intercepts: none; Symmetry: none

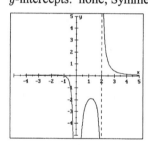

47. $y = \dfrac{x}{x^2+1}$

Vert: none; Horiz: $y = 0$

Slant: none; x-intercepts: $(0, 0)$

y-intercepts: $(0, 0)$; Symmetry: origin

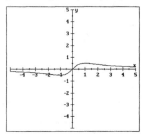

49. $y = \dfrac{3x^2}{x^2+1} = 3 + \dfrac{-3}{x^2+1}$

Vert: none; Horiz: $y = 3$

Slant: none; x-intercepts: $(0, 0)$

y-intercepts: $(0, 0)$; Symmetry: y-axis

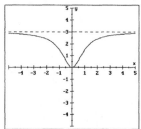

51. $y = \dfrac{x^2 - 2x - 8}{x - 1} = \dfrac{(x+2)(x-4)}{x-1}$

$\qquad = x - 1 + \dfrac{-9}{x-1}$

Vert: $x = 1$; Horiz: none; Slant: $y = x - 1$

x-intercepts: $(4, 0), (-2, 0)$

y-intercepts: $(0, 8)$; Symmetry: none

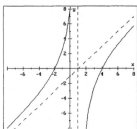

53. $y = \dfrac{x^3 + x^2 + 6x}{x^2 - 1} = \dfrac{x(x^2 + x + 6)}{(x+1)(x-1)}$

$\qquad = x + 1 + \dfrac{7x+1}{x^2-1}$

Vert: $x = -1$, $x = 1$; Horiz: none

Slant: $y = x + 1$; x-intercepts: $(0, 0)$

y-intercepts: $(0, 0)$; Symmetry: none

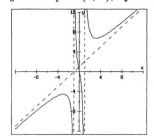

55. $y = \dfrac{x^2}{x} = x$ (if $x \neq 0$)

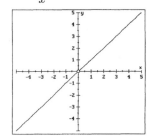

57. $y = \dfrac{x^3 + x}{x} = \dfrac{x(x^2+1)}{x} = x^2 + 1$

(if $x \neq 0$)

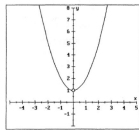

59. $y = \dfrac{x^2 - 2x + 1}{x - 1} = \dfrac{(x-1)(x-1)}{x-1}$
$= x - 1 \text{ (if } x \neq 1)$

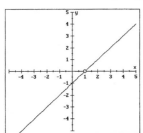

61. $y = \dfrac{x^3 - 1}{x - 1} = \dfrac{(x-1)(x^2 + x + 1)}{x - 1}$
$= x^2 + x + 1 \text{ (if } x \neq 1)$

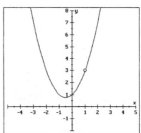

63. $c = f(x) = 1.25x + 700$

65. $\bar{c} = f(x) = \dfrac{1.25x + 700}{x}$

67. $c = 1.25(500) + 700 = \$1325$

69. $c = f(n) = 0.09n + 7.50$

71. $c = 0.09(775) + 7.50 = \$77.25$

73. $\bar{c} = \dfrac{0.09(1000) + 7.50}{1000} = \$0.0975$
$= 9.75¢$

75. **a.** $c(n) = 0.095n + 8.50$ **b.** $\bar{c}(n) = \dfrac{0.095n + 8.50}{n}$

c. $\bar{c}(850) = \dfrac{0.095(850) + 8.50}{850} = \$0.105 = 10.5¢$

77. **Answers may vary.**

79. $y = \dfrac{ax + b}{cx^2 + d} = \dfrac{\frac{ax+b}{x^2}}{\frac{cx^2+d}{x^2}} = \dfrac{\frac{ax}{x^2} + \frac{b}{x^2}}{\frac{cx^2}{x^2} + \frac{d}{x^2}} = \dfrac{\frac{a}{x} + \frac{b}{x^2}}{c + \frac{d}{x^2}}$
As x approaches $\pm\infty$, $y \approx \dfrac{0 + 0}{c + 0} = 0$. Thus the horizontal asymptote is $y = 0$.

81. $y = \dfrac{ax^2 + b}{cx^2 + d} = \dfrac{\frac{ax^2+b}{x^2}}{\frac{cx^2+d}{x^2}} = \dfrac{\frac{ax^2}{x^2} + \frac{b}{x^2}}{\frac{cx^2}{x^2} + \frac{d}{x^2}} = \dfrac{a + \frac{b}{x^2}}{c + \frac{d}{x^2}}$
As x approaches $\pm\infty$, $y \approx \dfrac{a + 0}{c + 0} = \dfrac{a}{c}$. Thus the horizontal asymptote is $y = \dfrac{a}{c}$.

83. **Answers may vary.**   **85.** **Answers may vary.**

87. $(2x^2 + 3x) + (x^2 - 2x) = 2x^2 + 3x + x^2 - 2x = 3x^2 + x$

89. $(5x + 2)(2x + 5) = 10x^2 + 25x + 4x + 10 = 10x^2 + 29x + 10$

91. $f(x + 1) = 3(x + 1) + 2 = 3x + 3 + 2 = 3x + 5$

Exercise 3.6 (page 303)

1. $f(x) + g(x)$ **3.** $f(x)g(x)$ **5.** $(-\infty, \infty)$

7. $g(f(x))$ **9.** identity

11. $(f + g)(x) = f(x) + g(x) = (2x + 1) + (3x - 2) = 5x - 1$; domain $= (-\infty, \infty)$

13. $(f \cdot g)(x) = f(x)g(x) = (2x + 1)(3x - 2) = 6x^2 - x - 2$; domain $= (-\infty, \infty)$

15. $(f - g)(x) = f(x) - g(x) = (x^2 + x) - (x^2 - 1) = x + 1$; domain $= (-\infty, \infty)$

17. $(f/g)(x) = \dfrac{f(x)}{g(x)} = \dfrac{x^2 + x}{x^2 - 1} = \dfrac{x(x + 1)}{(x + 1)(x - 1)} = \dfrac{x}{x - 1}$; domain $= (-\infty, -1) \cup (-1, 1) \cup (1, \infty)$

19. $(f + g)(2) = f(2) + g(2) = \left[(2)^2 - 1\right] + [3(2) - 2] = 3 + 4 = 7$

21. $(f - g)(0) = f(0) - g(0) = \left[(0)^2 - 1\right] - [3(0) - 2] = -1 - (-2) = 1$

23. $(f \cdot g)(2) = f(2) \cdot g(2) = \left[(2)^2 - 1\right] \cdot [3(2) - 2] = (3)(4) = 12$

25. $(f/g)\left(\frac{2}{3}\right) = \dfrac{f\left(\frac{2}{3}\right)}{g\left(\frac{2}{3}\right)} = \dfrac{\left[\left(\frac{2}{3}\right)^2 - 1\right]}{\left[3\left(\frac{2}{3}\right) - 2\right]} = \dfrac{-\frac{5}{9}}{0} \Rightarrow$ undefined

27. Let $f(x) = 3x^2$ and $g(x) = 2x$.
Then $(f + g)(x) = 3x^2 + 2x = h(x)$.

29. Let $f(x) = 3x^2$ and $g(x) = x^2 - 1$.
Then $(f/g)(x) = \dfrac{3x^2}{x^2 - 1} = h(x)$.

31. Let $f(x) = 3x^3$ and $g(x) = -x$.
Then $(f - g)(x) = 3x^3 + x$
$= x\left(3x^2 + 1\right) = h(x)$.

33. Let $f(x) = x + 9$ and $g(x) = x - 2$.
Then $(f \cdot g)(x) = (x + 9)(x - 2)$
$= x^2 + 7x - 18 = h(x)$.

35. $(f \circ g)(2) = f(g(2)) = f(5(2) - 2) = f(8) = 2(8) - 5 = 11$

37. $(f \circ f)\left(-\frac{1}{2}\right) = f\left(f\left(-\frac{1}{2}\right)\right) = f\left(2\left(-\frac{1}{2}\right) - 5\right) = f(-6) = 2(-6) - 5 = -17$

39. $(f \circ g)(-3) = f(g(-3)) = f(4(-3) + 4) = f(-8) = 3(-8)^2 - 2 = 190$

41. $(f \circ f)\left(\sqrt{3}\right) = f\left(f\left(\sqrt{3}\right)\right) = f\left(3\left(\sqrt{3}\right)^2 - 2\right) = f(7) = 3(7)^2 - 2 = 145$

43. The domain of $f \circ g$ is the set of all real numbers in the domain of $g(x)$ such that $g(x)$ is in the domain of $f(x)$. Domain of $g(x)$: $(-\infty, \infty)$. Domain of $f(x) = (-\infty, \infty)$. Thus, all values of $g(x)$ are in the domain of $f(x)$. $\boxed{\text{Domain of } f \circ g\text{: } (-\infty, \infty)}$
$(f \circ g)(x) = f(g(x)) = f(x + 1) = 3(x + 1) = 3x + 3$

45. The domain of $f \circ f$ is the set of all real numbers in the domain of $f(x)$ such that $f(x)$ is in the domain of $f(x)$. Domain of $f(x)$: $(-\infty, \infty)$. Thus, all values of $f(x)$ are in the domain of $f(x)$. $\boxed{\text{Domain of } f \circ f\text{: } (-\infty, \infty)}$ $(f \circ f)(x) = f(f(x)) = f(3x) = 3(3x) = 9x$

47. The domain of $g \circ f$ is the set of all real numbers in the domain of $f(x)$ such that $f(x)$ is in the domain of $g(x)$. Domain of $f(x)$: $(-\infty, \infty)$. Domain of $g(x) = (-\infty, \infty)$. Thus, all values of $f(x)$ are in the domain of $g(x)$. $\boxed{\text{Domain of } g \circ f\text{: } (-\infty, \infty)}$
$(g \circ f)(x) = g(f(x)) = g(x^2) = 2x^2$

49. The domain of $g \circ g$ is the set of all real numbers in the domain of $g(x)$ such that $g(x)$ is in the domain of $g(x)$. Domain of $g(x)$: $(-\infty, \infty)$. Thus, all values of $g(x)$ are in the domain of $g(x)$. $\boxed{\text{Domain of } g \circ g\text{: } (-\infty, \infty)}$ $(g \circ g)(x) = g(g(x)) = g(2x) = 2(2x) = 4x$

51. The domain of $f \circ g$ is the set of all real numbers in the domain of $g(x)$ such that $g(x)$ is in the domain of $f(x)$. Domain of $g(x)$: $(-\infty, \infty)$. Domain of $f(x) = [0, \infty)$. Thus, we must have $g(x) \geq 0 \Rightarrow x + 1 \geq 0 \Rightarrow x \geq -1$. $\boxed{\text{Domain of } f \circ g\text{: } [-1, \infty)}$
$(f \circ g)(x) = f(g(x)) = f(x + 1) = \sqrt{x + 1}$

53. The domain of $f \circ f$ is the set of all real numbers in the domain of $f(x)$ such that $f(x)$ is in the domain of $f(x)$. Domain of $f(x)$: $[0, \infty)$. Thus, we must have $f(x) \geq 0 \Rightarrow \sqrt{x} \geq 0$. This is true for all real values of x. $\boxed{\text{Domain of } f \circ f\text{: } [0, \infty)}$
$(f \circ f)(x) = f(f(x)) = f(\sqrt{x}) = \sqrt{\sqrt{x}} = \left((x)^{1/2}\right)^{1/2} = x^{1/4} = \sqrt[4]{x}$

55. The domain of $g \circ f$ is the set of all real numbers in the domain of $f(x)$ such that $f(x)$ is in the domain of $g(x)$. Domain of $f(x)$: $[-1, \infty)$. Domain of $g(x) = (-\infty, \infty)$. Thus, all values of $f(x)$ are in the domain of $g(x)$. $\boxed{\text{Domain of } g \circ f\text{: } [-1, \infty)}$
$(g \circ f)(x) = g(f(x)) = g\left(\sqrt{x + 1}\right) = \left(\sqrt{x + 1}\right)^2 - 1 = x$

57. The domain of $g \circ g$ is the set of all real numbers in the domain of $g(x)$ such that $g(x)$ is in the domain of $g(x)$. Domain of $g(x)$: $(-\infty, \infty)$. Thus, all values of $g(x)$ are in the domain of $g(x)$. $\boxed{\text{Domain of } g \circ g\text{: } (-\infty, \infty)}$ $(g \circ g)(x) = g(g(x)) = g(x^2 - 1) = (x^2 - 1)^2 - 1 = x^4 - 2x^2$

59. The domain of $f \circ g$ is the set of all real numbers in the domain of $g(x)$ such that $g(x)$ is in the domain of $f(x)$. Domain of $g(x)$: $(-\infty, 2) \cup (2, \infty)$. Domain of $f(x) = (-\infty, 1) \cup (1, \infty)$. Thus, we must have $g(x) \neq 1 \Rightarrow \dfrac{1}{x-2} \neq 1 \Rightarrow 1 \neq x - 2 \Rightarrow x \neq 3$

$\boxed{\text{Domain of } f \circ g: (-\infty, 2) \cup (2, 3) \cup (3, \infty)}$

$(f \circ g)(x) = f(g(x)) = f\left(\dfrac{1}{x-2}\right) = \dfrac{1}{\frac{1}{x-2} - 1} = \dfrac{1}{\frac{1}{x-2} - 1} \cdot \dfrac{x-2}{x-2} = \dfrac{x-2}{1 - (x-2)} = \dfrac{x-2}{3-x}$

61. The domain of $f \circ f$ is the set of all real numbers in the domain of $f(x)$ such that $f(x)$ is in the domain of $f(x)$. Domain of $f(x)$: $(-\infty, 1) \cup (1, \infty)$. Thus, we must have $f(x) \neq 1 \Rightarrow$
$\dfrac{1}{x-1} \neq 1 \Rightarrow 1 \neq x - 1 \Rightarrow x \neq 2$ $\boxed{\text{Domain of } f \circ f: (-\infty, 1) \cup (1, 2) \cup (2, \infty)}$
$(f \circ f)(x) = f(f(x)) = f\left(\dfrac{1}{x-1}\right) = \dfrac{1}{\frac{1}{x-1} - 1} = \dfrac{1}{\frac{1}{x-1} - 1} \cdot \dfrac{x-1}{x-1} = \dfrac{x-1}{1 - (x-1)} = \dfrac{x-1}{2-x}$

63. Let $f(x) = x - 2$ and $g(x) = 3x$.
Then $(f \circ g)(x) = f(g(x))$
$= f(3x) = 3x - 2.$

65. Let $f(x) = x - 2$ and $g(x) = x^2$.
Then $(f \circ g)(x) = f(g(x))$
$= f\left(x^2\right) = x^2 - 2.$

67. Let $f(x) = x^2$ and $g(x) = x - 2$.
Then $(f \circ g)(x) = f(g(x))$
$= f(x - 2) = (x - 2)^2.$

69. Let $f(x) = \sqrt{x}$ and $g(x) = x + 2$.
Then $(f \circ g)(x) = f(g(x))$
$= f(x + 2) = \sqrt{x + 2}.$

71. Let $f(x) = x + 2$ and $g(x) = \sqrt{x}$.
Then $(f \circ g)(x) = f(g(x))$
$= f\left(\sqrt{x}\right) = \sqrt{x} + 2.$

73. Let $f(x) = x$ and $g(x) = x$.
Then $(f \circ g)(x) = f(g(x))$
$= f(x) = x.$

75. **a.** $A = 17w$

b. $w^2 + 17^2 = d^2$
$w^2 = d^2 - 17^2$
$w = \sqrt{d^2 - 289}$

c. $A = 17w$
$= 17\sqrt{d^2 - 289}$

77. If the area is A and the length of a side is s, then $s^2 = A \Rightarrow s = \sqrt{A}$. Then $P = 4s = 4\sqrt{A}$.

79. $(f + f)(x) = f(x) + f(x) = 3x + 3x = 6x$
$f(x + x) = f(2x) = 3(2x) = 6x$

81. $(f \circ f)(x) = f(f(x)) = f\left(\dfrac{x-1}{x+1}\right) = \dfrac{\frac{x-1}{x+1} - 1}{\frac{x-1}{x+1} + 1} = \dfrac{(x+1)\left(\frac{x-1}{x+1} - 1\right)}{(x+1)\left(\frac{x-1}{x+1} + 1\right)} = \dfrac{x - 1 - (x+1)}{x - 1 + x + 1}$
$= \dfrac{-2}{2x} = -\dfrac{1}{x}$

83. **Answers may vary.**

85. **Answers may vary.**

87.
$$x = 3y - 7$$
$$x + 7 = 3y$$
$$\frac{x + 7}{3} = y$$

89.
$$x = \frac{y}{y + 3}$$
$$x(y + 3) = y$$
$$xy + 3x = y$$
$$xy - y = -3x$$
$$y(x - 1) = -3x$$
$$y = \frac{-3x}{x - 1} = \frac{3x}{1 - x}$$

Exercise 3.7 (page 311)

1. one-to-one

3. interchange

5. $y = 3x$
one-to-one

7. $y = x^2 + 3$
$x = 1$ and $x = -1$ both correspond to $y = 4$.
not one-to-one

9. $y = x^3 - x$
$x = 1$ and $x = -1$ both correspond to $y = 0$.
not one-to-one

11. $y = |x|$
$x = 1$ and $x = -1$ both correspond to $y = 1$.
not one-to-one

13. $y = 5$
$x = 1$ and $x = 2$ both correspond to $y = 5$.
not one-to-one

15. $y = (x - 2)^2$, $x \geq 2$
one-to-one

17. one-to-one

19. not one-to-one (not a function)

21. $(f \circ g)(x) = f(g(x)) = f\left(\frac{1}{5}x\right) = 5\left(\frac{1}{5}x\right) = x$

$(g \circ f)(x) = g(f(x)) = g(5x) = \frac{1}{5}(5x) = x$

23. $(f \circ g)(x) = f(g(x)) = f\left(\frac{1}{x - 1}\right) = \frac{\frac{1}{x-1} + 1}{\frac{1}{x-1}} = \frac{(x - 1)\left(\frac{1}{x-1} + 1\right)}{(x - 1)\frac{1}{x-1}} = \frac{1 + x - 1}{1} = x$

$(g \circ f)(x) = g(f(x)) = g\left(\frac{x + 1}{x}\right) = \frac{1}{\frac{x+1}{x} - 1} = \frac{x(1)}{x\left(\frac{x+1}{x} - 1\right)} = \frac{x}{x + 1 - x} = \frac{x}{1} = x$

25. $y = f(x) = 3x$
$x = 3y$
$\frac{x}{3} = y$
$f^{-1}(x) = \frac{x}{3}$

$(f \circ f^{-1})(x) = f\left(f^{-1}(x)\right)$
$= f\left(\frac{x}{3}\right)$
$= 3\left(\frac{x}{3}\right)$
$= x$

$\left(f^{-1} \circ f\right)(x) = f^{-1}(f(x))$
$= f^{-1}(3x)$
$= \frac{3x}{3}$
$= x$

27.
$$y = f(x) = 3x + 2$$
$$x = 3y + 2$$
$$x - 2 = 3y$$
$$\frac{x-2}{3} = y$$
$$f^{-1}(x) = \frac{x-2}{3}$$

$$\left(f \circ f^{-1}\right)(x) = f\left(f^{-1}(x)\right)$$
$$= f\left(\frac{x-2}{3}\right)$$
$$= 3\left(\frac{x-2}{3}\right) + 2$$
$$= x - 2 + 2 = x$$

$$\left(f^{-1} \circ f\right)(x) = f^{-1}(f(x))$$
$$= f^{-1}(3x + 2)$$
$$= \frac{(3x+2) - 2}{3}$$
$$= \frac{3x}{3} = x$$

29.
$$y = f(x) = \frac{1}{x+3}$$
$$x = \frac{1}{y+3}$$
$$x(y+3) = 1$$
$$y + 3 = \frac{1}{x}$$
$$f^{-1}(x) = \frac{1}{x} - 3$$

$$\left(f \circ f^{-1}\right)(x) = f\left(f^{-1}(x)\right)$$
$$= f\left(\frac{1}{x} - 3\right)$$
$$= \frac{1}{\frac{1}{x} - 3 + 3}$$
$$= \frac{1}{\frac{1}{x}}$$
$$= x$$

$$\left(f^{-1} \circ f\right)(x) = f^{-1}(f(x))$$
$$= f^{-1}\left(\frac{1}{x+3}\right)$$
$$= \frac{1}{\frac{1}{x+3}} - 3$$
$$= x + 3 - 3$$
$$= x$$

31.
$$y = f(x) = \frac{1}{2x} .$$
$$x = \frac{1}{2y}$$
$$x(2y) = 1$$
$$2xy = 1$$
$$y = \frac{1}{2x}$$
$$f^{-1}(x) = \frac{1}{2x}$$

$$\left(f \circ f^{-1}\right)(x) = f\left(f^{-1}(x)\right)$$
$$= f\left(\frac{1}{2x}\right)$$
$$= \frac{1}{2\left(\frac{1}{2x}\right)}$$
$$= \frac{1}{\frac{1}{x}}$$
$$= x$$

$$\left(f^{-1} \circ f\right)(x) = f^{-1}(f(x))$$
$$= f^{-1}\left(\frac{1}{2x}\right)$$
$$= \frac{1}{2\left(\frac{1}{2x}\right)}$$
$$= \frac{1}{\frac{1}{x}}$$
$$= x$$

33.
$$y = f(x) = 5x$$
$$x = 5y$$
$$\frac{x}{5} = y$$
$$f^{-1}(x) = \frac{1}{5}x$$

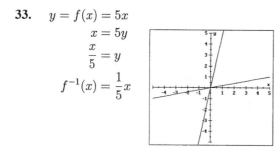

35.
$$y = f(x) = 2x - 4$$
$$x = 2y - 4$$
$$x + 4 = 2y$$
$$\frac{x+4}{2} = y$$
$$f^{-1}(x) = \frac{x+4}{2}$$

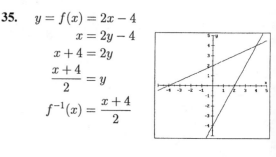

37.
$$x - y = 2$$
$$y - x = 2$$
$$y = x + 2$$
$$f^{-1}(x) = x + 2$$

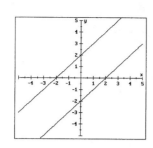

39.
$$2x + y = 4$$
$$2y + x = 4$$
$$2y = 4 - x$$
$$y = \frac{4 - x}{2}$$
$$f^{-1}(x) = \frac{4 - x}{2}$$

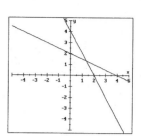

41.
$$y = \frac{1}{2x}$$
$$x = \frac{1}{2y}$$
$$2xy = 1$$
$$y = \frac{1}{2x}$$
$$f^{-1}(x) = \frac{1}{2x}$$

43.
$$y = \frac{x + 1}{x - 1}$$
$$x = \frac{y + 1}{y - 1}$$
$$x(y - 1) = y + 1$$
$$xy - x = y + 1$$
$$xy - y = x + 1$$
$$y(x - 1) = x + 1$$
$$f^{-1}(x) = \frac{x + 1}{x - 1}$$

45.
$$f(x) = x^2 - 3 \quad x \le 0$$
$$y = x^2 - 3 \quad x \le 0$$
$$x = y^2 - 3 \quad y \le 0$$
$$x + 3 = y^2 \quad y \le 0$$
$$\pm \sqrt{x + 3} = y \quad y \le 0$$
Thus, $f^{-1}(x) = -\sqrt{x + 3} \ (x \ge -3)$.

47.
$$f(x) = x^4 - 8 \quad x \ge 0$$
$$y = x^4 - 8 \quad x \ge 0$$
$$x = y^4 - 8 \quad y \ge 0$$
$$x + 8 = y^4 \quad y \ge 0$$
$$\pm \sqrt[4]{x + 8} = y \quad y \ge 0$$
Thus, $f^{-1}(x) = \sqrt[4]{x + 8} \ (x \ge -8)$.

49.
$$f(x) = \sqrt{4 - x^2} \quad 0 \le x \le 2$$
$$y = \sqrt{4 - x^2} \quad 0 \le x \le 2$$
$$x = \sqrt{4 - y^2} \quad 0 \le y \le 2$$
$$x^2 = 4 - y^2 \quad 0 \le y \le 2$$
$$y^2 = 4 - x^2 \quad 0 \le y \le 2$$
$$y = \pm \sqrt{4 - x^2} \quad 0 \le y \le 2$$
Thus, $f^{-1}(x) = \sqrt{4 - x^2} \ (0 \le x \le 2)$.

51.
$$f(x) = \frac{x}{x - 1}$$
Domain of $f = \boxed{(-\infty, 1) \cup (1, \infty)}$
$$f^{-1}(x) = \frac{x}{x - 1}$$
Range of f = Domain of f^{-1}
$$= \boxed{(-\infty, 1) \cup (1, \infty)}$$

53.
$$f(x) = \frac{1}{x} - 2$$
Domain of $f = \boxed{(-\infty, 0) \cup (0, \infty)}$

$$f^{-1}(x) = \frac{1}{x + 2}$$
Range of f = Domain of $f^{-1} = \boxed{(-\infty, -2) \cup (-2, \infty)}$

55.   **a.** $y = 0.75x + 8.50$

      b. $y = 0.75(4) + 8.50$
           $= \$11.50$

c. $y = 0.75x + 8.50$
          $x = 0.75y + 8.50$
     $x - 8.50 = 0.75y$
     $\dfrac{x - 8.50}{0.75} = y$

d. $y = \dfrac{x - 8.50}{0.75}$
      $= \dfrac{10 - 8.50}{0.75}$
      $= \dfrac{1.50}{0.75} = 2$

57.   **Answers may vary.**

59. $a \geq 0$

61. $16^{3/4} = \left(16^{1/4}\right)^3 = 8$

63. $(-8)^{2/3} = \left((-8)^{1/3}\right)^2 = 4$

65. $\left(\dfrac{64}{125}\right)^{-1/3} = \left(\dfrac{125}{64}\right)^{1/3} = \dfrac{5}{4}$

67. $49^{-1/2} = \dfrac{1}{49^{1/2}} = \dfrac{1}{7}$

Chapter 3 Summary   (page 315)

1.   **a.** $y = 3$
      Each value of x is paired
      with only one value of y.
      function

     b. $y + 5x^2 = 2$
          $y = -5x^2 + 2$
      Each value of x is paired
      with only one value of y.
      function

     c. $y^2 - x = 5$
          $y^2 = x + 5$
          $y = \pm\sqrt{x + 5}$
      Each value of x is paired
      with more than one value of y.
      not a function

     d. $y = |x| + x$
      Each value of x is paired
      with only one value of y.
      function

2.   **a.** $f(x) = y = 3x^2 - 5$: domain $= (-\infty, \infty)$
      $x = \pm\sqrt{\dfrac{y + 5}{3}}$: range $= [-5, \infty)$

     b. $f(x) = y = \dfrac{3x}{x - 5}$: domain $= (-\infty, 5) \cup (5, \infty)$
      $x = \dfrac{5y}{y - 3}$: range $= (-\infty, 3) \cup (3, \infty)$

     c. $f(x) = y = \sqrt{x - 1} \Rightarrow$ domain $= [1, \infty)$
      note: $\sqrt{x - 1} \geq 0$, so $y \geq 0 \Rightarrow$ range $= [0, \infty)$

     d. $f(x) = y = \sqrt{x^2 + 1} \Rightarrow x^2 + 1 \geq 0 \Rightarrow$ domain $= (-\infty, \infty)$
      note: $\sqrt{x^2 + 1} \geq 1$, so $y \geq 1 \Rightarrow$ range $= [1, \infty)$

3.   **a.** $f(x) = 5x - 2$
$f(2) = 5(2) - 2 = 8$
$f(-3) = 5(-3) - 2 = -17$
$f(0) = 5(0) - 2 = -2$

b. $f(x) = \dfrac{6}{x - 5}$

$f(2) = \dfrac{6}{2 - 5} = \dfrac{6}{-3} = -2$

$f(-3) = \dfrac{6}{-3 - 5} = \dfrac{6}{-8} = -\dfrac{3}{4}$

$f(0) = \dfrac{6}{0 - 5} = \dfrac{6}{-5} = -\dfrac{6}{5}$

c. $f(x) = |x - 2|$
$f(2) = |2 - 2| = |0| = 0$
$f(-3) = |-3 - 2| = |-5| = 5$
$f(0) = |0 - 2| = |-2| = 2$

d. $f(x) = \dfrac{x^2 - 3}{x^2 + 3}$

$f(2) = \dfrac{2^2 - 3}{2^2 + 3} = \dfrac{1}{7}$

$f(-3) = \dfrac{(-3)^2 - 3}{(-3)^2 + 3} = \dfrac{6}{12} = \dfrac{1}{2}$

$f(0) = \dfrac{0^2 - 3}{0^2 + 3} = \dfrac{-3}{3} = -1$

4.   **a.** $y = x^2 - x;\ a = 1, b = -1, c = 0$
vertex: $x = -\dfrac{b}{2a} = -\dfrac{-1}{2(1)} = \dfrac{1}{2}$
$y = x^2 - x = \left(\dfrac{1}{2}\right)^2 - \dfrac{1}{2} = -\dfrac{1}{4}$

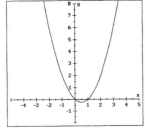

b. $y = x - x^2;\ a = -1, b = 1, c = 0$
vertex: $x = -\dfrac{b}{2a} = -\dfrac{1}{2(-1)} = \dfrac{1}{2}$
$y = x - x^2 = \dfrac{1}{2} - \left(\dfrac{1}{2}\right)^2 = \dfrac{1}{4}$

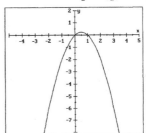

c. $y = x^2 - 3x - 4;\ a = 1, b = -3, c = -4$
vertex: $x = -\dfrac{b}{2a} = -\dfrac{-3}{2(1)} = \dfrac{3}{2}$
$y = x^2 - 3x - 4 = \left(\dfrac{3}{2}\right)^2 - 3\left(\dfrac{3}{2}\right) - 4$
$\qquad = -\dfrac{25}{4}$

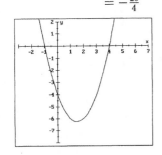

d. $y = 3x^2 - 8x - 3$
$a = 3, b = -8, c = -3$
vertex: $x = -\dfrac{b}{2a} = -\dfrac{-8}{2(3)} = \dfrac{4}{3}$
$y = 3x^2 - 8x - 3 = 3\left(\dfrac{4}{3}\right)^2 - 8\left(\dfrac{4}{3}\right) - 3$
$\qquad = -\dfrac{25}{3}$

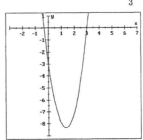

5.　**a.**　$3x^2 + y - 300 = 0$

$\qquad y = -3x^2 + 300$

$\qquad a = -3, b = 0, c = 300$

$\qquad x = -\frac{b}{2a} = -\frac{0}{2(-3)} = 0$

$\qquad y = -3(0)^2 + 300 = 300$

The maximum height is 300 units.

6.　Let the numbers be x and $1 - x$.

Product $= x(1 - x)$

$\qquad y = x - x^2 \colon a = -1, b = 1, c = 0$

vertex: $x = -\frac{b}{2a} = -\frac{1}{2(-1)} = \frac{1}{2}$

Both numbers are $\frac{1}{2}$.

7.　**a.**　$y = f(x) = x^3 - x$

$\qquad f(-x) = (-x)^3 - (-x)$

$\qquad\qquad = -x^3 + x$

$\qquad\qquad = -f(x) \Rightarrow$ odd

x-int.	y-int.
$x^3 - x = 0$	$y = 0^3 - 0$
$x(x^2 - 1) = 0$	$y = 0$
$x(x + 1)(x - 1) = 0$	$(0, 0)$
$x = 0, x = -1, x = 1$	
$(0, 0), (-1, 0), (1, 0)$	

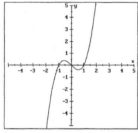

b.　$f(x) = x^2 - 4x$

$\qquad f(-x) = (-x)^2 - 4(-x)$

$\qquad\qquad = x^2 + 4x$

$\qquad \Rightarrow$ neither even nor odd

x-int.	y-int.
$x^2 - 4x = 0$	$y = 0^2 - 4(0)$
$x(x - 4) = 0$	$y = 0$
$x = 0, x = 4$	$(0, 0)$
$(0, 0), (4, 0)$	

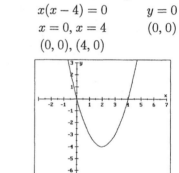

c.　$f(x) = x^3 - x^2$

$\qquad f(-x) = (-x)^3 - (-x)^2$

$\qquad\qquad = -x^3 - x^2$

$\qquad \Rightarrow$ neither even nor odd

x-int.	y-int.
$x^3 - x^2 = 0$	$y = 0^3 - 0^2$
$x^2(x - 1) = 0$	$y = 0$
$x = 0, x = 1$	$(0, 0)$
$(0, 0), (1, 0)$	

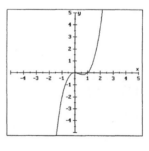

d. $y = f(x) = 1 - x^4$

$f(-x) = 1 - (-x)^4$

$\quad\quad = 1 - x^4 = f(x) \Rightarrow$ even

x-int.	y-int.
$1 - x^4 = 0$	$y = 1 - 0^4$
$(1 + x^2)(1 - x^2) = 0$	$y = 1$
$x^2 = -1, \; x^2 = 1$	$(0, 1)$
$x = -1, \; x = 1$	
$(-1, 0), (1, 0)$	

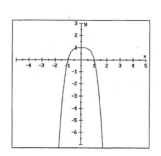

8. **a.** inc: $(-\infty, 0)$; dec: $(0, \infty)$

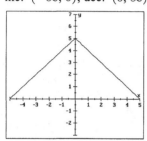

b. inc: $(-\infty, 0)$; const: $(0, \infty)$

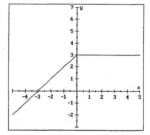

9. **a.** $y = \sqrt{x + 2} + 3$

Shift $y = \sqrt{x}$ U 3, L 2

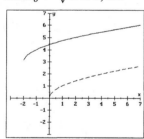

b. $y = |x - 4| + 2$

Shift $y = |x|$ U 2, R 4

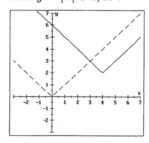

10. **a.** $y = \dfrac{1}{3}x^3$: Stretch

$y = x^3$ vert. by a factor of $\frac{1}{3}$

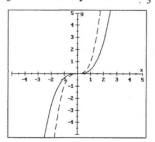

b. $y = (-5x)^3$: Stretch $y = x^3$ hor.

by a factor of $\frac{1}{5}$. Reflect about y.

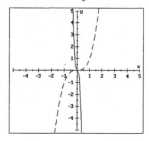

11. **a.** $y = \dfrac{x}{(x-1)^2}$

Vert: $x = 1$; Horiz: $y = 0$

Slant: none; x-intercepts: $(0, 0)$

y-intercepts: $(0, 0)$; Symmetry: none

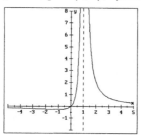

b. $y = \dfrac{(x-1)^2}{x}$: Vert: $x = 0$

Horiz: none; Slant: $y = x - 2$

x-intercepts: $(1, 0)$

y-intercepts: none; Symmetry: none

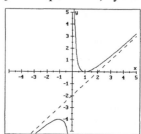

c. $y = \dfrac{x^2 - x - 2}{x^2 + x - 2} = \dfrac{(x+1)(x-2)}{(x+2)(x-1)}$

Vert: $x = -2$, $x = 1$; Horiz: $y = 1$

Slant: none; x-intercepts: $(-1, 0)$, $(2, 0)$

y-intercepts: $(0, 1)$; Symmetry: none

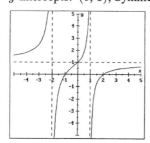

d. $y = \dfrac{x^3 + x}{x^2 - 4} = \dfrac{x(x^2 + 1)}{(x+2)(x-2)}$

Vert: $x = -2$, $x = 2$;

Slant: $y = x$;

x-int: $(0, 0)$ y-int: $(0, 0)$;

Symmetry: none

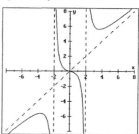

12. **a.** $(f + g)(x) = f(x) + g(x) = (x^2 - 1) + (2x + 1) = x^2 + 2x$

b. $(f \cdot g)(x) = f(x)g(x) = (x^2 - 1)(2x + 1) = 2x^3 + x^2 - 2x - 1$

c. $(f - g)(x) = f(x) - g(x) = (x^2 - 1) - (2x + 1) = x^2 - 2x - 2$

d. $(f/g)(x) = \dfrac{f(x)}{g(x)} = \dfrac{x^2 - 1}{2x + 1}$

e. $(f \circ g)(x) = f(g(x)) = f(2x + 1) = (2x + 1)^2 - 1 = 4x^2 + 4x + 1 - 1 = 4x^2 + 4x$

f. $(g \circ f)(x) = g(f(x)) = g(x^2 - 1) = 2(x^2 - 1) + 1 = 2x^2 - 2 + 1 = 2x^2 - 1$

13. a.
$$y = f(x) = 7x - 1$$
$$x = 7y - 1$$
$$x + 1 = 7y$$
$$\frac{x+1}{7} = y$$
$$f^{-1}(x) = \frac{x+1}{7}$$

b.
$$y = f(x) = \frac{1}{2-x}$$
$$x = \frac{1}{2-y}$$
$$x(2-y) = 1$$
$$2 - y = \frac{1}{x}$$
$$2 - \frac{1}{x} = y$$
$$f^{-1}(x) = 2 - \frac{1}{x}$$

c.
$$y = f(x) = \frac{x}{1-x}$$
$$x = \frac{y}{1-y}$$
$$x(1-y) = y$$
$$x - xy = y$$
$$x = xy + y$$
$$x = y(x+1)$$
$$\frac{x}{x+1} = y$$
$$f^{-1}(x) = \frac{x}{x+1}$$

d.
$$y = f(x) = \frac{3}{x^3}$$
$$x = \frac{3}{y^3}$$
$$xy^3 = 3$$
$$y^3 = \frac{3}{x}$$
$$y = \sqrt[3]{\frac{3}{x}}$$
$$f^{-1}(x) = \sqrt[3]{\frac{3}{x}}$$

14.
$$y = \frac{2x+3}{5x-10}$$
$$x = \frac{2y+3}{5y-10}$$
$$x(5y - 10) = 2y + 3$$
$$5xy - 10x = 2y + 3$$
$$5xy - 2y = 10x + 3$$
$$y(5x - 2) = 10x + 3$$
$$y = \frac{10x+3}{5x-2}$$

Range of f = Domain of f^{-1}
$$= \boxed{\left(-\infty, \tfrac{2}{5}\right) \cup \left(\tfrac{2}{5}, \infty\right)}$$

Chapter 3 Test (page 319)

1. $f(x) = \dfrac{3}{x-5}$: domain $= (-\infty, 5) \cup (5, \infty)$

$x = \dfrac{3}{x} + 5$: range $= (-\infty, 0) \cup (0, \infty)$

2. $f(x) = \sqrt{x+3}$: domain $= [-3, \infty)$
Since $\sqrt{x+3} \geq 0$, range $= [0, \infty)$.

3. $f(-1) = \dfrac{-1}{-1-1} = \dfrac{-1}{-2} = \dfrac{1}{2}$

$f(2) = \dfrac{2}{2-1} = \dfrac{2}{1} = 2$

4. $f(-1) = \sqrt{-1+7} = \sqrt{6}$
$f(2) = \sqrt{2+7} = \sqrt{9} = 3$

5. $y = 3(x - 7)^2 - 3$
Vertex: $(7, -3)$

6. $y = x^2 - 2x - 3$; $a = 1, b = -2, c = -3$
vertex: $x = -\frac{b}{2a} = -\frac{-2}{2(1)} = 1$
$y = x^2 - 2x - 3 = 1^2 - 2(1) - 3 = -4$

7. $y = 3x^2 - 24x + 38$
$a = 3, b = -24, c = 38$
vertex: $x = -\frac{b}{2a} = -\frac{-24}{2(3)} = 4$
$y = 3x^2 - 24x + 38 = 3(4)^2 - 24(4) + 38$
$= -10$

8. $y = 5 - 4x - x^2$; $a = -1, b = -4, c = 5$
vertex: $x = -\frac{b}{2a} = -\frac{-4}{2(-1)} = -2$
$y = 5 - 4x - x^2 = 5 - 4(-2) - (-2)^2$
$= 9$

9. $y = f(x) = x^4 - x^2$
$f(-x) = (-x)^4 - (-x)^2$
$\quad = x^4 - x^2 = f(x) \Rightarrow$ even

x-int.	y-int.
$x^4 - x^2 = 0$	$y = 0^4 - 0^2$
$x^2(x^2 - 1) = 0$	$y = 0$
$x^2 = 0, x^2 = 1$	$(0, 0)$

$x = 0, x = -1, x = 1$
$(0, 0), (-1, 0), (1, 0)$

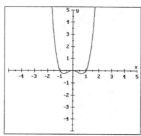

10. $y = f(x) = x^5 - x^3$
$f(-x) = (-x)^5 - (-x)^3$
$\quad = -x^5 - (-x)^3$
$\quad = -x^5 + x^3 = -f(x) \Rightarrow$ odd

x-int.	y-int.
$x^5 - x^3 = 0$	$y = 0^5 - 0^3$
$x^3(x^2 - 1) = 0$	$y = 0$
$x^3 = 0, x^2 = 1$	$(0, 0)$

$x = 0, x = -1, x = 1$
$(0, 0), (-1, 0), (1, 0)$

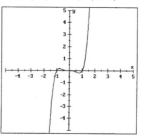

11. $h = 100t - 16t^2$: $a = -16, b = 100, c = 0$
$x = -\frac{b}{2a} = -\frac{100}{2(-16)} = \frac{25}{8}$ seconds

12. $h = 100t - 16t^2$: $a = -16, b = 100, c = 0$
From **#11**, $x = \frac{25}{8}$.
$y = 100\left(\frac{25}{8}\right) - 16\left(\frac{25}{8}\right)^2 = \frac{625}{4}$ feet

13. The roadway is at $y = 0$, so the distance to the lowest point will be the y-coord. of the vertex:
$$x^2 - 2500y + 25000 = 0$$
$$x^2 + 25000 = 2500y$$
$$\frac{1}{2500}x^2 + 10 = y$$
$$y = c - \frac{b^2}{4a} = 10 - \frac{0^2}{4\left(\frac{1}{2500}\right)} = 10$$
The lowest point is 10 ft above.

14. Let $x = \pm 500$:
$$x^2 - 2500y + 25000 = 0$$
$$(\pm 500)^2 - 2500y + 25000 = 0$$
$$250,000 - 2500y + 25,000 = 0$$
$$275,000 = 2500y$$
$$110 = y$$
The cable attaches to the vertical pillars 110 feet above the roadway.

15. $y = (x - 3)^2 + 1$
Shift $y = x^2$ U 1, R 3

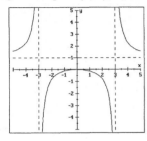

16. $y = \sqrt{x - 1} + 5$
Shift $y = \sqrt{x}$ U 5, R 1

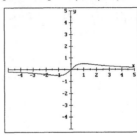

17. $y = \dfrac{x - 1}{x^2 - 9} = \dfrac{x - 1}{(x + 3)(x - 3)}$
Vert: $x = -3$, $x = 3$; Horiz: $y = 0$

18. $y = \dfrac{x^2 - 5x - 14}{x - 3} = \dfrac{(x - 7)(x + 2)}{(x - 3)}$
$$= x - 2 + \tfrac{-20}{x-3}$$
Vert: $x = 3$; Slant: $y = x - 2$

19. $y = \dfrac{x^2}{x^2 - 9} = \dfrac{x^2}{(x + 3)(x - 3)}$
Vert: $x = -3$, $x = 3$; Horiz: $y = 1$
Slant: none; x-intercepts: $(0, 0)$
y-intercepts: $(0, 0)$; Symmetry: y-axis

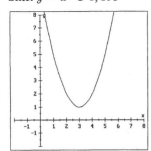

20. $y = \dfrac{x}{x^2 + 1}$
Vert: none; Horiz: $y = 0$
Slant: none; x-intercepts: $(0, 0)$
y-intercepts: $(0, 0)$; Symmetry: origin

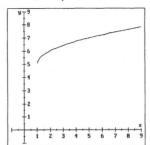

21. $y = \dfrac{2x^2 - 3x - 2}{x - 2} = \dfrac{(2x+1)(x-2)}{x-2}$
$$= 2x + 1 \text{ (if } x \neq 2)$$

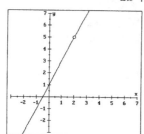

22. $y = \dfrac{x}{x^2 - x} = \dfrac{x}{x(x-1)} = \dfrac{1}{x-1}$ (if $x \neq 0$

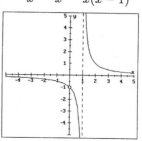

23. $(f+g)(x) = f(x) + g(x) = (3x) + (x^2 + 2) = x^2 + 3x + 2$

24. $(g \circ f)(x) = g(f(x)) = g(3x) = (3x)^2 + 2 = 9x^2 + 2$

25. $(f/g)(x) = \dfrac{f(x)}{g(x)} = \dfrac{3x}{x^2 + 2}$

26. $(f \circ g)(x) = f(g(x)) = f(x^2 + 2) = 3(x^2 + 2) = 3x^2 + 6$

27.
$$y = \frac{x+1}{x-1}$$
$$x = \frac{y+1}{y-1}$$
$$x(y-1) = y + 1$$
$$xy - x = y + 1$$
$$xy - y = x + 1$$
$$y(x-1) = x + 1$$
$$f^{-1}(x) = \frac{x+1}{x-1}$$

28.
$$y = x^3 - 3$$
$$x = y^3 - 3$$
$$x + 3 = y^3$$
$$\sqrt[3]{x+3} = y$$
$$f^{-1}(x) = \sqrt[3]{x+3}$$

29. $f(x) = \dfrac{3}{x} - 2;\ f^{-1}(x) = \dfrac{3}{x+2}$
Range of f = Domain of f^{-1}
$$= \boxed{(-\infty, -2) \cup (-2, \infty)}$$

30. $f(x) = \dfrac{3x - 1}{x - 3};\ f^{-1}(x) = \dfrac{3x - 1}{x - 3}$
Range of f = Domain of f^{-1}
$$= \boxed{(-\infty, 3) \cup (3, \infty)}$$

Cumulative Review Exercises (page 320)

1.

$$5x - 3y = 15 \qquad 5x - 3y = 15$$
$$5x - 3(0) = 15 \qquad 5(0) - 3y = 15$$
$$5x = 15 \qquad -3y = 15$$
$$x = 3 \qquad y = -5$$
$$(3, 0) \qquad (0, -5)$$

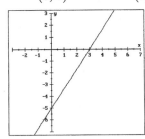

2.

$$3x + 2y = 12 \qquad 3x + 2y = 12$$
$$3x + 2(0) = 12 \qquad 3(0) + 2y = 12$$
$$3x = 12 \qquad 2y = 12$$
$$x = 4 \qquad y = 6$$
$$(4, 0) \qquad (0, 6)$$

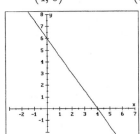

3. **a.** $d = \sqrt{(x_2 - x_1)^2 + (y_2 - y_1)^2} = \sqrt{(-2 - 3)^2 + \left(\frac{7}{2} - \left(-\frac{1}{2}\right)\right)^2} = \sqrt{25 + 16} = \sqrt{41}$

b. $x = \dfrac{x_1 + x_2}{2} = \dfrac{-2 + 3}{2} = \dfrac{1}{2}; \; y = \dfrac{\frac{7}{2} + \left(-\frac{1}{2}\right)}{2} = \dfrac{\frac{6}{2}}{2} = \dfrac{3}{2} \quad \left(\dfrac{1}{2}, \dfrac{3}{2}\right)$

c. $m = \dfrac{-\frac{1}{2} - \frac{7}{2}}{3 - (-2)} = \dfrac{-\frac{8}{2}}{5} = -\dfrac{4}{5}$

4. **a.** $d = \sqrt{(x_2 - x_1)^2 + (y_2 - y_1)^2} = \sqrt{(3 - (-7))^2 + (7 - 3)^2} = \sqrt{100 + 16} = \sqrt{116} = 2\sqrt{29}$

b. $x = \dfrac{x_1 + x_2}{2} = \dfrac{3 + (-7)}{2} = \dfrac{-4}{2} = -2; \; y = \dfrac{7 + 3}{2} = \dfrac{10}{2} = 5 \quad (-2, 5)$

c. $m = \dfrac{3 - 7}{-7 - 3} = \dfrac{-4}{-10} = \dfrac{2}{5}$

5. $m = \dfrac{y_2 - y_1}{x_2 - x_1} = \dfrac{-7 - 5}{3 - (-3)} = \dfrac{-12}{6} = -2$

$y - y_1 = m(x - x_1)$
$y - 5 = -2(x + 3)$
$y = -2x - 1$

6. $y - y_1 = m(x - x_1)$

$y - \dfrac{5}{2} = \dfrac{7}{2}\left(x - \dfrac{3}{2}\right)$

$y = \dfrac{7}{2}x - \dfrac{21}{4} + \dfrac{5}{2}$

$y = \dfrac{7}{2}x - \dfrac{11}{4}$

7. $3x - 5y = 7$ $\qquad$ $y - y_1 = m(x - x_1)$

$\qquad -5y = -3x + 7$ $\qquad$ $y - 3 = \frac{3}{5}(x + 5)$

$\qquad y = \frac{3}{5}x - \frac{7}{5}$ $\qquad$ $y - 3 = \frac{3}{5}x + 3$

$m = \frac{3}{5}$ $\qquad\qquad$ $\boxed{y = \frac{3}{5}x + 6}$

Use $m = \frac{3}{5}$.

8. $x - 4y = 12$ $\qquad$ $y - y_1 = m(x - x_1)$

$\qquad -4y = -x + 12$ $\qquad$ $y - 0 = -4(x - 0)$

$\qquad y = \frac{1}{4}x - 3$ $\qquad$ $\boxed{y = -4x}$

$m = \frac{1}{4}$

Use $m = -4$.

9. $x^2 = y - 2$

symmetry: y-axis

x-int: none, y-int: $(2, 0)$

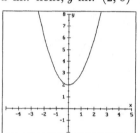

10. $y^2 = x - 2$

symmetry: x-axis

x-int: $(2, 0)$, y-int: none

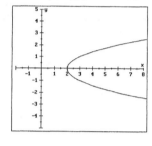

11. $x^2 + y^2 = 100$

Circle: $C(0, 0)$; $r = 10$

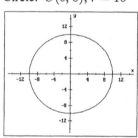

12. $x^2 - 2x + y^2 = 8$

$(x - 1)^2 + y^2 = 9$

Circle: $C(1, 0)$; $r = 3$

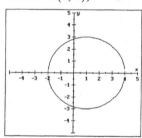

13. $\dfrac{x - 2}{x} = \dfrac{x - 6}{5}$

$5(x - 2) = x(x - 6)$

$5x - 10 = x^2 - 6x$

$0 = x^2 - 11x + 10$

$0 = (x - 10)(x - 1)$

$x = 10$ or $x = 1$

14. $\dfrac{x + 2}{x - 6} = \dfrac{3x + 1}{2x - 11}$

$(x + 2)(2x - 11) = (3x + 1)(x - 6)$

$2x^2 - 7x - 22 = 3x^2 - 17x - 6$

$0 = x^2 - 10x + 16$

$0 = (x - 8)(x - 2)$

$x = 8$ or $x = 2$

15. $m = \dfrac{54 - 37}{4 - 2} = \dfrac{17}{2} = 8.5$

$y = mx + b \qquad y = 8.5x + 20$

$y = 8.5x + b \qquad y = 8.5(5) + 20$

$37 = 8.5(2) + b \qquad y = 62.50$

$37 = 17 + b \qquad$ It will cost \$62.50.

$20 = b$

16.
$$E = ks^2$$

$E = k(50)^2 \qquad E = k(20)^2$

$E = 2500k \qquad E = 400k$

$\dfrac{50 \text{ mph } E}{20 \text{ mph } E} = \dfrac{2500k}{400k} = \dfrac{25}{4}$

17. $y = 3x - 1$: Each value of x is paired with only one value of y. $\Rightarrow$ **function**

18. $y = x^2 + 3$: Each value of x is paired with only one value of y. $\Rightarrow$ **function**

19. $y = \dfrac{1}{x - 2}$: Each value of x is paired with only one value of y. $\Rightarrow$ **function**

20. $y^2 = 4x$

$y = \pm\sqrt{4x}$: Each value of x is paired with more than one value of y. $\Rightarrow$ **not a function**

21. $y = f(x) = x^2 + 5$: domain $= (-\infty, \infty)$; $x = \pm\sqrt{y - 5}$: range $= [5, \infty)$

22. $y = f(x) = \dfrac{7}{x + 2}$: domain $= (-\infty, -2) \cup (-2, \infty)$; $x = \dfrac{7}{y} - 2$: range $= (-\infty, 0) \cup (0, \infty)$

23. $y = f(x) = -\sqrt{x - 2} \Rightarrow$ domain $= [2, \infty)$; $\sqrt{x - 2} \geq 0$, so $-\sqrt{x - 2} \leq 0 \Rightarrow$ range $= (-\infty, 0]$

24. $y = f(x) = \sqrt{x + 4} \Rightarrow$ domain $= [-4, \infty)$; $\sqrt{x + 4} \geq 0$, so $y \geq 0 \Rightarrow$ range $= [0, \infty)$

25. $y = x^2 + 5x - 6$; $a = 1, b = 5, c = -6$

vertex: $x = -\dfrac{b}{2a} = -\dfrac{5}{2(1)} = -\dfrac{5}{2}$

$y = x^2 + 5x - 6 = \left(-\dfrac{5}{2}\right)^2 + 5\left(-\dfrac{5}{2}\right) - 6$

$= -\dfrac{49}{4}$

26. $y = -x^2 + 5x + 6$; $a = -1, b = 5, c = 6$

vertex: $x = -\dfrac{b}{2a} = -\dfrac{5}{2(-1)} = \dfrac{5}{2}$

$y = -x^2 + 5x + 6 = -\left(\dfrac{5}{2}\right)^2 + 5\left(\dfrac{5}{2}\right) + 6$

$= \dfrac{49}{4}$

27. $y = x^2 - 4$: Shift $y = x^2$ D 4.

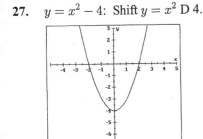

28. $y = -x^2 + 4$: Reflect $y = x^2$ about x and shift U 4.

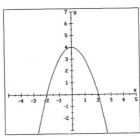

29. $y = f(x) = x^3 + x$

$f(-x) = (-x)^3 + (-x)$

$\qquad = -x^3 - x$

$\qquad = -f(x) \Rightarrow$ odd

x-int.	y-int.
$x^3 + x = 0$	$y = 0^3 + 0$
$x(x^2 + 1) = 0$	$y = 0$
$x = 0$	$(0, 0)$
$(0, 0)$	

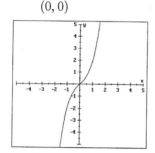

30. $f(x) = -x^4 + 2x^2 + 1$

$f(-x) = -(-x)^4 + 2(-x)^2 + 1$

$\qquad = -x^4 + 2x^2 + 1 \Rightarrow$ even

x-int.	y-int.
$-x^4 + 2x^2 + 1 = 0$	$f(0) = 1$
$x^4 - 2x^2 - 1 = 0$	$y = 1$
not rational numbers	$(0, 1)$

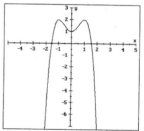

31. $y = \dfrac{x}{x - 3} = 1 + \dfrac{3}{x - 3}$

Vertical: $x = 3$; Horizontal: $y = 1$

Slant: none; x-intercepts: $(0, 0)$

y-intercepts: $(0, 0)$; Symmetry: none

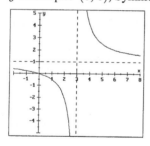

32. $y = \dfrac{x^2 - 1}{x^2 - 9} = \dfrac{(x + 1)(x - 1)}{(x + 3)(x - 3)} = 1 + \frac{8}{x^2 - 9}$

Vertical: $x = -3$, $x = 3$; Horizontal: $y = 1$

Slant: none; x-intercepts: $(-1, 0)$, $(1, 0)$

y-intercepts: $\left(0, \frac{1}{9}\right)$; Symmetry: y-axis

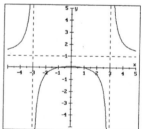

33. $(f + g)(x) = f(x) + g(x) = (3x - 4) + (x^2 + 1) = x^2 + 3x - 3$; domain $= (-\infty, \infty)$

34. $(f - g)(x) = f(x) - g(x) = (3x - 4) - (x^2 + 1) = -x^2 + 3x - 5$; domain $= (-\infty, \infty)$

35. $(f \cdot g)(x) = f(x)g(x) = (3x - 4)(x^2 + 1) = 3x^3 - 4x^2 + 3x - 4$; domain $= (-\infty, \infty)$

36. $(f/g)(x) = \frac{f(x)}{g(x)} = \frac{3x-4}{x^2+1}$; domain $= (-\infty, \infty)$

37. $(f \circ g)(2) = f(g(2)) = f((2^2 + 1)) = f(5) = 3(5) - 4 = 11$

38. $(g \circ f)(2) = g(f(2)) = g(3(2) - 4) = g(2) = 2^2 + 1 = 5$

39. $(f \circ g)(x) = f(g(x)) = f(x^2 + 1) = 3(x^2 + 1) - 4 = 3x^2 - 1$

40. $(g \circ f)(x) = g(f(x)) = g(3x - 4) = (3x - 4)^2 + 1 = 9x^2 - 24x + 16 + 1 = 9x^2 - 24x + 17$

41.
$$y = 3x + 2$$
$$x = 3y + 2$$
$$x - 2 = 3y$$
$$\frac{x-2}{3} = y$$
$$f^{-1}(x) = \frac{x-2}{3}$$

42.
$$y = \frac{1}{x-3}$$
$$x = \frac{1}{y-3}$$
$$x(y - 3) = 1$$
$$y - 3 = \frac{1}{x}$$
$$y = \frac{1}{x} + 3$$
$$f^{-1}(x) = \frac{1}{x} + 3$$

43.
$$y = x^2 + 5$$
$$x = y^2 + 5$$
$$x - 5 = y^2$$
$$\pm\sqrt{x - 5} = y$$
$$\sqrt{x - 5} = y \ (y \geq 0)$$
$$f^{-1}(x) = \sqrt{x - 5}$$

44.
$$3x - y = 1$$
$$3y - x = 1$$
$$3y = x + 1$$
$$y = \frac{x+1}{3}$$
$$f^{-1}(x) = \frac{x+1}{3}$$

45. $y = kwz$

46. $y = \dfrac{kx}{t^2}$

Exercise 4.1 (page 333)

1. exponential

3. $(-\infty, \infty)$

5. $(0, \infty)$

7. asymptote

9. increasing

11. 2.72

13. increasing

15. $4^{\sqrt{3}} \approx 11.0357$

17. $7^\pi \approx 451.8079$

19. $5^{\sqrt{2}}5^{\sqrt{2}} = 5^{\sqrt{2}+\sqrt{2}} = 5^{2\sqrt{2}} = \left(5^2\right)^{\sqrt{2}}$
$$= 25^{\sqrt{2}}$$

21. $\left(a^{\sqrt{8}}\right)^{\sqrt{2}} = a^{\sqrt{8}\cdot\sqrt{2}} = a^{\sqrt{16}} = a^4=$

23.
$$y = 3^x$$
points: $(0, 1), (1, 3)$

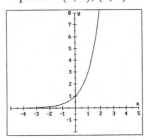

25.
$$y = \left(\frac{1}{5}\right)^x$$
points: $(0, 1), \left(1, \frac{1}{5}\right)$

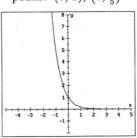

27.
$$y = \left(\frac{3}{4}\right)^x$$
points: $(0, 1)$, $\left(1, \frac{3}{4}\right)$

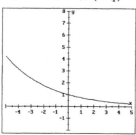

29.
$$y = (1.5)^x$$
points: $(0, 1)$, $(1, 1.5)$

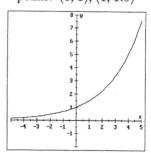

31. The graph passes through $(0, 1)$ and has the x-axis as an asymptote. YES

33. The graph does not pass through $(0, 1)$. NO

35. The graph passes through $(0, 1)$ and has the x-axis as an asymptote, so it could be an exponential function. It passes through the point $\left(1, \frac{1}{2}\right) = (1, b)$. $b = \frac{1}{2}$

37. The graph does not pass through $(0, 1)$. It is not an exponential function.

39. The graph passes through $(0, 1)$ and has the x-axis as an asymptote, so it could be an exponential function. It passes through the point $(1, 2) = (1, b)$. $b = 2$

41. The graph passes through $(0, 1)$ and has the x-axis as an asymptote, so it could be an exponential function.
$$y = b^x$$
$$e^2 = b^2$$
$$e = b$$

43.
$$y = 3^x - 1$$
Shift $y = 3^x$ D1.

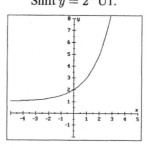

45.
$$y = 2^x + 1$$
Shift $y = 2^x$ U1.

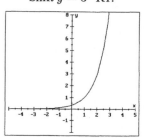

47.
$$y = 3^{x-1}$$
Shift $y = 3^x$ R1.

49.
$$y = 3^{x+1}$$
Shift $y = 3^x$ L1.

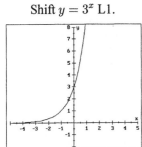

51.
$$y = e^x - 4$$
Shift $y = e^x$ D4.

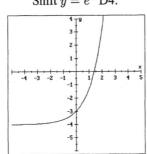

53.
$$y = e^{x-2}$$
Shift $y = e^x$ R2.

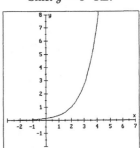

55.
$$y = 2^{x+1} - 2$$
Shift $y = 2^x$ L1, D2.

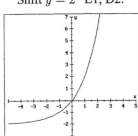

57.
$$y = 3^{x-2} + 1$$
Shift $y = 3^x$ R2, U1.

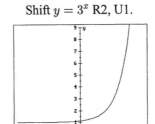

59.
$$y = 5(2^x)$$

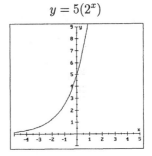

61.
$$y = 3^{-x}$$

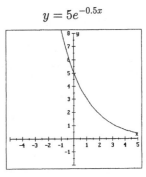

63.
$$y = 2e^x$$

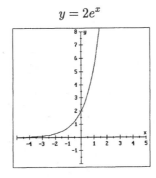

65.
$$y = 5e^{-0.5x}$$

67. $A = P\left(1 + \dfrac{r}{k}\right)^{kt} = 500\left(1 + \dfrac{0.08}{4}\right)^{4(10)} \approx \$1104.02$

69. 5% interest:

$A = P\left(1 + \dfrac{r}{k}\right)^{kt}$

$= 500\left(1 + \dfrac{0.05}{2}\right)^{2(5)}$

$\approx \$640.04$

$5\frac{1}{2}\%$ interest:

$A = P\left(1 + \dfrac{r}{k}\right)^{kt}$

$= 500\left(1 + \dfrac{0.055}{2}\right)^{2(5)}$

$\approx \$655.83$

Difference $= 655.83 - 640.04$

$= \$15.79$ more

171

71. $A = A_0\left(1 + \dfrac{r}{360}\right)^{365t}$

$= 1000\left(1 + \dfrac{0.07}{360}\right)^{365(5)}$

$\approx \$1425.93$

73. $A = P\left(1 + \dfrac{r}{k}\right)^{kt}$

$= 1500\left(1 + \dfrac{0.21}{12}\right)^{12(1)}$

$\approx \$1847.16$

75. $A = Pe^{rt} = 2000e^{0.08(15)} \approx \$6640.23$

77. Continuous:

$A = Pe^{rt} = 30,000e^{0.08(20)}$

$\approx \$148,590.97$

Annually:

$A = P\left(1 + \dfrac{r}{k}\right)^{kt} = 30,000\left(1 + \dfrac{0.08}{1}\right)^{1(20)} \approx \$139,828.71$

79. $A = P\left(1 + \dfrac{r}{k}\right)^{kt}$

$P = A(1 + r)^n$

$\dfrac{P}{(1 + r)^n} = A$

$P(1 + r)^{-n} = A$

81. $5^{3t} = k^t$

$\left(5^3\right)^t = k^t$

$125^t = k^t$

$125 = k$

83. $x^2 + 9x^4 = x^2(1 + 9x^2)$

85. $x^2 + x - 12 = (x + 4)(x - 3)$

Exercise 4.2 (page 340)

1. birth, death

3. $A = A_0 2^{-t/h} = 50 \cdot 2^{-100/(12.4)}$

≈ 0.1868 grams

5. $A = A_0 2^{-t/h} = 1000 \cdot 2^{-200/(30.17)}$

≈ 10 kg

7. $A = A_0 2^{-t/h} = A_0 \cdot 2^{-60/40}$

$\approx A_0(0.354)$

About 35.4% will remain.

9. $A = A_0 2^{-t/h}$

$= 1 \cdot 2^{-12/(4.5)}$

≈ 0.1575 unit

11. $P = P_0 2^{t/2}$

$= 10,000 \cdot 2^{5/2}$

$\approx 56,570$ fish

13. $I = I_0 k^x$

$= 8(0.5)^2$

$= 2$ lumens

15. $T = 40 + 60(0.75)^t$

$= 40 + 60(0.75)^{3.5}$

$\approx 61.9°$ C

17. $P = 173e^{0.03t}$

$= 173e^{0.03(20)}$

≈ 315

19. $P = P_0 e^{0.27t}$

$= 2e^{0.27(12)}$

≈ 51 cases

21. $P = P_0 e^{kt}$

$= 6e^{0.019(30)}$

≈ 10.6 billion

23. $P = P_0 e^{kt}$

$= 6e^{0.019(50)}$

≈ 15.5 billion

$\dfrac{15.5}{6} \approx$ a factor of 2.6

25. $P = e^{-0.3t}$

$= e^{-0.3(24)}$

≈ 0.0007

$= 0.07\%$

27. Let $t = 0$:
$$x = 0.08(1 - e^{-0.1t})$$
$$= 0.08(1 - e^{-0.1(0)})$$
$$= 0.08(1 - e^0) = 0.08(1 - 1) = 0$$

29.
$$P = \frac{1,200,000}{1 + (1200 - 1)e^{-0.4t}}$$
$$= \frac{1,200,000}{1 + (1200 - 1)e^{-0.4(8)}}$$
$$\approx 24,060 \text{ people}$$

31.
$$w = 1.54e^{0.503n}$$
$$= 1.54e^{0.503(5)}$$
$$\approx 19.0 \text{ mm}$$

33.
$$v = 50(1 - e^{-0.2t})$$
$$= 50(1 - e^{-0.2(20)})$$
$$\approx 49 \text{ meters/second}$$

35.

Males	Females
$P = P_0 e^{kt}$	$P = P_0 e^{kt}$
$= 133 e^{0.01t}$	$= 139 e^{0.01t}$
$= 133 e^{0.01(20)}$	$= 139 e^{0.01(20)}$
≈ 162.4 million	≈ 169.8 million

There will be about 7 million more females.

37. Find where these graphs meet:
$$y = 1000 e^{0.02t}, \quad y = 31x + 2000$$

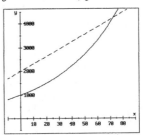

It will take about 72.2 years.

39.
$$1 + 1 + \frac{1}{2} + \frac{1}{2 \cdot 3} + \frac{1}{2 \cdot 3 \cdot 4}$$
$$+ \frac{1}{2 \cdot 3 \cdot 4 \cdot 5} \approx 2.71\overline{6}$$
$$e \approx 2.718: \text{ accurate to 2 places}$$

41.
$$P = \frac{1,200,000}{1 + 1199 e^{-0.4t}}$$

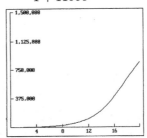

43. $2^3 = x$
$8 = x$

45. $x^3 = 27$
$x = 3$

47. $x^{-3} = \dfrac{1}{8}$
$\dfrac{1}{x^3} = \dfrac{1}{8}$
$x = 2$

49. $9^{1/2} = x$
$3 = x$

Exercise 4.3 (page 350)

1. $x = b^y$

3. range

5. inverse

7. exponent

9. $(b, 1), (1, 0)$

11. $\log_e x$

13. $(-\infty, \infty)$

15. 10

17. $\log_3 81 = 4$
$3^4 = 81$

19. $\log_{1/2} \dfrac{1}{8} = 3$
$\left(\dfrac{1}{2}\right)^3 = \dfrac{1}{8}$

21. $\log_4 \dfrac{1}{64} = -3$
$4^{-3} = \dfrac{1}{64}$

23. $\log_\pi \pi = 1$
$\pi^1 = \pi$

25. $8^2 = 64$
$\log_8 64 = 2$

27. $4^{-2} = \dfrac{1}{16}$
$\log_4 \dfrac{1}{16} = -2$

29. $\left(\dfrac{1}{2}\right)^{-5} = 32$
$\log_{1/2} 32 = -5$

31. $x^y = z$
$\log_x z = y$

33. $\log_2 8 = x$
$2^x = 8$
$x = 3$

35. $\log_4 64 = x$
$4^x = 64$
$x = 3$

37. $\log_{1/2} \dfrac{1}{8} = x$
$\left(\dfrac{1}{2}\right)^x = \dfrac{1}{8}$
$x = 3$

39. $\log_9 3 = x$
$9^x = 3$
$x = \dfrac{1}{2}$

41. $\log_{1/2} 8 = x$
$\left(\dfrac{1}{2}\right)^x = 8$
$x = -3$

43. $\log_8 x = 2$
$8^2 = x$
$64 = x$

45. $\log_7 x = 1$
$7^1 = x$
$7 = x$

47. $\log_{25} x = \dfrac{1}{2}$
$25^{1/2} = x$
$5 = x$

49. $\log_5 x = -2$
$5^{-2} = x$
$\dfrac{1}{5^2} = x$
$\dfrac{1}{25} = x$

51. $\log_{36} x = -\dfrac{1}{2}$
$36^{-1/2} = x$
$\dfrac{1}{36^{1/2}} = x$
$\dfrac{1}{6} = x$

53. $\log_x 5^3 = 3$
$x^3 = 5^3$
$x = 5$

55. $\log_x \dfrac{9}{4} = 2$
$x^2 = \dfrac{9}{4}$
$x = \dfrac{3}{2}$

57. $\log_x \dfrac{1}{64} = -3$
$x^{-3} = \dfrac{1}{64}$
$\dfrac{1}{x^3} = \dfrac{1}{4^3}$
$x = 4$

59. $\log_x \dfrac{9}{4} = -2$
$x^{-2} = \dfrac{9}{4}$
$\left(x^{-2}\right)^{-1} = \left(\dfrac{9}{4}\right)^{-1}$
$x^2 = \dfrac{4}{9}$
$x = \dfrac{2}{3}$

61. From the definition:
$2^{\log_2 5} = 5$

63. From the definition:
$x^{\log_4 6} = 6 \Rightarrow x = 4$

65. $\log 3.25 \approx 0.5119$

67. $\log 0.00467 \approx -2.3307$

69. $\ln 45.7 \approx 3.8221$

71. $\ln \frac{2}{3} \approx -0.4055$

73. $\ln 35.15 \approx 3.5596$

SECTION 4.3

75. $\ln 7.896 \approx 2.0664$

77. $\log (\ln 1.7) \approx -0.2752$

79. $\ln (\log 0.1)$: undefined

81. $\log y = 1.4023$
$\qquad y \approx 25.2522$

83. $\log y = -3.71$
$\qquad y \approx 1.9498 \times 10^{-4}$

85. $\ln y = 1.4023$
$\qquad y \approx 4.0645$

87. $\ln y = 4.24$
$\qquad y \approx 69.4079$

89. $\ln y = -3.71$
$\qquad y \approx 0.0245$

91. $\log y = \ln 8$
$\qquad y \approx 120.0719$

93. The graph passes through the point $(b, 1) = (2, 1)$.
$\Rightarrow b = 2$

95. $y = \log_b x$
$b^y = x$
$b^{-1} = \frac{1}{2}$
$\left(b^{-1}\right)^{-1} = \left(\frac{1}{2}\right)^{-1}$
$b = 2$

97. $y = \log_3 x$
points: $(1, 0), (3, 1)$

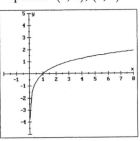

99. $y = \log_{1/3} x$
points: $(1, 0), \left(\frac{1}{3}, 1\right)$

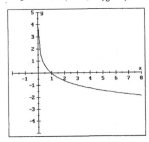

101. $y = 2 + \log_2 x$
Shift $y = \log_2 x$ U2.

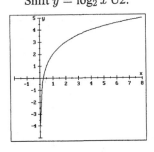

103. $y = \log_3 (x + 2)$
Shift $y = \log_3 x$ L2.

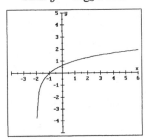

105. $y = \log (3x)$

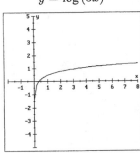

107. $y = \log (-x)$

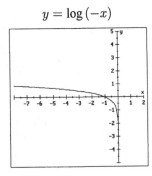

109. $y = \ln \left(\frac{1}{2} x\right)$

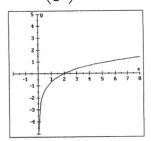

111. $y = \ln(-x)$

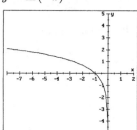

113. Answers may vary.

115. Answers may vary.

117. $y = x^2 + 7x + 3$; $a = 1$, $b = 7$, $c = 3$
$x = -\frac{b}{2a} = -\frac{7}{2(1)} = -\frac{7}{2}$
$y = x^2 + 7x + 3$
$= \left(-\frac{7}{2}\right)^2 + 7\left(-\frac{7}{2}\right) + 3 = -\frac{37}{4}$
Vertex: $\left(-\frac{7}{2}, -\frac{37}{4}\right)$

119. Let $x = $ width. Then $\dfrac{3400 - 4x}{2} = $ length.
$A = x\left(\dfrac{3400 - 4x}{2}\right) = x(1700 - 2x)$
$= 1700x - 2x^2$
$= -2x^2 + 1700x$
$x = -\dfrac{b}{2a} = -\dfrac{1700}{2(-2)} = 425$
The dimensions are 425 feet by 850 feet.

121. $y = 5x - 8$: $m = 5$
Use $m = 5$:
$y - y_1 = m(x - x_1)$
$y - 0 = 5(x - 0)$
$y = 5x$

Exercise 4.4 (page 356)

1. $20 \log \dfrac{E_O}{E_I}$

3. $t = -\dfrac{1}{k} \ln\left(1 - \dfrac{C}{M}\right)$

5. $E = RT \ln\left(\dfrac{V_f}{V_i}\right)$

7. db gain $= 20 \log \dfrac{E_0}{E_I}$
$= 20 \log \dfrac{17}{0.03}$
≈ 55 db

9. db gain $= 20 \log \dfrac{E_0}{E_I}$
$= 20 \log \dfrac{20}{0.71}$
≈ 29 db

11. $R = \log \dfrac{A}{P}$
$= \log \dfrac{5000}{0.2}$
≈ 4.4

13. $R = \log \dfrac{A}{P}$
$= \log \dfrac{2500}{\frac{1}{4}}$
$= 4$

15. $t = -\dfrac{1}{k} \ln\left(1 - \dfrac{C}{M}\right)$
$t = -\dfrac{1}{0.116} \ln\left(1 - \dfrac{0.9M}{M}\right)$
$t = -\dfrac{1}{0.116} \ln(1 - 0.9) \approx 19.8$ hours

17. $t = \dfrac{\ln 2}{r} = \dfrac{\ln 2}{0.12} \approx 5.8$ years

19. $t = \dfrac{\ln 3}{r} = \dfrac{\ln 3}{0.12} \approx 9.2$ years

21. $E = RT \ln\left(\dfrac{V_f}{V_i}\right)$

$E = (8.314)(400) \ln\left(\dfrac{3V_i}{V_i}\right)$

$E = (8.314)(400) \ln(3)$

$E \approx 3654$ joules

23. $r = \dfrac{1}{t} \ln \dfrac{P}{P_0}$

$= \dfrac{1}{7} \ln \dfrac{10,400,000}{10,400}$

$\approx 0.99 \Rightarrow$ about 99% per year

25. $n = \dfrac{\log V - \log C}{\log\left(1 - \frac{2}{N}\right)}$

$= \dfrac{\log 8000 - \log 37,000}{\log\left(1 - \frac{2}{5}\right)}$

≈ 3 years old

27. $n = \dfrac{\log\left[\frac{Ar}{P} + 1\right]}{\log(1 + r)}$

$= \dfrac{\log\left[\frac{20000(0.12)}{1000} + 1\right]}{\log(1 + 0.12)}$

≈ 10.8 years

29. $V = ER_1 \ln \dfrac{R_2}{R_1}(400,000)(0.25) \ln\left(\dfrac{2}{0.25}\right) \approx 208,000$ V

31. $f(x) = 12 \Rightarrow \ln x = 12 \Rightarrow x \approx 162755 \Rightarrow$ You would go out 162,755 cm.

$162,755 \text{ cm} = \dfrac{162,755 \text{ cm}}{1} \cdot \dfrac{1 \text{ in}}{2.54 \text{ cm}} \cdot \dfrac{1 \text{ ft}}{12 \text{ in}} \cdot \dfrac{1 \text{ mi}}{5280 \text{ ft}} \approx 1 \text{ mile}$

33. Set $x = 0$:

$y = \dfrac{1}{1 + e^{-2(0)}}$

$= \dfrac{1}{1 + e^0}$

$= \dfrac{1}{1 + 1} = \dfrac{1}{2}$

y-intercept: $\left(0, \dfrac{1}{2}\right)$

35. $y = mx + b$

$y = 7x + 3$

37. Vertical through $(2, 3)$

$x = 2$

39. $\dfrac{2(x + 2) - 1}{4x^2 - 9} = \dfrac{2x + 4 - 1}{(2x + 3)(2x - 3)} = \dfrac{2x + 3}{(2x + 3)(2x - 3)} = \dfrac{1}{2x - 3}$

41. $\dfrac{x^2 + 3x + 2}{3x + 9} \cdot \dfrac{x + 3}{x^2 - 4} = \dfrac{(x + 2)(x + 1)}{3(x + 3)} \cdot \dfrac{x + 3}{(x + 2)(x - 2)} = \dfrac{x + 1}{3(x - 2)}$

Exercise 4.5 (page 365)

1. 0

3. M, N

5. x, y

7. x

9. $\neq$

11. $\log_4 1 = 0$

13. $\log_4 4^7 = 7 \log_4 4 = 7$

15. $5^{\log_5 10} = 10$ **17.** $\log_5 5 = 1$ **19-23. Answers may vary.**

25. $\log_b 2xy = \log_b 2 + \log_b x + \log_b y$

27. $\log_b \dfrac{2x}{y} = \log_b 2x - \log_b y$
$$= \log_b 2 + \log_b x - \log_b y$$

29. $\log_b x^2 y^3 = \log_b x^2 + \log_b y^3$
$$= 2\log_b x + 3\log_b y$$

31. $\log_b (xy)^{1/3} = \dfrac{1}{3}\log_b xy$
$$= \dfrac{1}{3}(\log_b x + \log_b y)$$

33. $\log_b x\sqrt{z} = \log_b xz^{1/2} = \log_b x + \log_b z^{1/2} = \log_b x + \frac{1}{2}\log_b z$

35. $\log_b \dfrac{\sqrt[3]{x}}{\sqrt[3]{yz}} = \log_b \sqrt[3]{x} - \log_b \sqrt[3]{yz} = \log_b x^{1/3} - \log_b (yz)^{1/3} = \dfrac{1}{3}\log_b x - \dfrac{1}{3}\log_b yz$
$$= \dfrac{1}{3}\log_b x - \dfrac{1}{3}(\log_b y + \log_b z)$$
$$= \dfrac{1}{3}\log_b x - \dfrac{1}{3}\log_b y - \dfrac{1}{3}\log_b z$$

37. $\log_b (x+1) - \log_b x = \log_b \dfrac{x+1}{x}$

39. $2\log_b x + \dfrac{1}{3}\log_b y = \log_b x^2 + \log_b y^{1/3} = \log_b x^2 y^{1/3} = \log_b x^2 \sqrt[3]{y}$

41. $-3\log_b x - 2\log_b y + \dfrac{1}{2}\log_b z = \log_b x^{-3} + \log_b y^{-2} + \log_b z^{1/2} = \log_b x^{-3}y^{-2}\sqrt{z} = \log_b \dfrac{\sqrt{z}}{x^3 y^2}$

43. $\log_b \left(\dfrac{x}{z}+x\right) - \log_b \left(\dfrac{y}{z}+y\right) = \log_b \dfrac{\frac{x}{z}+x}{\frac{y}{z}+y} = \log_b \dfrac{z\left(\frac{x}{z}+x\right)}{z\left(\frac{y}{z}+y\right)} = \log_b \dfrac{x+xz}{y+yz} = \log_b \dfrac{x}{y}$

45. $\log_b ab = \log_b a + \log_b b = \log_b a + 1$
   TRUE

47. $\log_b 0$ is undefined.
   FALSE

49. $\log_b (x \cdot y) = \log_b x + \log_b y$, so
$\log_b (x+y) \neq \log_b x + \log_b y$
   TRUE (unless $x \cdot y = x + y$)

51. If $\log_a b = c$, then $\log_b a = c$
   FALSE

53. $\log_7 7^7 = 7 \Rightarrow 7^7 = 7^7$
   TRUE

55. $-\log_b x = \log_b x^{-1} = \log_b \dfrac{1}{x}$
   FALSE

57. $\log_b \left(\dfrac{A}{B} \right) = \log_b A - \log_b B$, so

$\dfrac{\log_b (A)}{\log_b (B)} \neq \log_b A - \log_b B$

FALSE

59. $\log_b \dfrac{1}{5} = \log_b 5^{-1} = -\log_b 5$

TRUE

61. $\dfrac{1}{3} \log_b a^3 = \dfrac{1}{3} \cdot 3 \log_b a = \log_b a$

TRUE

63. Let $\log_{1/b} y = c$.

Then $\left(\dfrac{1}{b} \right)^c = y$.

$\left(\left(\dfrac{b}{1} \right)^{-1} \right)^c = y$

$(b)^{-c} = y \Rightarrow \log_b y = -c$.

$\log_{1/b} y + \log_b y = c + (-c) = 0$.

TRUE

65. $\log_{10} 28 = \log_{10} (4 \cdot 7)$

$= \log_{10} 4 + \log_{10} 7$

$= 0.6021 + 0.8451 = 1.4472$

67. $\log_{10} 2.25 = \log_{10} \left(\dfrac{9}{4} \right)$

$= \log_{10} 9 - \log_{10} 4$

$= 0.9542 - 0.6021 = 0.3521$

69. $\log_{10} \left(\dfrac{63}{4} \right) = \log_{10} 63 - \log_{10} 4$

$= \log_{10} (7 \cdot 9) - \log_{10} 4$

$= \log_{10} 7 + \log_{10} 9 - \log_{10} 4$

$= 0.8451 + 0.9542 - 0.6021$

$= 1.1972$

71. $\log_{10} 252 = \log_{10} (4 \cdot 63)$

$= \log_{10} (4 \cdot 7 \cdot 9)$

$= \log_{10} 4 + \log_{10} 7 + \log_{10} 9$

$= 0.6021 + 0.8451 + 0.9542$

$= 2.4014$

73. $\log_{10} 112 = \log_{10} (4 \cdot 28)$

$= \log_{10} (4 \cdot 7 \cdot 4)$

$= \log_{10} 4 + \log_{10} 7 + \log_{10} 4$

$= 0.6021 + 0.8451 + 0.6021$

$= 2.0493$

75. $\log_{10} \left(\dfrac{144}{49} \right) = \log_{10} 144 - \log_{10} 49$

$= \log_{10} (4 \cdot 4 \cdot 9) - \log_{10} (7 \cdot 7)$

$= \log_{10} 4 + \log_{10} 4 + \log_{10} 9 - \log_{10} 7 - \log_{10} 7$

$= 0.6021 + 0.6021 + 0.9542 - 0.8451 - 0.8451$

$= 0.4682$

77. $\log_3 7 = \dfrac{\log_{10} 7}{\log_{10} 3} \approx 1.7712$

79. $\log_\pi 3 = \dfrac{\log_{10} 3}{\log_{10} \pi} \approx 0.9597$

81. $\log_3 8 = \dfrac{\log_e 8}{\log_e 3} \approx 1.8928$

83. $\log_{\sqrt{2}} \sqrt{5} = \dfrac{\log_e \sqrt{5}}{\log_e \sqrt{2}} \approx 2.3219$

85. $\text{pH} = -\log [\text{H}^+]$
$\quad\quad = -\log \left(1.7 \times 10^{-5}\right)$
$\quad\quad \approx 4.77$

87.
$\quad\quad \text{pH} = -\log [\text{H}^+] \quad\quad\quad\quad \text{pH} = -\log [\text{H}^+]$
$\quad\quad 2.9 = -\log [\text{H}^+] \quad\quad\quad\quad 3.3 = -\log [\text{H}^+]$
$\quad\quad -2.9 = \log [\text{H}^+] \quad\quad\quad\quad -3.3 = \log [\text{H}^+]$
$\quad 1.26 \times 10^{-3} \approx [\text{H}^+] \quad\quad 5.01 \times 10^{-4} \approx [\text{H}^+]$
The hydrogen ion concentration can range from 5.01×10^{-4} to 1.26×10^{-3}.

89. db gain $= 10 \log \dfrac{P_O}{P_I}$
$\quad\quad\quad\quad = 10 \log \dfrac{40}{\frac{1}{2}}$
$\quad\quad\quad\quad \approx 19 \text{ db}$

91. $L = k \ln I$
$4L = 4k \ln I = k \cdot 4 \ln I = k \ln I^4$
The original intensity must be raised to the fourth power.

93. $E = 8300 \ln V$
$2E = 2 \cdot 8300 \ln V = 8300 \ln V^2$
The original volume is squared.

95. Let $\log_b M = x$ and $\log_b N = y$.
Then $b^x = M$ and $b^y = N$.
$\dfrac{M}{N} = \dfrac{b^x}{b^y} = b^{x-y}$.
So $\log_b \dfrac{M}{N} = x - y$, or
$\log_b \dfrac{M}{N} = \log_b M - \log_b N$.

97. $e^{x \ln a} = e^{\ln a^x} = a^x$

99. $\ln (e^x) = x \ln e = x(1) = x$

101. $\log (0.9) < 0$, so $\ln(\log(0.9))$ is undefined.

103. Answers may vary.

105. $y = 3x - 1 \Rightarrow$ function

107. $y^2 = 4x$
$y = \pm \sqrt{4x} \Rightarrow$ not a function

109. $f(x) = x^2 - 4$; domain $= (-\infty, \infty)$

111. $f(x) = \sqrt{x^2 + 4}$; domain $= (-\infty, \infty)$

Exercise 4.6 (page 373)

1. exponential

3. $A_0 2^{-t/h}$

5. $\quad 4^x = 5$
$\quad \log 4^x = \log 5$
$\quad x \log 4 = \log 5$
$\quad\quad x = \dfrac{\log 5}{\log 4}$
$\quad\quad x \approx 1.1610$

7.
$$13^{x-1} = 2$$
$$\log 13^{x-1} = \log 2$$
$$(x-1)\log 13 = \log 2$$
$$x\log 13 - \log 13 = \log 2$$
$$x\log 13 = \log 2 + \log 13$$
$$x = \frac{\log 2 + \log 13}{\log 13}$$
$$x \approx 1.2702$$

9.
$$2^{x+1} = 3^x$$
$$\log 2^{x+1} = \log 3^x$$
$$(x+1)\log 2 = x\log 3$$
$$x\log 2 + \log 2 = x\log 3$$
$$x\log 2 - x\log 3 = -\log 2$$
$$x(\log 2 - \log 3) = -\log 2$$
$$x = \frac{-\log 2}{\log 2 - \log 3}$$
$$x \approx 1.7095$$

11.
$$2^x = 3^x$$
$$\log 2^x = \log 3^x$$
$$x\log 2 = x\log 3$$
$$x\log 2 - x\log 3 = 0$$
$$x(\log 2 - \log 3) = 0$$
$$x = \frac{0}{\log 2 - \log 3} = 0$$

13.
$$7^{x^2} = 10$$
$$\log 7^{x^2} = \log 10$$
$$x^2 \log 7 = \log 10$$
$$x^2 = \frac{\log 10}{\log 7}$$
$$x = \pm\sqrt{\frac{\log 10}{\log 7}}$$
$$x \approx \pm 1.0878$$

15.
$$8^{x^2} = 9^x$$
$$\log 8^{x^2} = \log 9^x$$
$$x^2 \log 8 = x\log 9$$
$$x^2 \log 8 - x\log 9 = 0$$
$$x(x\log 8 - \log 9) = 0$$
$$x = 0 \quad \textbf{or} \quad x\log 8 - \log 9 = 0$$
$$x = 0 \qquad\qquad x\log 8 = \log 9$$
$$x = 0 \qquad\qquad x = \frac{\log 9}{\log 8}$$
$$x = 0 \qquad\qquad x \approx 1.0566$$

17.
$$2^{x^2-2x} = 8$$
$$2^{x^2-2x} = 2^3$$
$$x^2 - 2x = 3$$
$$x^2 - 2x - 3 = 0$$
$$(x-3)(x+1) = 0$$
$$x - 3 = 0 \quad \textbf{or} \quad x + 1 = 0$$
$$x = 3 \qquad\qquad x = -1$$

19.
$$3^{x^2+4x} = \frac{1}{81}$$
$$3^{x^2+4x} = 3^{-4}$$
$$x^2 + 4x = -4$$
$$x^2 + 4x + 4 = 0$$
$$(x+2)(x+2) = 0$$
$$x + 2 = 0 \quad \textbf{or} \quad x + 2 = 0$$
$$x = -2 \qquad\qquad x = -2$$

21.
$$4^{x+2} - 4^x = 15$$
$$4^x 4^2 - 4^x = 15$$
$$16 \cdot 4^x - 4^x = 15$$
$$15 \cdot 4^x = 15$$
$$4^x = 1$$
$$x = 0$$

23.
$$2(3^x) = 6^{2x}$$
$$\log [2(3^x)] = \log 6^{2x}$$
$$\log 2 + \log 3^x = 2x \log 6$$
$$\log 2 + x \log 3 = 2x \log 6$$
$$x \log 3 - 2x \log 6 = -\log 2$$
$$x(\log 3 - 2 \log 6) = -\log 2$$
$$x = \frac{-\log 2}{\log 3 - 2 \log 6}$$
$$x \approx 0.2789$$

25.
$$2^{2x} - 10(2^x) + 16 = 0$$
$$(2^x - 2)(2^x - 8) = 0$$
$$2^x - 2 = 0 \quad \text{or} \quad 2^x - 8 = 0$$
$$2^x = 2 \qquad\qquad 2^x = 8$$
$$x = 1 \qquad\qquad x = 3$$

27.
$$2^{2x+1} - 2^x = 1$$
$$2^{2x}2^1 - 2^x - 1 = 0$$
$$2(2^{2x}) - 2^x - 1 = 0$$
$$(2(2^x) + 1)(2^x - 1) = 0$$
$$2(2^x) + 1 = 0 \quad \text{or} \quad 2^x - 1 = 0$$
$$2(2^x) = -1 \qquad\qquad 2^x = 1$$
$$2^x = -\tfrac{1}{2} \qquad\qquad x = 0$$
impossible

29.
$$\log (2x - 3) = \log (x + 4)$$
$$2x - 3 = x + 4$$
$$x = 7$$

31.
$$\log \frac{4x + 1}{2x + 9} = 0$$
$$10^0 = \frac{4x + 1}{2x + 9}$$
$$1 = \frac{4x + 1}{2x + 9}$$
$$2x + 9 = 4x + 1$$
$$8 = 2x$$
$$4 = x$$

33.
$$\log x^2 = 2$$
$$10^2 = x^2$$
$$100 = x^2$$
$$\pm \sqrt{100} = x$$
$$\pm 10 = x$$

35.
$$\log x + \log (x - 48) = 2$$
$$\log x(x - 48) = 2$$
$$x(x - 48) = 10^2$$
$$x^2 - 48x - 100 = 0$$
$$(x - 50)(x + 2) = 0$$
$$x - 50 = 0 \quad \text{or} \quad x + 2 = 0$$
$$x = 50 \qquad\qquad x = -2: \text{extraneous}$$

37.
$$\log x + \log (x - 15) = 2$$
$$\log x(x - 15) = 2$$
$$x(x - 15) = 10^2$$
$$x^2 - 15x - 100 = 0$$
$$(x - 20)(x + 5) = 0$$
$$x - 20 = 0 \quad \text{or} \quad x + 5 = 0$$
$$x = 20 \qquad\qquad x = -5$$
$$\text{extraneous}$$

39.
$$\log(x+90) = 3 - \log x$$
$$\log x + \log(x+90) = 3$$
$$\log x(x+90) = 3$$
$$x(x+90) = 10^3$$
$$x^2 + 90x - 1000 = 0$$
$$(x-10)(x+100) = 0$$
$$x - 10 = 0 \quad \textbf{or} \quad x + 100 = 0$$
$$x = 10 \qquad\qquad x = -100$$
$$\text{extraneous}$$

41. $\log(x-1) - \log 6 = \log(x-2) - \log x$
$$\log \tfrac{x-1}{6} = \log \tfrac{x-2}{x}$$
$$\tfrac{x-1}{6} = \tfrac{x-2}{x}$$
$$x(x-1) = 6(x-2)$$
$$x^2 - x = 6x - 12$$
$$x^2 - 7x + 12 = 0$$
$$(x-3)(x-4) = 0$$
$$x - 3 = 0 \quad \textbf{or} \quad x - 4 = 0$$
$$x = 3 \qquad\qquad x = 4$$

43. $\log x^2 = (\log x)^2$
$$2\log x = (\log x)^2$$
$$0 = (\log x)^2 - 2\log x$$
$$0 = \log x(\log x - 2)$$
$$\log x = 0 \quad \textbf{or} \quad \log x - 2 = 0$$
$$x = 1 \qquad\qquad \log x = 2$$
$$x = 1 \qquad\qquad x = 100$$

45. $\dfrac{\log(3x-4)}{\log x} = 2$
$$\log(3x-4) = 2\log x$$
$$\log(3x-4) = \log x^2$$
$$3x - 4 = x^2$$
$$0 = x^2 - 3x + 4$$
No real solutions

47. $\dfrac{\log(5x+6)}{2} = \log x$
$$\log(5x+6) = 2\log x$$
$$\log(5x+6) = \log x^2$$
$$5x + 6 = x^2$$
$$0 = x^2 - 5x - 6$$
$$0 = (x-6)(x+1)$$
$$x - 6 = 0 \quad \textbf{or} \quad x + 1 = 0$$
$$x = 6 \qquad\qquad x = -1$$
$$\text{extraneous}$$

49. $\log_3 x = \log_3\left(\dfrac{1}{x}\right) + 4$
$$\log_3 x = \log_3 x^{-1} + 4$$
$$\log_3 x = -\log_3 x + 4$$
$$2\log_3 x = 4$$
$$\log_3 x = 2$$
$$x = 9$$

51.
$$2\log_2 x = 3 + \log_2(x-2)$$
$$\log_2 x^2 - \log_2(x-2) = 3$$
$$\log_2 \tfrac{x^2}{x-2} = 3$$
$$\tfrac{x^2}{x-2} = 2^3$$
$$x^2 = 8(x-2)$$
$$x^2 - 8x + 16 = 0$$
$$(x-4)(x-4) = 0$$
$$x - 4 = 0 \quad \textbf{or} \quad x - 4 = 0$$
$$x = 4 \qquad\qquad x = 4$$

53. $\log(7y+1) = 2\log(y+3) - \log 2$
$$\log(7y+1) = \log \tfrac{(y+3)^2}{2}$$
$$7y + 1 = \tfrac{y^2+6y+9}{2}$$
$$14y + 2 = y^2 + 6y + 9$$
$$0 = y^2 - 8y + 7$$
$$0 = (y-7)(y-1)$$
$$y - 7 = 0 \quad \textbf{or} \quad y - 1 = 0$$
$$y = 7 \qquad\qquad y = 1$$

55. Graph $y = \log x + \log (x - 15)$ and $y = 2$ and find the x-coordinate of the point(s) of intersection: $x = 20$

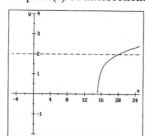

57. Graph $y = 2^{x+1}$ and $y = 7$ and find the x-coordinate of the point(s) of intersection: $x \approx 1.81$

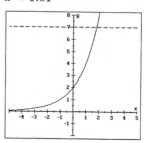

59.
$$A = A_0 2^{-t/h}$$
$$0.75 A_0 = A_0 \cdot 2^{-t/(12.4)}$$
$$0.75 = 2^{-t/12.4}$$
$$\log (0.75) = \log \left(2^{-t/12.4}\right)$$
$$\log (0.75) = -\frac{t}{12.4} \log 2$$
$$-\frac{12.4 \log (0.75)}{\log 2} = t$$
$$5.1 \text{ years} \approx t$$

61.
$$A = A_0 2^{-t/h}$$
$$0.20 A_0 = A_0 \cdot 2^{-t/(18.4)}$$
$$0.20 = 2^{-t/18.4}$$
$$\log (0.20) = \log \left(2^{-t/18.4}\right)$$
$$\log (0.20) = -\frac{t}{18.4} \log 2$$
$$-\frac{18.4 \log (0.20)}{\log 2} = t$$
$$42.7 \text{ days} \approx t$$

63.
$$A = A_0 2^{-t/h}$$
$$0.70 A_0 = A_0 \cdot 2^{-t/5700}$$
$$0.70 = 2^{-t/5700}$$
$$\log (0.70) = \log \left(2^{-t/5700}\right)$$
$$\log (0.70) = -\frac{t}{5700} \log 2$$
$$-\frac{5700 \log (0.70)}{\log 2} = t$$
$$2900 \text{ years} \approx t$$

65.
$$A = P\left(1 + \frac{r}{k}\right)^{kt}$$
$$800 = 500\left(1 + \frac{0.085}{2}\right)^{2t}$$
$$\frac{8}{5} = (1.0425)^{2t}$$
$$\log \left(\frac{8}{5}\right) = \log (1.0425)^{2t}$$
$$\log 8 - \log 5 = 2t \log (1.0425)$$
$$\frac{\log 8 - \log 5}{2 \log (1.0425)} = t$$
$$5.6 \text{ years} \approx t$$

67.
$$A = P\left(1 + \frac{r}{k}\right)^{kt}$$
$$2100 = 1300\left(1 + \frac{0.09}{4}\right)^{4t}$$
$$\frac{21}{13} = (1.0225)^{4t}$$
$$\log\left(\frac{21}{13}\right) = \log(1.0225)^{4t}$$
$$\log 21 - \log 13 = 4t\log(1.0225)$$
$$\frac{\log 21 - \log 13}{4\log(1.0225)} = t$$
$$5.4 \text{ years} \approx t$$

69.
$$A = Pe^{rt}$$
$$2P = Pe^{rt}$$
$$2 = e^{rt}$$
$$\ln 2 = \ln e^{rt}$$
$$\ln 2 = rt$$
$$\frac{\ln 2}{r} = t$$
$$\frac{0.70}{r} \approx t$$
$$\frac{70}{(100 \cdot r)\%} \approx t$$

71.
$$I = I_0 e^{kd}$$
$$0.70 I_0 = I_0 e^{k(6)}$$
$$0.70 = e^{6k}$$
$$\ln 0.70 = \ln e^{6k}$$
$$\ln 0.70 = 6k$$
$$\frac{\ln 0.70}{6} = k$$

$$I = I_0 e^{kd}$$
$$0.20 I_0 = I_0 e^{\frac{\ln 0.70}{6}d}$$
$$0.20 = e^{\frac{\ln 0.70}{6}d}$$
$$\ln 0.20 = \ln e^{\frac{\ln 0.70}{6}d}$$
$$\ln 0.20 = \frac{\ln 0.70}{6}d$$
$$\frac{6\ln 0.20}{\ln 0.70} = d$$
$$d \approx 27 \text{ meters}$$

73.
$$T = 60 + 40e^{kt}$$
$$90 = 60 + 40e^{k(3)}$$
$$30 = 40e^{3k}$$
$$0.75 = e^{3k}$$
$$\ln(0.75) = \ln e^{3k}$$
$$\ln(0.75) = 3k$$
$$\frac{\ln(0.75)}{3} = k$$

75.
$$T = 300 - 300e^{kt}$$
$$100 = 300 - 300e^{k(5)}$$
$$-200 = -300e^{5k}$$
$$\frac{2}{3} = e^{5k}$$
$$\ln\left(\frac{2}{3}\right) = \ln e^{5k}$$
$$\ln\left(\frac{2}{3}\right) = 5k$$
$$\frac{\ln\left(\frac{2}{3}\right)}{5} = k$$

77. **Answers may vary.**

79.
$$P = P_0 e^{rt}$$
$$3P_0 = P_0 e^{rt}$$
$$3 = e^{rt}$$
$$\ln 3 = \ln e^{rt}$$
$$\ln 3 = rt$$
$$\frac{\ln 3}{r} = t$$

81.
$$y = 3x + 2$$
$$x = 3y + 2$$
$$x - 2 = 3y$$
$$\frac{x - 2}{3} = y$$

83.
$$(f \circ g)(2) = f(g(2))$$
$$= f(2^2)$$
$$= f(4)$$
$$= 5(4) - 1$$
$$= 19$$

85.
$$(f \circ g)(x) = f(g(x))$$
$$= f(x^2)$$
$$= 5x^2 - 1$$

Chapter 4 Summary (page 377)

1. **a.** $5^{\sqrt{2}}5^{\sqrt{2}} = 5^{\sqrt{2}+\sqrt{2}} = 5^{2\sqrt{2}}$ **b.** $\left(2^{\sqrt{5}}\right)^{\sqrt{2}} = 2^{\sqrt{5}\cdot\sqrt{2}} = 2^{\sqrt{10}}$

2. **a.** $y = 3^x$: $(0, 1)$, $(1, 3)$ **b.** $y = \left(\dfrac{1}{3}\right)^x$: $(0, 1)$, $\left(1, \dfrac{1}{3}\right)$

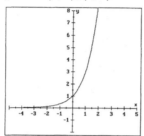 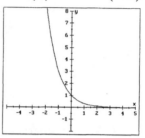

3. $f(x) = 6^x$: goes through $(0, 1)$ and $(1, 6)$ **4.** $y = b^x$: domain $= (-\infty, \infty)$; range $= (0, \infty)$

$p = 1, q = 6$

5. **a.** $y = \left(\dfrac{1}{2}\right)^x - 2$ **b.** $y = \left(\dfrac{1}{2}\right)^{x+2}$

Shift $y = \left(\dfrac{1}{2}\right)^x$ down 2:   Shift $y = \left(\dfrac{1}{2}\right)^x$ left 2:

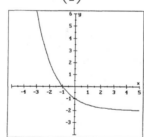 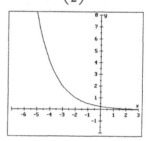

6. **a.** $y = e^x + 1$

Shift $y = e^x$ up 1:

b. $y = e^{x-3}$

Shift $y = e^x$ right 3:

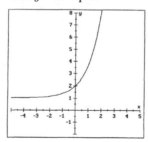 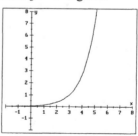

7. If $\frac{1}{3}$ decays in 20 years, $\frac{2}{3}$ is left.

$$A = A_0 2^{-t/h}$$
$$\tfrac{2}{3} A_0 = A_0 \cdot 2^{-20/h}$$
$$\tfrac{2}{3} = 2^{-20/h}$$
$$\log\left(\tfrac{2}{3}\right) = \log\left(2^{-20/h}\right)$$
$$\log\left(\tfrac{2}{3}\right) = -\tfrac{20}{h} \log 2$$
$$h \log\left(\tfrac{2}{3}\right) = -20 \log 2$$
$$h = \frac{-20 \log 2}{\log(2/3)}$$
$$h \approx 34.2 \text{ years}$$

8. $$A = P\left(1 + \frac{r}{k}\right)^{kt}$$
$$= 10{,}500\left(1 + \frac{0.09}{4}\right)^{4(60)}$$
$$\approx \$2{,}189{,}703.45$$

9. $$I = I_0 k^x$$
$$= 14(0.7)^{12}$$
$$\approx 0.19 \text{ lumens}$$

10. $$A = Pe^{rt}$$
$$= 10{,}500 e^{0.09(60)}$$
$$\approx \$2{,}324{,}767.37$$

11. $$P = P_0 e^{kt}$$
$$P = 275{,}000{,}000 e^{0.015(50)}$$
$$P \approx 582{,}000{,}000 \text{ people}$$

12. domain $= (0, \infty)$
range $= (-\infty, \infty)$

13. **a.** $\log_3 9 = ?$
$$3^? = 9$$
$$\log_3 9 = 2$$

   b. $\log_9 \dfrac{1}{3} = ?$
$$9^? = \frac{1}{3}$$
$$\log_9 \frac{1}{3} = -\frac{1}{2}$$

   c. $\log_x 1 = ?$
$$x^? = 1$$
$$\log_x 1 = 0$$

   d. $\log_5 0.04 = ?$
$$5^? = 0.04 = \frac{1}{25}$$
$$\log_5 0.04 = -2$$

   e. $\log_a \sqrt{a} = ?$
$$a^? = \sqrt{a}$$
$$\log_a \sqrt{a} = \frac{1}{2}$$

   f. $\log_a \sqrt[3]{a} = ?$
$$a^? = \sqrt[3]{a}$$
$$\log_a \sqrt[3]{a} = \frac{1}{3}$$

14. **a.** $\log_2 x = 5$
$$2^5 = x$$
$$32 = x$$

   b. $\log_{\sqrt{3}} x = 4$
$$\left(\sqrt{3}\right)^4 = x$$
$$9 = x$$

c. $\log_{\sqrt{2}} x = 6$

$\left(\sqrt{2}\right)^6 = x$

$8 = x$

d. $\log_{0.1} 10 = x$

$(0.1)^x = 10$

$\left(\dfrac{1}{10}\right)^x = 10$

$x = -1$

e. $\log_x 2 = -\dfrac{1}{3}$

$x^{-1/3} = 2$

$\left(x^{-1/3}\right)^{-3} = 2^{-3}$

$x = \dfrac{1}{8}$

f. $\log_x 32 = 5$

$x^5 = 32$

$x = 2$

g. $\log_{0.25} x = -1$

$(0.25)^{-1} = x$

$\left(\tfrac{1}{4}\right)^{-1} = x$

$4 = x$

h. $\log_{0.125} x = -\dfrac{1}{3}$

$(0.125)^{-1/3} = x$

$\left(\tfrac{1}{8}\right)^{-1/3} = x$

$2 = x$

i. $\log_{\sqrt{2}} 32 = x$

$\left(\sqrt{2}\right)^x = 32$

$\left(2^{1/2}\right)^x = 2^5$

$\dfrac{1}{2}x = 5$

$x = 10$

j. $\log_{\sqrt{5}} x = -4$

$\left(\sqrt{5}\right)^{-4} = x$

$\dfrac{1}{25} = x$

k. $\log_{\sqrt{3}} 9\sqrt{3} = x$

$\left(\sqrt{3}\right)^x = 9\sqrt{3}$

$\left(3^{1/2}\right)^x = 3^{5/2}$

$\dfrac{1}{2}x = \dfrac{5}{2}$

$x = 5$

l. $\log_{\sqrt{5}} 5\sqrt{5} = x$

$\left(\sqrt{5}\right)^x = 5\sqrt{5}$

$\left(5^{1/2}\right)^x = 5^{3/2}$

$\dfrac{1}{2}x = \dfrac{3}{2}$

$x = 3$

15. a. $y = \log(x - 2)$
Shift $y = \log x$ right 2:

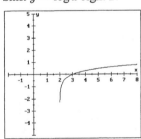

b. $y = 3 + \log x$
Shift $y = \log x$ up 3:

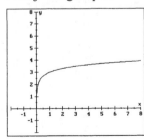

16. **a.** $y = 4^x; y = \log_4 x$

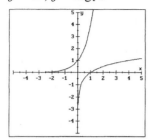

b. $y = \left(\dfrac{1}{3}\right)^x; y = \log_{1/3} x$

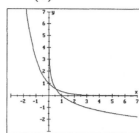

17. **a.** $\ln 452 \approx 6.1137$

b. $\ln(\log 7.85) \approx -0.1111$

18. **a.** $\ln x = 2.336$
$x \approx 10.3398$

b. $\ln x = \log 8.8$
$x \approx 2.5715$

19. **a.** $y = 1 + \ln x$
Shift $y = \ln x$ up 1:

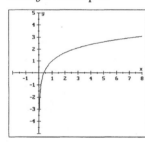

b. $y = \ln(x + 1)$
Shift $y = \ln x$ left 1:

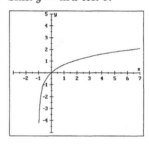

20. db gain $= 20 \log \dfrac{E_0}{E_I}$

$= 20 \log \dfrac{18}{0.04}$

≈ 53 db gain

21. $R = \log \dfrac{A}{P}$

$= \log \dfrac{7500}{0.3}$

≈ 4.4

22. $t = -\dfrac{1}{k} \ln\left(1 - \dfrac{C}{M}\right)$

$t = -\dfrac{1}{0.17} \ln\left(1 - \dfrac{0.8M}{M}\right)$

$t = -\dfrac{1}{0.17} \ln(1 - 0.8) \approx 9.5$ hours

23. $t = \dfrac{\ln 2}{r} = \dfrac{\ln 2}{0.03} \approx 23$ years

24. $E = RT \ln\left(\dfrac{V_f}{V_i}\right) = (8.314)(350) \ln\left(\dfrac{2V_i}{V_i}\right) = (8.314)(350) \ln(2) \approx 2017$ joules

25. **a.** $\log_7 1 = 0$

b. $\log_7 7 = 1$

    c. $\log_7 7^3 = 3$ **d.** $7^{\log_7 4} = 4$

26. **a.** $\ln e^4 = 4 \ln e = 4$ **b.** $\ln 1 = 0$

    c. $10^{\log_{10} 7} = 7$ **d.** $e^{\log_e 3} = 3$

    e. $\log_b b^4 = 4 \log_b b = 4$ **f.** $\ln e^9 = 9 \ln e = 9$

27. **a.** $\log_b \dfrac{x^2 y^3}{z^4} = \log_b \left(x^2 y^3 \right) - \log_b \left(z^4 \right) = \log_b x^2 + \log_b y^3 - \log_b z^4$

$$= 2 \log_b x + 3 \log_b y - 4 \log_b z$$

    b. $\log_b \sqrt{\dfrac{x}{yz^2}} = \log_b \left(\dfrac{x}{yz^2} \right)^{1/2} = \dfrac{1}{2} \log_b \left(\dfrac{x}{yz^2} \right) = \dfrac{1}{2} \left(\log_b x - \log_b yz^2 \right)$

$$= \dfrac{1}{2} \left(\log_b x - \log_b y - 2 \log_b z \right)$$

28. **a.** $3 \log_b x - 5 \log_b y + 7 \log_b z = \log_b x^3 + \log_b y^{-5} + \log_b z^7 = \log_b \dfrac{x^3 z^7}{y^5}$

    b. $\dfrac{1}{2} \left(\log_b x + 3 \log_b y \right) - 7 \log_b z = \dfrac{1}{2} \left(\log_b x + \log_b y^3 \right) + \log_b z^{-7} = \dfrac{1}{2} \log_b xy^3 + \log_b z^{-7}$

$$= \log_b \sqrt{xy^3} + \log_b z^{-7}$$

$$= \log_b \dfrac{\sqrt{xy^3}}{z^7}$$

29. **a.** $\log abc = \log a + \log b + \log c = 0.6 + 0.36 + 2.4 = 3.36$

    b. $\log a^2 b = \log a^2 + \log b = 2 \log a + \log b = 2(0.6) + 0.36 = 1.56$

    c. $\log \dfrac{ac}{b} = \log a + \log c - \log b = 0.6 + 2.4 - 0.36 = 2.64$

    d. $\log \dfrac{a^2}{c^3 b^2} = \log a^2 - \log c^3 b^2 = \log a^2 - \log c^3 - \log b^2$

$$= 2 \log a - 3 \log c - 2 \log b$$

$$= 2(0.6) - 3(2.4) - 2(0.36) = -6.72$$

30. $\log_5 17 = \dfrac{\log 17}{\log 5} \approx 1.7604$ **31.** $\text{pH} = -\log \left[\text{H}^+ \right]$

$$3.1 = -\log \left[\text{H}^+ \right]$$
$$-3.1 = \log \left[\text{H}^+ \right]$$
$$7.94 \times 10^{-4} \approx \left[\text{H}^+ \right]$$

32. $L = k \ln I$

$k \ln \dfrac{I}{2} = k \left(\ln I - \ln 2 \right) = k \ln I - k \ln 2 \Rightarrow$ The loudness decreases by $k \ln 2$.

33. **a.**
$$3^x = 7$$
$$\log 3^x = \log 7$$
$$x \log 3 = \log 7$$
$$x = \frac{\log 7}{\log 3}$$
$$x \approx 1.7712$$

b.
$$5^{x+2} = 625$$
$$\log 5^{x+2} = \log 625$$
$$(x + 2)\log 5 = \log 625$$
$$x \log 5 + 2 \log 5 = \log 625$$
$$x \log 5 = \log 625 - 2 \log 5$$
$$x = \frac{\log 625 - 2 \log 5}{\log 5}$$
$$x = 2$$

c.
$$2^x = 3^{x-1}$$
$$\log 2^x = \log 3^{x-1}$$
$$x \log 2 = (x - 1)\log 3$$
$$x \log 2 = x \log 3 - \log 3$$
$$x \log 2 - x \log 3 = -\log 3$$
$$x(\log 2 - \log 3) = -\log 3$$
$$x = \frac{-\log 3}{\log 2 - \log 3}$$
$$x \approx 2.7095$$

d.
$$2^{x^2+4x} = \frac{1}{8}$$
$$2^{x^2+4x} = 2^{-3}$$
$$x^2 + 4x = -3$$
$$x^2 + 4x + 3 = 0$$
$$(x + 1)(x + 3) = 0$$
$$x + 1 = 0 \quad \text{or} \quad x + 3 = 0$$
$$x = -1 \qquad\qquad x = -3$$

34. **a.**
$$\log x + \log (29 - x) = 2$$
$$\log x(29 - x) = 2$$
$$x(29 - x) = 10^2$$
$$-x^2 + 29x - 100 = 0$$
$$x^2 - 29x + 100 = 0$$
$$(x - 25)(x - 4) = 0$$
$$x - 25 = 0 \quad \text{or} \quad x - 4 = 0$$
$$x = 25 \qquad\qquad x = 4$$

b.
$$\log_2 x + \log_2 (x - 2) = 3$$
$$\log_2 x(x - 2) = 3$$
$$x(x - 2) = 2^3$$
$$x^2 - 2x - 8 = 0$$
$$(x - 4)(x + 2) = 0$$
$$x - 4 = 0 \quad \text{or} \quad x + 2 = 0$$
$$x = 4 \qquad\qquad x = -2$$
$$\text{extraneous}$$

c.
$$\log_2 (x + 2) + \log_2 (x - 1) = 2$$
$$\log_2 (x + 2)(x - 1) = 2$$
$$(x + 2)(x - 1) = 2^2$$
$$x^2 + x - 2 = 4$$
$$x^2 + x - 6 = 0$$
$$(x - 2)(x + 3) = 0$$
$$x - 2 = 0 \quad \text{or} \quad x + 3 = 0$$
$$x = 2 \qquad\qquad x = -3$$
$$\text{extraneous}$$

d.
$$\frac{\log (7x - 12)}{\log x} = 2$$
$$\log (7x - 12) = 2 \log x$$
$$\log (7x - 12) = \log x^2$$
$$7x - 12 = x^2$$
$$0 = x^2 - 7x + 12$$
$$0 = (x - 3)(x - 4)$$
$$x = 3, x = 4$$

e.
$$\log x + \log (x - 5) = \log 6$$
$$\log x(x - 5) = \log 6$$
$$x(x - 5) = 6$$
$$x^2 - 5x = 6$$
$$x^2 - 5x - 6 = 0$$
$$(x - 6)(x + 1) = 0$$
$$x - 6 = 0 \quad \textbf{or} \quad x + 1 = 0$$
$$x = 6 \qquad\qquad x = -1$$
$$\text{extraneous}$$

f.
$$\log 3 - \log (x - 1) = -1$$
$$\log \frac{3}{x - 1} = -1$$
$$\frac{3}{x - 1} = 10^{-1}$$
$$\frac{3}{x - 1} = \frac{1}{10}$$
$$30 = x - 1$$
$$31 = x$$

g.
$$e^{x \ln 2} = 9$$
$$\ln e^{x \ln 2} = \ln 9$$
$$x \ln 2 \ln e = \ln 9$$
$$x \ln 2 = \ln 9$$
$$x = \frac{\ln 9}{\ln 2} \approx 3.1699$$

h.
$$\ln x = \ln (x - 1)$$
$$x = x - 1$$
$$\text{no solution}$$

i.
$$\ln x = \ln (x - 1) + 1$$
$$\ln x - \ln (x - 1) = 1$$
$$\ln \frac{x}{x - 1} = 1$$
$$\frac{x}{x - 1} = e^1$$
$$\frac{x}{x - 1} = \frac{e}{1}$$
$$x = e(x - 1)$$
$$\frac{e}{e - 1} = x, \text{ or } x \approx 1.5820$$

j. Note:
$$\log_{10} x = \frac{\ln x}{\ln 10}$$
$$\ln x = \log_{10} x$$
$$\ln x = \frac{\ln x}{\ln 10}$$
$$\ln x \ln 10 = \ln x$$
$$\ln x \ln 10 - \ln x = 0$$
$$\ln x (\ln 10 - 1) = 0$$
$$\ln x = 0 \Rightarrow x = 1$$

35.
$$A = A_0 2^{-t/h}$$
$$\frac{2}{3} A_0 = A_0 \cdot 2^{-t/5700}$$
$$\frac{2}{3} = 2^{-t/5700}$$
$$\log (2/3) = \log \left(2^{-t/5700}\right)$$
$$\log (2/3) = -\frac{t}{5700} \log 2$$
$$-\frac{5700 \log (2/3)}{\log 2} = t \Rightarrow t \approx 3300 \text{ years}$$

Chapter 4 Test (page 382)

1. $y = 2^x + 1 \Rightarrow$ Shift $y = 2^x$ up 1.

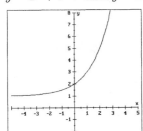

2. $y = e^{x-2} \Rightarrow$ Shift $y = e^x$ right 2.

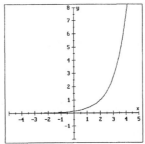

3. $A = 3(2)^{-6} = 3 \cdot \frac{1}{64} = \frac{3}{64}$ gram

4. $A = 1000\left(1 + \frac{0.06}{2}\right)^{2(1)} \approx \$1060.90$

5. $A = 2000e^{0.08(10)} \approx \$4451.08$

6. $\log_7 343 = \log_7 7^3 = 3$

7. $\log_3 \dfrac{1}{27} = \log_3 3^{-3} = -3$

8. $\log_{10} 10^{12} + 10^{\log_{10} 5} = 12 + 5 = 17$

9. $\log_{3/2} \dfrac{9}{4} = \log_{3/2}\left(\frac{3}{2}\right)^2 = 2$

10. $\log_{2/3} \dfrac{27}{8} = \log_{2/3}\left(\frac{2}{3}\right)^{-3} = -3$

11. $y = \log(x - 1) \Rightarrow$ Shift $y = \log x$ right 1.

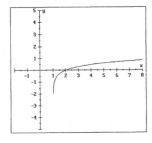

12. $y = 2 + \ln x \Rightarrow$ Shift $y = \ln x$ up 2.

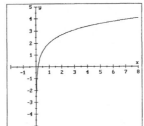

13. $\log a^2 bc^3 = \log a^2 + \log b + \log c^3 = 2\log a + \log b + 3\log c$

14. $\ln\sqrt{\dfrac{a}{b^2 c}} = \ln\left(\dfrac{a}{b^2 c}\right)^{1/2} = \dfrac{1}{2}\ln\dfrac{a}{b^2 c} = \dfrac{1}{2}\left(\ln a - \ln b^2 - \ln c\right) = \dfrac{1}{2}\left(\ln a - 2\ln b - \ln c\right)$

15. $\dfrac{1}{2}\log(a + 2) + \log b - 2\log c = \log\sqrt{a + 2} + \log b - \log c^2 = \log\dfrac{b\sqrt{a + 2}}{c^2}$

16. $\dfrac{1}{3}(\log a - 2\log b) - \log c = \dfrac{1}{3}\log\dfrac{a}{b^2} - \log c = \log\sqrt[3]{\dfrac{a}{b^2}} - \log c = \log\dfrac{\sqrt[3]{\frac{a}{b^2}}}{c}$

17. $\log 24 = \log 8 \cdot 3 = \log 2^3 \cdot 3 = 3\log 2 + \log 3 = 3(0.3010) + 0.4771 = 1.3801$

18. $\log \dfrac{8}{3} = \log \dfrac{2^3}{3} = 3\log 2 - \log 3 = 3(0.3010) - 0.4771 = 0.4259$

19. $\log_7 3 = \dfrac{\log 3}{\log 7} \text{ or } \dfrac{\ln 3}{\ln 7}$

20. $\log_\pi e = \dfrac{\log e}{\log \pi} \text{ or } \dfrac{\ln e}{\ln \pi}$

21. $\log_a ab = \log_a a + \log_a b = 1 + \log_a b$
    TRUE

22. $\log \dfrac{a}{b} = \log a - \log b$
    FALSE

23. $\begin{aligned} \text{pH} &= -\log [\text{H}^+] \\ &= -\log \left(3.7 \times 10^{-7}\right) \\ &\approx 6.4 \end{aligned}$

24. $\begin{aligned} \text{db gain} &= 20 \log \dfrac{E_O}{E_I} \\ &= 20 \log \dfrac{60}{0.3} \\ &\approx 46 \text{ db gain} \end{aligned}$

25.
$$3^{x-1} = 100^x$$
$$\log 3^{x-1} = \log 100^x$$
$$(x-1)\log 3 = x \log 100$$
$$x \log 3 - \log 3 = x \log 100$$
$$x \log 3 - x \log 100 = \log 3$$
$$x(\log 3 - \log 100) = \log 3$$
$$x = \frac{\log 3}{\log 3 - \log 100}$$
$$x \approx -0.3133$$

26.
$$3^{x^2-2x} = 27$$
$$3^{x^2-2x} = 3^3$$
$$x^2 - 2x = 3$$
$$x^2 - 2x - 3 = 0$$
$$(x-3)(x+1) = 0$$
$$x - 3 = 0 \quad \text{or} \quad x + 1 = 0$$
$$x = 3 \qquad\qquad x = -1$$

27.
$$\log (5x + 2) = \log (2x + 5)$$
$$5x + 2 = 2x + 5$$
$$3x = 3$$
$$x = 1$$

28.
$$\log x + \log (x - 9) = 1$$
$$\log x(x - 9) = 1$$
$$x(x - 9) = 10^1$$
$$x^2 - 9x - 10 = 0$$
$$(x - 10)(x + 1) = 0$$
$$x - 10 = 0 \quad \text{or} \quad x + 1 = 0$$
$$x = 10 \qquad\qquad x = -1: \text{ extraneous}$$

Exercise 5.1 (page 393)

1. whole **3.** any **5.** factor

7.
$$P(1) = 2(1)^4 - 2(1)^3 + 5(1)^2 - 1$$
$$= 2(1) - 2(1) + 5(1) - 1$$
$$= 2 - 2 + 5 - 1 = \boxed{4}$$

$$
\begin{array}{r}
2x^3 + 5x + 5 \\
x-1\;\overline{\smash{\big)}\;2x^4 - 2x^3 + 5x^2 + 0x - 1} \\
\underline{2x^4 - 2x^3} \\
0 + 5x^2 \\
\underline{5x^2 - 5x} \\
5x - 1 \\
\underline{5x - 5} \\
\boxed{4}
\end{array}
$$

9.
$$P(-2) = 2(-2)^4 - 2(-2)^3 + 5(-2)^2 - 1$$
$$= 2(16) - 2(-8) + 5(4) - 1$$
$$= 32 + 16 + 20 - 1 = \boxed{67}$$

$$
\begin{array}{r}
2x^3 - 6x^2 + 17x - 34 \\
x+2\;\overline{\smash{\big)}\;2x^4 - 2x^3 + 5x^2 + 0x - 1} \\
\underline{2x^4 + 4x^3} \\
-6x^3 + 5x^2 \\
\underline{-6x^3 - 12x^2} \\
17x^2 + 0x \\
\underline{17x^2 + 34x} \\
-34x - 1 \\
\underline{-34x - 68} \\
\boxed{67}
\end{array}
$$

11. remainder $= P(12) = 3(12)^4 + 5(12)^3 - 4(12)^2 - 2(12) + 1$
$$= 3(20{,}736) + 5(1728) - 4(144) - 2(12) + 1$$
$$= 62{,}208 + 8640 - 576 - 24 + 1 = 70{,}249$$

13. remainder $= P(-3.25) = 3(-3.25)^4 + 5(-3.25)^3 - 4(-3.25)^2 - 2(-3.25) + 1$
$$= 3(111.56640625) + 5(-34.328125) - 4(10.5625) - 2(-3.25) + 1$$
$$= 334.69921875 - 171.640625 - 42.25 + 6.5 + 1 = 128.3085938$$

15. $(x - 1)$ is a factor if $P(1) = 0$.
$P(1) = 1^7 - 1 = 0$: true

17. $(x - 1)$ is a factor if $P(1) = 0$.
$P(1) = 3(1)^5 + 4(1)^2 - 7 = 0$: true

19. $(x + 3)$ is a factor if $P(-3) = 0$. $P(-3) = 2(-3)^3 - 2(-3)^2 + 1 = -71$: false

21. $(x - 1)$ is a factor if $P(1) = 0$. $P(1) = (1)^{1984} - (1)^{1776} + (1)^{1492} - (1)^{1066} = 0$: true

23. $x = -1$ is a solution, so $(x + 1)$ is a factor.
Use synthetic division to divide by $(x + 1)$.

$$
\begin{array}{r|rrrr}
-1 & 1 & 3 & -13 & -15 \\
   &   & -1 & -2 & 15 \\
\hline
   & 1 & 2 & -15 & 0
\end{array}
$$

$$x^3 + 3x^2 - 13x - 15 = 0$$
$$(x + 1)(x^2 + 2x - 15) = 0$$
$$(x + 1)(x + 5)(x - 3) = 0$$
Solution set: $\{-1, -5, 3\}$

25. $x = 1$ is a solution, so $(x - 1)$ is a factor.
Use synthetic division to divide by $(x - 1)$.

$$\begin{array}{r|rrrrr} 1 & 1 & -2 & -2 & 6 & -3 \\ & & 1 & -1 & -3 & 3 \\ \hline & 1 & -1 & -3 & 3 & 0 \end{array}$$

$x^4 - 2x^3 - 2x^2 + 6x - 3 = 0$
$(x - 1)(x^3 - x^2 - 3x + 3) = 0$

Use the fact that $x = 1$ is a double root
and divide the depressed polynomial by $(x - 1)$:

$$\begin{array}{r|rrrr} 1 & 1 & -1 & -3 & 3 \\ & & 1 & 0 & -3 \\ \hline & 1 & 0 & -3 & 0 \end{array}$$

$(x - 1)(x^3 - x^2 - 3x + 3) = 0$
$(x - 1)(x - 1)(x^2 - 3) = 0$

Solution set: $\left\{1, 1, \sqrt{3}, -\sqrt{3}\right\}$

27. $x = 2$ is a solution, so $(x - 2)$ is a factor.
Use synthetic division to divide by $(x - 2)$.

$$\begin{array}{r|rrrrr} 2 & 1 & -5 & 7 & -5 & 6 \\ & & 2 & -6 & 2 & -6 \\ \hline & 1 & -3 & 1 & -3 & 0 \end{array}$$

$x^4 - 5x^3 + 7x^2 - 5x + 6 = 0$
$(x - 2)(x^3 - 3x^2 + x - 3) = 0$

$x = 3$ is a root, so $(x - 3)$ is a factor.
Use synthetic division to divide by $(x - 3)$.

$$\begin{array}{r|rrrr} 3 & 1 & -3 & 1 & -3 \\ & & 3 & 0 & 3 \\ \hline & 1 & 0 & 1 & 0 \end{array}$$

$(x - 2)(x^3 - 3x^2 + x - 3) = 0$
$(x - 2)(x - 3)(x^2 + 1) = 0$
$x^2 + 1 = 0 \Rightarrow x^2 = -1 \Rightarrow x = \pm i$
Solution set: $\{2, 3, \pm i\}$

29. $(x - 4)(x - 5) = x^2 - 9x + 20$

31. $(x - 1)(x - 1)(x - 1) = (x^2 - 2x + 1)(x - 1) = x^3 - 3x^2 + 3x - 1$

33. $(x - 2)(x - 4)(x - 5) = (x^2 - 6x + 8)(x - 5) = x^3 - 11x^2 + 38x - 40$

35. $(x - 1)(x + 1)\left(x - \sqrt{2}\right)\left(x + \sqrt{2}\right) = (x^2 - 1)(x^2 - 2) = x^4 - 3x^2 + 2$

37. $\left(x - \sqrt{2}\right)(x - i)(x + i) = \left(x - \sqrt{2}\right)(x^2 - i^2) = \left(x - \sqrt{2}\right)(x^2 + 1) = x^3 - \sqrt{2}x^2 + x - \sqrt{2}$

39. $(x - 0)[x - (1 + i)][x - (1 - i)] = x\left[x^2 - (1 - i)x - (1 + i)x + (1 + i)(1 - i)\right]$
$$= x\left[x^2 - x + ix - x - ix + 1 - i^2\right]$$
$$= x\left[x^2 - 2x + 2\right] = x^3 - 2x^2 + 2x$$

41.
$$\begin{array}{r|rrrr} 1 & 3 & -2 & -6 & -4 \\ & & 3 & 1 & -5 \\ \hline & 3 & 1 & -5 & -9 \end{array}$$

$P(x) = 3x^3 - 2x^2 - 6x - 4$
$\quad = (x - 1)(3x^2 + x - 5) - 9$

43.
$$\begin{array}{r|rrrr} 3 & 3 & -2 & -6 & -4 \\ & & 9 & 21 & 45 \\ \hline & 3 & 7 & 15 & 41 \end{array}$$

$P(x) = 3x^3 - 2x^2 - 6x - 4$
$\quad = (x - 3)(3x^2 + 7x + 15) + 41$

45.
$$\begin{array}{r|rrrr} -1 & 3 & -2 & -6 & -4 \\ & & -3 & 5 & 1 \\ \hline & 3 & -5 & -1 & -3 \end{array}$$
$$P(x) = 3x^3 - 2x^2 - 6x - 4$$
$$= (x+1)(3x^2 - 5x - 1) - 3$$

47.
$$\begin{array}{r|rrrr} -3 & 3 & -2 & -6 & -4 \\ & & -9 & 33 & -81 \\ \hline & 3 & -11 & 27 & -85 \end{array}$$
$$P(x) = 3x^3 - 2x^2 - 6x - 4$$
$$= (x+3)(3x^2 - 11x + 27) - 85$$

49.
$$\begin{array}{r|rrrr} 1 & 1 & 1 & 1 & -3 \\ & & 1 & 2 & 3 \\ \hline & 1 & 2 & 3 & 0 \end{array}$$
$$x^2 + 2x + 3$$

51.
$$\begin{array}{r|rrrr} -1 & 7 & -3 & -5 & 1 \\ & & -7 & 10 & -5 \\ \hline & 7 & -10 & 5 & -4 \end{array}$$
$$7x^2 - 10x + 5 + \frac{-4}{x+1}$$

53.
$$\begin{array}{r|rrrrr} 3 & 4 & -3 & 0 & -1 & 5 \\ & & 12 & 27 & 81 & 240 \\ \hline & 4 & 9 & 27 & 80 & 245 \end{array}$$
$$4x^3 + 9x^2 + 27x + 80 + \frac{245}{x-3}$$

55.
$$\begin{array}{r|rrrrrr} 4 & 3 & 0 & 0 & 0 & -768 & 0 \\ & & 12 & 48 & 192 & 768 & 0 \\ \hline & 3 & 12 & 48 & 192 & 0 & 0 \end{array}$$
$$3x^4 + 12x^3 + 48x^2 + 192x$$

57.
$$\begin{array}{r|rrrr} 2 & 5 & 2 & -1 & 1 \\ & & 10 & 24 & 46 \\ \hline & 5 & 12 & 23 & 47 \end{array} \quad \boxed{P(2) = 47}$$

59.
$$\begin{array}{r|rrrr} -5 & 5 & 2 & -1 & 1 \\ & & -25 & 115 & -570 \\ \hline & 5 & -23 & 114 & -569 \end{array}$$
$$\boxed{P(-5) = -569}$$

61.
$$\begin{array}{r|rrrrr} \frac{1}{2} & 2 & 0 & -1 & 0 & 2 \\ & & 1 & \frac{1}{2} & -\frac{1}{4} & -\frac{1}{8} \\ \hline & 2 & 1 & -\frac{1}{2} & -\frac{1}{4} & \frac{15}{8} \end{array}$$
$$\boxed{P\left(\tfrac{1}{2}\right) = \tfrac{15}{8}}$$

63.
$$\begin{array}{r|rrrrr} i & 2 & 0 & -1 & 0 & 2 \\ & & 2i & -2 & -3i & 3 \\ \hline & 2 & 2i & -3 & -3i & 5 \end{array}$$
$$\boxed{P(i) = 5}$$

65.
$$\begin{array}{r|rrrrr} 1 & 1 & -8 & 14 & 8 & -15 \\ & & 1 & -7 & 7 & 15 \\ \hline & 1 & -7 & 7 & 15 & 0 \end{array}$$
$$\boxed{P(1) = 0}$$

67.
$$\begin{array}{r|rrrrr} -3 & 1 & -8 & 14 & 8 & -15 \\ & & -3 & 33 & -141 & 399 \\ \hline & 1 & -11 & 47 & -133 & 384 \end{array}$$
$$\boxed{P(-3) = 384}$$

69.
$$\begin{array}{r|rrrrrr} i & 1 & 0 & -1 & -8 & 0 & 8 \\ & & i & -1 & -2i & 2-8i & 8+2i \\ \hline & 1 & i & -2 & -8-2i & 2-8i & 16+2i \end{array} \quad \boxed{P(i) = 16 + 2i}$$

71.
$$\begin{array}{r|rrrrrr} -2i & 1 & 0 & -1 & -8 & 0 & 8 \\ & & -2i & -4 & 10i & 20+16i & 32-40i \\ \hline & 1 & -2i & -5 & -8+10i & 20+16i & 40-40i \end{array} \quad \boxed{P(-2i) = 40 - 40i}$$

73.
$$P(0) = 0$$
$$a_n(0)^n + a_{n-1}(0)^{n-1} + \cdots + a_1(0) + a_0 = 0$$
$$0 + 0 + \cdots + 0 + a_0 = 0$$
$$a_0 = 0$$

75. $P(2) = 0 \Rightarrow (x - 2)$ is a factor. $P(-2) = 0 \Rightarrow (x + 2)$ is a factor. The product of two factors will also be a factor, so $(x - 2)(x + 2) = x^2 - 4$ is a factor of the polynomial $P(x)$.

77. $P(3, -2)$: QIV   **79.** $R(8, \pi)$: QI

81. $d = \sqrt{(3 - (-5))^2 + (-3 - 3)^2} = \sqrt{(8)^2 + (-6)^2} = \sqrt{64 + 36} = \sqrt{100} = 10$

83. $m = \dfrac{-3 - 5}{-5 - 3} = \dfrac{-8}{-8} = 1$

Exercise 5.2 (page 400)

1. zero   **3.** conjugate   **5.** $(-x)^3 - (-x)^2 - 4$
$$= -x^3 - x^2 - 4$$
$$\text{0 variations}$$

7. $7x^4 + 5x^3 - 2x + 1 \Rightarrow 7(-x)^4 + 5(-x)^3 - 2(-x) + 1$
$$\Rightarrow 7x^4 - 5x^3 + 2x + 1 \Rightarrow 2 \text{ variations} \Rightarrow \text{at most 2 negative roots}$$

9. lower bound

11.
$$x^{10} = 1$$
$$x^{10} - 1 = 0 \Rightarrow 10 \text{ roots}$$

13.
$$3x^4 - 4x^2 - 2x = -7$$
$$3x^4 - 4x^2 - 2x + 7 = 0 \Rightarrow 4 \text{ roots}$$

15. $x(3x^4 - 2) = 12x$
$$3x^5 - 14x = 0$$
$$5 \text{ total roots} \Rightarrow \boxed{4 \text{ other roots}}$$

17. If $-i$ is a root, then i is a root also:
$$(x - 3)(x + i)(x - i) = 0$$
$$(x - 3)(x^2 - i^2) = 0$$
$$(x - 3)(x^2 + 1) = 0$$
$$x^3 - 3x^2 + x - 3 = 0$$

19. If $2 + i$ is a root, then $2 - i$ is a root also:
$$(x - 2)[x - (2 + i)][x - (2 - i)] = 0$$
$$(x - 2)\left[x^2 - (2 - i)x - (2 + i)x + (2 + i)(2 - i)\right] = 0$$
$$(x - 2)\left[x^2 - 2x + ix - 2x - ix + 4 - i^2\right] = 0$$
$$(x - 2)\left[x^2 - 4x + 5\right] = 0$$
$$x^3 - 6x^2 + 13x - 10 = 0$$

21. If i is a root, then $-i$ is a root also:
$$(x-3)(x-2)(x-i)(x+i) = 0$$
$$\left(x^2 - 5x + 6\right)\left(x^2 - i^2\right) = 0$$
$$\left(x^2 - 5x + 6\right)\left(x^2 + 1\right) = 0$$
$$x^4 - 5x^3 + 7x^2 - 5x + 6 = 0$$

23. If i and $1 - i$ are roots, then $-i$ and $1 + i$ are roots also:
$$(x-i)(x+i)[x-(1-i)][x-(1+i)] = 0$$
$$\left(x^2 - i^2\right)\left[x^2 - (1+i)x - (1-i)x + (1-i)(1+i)\right] = 0$$
$$\left(x^2 + 1\right)\left[x^2 - x - ix - x + ix + 1 - i^2\right] = 0$$
$$\left(x^2 + 1\right)\left[x^2 - 2x + 2\right] = 0$$
$$x^4 - 2x^3 + 3x^2 - 2x + 2 = 0$$

25. $P(x) = 3x^3 + 5x^2 - 4x + 3$
2 sign variations $\Rightarrow$ 2 or 0 positive roots
$$P(-x) = 3(-x)^3 + 5(-x)^2 - 4(-x) + 3$$
$$= -3x^3 + 5x^2 + 4x + 3$$
1 sign variation $\Rightarrow$ 1 negative root

# pos	# neg	# nonreal
2	1	0
0	1	2

27. $P(x) = 2x^3 + 7x^2 + 5x + 5$
0 sign variations $\Rightarrow$ 0 positive roots
$$P(-x) = 2(-x)^3 + 7(-x)^2 + 5(-x) + 3$$
$$= -2x^3 + 7x^2 - 5x + 3$$
3 sign variations $\Rightarrow$ 3 or 1 negative roots

# pos	# neg	# nonreal
0	3	0
0	1	2

29. $P(x) = 8x^4 + 5$
0 sign variations $\Rightarrow$ 0 positive roots
$$P(-x) = 8(-x)^4 + 5$$
$$= 8x^4 + 5$$
0 sign variations $\Rightarrow$ 0 negative roots

# pos	# neg	# nonreal
0	0	4

31. $P(x) = x^4 + 8x^2 - 5x - 10$
1 sign variation $\Rightarrow$ 1 positive root
$$P(-x) = (-x)^4 + 8(-x)^2 - 5(-x) - 10$$
$$= x^4 + 8x^2 + 5x - 10$$
1 sign variation $\Rightarrow$ 1 negative root

# pos	# neg	# nonreal
1	1	2

33. $P(x) = -x^{10} - x^8 - x^6 - x^4 - x^2 - 1$: 0 sign variations $\Rightarrow$ 0 positive roots
$$P(-x) = -(-x)^{10} - (-x)^8 - (-x)^6 - (-x)^4 - (-x)^2 - 1$$
$$= -x^{10} - x^8 - x^6 - x^4 - x^2 - 1$$: 0 sign variations $\Rightarrow$ 0 negative roots

# pos	# neg	# nonreal
0	0	10

35. $P(x) = x^9 + x^7 + x^5 + x^3 + x = x\left(x^8 + x^6 + x^4 + x^2 + 1\right)$: 0 sign variations $\Rightarrow$ 0 positive roots
$$P(-x) = (-x)\left[(-x)^8 + (-x)^6 + (-x)^4 + (-x)^2 + 1\right]$$
$$= -x\left[x^8 + x^6 + x^4 + x^2 + 1\right]$$: 0 sign variations $\Rightarrow$ 0 negative roots

# pos	# neg	# zero	# nonreal
0	0	1	8

37. $P(x) = -2x^4 - 3x^2 + 2x + 3$: 1 sign variation $\Rightarrow$ 1 positive root

$P(-x) = -2(-x)^4 - 3(-x)^2 + 2(-x) + 3$

$\quad\quad = -2x^4 - 3x^2 - 2x + 3$: 1 sign variation $\Rightarrow$ 1 negative root

# pos	# neg	# nonreal
1	1	2

39. $\quad\quad P(x) = x^2 - 2x - 4$

$$\begin{array}{r|rrr} 4 & 1 & -2 & -4 \\ & & 4 & 8 \\ \hline & 1 & 2 & 4 \end{array}$$

Upper bound: 4

$$\begin{array}{r|rrr} -2 & 1 & -2 & -4 \\ & & -2 & 8 \\ \hline & 1 & -4 & 4 \end{array}$$

Lower bound: -2

41. $\quad P(x) = 18x^2 - 6x - 1$

$$\begin{array}{r|rrr} 1 & 18 & -6 & -1 \\ & & 18 & 12 \\ \hline & 18 & 12 & 11 \end{array}$$

Upper bound: 1

$$\begin{array}{r|rrr} -1 & 18 & -6 & -1 \\ & & -18 & 24 \\ \hline & 18 & -24 & 23 \end{array}$$

Lower bound: -1

43. $\quad\quad P(x) = 6x^3 - 13x^2 - 110x$

$$\begin{array}{r|rrrr} 6 & 6 & -13 & -110 & 0 \\ & & 36 & 138 & 168 \\ \hline & 6 & 23 & 28 & 168 \end{array}$$

Upper bound: 6

$$\begin{array}{r|rrrr} -4 & 6 & -13 & -110 & 0 \\ & & -24 & 148 & -152 \\ \hline & 6 & -37 & 38 & -152 \end{array}$$

Lower bound: -4

45. $\quad P(x) = x^5 + x^4 - 8x^3 - 8x^2 + 15x + 15$

$$\begin{array}{r|rrrrrr} 3 & 1 & 1 & -8 & -8 & 15 & 15 \\ & & 3 & 12 & 12 & 12 & 81 \\ \hline & 1 & 4 & 4 & 4 & 27 & 96 \end{array}$$

Upper bound: 3

$$\begin{array}{r|rrrrrr} -4 & 1 & 1 & -8 & -8 & 15 & 15 \\ & & -4 & 12 & -16 & 96 & -444 \\ \hline & 1 & -3 & 4 & -24 & 111 & -429 \end{array}$$

Lower bound: -4

47. $\quad\quad\quad P(x) = 3x^5 - 11x^4 - 2x^3 + 38x^2 - 21x - 15$

$$\begin{array}{r|rrrrrr} 4 & 3 & -11 & -2 & 38 & -21 & -15 \\ & & 12 & 4 & 8 & 184 & 652 \\ \hline & 3 & 1 & 2 & 46 & 163 & 637 \end{array}$$

Upper bound: 4

$$\begin{array}{r|rrrrrr} -2 & 3 & -11 & -2 & 38 & -21 & -15 \\ & & -6 & 34 & -64 & 52 & -62 \\ \hline & 3 & -17 & 32 & -26 & 31 & -77 \end{array}$$

Lower bound: -2

49. **Answers may vary.**

51. The number of nonreal roots must occur in conjugate pairs, so the number of nonreal roots will always be even. Since a polynomial of odd degree has an odd number of roots, at least one root must not be nonreal. Thus, at least one root of such a polynomial will be real.

53. k units to the right

55. reflected about the y-axis

57. stretched vertically by a factor of k

Exercise 5.3 (page 408)

1. -7

3. root

5.

$$x^3 - 5x^2 - x + 5 = 0$$

Possible rational roots: $\pm 1, \pm 5$

Descartes' Rule of Signs:

# pos	# neg	# nonreal
2	1	0
0	1	2

Test $x = -1$:

```
-1 | 1  -5  -1   5
   |     -1   6  -5
   ----------------
     1  -6   5   0
```

Factor out $(x + 1)$:

$$x^3 - 5x^2 - x + 5 = 0$$
$$(x + 1)(x^2 - 6x + 5) = 0$$
$$(x + 1)(x - 5)(x - 1) = 0$$

Solution set: $\{-1, 5, 1\}$

7.

$$x^3 - 2x^2 - 9x + 18 = 0$$

Possible rational roots:

$$\pm 1, \pm 2, \pm 3, \pm 6, \pm 9, \pm 18$$

Descartes' Rule of Signs:

# pos	# neg	# nonreal
2	1	0
0	1	2

Test $x = 2$:

```
2 | 1  -2  -9   18
  |      2   0  -18
  -----------------
    1   0  -9    0
```

Factor out $(x - 2)$:

$$x^3 - 2x^2 - 9x + 18 = 0$$
$$(x - 2)(x^2 - 9) = 0$$
$$(x - 2)(x + 3)(x - 3) = 0$$

Solution set: $\{2, -3, 3\}$

9.

$$x^3 - 2x^2 - x + 2 = 0$$

Possible rational roots

$$\pm 1, \pm 2$$

Descartes' Rule of Signs

# pos	# neg	# nonreal
2	1	0
0	1	2

Test $x = -1$:

```
-1 | 1  -2  -1   2
   |     -1   3  -2
   ---------------
     1  -3   2   0
```

$$x^3 - 2x^2 - x + 2 = 0$$
$$(x + 1)(x^2 - 3x + 2) = 0$$
$$(x + 1)(x - 1)(x - 2) = 0$$

Solution set: $\{-1, 1, 2\}$

11.

Possible rational roots

$$\pm 1, \pm 2, \pm 3, \pm 4,$$
$$\pm 6, \pm 8, \pm 12, \pm 24$$

Descartes' Rule of Signs

# pos	# neg	# nonreal
4	0	0
2	0	2
0	0	4

$$x^4 - 10x^3 + 35x^2 - 50x + 24 = 0$$
$$(x - 1)(x^3 - 9x^2 + 26x - 24) = 0$$
$$(x - 1)(x - 2)(x^2 - 7x + 12) = 0$$
$$(x - 1)(x - 2)(x - 3)(x - 4) = 0$$

Solution set: $\{1, 2, 3, 4\}$

Test $x = 1$:

```
1 | 1  -10   35  -50   24
  |       1   -9   26  -24
  -----------------------
    1   -9   26  -24    0
```

Test $x = 2$:

```
2 | 1  -9   26  -24
  |      2  -14   24
  -----------------
    1  -7   12    0
```

13.

Possible rational roots
$\pm 1, \pm 2, \pm 3, \pm 5,$
$\pm 6, \pm 10, \pm 15,$
± 30

Descartes' Rule of Signs

# pos	# neg	# nonreal
2	2	0
2	0	2
0	2	2
0	0	4

$$x^4 + 3x^3 - 13x^2 - 9x + 30 = 0$$
$$(x - 2)(x^3 + 5x^2 - 3x - 15) = 0$$
$$(x - 2)(x + 5)(x^2 - 3) = 0$$
$x^2 - 3$ does not factor rationally.
Rational solutions: $\{2, -5\}$

Test $x = 2$:

$$
\begin{array}{r|rrrrr}
2 & 1 & 3 & -13 & -9 & 30 \\
  &   & 2 & 10 & -6 & -30 \\
\hline
  & 1 & 5 & -3 & -15 & 0
\end{array}
$$

Test $x = -5$:

$$
\begin{array}{r|rrrr}
-5 & 1 & 5 & -3 & -15 \\
   &   & -5 & 0 & 15 \\
\hline
   & 1 & 0 & -3 & 0
\end{array}
$$

15.

Possible rat. roots
$\pm 1, \pm 2, \pm 3,$
$\pm 4, \pm 6, \pm 12$

Descartes' Rule of Signs

# pos	# neg	# nonreal
2	3	0
2	1	2
0	3	2
0	1	4

$$x^5 + 3x^4 - 5x^3 - 15x^2 + 4x + 12 = 0$$
$$(x + 1)(x^4 + 2x^3 - 7x^2 - 8x + 12) = 0$$
$$(x + 1)(x - 1)(x^3 + 3x^2 - 4x - 12) = 0$$
$$(x + 1)(x - 1)(x - 2)(x^2 + 5x + 6) = 0$$
$$(x + 1)(x - 1)(x - 2)(x + 2)(x + 3) = 0$$
Solution set: $\{-1, 1, 2, -2, -3\}$

Test $x = -1$:

$$
\begin{array}{r|rrrrrr}
-1 & 1 & 3 & -5 & -15 & 4 & 12 \\
   &   & -1 & -2 & 7 & 8 & -12 \\
\hline
   & 1 & 2 & -7 & -8 & 12 & 0
\end{array}
$$

Test $x = 1$:

$$
\begin{array}{r|rrrrr}
1 & 1 & 2 & -7 & -8 & 12 \\
  &   & 1 & 3 & -4 & -12 \\
\hline
  & 1 & 3 & -4 & -12 & 0
\end{array}
$$

Test $x = 2$:

$$
\begin{array}{r|rrrr}
2 & 1 & 3 & -4 & -12 \\
  &   & 2 & 10 & 12 \\
\hline
  & 1 & 5 & 6 & 0
\end{array}
$$

17. First, factor out the common factor of x: $\quad x^7 - 12x^5 + 48x^3 - 64x = x(x^6 - 12x^4 + 48x^2 - 64)$

Possible rat. roots

$$\pm 1, \pm 2, \pm 4, \pm 8, \pm 16, \pm 32, \pm 64$$

Descartes' Rule of Signs

# pos	# neg	# zero	# nonreal
3	3	1	0
3	1	1	2
1	3	1	2
1	1	1	4

Test $x = 2$:

$$
\begin{array}{r|rrrrrrr}
2 & 1 & 0 & -12 & 0 & 48 & 0 & -64 \\
  &   & 2 & 4 & -16 & -32 & 32 & 64 \\
\hline
  & 1 & 2 & -8 & -16 & 16 & 32 & 0
\end{array}
$$

Test $x = 2$:

$$
\begin{array}{r|rrrrrr}
2 & 1 & 2 & -8 & -16 & 16 & 32 \\
  &   & 2 & 8 & 0 & -32 & -32 \\
\hline
  & 1 & 4 & 0 & -16 & -16 & 0
\end{array}
$$

Test $x = 2$:

$$
\begin{array}{r|rrrrr}
2 & 1 & 4 & 0 & -16 & -16 \\
  &   & 2 & 12 & 24 & 16 \\
\hline
  & 1 & 6 & 12 & 8 & 0
\end{array}
$$

Test $x = -2$:

$$
\begin{array}{r|rrrr}
-2 & 1 & 6 & 12 & 8 \\
   &   & -2 & -8 & -8 \\
\hline
   & 1 & 4 & 4 & 0
\end{array}
$$

continued on next page...

17. continued

$$x^7 - 12x^5 + 48x^3 - 64x = 0$$
$$x(x^6 - 12x^4 + 48x^2 - 64) = 0$$
$$x(x-2)(x^5 + 2x^4 - 8x^3 - 16x^2 + 16x + 32) = 0$$
$$x(x-2)(x-2)(x^4 + 4x^3 - 16x - 16) = 0$$
$$x(x-2)(x-2)(x-2)(x^3 + 6x^2 + 12x + 8) = 0$$
$$x(x-2)(x-2)(x-2)(x+2)(x^2 + 4x + 4) = 0$$
$$x(x-2)(x-2)(x-2)(x+2)(x+2)(x+2) = 0 \Rightarrow \text{Sol. set} = \{0, 2, 2, 2, -2, -2, -2\}$$

19.

Possible rational roots

$$\pm 1, \pm 2, \pm 4, \pm 8$$
$$\pm \tfrac{1}{3}, \pm \tfrac{2}{3}, \pm \tfrac{4}{3}, \pm \tfrac{8}{3}$$

Descartes' Rule of Signs

# pos	# neg	# nonreal
3	0	0
1	0	2

$3x^3 - 2x^2 + 12x - 8 = 0$
$(x - \tfrac{2}{3})(3x^2 + 12) = 0$
$3x^2 + 12$ does not factor rationally.
Rational solutions: $\left\{\tfrac{2}{3}\right\}$

Test $x = \tfrac{2}{3}$:

$$\begin{array}{r|rrrr}
\tfrac{2}{3} & 3 & -2 & 12 & -8 \\
& & 2 & 0 & 8 \\
\hline
& 3 & 0 & 12 & 0
\end{array}$$

21.

Possible rational roots

$$\pm 1, \pm 2, \pm 3, \pm 4,$$
$$\pm 6, \pm 12, \pm \tfrac{1}{3},$$
$$\pm \tfrac{2}{3}, \pm \tfrac{4}{3}$$

Descartes' Rule of Signs

# pos	# neg	# nonreal
3	1	0
1	1	2

$3x^4 - 14x^3 + 11x^2 + 16x - 12 = 0$
$(x+1)(3x^3 - 17x^2 + 28x - 12) = 0$
$(x+1)(x-2)(3x^2 - 11x + 6) = 0$
$(x+1)(x-2)(3x-2)(x-3) = 0$
Solution set: $\left\{-1, 2, \tfrac{2}{3}, 3\right\}$

Test $x = -1$:

$$\begin{array}{r|rrrrr}
-1 & 3 & -14 & 11 & 16 & -12 \\
& & -3 & 17 & -28 & 12 \\
\hline
& 3 & -17 & 28 & -12 & 0
\end{array}$$

Test $x = 2$:

$$\begin{array}{r|rrrr}
2 & 3 & -17 & 28 & -12 \\
& & 6 & -22 & 12 \\
\hline
& 3 & -11 & 6 & 0
\end{array}$$

23.

Possible rational roots

$$\pm 1, \pm 3, \pm \tfrac{1}{2}, \pm \tfrac{3}{2},$$
$$\pm \tfrac{1}{3}, \pm \tfrac{1}{4}, \pm \tfrac{3}{4}, \pm \tfrac{1}{6},$$
$$\pm \tfrac{1}{12}$$

Descartes' Rule of Signs

# pos	# neg	# nonreal
3	1	0
1	1	2

$12x^4 + 20x^3 - 41x^2 + 20x - 3 = 0$
$(x+3)(12x^3 - 16x^2 + 7x - 1) = 0$
$(x+3)(x - \tfrac{1}{2})(12x^2 - 10x + 2) = 0$
$(x+3)(x - \tfrac{1}{2})(2x-1)(6x-2) = 0$
Solution set: $\left\{-3, \tfrac{1}{2}, \tfrac{1}{2}, \tfrac{1}{3}\right\}$

Test $x = -3$:

$$\begin{array}{r|rrrrr}
-3 & 12 & 20 & -41 & 20 & -3 \\
& & -36 & 48 & -21 & 3 \\
\hline
& 12 & -16 & 7 & -1 & 0
\end{array}$$

Test $x = \tfrac{1}{2}$:

$$\begin{array}{r|rrrr}
\tfrac{1}{2} & 12 & -16 & 7 & -1 \\
& & 6 & -5 & 1 \\
\hline
& 12 & -10 & 2 & 0
\end{array}$$

25.

Possible rat. roots
$\pm 1, \pm 2, \pm 3,$
$\pm 4, \pm 6, \pm 12,$
$\pm\frac{1}{2}, \pm\frac{3}{2}, \pm\frac{1}{3},$
$\pm\frac{2}{3}, \pm\frac{4}{3}, \pm\frac{1}{6}$

Descartes' Rule of Signs

# pos	# neg	# nonreal
3	2	0
3	0	2
1	2	2
1	0	4

$6x^5 - 7x^4 - 48x^3 + 81x^2 - 4x - 12 = 0$
$(x - 2)(6x^4 + 5x^3 - 38x^2 + 5x + 6) = 0$
$(x - 2)(x - 2)(6x^3 + 17x^2 - 4x - 3) = 0$
$(x - 2)(x - 2)(x + 3)(6x^2 - x - 1) = 0$
$(x - 2)(x - 2)(x + 3)(2x - 1)(3x + 1) = 0$
Solution set: $\left\{2, 2, -3, \frac{1}{2}, -\frac{1}{3}\right\}$

Test $x = 2$:

$$
\begin{array}{r|rrrrrr}
2 & 6 & -7 & -48 & 81 & -4 & -12 \\
  &   & 12 & 10 & -76 & 10 & 12 \\
\hline
  & 6 & 5 & -38 & 5 & 6 & 0
\end{array}
$$

Test $x = 2$:

$$
\begin{array}{r|rrrrr}
2 & 6 & 5 & -38 & 5 & 6 \\
  &   & 12 & 34 & -8 & -6 \\
\hline
  & 6 & 17 & -4 & -3 & 0
\end{array}
$$

Test $x = -3$:

$$
\begin{array}{r|rrrr}
-3 & 6 & 17 & -4 & -3 \\
   &   & -18 & 3 & 3 \\
\hline
   & 6 & -1 & -1 & 0
\end{array}
$$

27.

Possible rational roots
$\pm 1, \pm 2, \pm 3, \pm 6,$
$\pm 9, \pm 18, \pm\frac{1}{2}, \pm\frac{3}{2}$
$\pm\frac{9}{2}, \pm\frac{1}{3}, \pm\frac{2}{3}, \pm\frac{1}{5},$
$\pm\frac{2}{5}, \pm\frac{3}{5}, \pm\frac{6}{5}, \pm\frac{9}{5},$
$\pm\frac{18}{5}, \pm\frac{1}{6}, \pm\frac{1}{10}, \pm\frac{3}{10},$
$\pm\frac{9}{10}, \pm\frac{1}{15}, \pm\frac{2}{15}, \pm\frac{1}{30}$

Descartes' Rule of Signs

# pos	# neg	# nonreal
2	1	0
0	1	2

Test $x = -\frac{3}{5}$:

$$
\begin{array}{r|rrrr}
-\frac{3}{5} & 30 & -47 & -9 & 18 \\
             &   & -18 & 39 & -18 \\
\hline
             & 30 & -65 & 30 & 0
\end{array}
$$

$30x^3 - 47x^2 - 9x + 18 = 0$
$\left(x + \frac{3}{5}\right)(30x^2 - 65x + 30) = 0$
$\left(x + \frac{3}{5}\right)(3x - 2)(10x - 15) = 0$
Solution set: $\left\{-\frac{3}{5}, \frac{2}{3}, \frac{3}{2}\right\}$

29.

Possible rational roots
$\pm 1, \pm 2, \pm 3, \pm 4,$
$\pm 6, \pm 8, \pm 12, \pm 24$
$\pm\frac{1}{3}, \pm\frac{2}{3}, \pm\frac{4}{3}, \pm\frac{8}{3},$
$\pm\frac{1}{5}, \pm\frac{2}{5}, \pm\frac{3}{5}, \pm\frac{4}{5},$
$\pm\frac{6}{5}, \pm\frac{8}{5}, \pm\frac{12}{5}, \pm\frac{24}{5},$
$\pm\frac{1}{15}, \pm\frac{2}{15}, \pm\frac{4}{15}, \pm\frac{8}{15}$

Descartes' Rule of Signs

# pos	# neg	# nonreal
2	1	0
0	1	2

Test $x = -\frac{3}{5}$:

$$
\begin{array}{r|rrrr}
-\frac{3}{5} & 15 & -61 & -2 & 24 \\
             &   & -9 & 42 & -24 \\
\hline
             & 15 & -70 & 40 & 0
\end{array}
$$

$15x^3 - 61x^2 - 2x + 24 = 0$
$\left(x + \frac{3}{5}\right)(15x^2 - 70x + 40) = 0$
$\left(x + \frac{3}{5}\right)(3x - 2)(5x - 20) = 0$
Solution set: $\left\{-\frac{3}{5}, \frac{2}{3}, 4\right\}$

31.

Possible rational roots
$\pm 1, \pm 2, \pm 3, \pm 6,$
$\pm 9, \pm 18, \pm\frac{1}{2}, \pm\frac{3}{2}$
$\pm\frac{9}{2}, \pm\frac{1}{3}, \pm\frac{2}{3}, \pm\frac{1}{4},$
$\pm\frac{3}{4}, \pm\frac{9}{4}, \pm\frac{1}{5}, \pm\frac{2}{5},$
$\pm\frac{3}{5}, \pm\frac{6}{5}, \pm\frac{9}{5}, \pm\frac{18}{5},$
$\pm\frac{1}{10}, \pm\frac{3}{10}, \pm\frac{9}{10}, \pm\frac{1}{20}$

Descartes' Rule of Signs

# pos	# neg	# nonreal
2	1	0
0	1	2

Test $x = -\frac{1}{2}$:

$$
\begin{array}{r|rrrr}
-\frac{1}{2} & 20 & -44 & 9 & 18 \\
             &   & -10 & 27 & -18 \\
\hline
             & 20 & -54 & 36 & 0
\end{array}
$$

$20x^3 - 44x^2 + 9x + 18 = 0$
$\left(x + \frac{1}{2}\right)(20x^2 - 54x + 36) = 0$
$\left(x + \frac{1}{2}\right)(4x - 6)(5x - 6) = 0$
Solution set: $\left\{-\frac{1}{2}, \frac{3}{2}, \frac{6}{5}\right\}$

33. If $(1+i)$ is a root, then so is $(1-i)$, and $x - (1+i)$ and $x - (1-i)$ are factors.

Then $[x - (1+i)][x - (1-i)] = x^2 - 2x + 2$ is a factor. Divide it out:

$$\begin{array}{r} x - 3 \\ x^2 - 2x + 2 \overline{\smash{\big)}\ x^3 - 5x^2 + 8x - 6} \\ \underline{x^3 - 2x^2 + 2x} \\ -3x^2 + 6x - 6 \\ \underline{-3x^2 + 6x - 6} \\ 0 \end{array}$$

$x^3 - 5x^2 + 8x - 6 = 0$

$(x^2 - 2x + 2)(x - 3) = 0$

Solution set: $\{1 + i, 1 - i, 3\}$

35. If $(1+i)$ is a root, then so is $(1-i)$, and $x - (1+i)$ and $x - (1-i)$ are factors.

Then $[x - (1+i)][x - (1-i)] = x^2 - 2x + 2$ is a factor. Divide it out:

$$\begin{array}{r} x^2 - 9 \\ x^2 - 2x + 2 \overline{\smash{\big)}\ x^4 - 2x^3 - 7x^2 + 18x - 18} \\ \underline{x^4 - 2x^3 + 2x^2} \\ -9x^2 + 18x - 18 \\ \underline{-9x^2 + 18x - 18} \\ 0 \end{array}$$

$x^4 - 2x^3 - 7x^2 + 18x - 18 = 0$

$(x^2 - 2x + 2)(x^2 - 9) = 0$

$(x^2 - 2x + 2)(x + 3)(x - 3) = 0$

Solution set: $\{1 + i, 1 - i, -3, 3\}$

37.

Possible rational roots
$\pm 1, \pm 2, \pm 3, \pm 6,$
$\pm \frac{1}{3}, \pm \frac{2}{3}$

Descartes' Rule of Signs

# pos	# neg	# nonreal
1	2	0
1	0	2

$x^3 - \dfrac{4}{3}x^2 - \dfrac{13}{3}x - 2 = 0$

$3x^3 - 4x^2 - 13x - 6 = 0$

$(x + 1)(3x^2 - 7x - 6) = 0$

$(x + 1)(3x + 2)(x - 3) = 0$

Solution set: $\left\{-1, -\frac{2}{3}, 3\right\}$

Test $x = -1$:

$$\begin{array}{r|rrrr} -1 & 3 & -4 & -13 & -6 \\ & & -3 & 7 & 6 \\ \hline & 3 & -7 & -6 & 0 \end{array}$$

39. $x^{-5} - 8x^{-4} + 25x^{-3} - 38x^{-2} + 28x^{-1} - 8 = 0$

$x^5\left(x^{-5} - 8x^{-4} + 25x^{-3} - 38x^{-2} + 28x^{-1} - 8\right) = x^5(0)$

$1 - 8x + 25x^2 - 38x^3 + 28x^4 - 8x^5 = 0$

Possible rat. roots
$\pm 1, \pm \frac{1}{2}, \pm \frac{1}{4},$
$\pm \frac{1}{8}$

Descartes' Rule of Signs

# pos	# neg	# nonreal
5	0	0
3	0	2
1	0	4

$8x^5 - 28x^4 + 38x^3 - 25x^2 + 8x - 1 = 0$

$(x - 1)\left(8x^4 - 20x^3 + 18x^2 - 7x + 1\right) = 0$

Test $x = 1$:

$$\begin{array}{r|rrrrrr} 1 & 8 & -28 & 38 & -25 & 8 & -1 \\ & & 8 & -20 & 18 & -7 & 1 \\ \hline & 8 & -20 & 18 & -7 & 1 & 0 \end{array}$$

continued on next page...

39. **continued...**

Test $x = 1$:

$$
\begin{array}{r|rrrrr}
1 & 8 & -20 & 18 & -7 & 1 \\
  &   & 8 & -12 & 6 & -1 \\
\hline
  & 8 & -12 & 6 & -1 & 0
\end{array}
$$

Test $x = \frac{1}{2}$:

$$
\begin{array}{r|rrrr}
\frac{1}{2} & 8 & -12 & 6 & -1 \\
 & & 4 & -4 & 1 \\
\hline
 & 8 & -8 & 2 & 0
\end{array}
$$

$(x-1)(x-1)\left(8x^3 - 12x^2 + 6x - 1\right) = 0$

$(x-1)(x-1)\left(x - \frac{1}{2}\right)\left(8x^2 - 8x + 2\right) = 0$

$(x-1)(x-1)\left(x - \frac{1}{2}\right)(4x-2)(2x-1) = 0 \Rightarrow$ Solution set: $\left\{1, 1, \frac{1}{2}, \frac{1}{2}, \frac{1}{2}\right\}$

41. Let $x = R_1$. Then $x + 10 = R_2$ and $x + 50 = R_3$.

$$\frac{1}{R} = \frac{1}{R_1} + \frac{1}{R_2} + \frac{1}{R_3}$$

$$\frac{1}{6} = \frac{1}{x} + \frac{1}{x+10} + \frac{1}{x+50}$$

$$6x(x+10)(x+50) \cdot \frac{1}{6} = 6x(x+10)(x+50)\left(\frac{1}{x} + \frac{1}{x+10} + \frac{1}{x+50}\right)$$

$$x(x+10)(x+50) = 6(x+10)(x+50) + 6x(x+50) + 6x(x+10)$$

$$x^3 + 60x^2 + 500x = 6x^2 + 360x + 3000 + 6x^2 + 300x + 6x^2 + 60x$$

$$x^3 + 42x^2 - 220x - 3000 = 0$$

$$(x-10)\left(x^2 + 52x + 300\right) = 0$$

$$
\begin{array}{r|rrrr}
10 & 1 & 42 & -220 & -3000 \\
 & & 10 & 520 & 3000 \\
\hline
 & 1 & 52 & 300 & 0
\end{array}
$$

Use the quadratic formula on the second factor. The two solutions from that factor are negative. The only solution that makes sense is $x = 10$. The resistances are 10, 20 and 60 ohms.

43. **Answers may vary.**

45. A point on the parabola has coordinates $\left(x, 16 - x^2\right)$. $A = 2x\left(16 - x^2\right) = 32x - 2x^3$.

$$32x - 2x^3 = 42$$

$$-2x^3 + 32x - 42 = 0$$

$$x^3 - 16x + 21 = 0$$

$$(x-3)\left(x^2 + 3x - 7\right) = 0$$

Using the quadratic formula on the second factor

yields the solutions $x = \dfrac{-3 \pm \sqrt{37}}{2} \approx 1.54$ or -4.54.

Test $x = 3$:

$$
\begin{array}{r|rrrr}
3 & 1 & 0 & -16 & 21 \\
 & & 3 & 9 & -21 \\
\hline
 & 1 & 3 & -7 & 0
\end{array}
$$

The only solutions that make sense are $x = 3$ or $x = 1.54$.

Points: $(3, 7)$ or $(1.54, 13.63)$

47. $\sqrt{72a^3b^5c} = \sqrt{36a^2b^4}\sqrt{2abc} = 6ab^2\sqrt{2abc}$

49. $\sqrt{18a^2b} + a\sqrt{50b} = \sqrt{9a^2}\sqrt{2b} + a\sqrt{25}\sqrt{2b} = 3a\sqrt{2b} + 5a\sqrt{2b} = 8a\sqrt{2b}$

51. $\dfrac{2}{\sqrt{3}-1} = \dfrac{2\left(\sqrt{3}+1\right)}{\left(\sqrt{3}-1\right)\left(\sqrt{3}+1\right)} = \dfrac{2\left(\sqrt{3}+1\right)}{\left(\sqrt{3}\right)^2 - 1^2} = \dfrac{2\left(\sqrt{3}+1\right)}{3-1} = \dfrac{2\left(\sqrt{3}+1\right)}{2} = \sqrt{3}+1$

Exercise 5.4 (page 413)

1. $P(a)$ and $P(b)$

3. x_l and x_r

5. $P(x) = 2x^2 + x - 3$
$P(-2) = 3; P(-1) = -2$
Thus, there is a root between -2 and -1.

7. $P(x) = 3x^3 - 11x^2 - 14x$
$P(4) = -40; P(5) = 30$
Thus, there is a root between 4 and 5.

9. $P(x) = x^4 - 8x^2 + 15$
$P(1) = 8; P(2) = -1$
Thus, there is a root between 1 and 2.

11.
$$30x^3 + 10 = 61x^2 + 39x$$
$30x^3 - 61x^2 - 39x + 10 = 0$
$P(x) = 30x^3 - 61x^2 - 39x + 10$
$P(2) = -72; P(3) = 154$
Thus, there is a root between 2 and 3.

13.
$$30x^3 + 10 = 61x^2 + 39x$$
$30x^3 - 61x^2 - 39x + 10 = 0$
$P(x) = 30x^3 - 61x^2 - 39x + 10$
$P(0) = 10; P(1) = -60$
Thus, there is a root between 0 and 1.

15. $P(x) = x^2 - 3$. Note that $P(1) = -2$ and $P(2) = 1$. There is a root between 1 and 2.

STEP	x_l	c	x_r	$P(x_l)$	$P(x_c)$	$P(x_r)$
0	1	1.5	2	−2.0	−0.75	1.0
1	1.5	1.75	2	−0.75	0.0625	1.0
2	1.5	1.625	1.75	−0.75	−0.3594	0.0625
3	1.625	1.6875	1.75	−0.3594	−0.1523	0.0625
4	1.6875	1.71875	1.75	−0.1523	−0.0459	0.0625
5	1.71875	1.734375	1.75			

To the nearest tenth, the solution is $x = 1.7$.

17. $P(x) = x^2 - 5$. Note that $P(-3) = 4$ and $P(-2) = -1$. There is a root between -3 and -2.

STEP	x_l	c	x_r	$P(x_l)$	$P(x_c)$	$P(x_r)$
0	−3	−2.5	−2	4.0	1.25	−1.0
1	−2.5	−2.25	−2	1.25	0.0625	−1.0
2	−2.25	−2.125	−2	0.0625	−0.4844	−1.0
3	−2.25	−2.1875	−2.125	0.0625	−0.2148	−0.4844
4	−2.25	−2.21875	−2.1875	0.0625	−0.0771	−0.2148
5	−2.25	−2.234375	−2.21875			

To the nearest tenth, the solution is $x = -2.2$.

SECTION 5.4

19. $P(x) = x^3 - x^2 - 2$. Note that $P(1) = -2$ and $P(2) = 2$. There is a root between 1 and 2.

STEP	x_l	c	x_r	$P(x_l)$	$P(x_c)$	$P(x_r)$
0	1	1.5	2	−2.0	−0.875	2.0
1	1.5	1.75	2	−0.875	0.2969	2.0
2	1.5	1.625	1.75	−0.875	−0.3496	0.2969
3	1.625	1.6875	1.75	−0.3496	−0.0422	0.2969
4	1.6875	1.71875	1.75	−0.0422	0.1233	0.2969
5	1.6875	1.703125	1.71875			

To the nearest tenth, the solution is $x = 1.7$.

21. $P(x) = 3x^4 + 3x^3 - x^2 - 4x - 4$. Note that $P(-2) = 24$ and $P(-1) = -1$.
There is a root between -2 and -1.

STEP	x_l	c	x_r	$P(x_l)$	$P(x_c)$	$P(x_r)$
0	−2	−1.5	−1	24.0	4.8125	−1.0
1	−1.5	−1.25	−1	4.8125	0.9023	−1.0
2	−1.25	−1.125	−1	0.9023	−0.2317	−1.0
3	−1.25	−1.1875	−1.125	0.9023	0.2818	−0.2317
4	−1.1875	−1.15625	−1.125	0.2818	0.0127	−0.2317
5	−1.15625	−1.140625	−1.125	0.0127	−0.1125	−0.2317
6	−1.15625	−1.1484375	−1.140625	0.0127	−0.0507	−0.1125
7	−1.15625	−1.15234375	−1.1484375	0.0127	−0.0192	−0.0507
8	−1.15625	−1.154296875	−1.15234375			

To the nearest tenth, the solution is $x = -1.2$.

23. Find the x-coordinates of any x-intercepts:

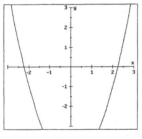

$\{-2.2, 2.2\}$

25. Find the x-coordinates of any x-intercepts:

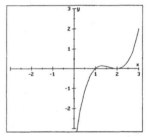

$\{1, 2\}$: Bisection fails for $x = 2$ since the solution cannot be bracketed by a negative y-coordinate and a positive y-coordinate.

27. A point on the graph has coordinates (x, x^3). Then the distance to the origin is found by the distance formula:

$$d = \sqrt{(x-0)^2 + (x^3-0)^2} = \sqrt{x^2 + x^6} = 1$$
$$\sqrt{x^2 + x^6} = 1$$
$$x^2 + x^6 = 1$$

$x^2 + x^6 - 1 = 0$: Graph and find intercepts.
The x-coordinates are about 0.83 and -0.83.
Find the y-coordinates using $y = x^3$.
The points are $(0.83, 0.56)$ and $(-0.83, -0.56)$.

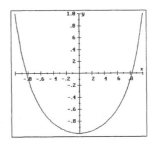

29. **Answers may vary.**

31. $y = \dfrac{5x-3}{x^2-4} = \dfrac{5x-3}{(x+2)(x-2)}$
vertical: $x = -2, x = 2$
horizontal: $y = 0$
slant: none

33. $y = \dfrac{x^2}{x-2} = x + 2 + \dfrac{4}{x-2}$
vertical: $x = 2$
horizontal: none
slant: $y = x + 2$

Chapter 5 Summary (page 416)

1. **a.** $P(1) = 4(1)^4 + 2(1)^3 - 3(1)^2 - 2 = 1 \Rightarrow$ The remainder is 1.

b. $P(2) = 4(2)^4 + 2(2)^3 - 3(2)^2 - 2 = 66 \Rightarrow$ The remainder is 66.

c. $P(-3) = 4(-3)^4 + 2(-3)^3 - 3(-3)^2 - 2 = 241 \Rightarrow$ The remainder is 241.

d. $P(-2) = 4(-2)^4 + 2(-2)^3 - 3(-2)^2 - 2 = 34 \Rightarrow$ The remainder is 34.

2. **a.** $P(2) = (2)^3 + 4(2)^2 - 2(2) + 4 = 24 \Rightarrow$ The remainder is 24. $\Rightarrow$ not a factor

b. $P(-3) = 2(-3)^4 + 10(-3)^3 + 4(-3)^2 + 7(-3) + 21 = -83 \Rightarrow$ The remainder is -83.
$\Rightarrow$ not a factor

c. $P(5) = (5)^5 - 3125 = 0 \Rightarrow$ The remainder is 0. $\Rightarrow$ factor

d. $P(6) = (6)^5 - 6(6)^4 - 4(6) + 24 = 0 \Rightarrow$ The remainder is 0. $\Rightarrow$ factor

3. **a.** $2(x+1)(x-2)\left(x-\frac{3}{2}\right) = 2(x^2 - x - 2)\left(x - \frac{3}{2}\right) = 2\left(x^3 - \frac{5}{2}x^2 - \frac{1}{2}x + 3\right)$
$$= 2x^3 - 5x^2 - x + 6$$

b. $2(x-1)(x+3)\left(x-\frac{1}{2}\right) = 2(x^2 + 2x - 3)\left(x - \frac{1}{2}\right) = 2\left(x^3 + \frac{3}{2}x^2 - 4x + \frac{3}{2}\right)$
$$= 2x^3 + 3x^2 - 8x + 3$$

c. $(x-2)(x+5)(x-i)(x+i) = \left(x^2 + 3x - 10\right)\left(x^2 - i^2\right) = \left(x^2 + 3x - 10\right)\left(x^2 + 1\right)$
$$= x^4 + 3x^3 - 9x^2 + 3x - 10$$

d. $(x+3)(x-2)(x-i)(x+i) = \left(x^2 + x - 6\right)\left(x^2 - i^2\right) = \left(x^2 + x - 6\right)\left(x^2 + 1\right)$
$$= x^4 + x^3 - 5x^2 + x - 6$$

4. **a.**

$$
\begin{array}{r|rrrrr}
3 & 3 & 0 & 2 & 3 & 7 \\
 & & 9 & 27 & 87 & 270 \\
\hline
 & 3 & 9 & 29 & 90 & 277
\end{array}
$$

quotient: $3x^3 + 9x^2 + 29x + 90$
remainder: 277

b.

$$
\begin{array}{r|rrrrr}
2 & 2 & 0 & -3 & 3 & -1 \\
 & & 4 & 8 & 10 & 26 \\
\hline
 & 2 & 4 & 5 & 13 & 25
\end{array}
$$

quotient: $2x^3 + 4x^2 + 5x + 13$
remainder: 25

c.

$$
\begin{array}{r|rrrrrr}
-2 & 5 & -4 & 3 & -2 & 1 & -1 \\
 & & -10 & 28 & -62 & 128 & -258 \\
\hline
 & 5 & -14 & 31 & -64 & 129 & -259
\end{array}
$$

quotient: $5x^4 - 14x^3 + 31x^2 - 64x + 129$
remainder: -259

d.

$$
\begin{array}{r|rrrrrr}
-1 & 4 & 2 & -1 & 3 & 2 & 1 \\
 & & -4 & 2 & -1 & -2 & 0 \\
\hline
 & 4 & -2 & 1 & 2 & 0 & 1
\end{array}
$$

quotient: $4x^4 - 2x^3 + x^2 + 2x$
remainder: 1

5. **a.** $3x^6 - 4x^5 + 3x + 2 = 0$
6 roots

b. $2x^6 - 5x^4 + 5x^3 - 4x + x - 12 = 0$
6 roots

c. $3x^{65} - 4x^{50} + 3x^{17} + 2x = 0$
65 roots

d. $x^{1984} - 12 = 0$
1984 roots

6. **a.** $2 - i$ is also a root.

b. $-i = 0 - i$, so $0 + i = i$ is also a root.

7. **a.** $P(x) = 3x^4 + 2x^3 - 4x + 2$: 2 sign variations $\Rightarrow$ 2 or 0 positive roots
$P(-x) = 3(-x)^4 + 2(-x)^3 - 4(-x) + 2$
$$= 3x^4 - 2x^3 + 4x + 2: \text{ 2 sign variations} \Rightarrow \text{2 or 0 negative roots}$$

# pos	# neg	# nonreal
2	2	0
2	0	2
0	2	2
0	0	4

b. $P(x) = 2x^4 - 3x^3 + 5x^2 + x - 5$: 3 sign variations $\Rightarrow$ 3 or 1 positive roots
$P(-x) = 2(-x)^4 - 3(-x)^3 + 5(-x)^2 + (-x) - 5$
$$= 2x^4 + 3x^3 + 5x^2 - x - 5: \text{ 1 sign variation} \Rightarrow \text{1 negative root}$$

# pos	# neg	# nonreal
3	1	0
1	1	2

c.
$$4x^5 + 3x^4 + 2x^3 + x^2 + x = 7$$
$$4x^5 + 3x^4 + 2x^3 + x^2 + x - 7 = 0$$
$P(x) = 4x^5 + 3x^4 + 2x^3 + x^2 + x - 7$: 1 sign variation $\Rightarrow$ 1 positive root
$P(-x) = 4(-x)^5 + 3(-x)^4 + 2(-x)^3 + (-x)^2 + (-x) - 7$
$\qquad = -4x^5 + 3x^4 - 2x^3 + x^2 - x - 7$: 4 sign variations $\Rightarrow$ 4 or 2 or 0 negative roots

# pos	# neg	# nonreal
1	4	0
1	2	2
1	0	4

d. $P(x) = 3x^7 - 4x^5 + 3x^3 + x - 4$: 3 sign variations $\Rightarrow$ 3 or 1 positive roots
$P(-x) = 3(-x)^7 - 4(-x)^5 + 3(-x)^3 + (-x) - 4$
$\qquad = -3x^7 + 4x^5 - 3x^3 - x - 4$: 2 sign variations $\Rightarrow$ 2 or 0 negative roots

# pos	# neg	# nonreal
3	2	2
3	0	4
1	2	4
1	0	6

e. $P(x) = x^4 + x^2 + 24{,}567$: 0 sign variations $\Rightarrow$ 0 positive roots
$P(-x) = (-x)^4 + (-x)^2 + 24{,}567$
$\qquad = x^4 + x^2 + 24{,}567$: 0 sign variations $\Rightarrow$ 0 negative roots

# pos	# neg	# nonreal
0	0	4

f. $P(x) = -x^7 - 5$: 0 sign variations $\Rightarrow$ 0 positive roots
$P(-x) = -(-x)^7 - 5$
$\qquad = x^7 - 5$: 1 sign variation $\Rightarrow$ 1 negative root

# pos	# neg	# nonreal
0	1	6

8. a. $P(x) = 5x^3 - 4x^2 - 2x + 4$

$$
\begin{array}{r|rrrr}
2 & 5 & -4 & -2 & 4 \\
  &   & 10 & 12 & 20 \\
\hline
  & 5 & 6 & 10 & 24
\end{array}
$$

Upper bound: 2

$$
\begin{array}{r|rrrr}
-1 & 5 & -4 & -2 & 4 \\
   &   & -5 & 9 & -7 \\
\hline
   & 5 & -9 & 7 & -3
\end{array}
$$

Lower bound: -1

b. $P(x) = x^4 + 3x^3 - 5x^2 - 9x + 1$

$$
\begin{array}{r|rrrrr}
2 & 1 & 3 & -5 & -9 & 1 \\
  &   & 2 & 10 & 10 & 2 \\
\hline
  & 1 & 5 & 5 & 1 & 3
\end{array}
$$

Upper bound: 2

$$
\begin{array}{r|rrrrr}
-5 & 1 & 3 & -5 & -9 & 1 \\
   &   & -5 & 10 & -25 & 170 \\
\hline
   & 1 & -2 & 5 & -34 & 171
\end{array}
$$

Lower bound: -5

9. a.

Possible rational roots

$\pm 1, \pm 2, \pm 3, \pm 5,$
$\pm 6, \pm 10, \pm 15, \pm 30,$
$\pm \frac{1}{2}, \pm \frac{3}{2}, \pm \frac{5}{2}, \pm \frac{15}{2}$

Test $x = -2$:

$$\begin{array}{r|rrrr}
-2 & 2 & 17 & 41 & 30 \\
   &   & -4 & -26 & -30 \\
\hline
   & 2 & 13 & 15 & 0
\end{array}$$

Descartes' Rule of Signs

# pos	# neg	# nonreal
0	3	0
0	1	2

$2x^3 + 17x^2 + 41x + 30 = 0$
$(x + 2)(2x^2 + 13x + 15) = 0$
$(x + 2)(2x + 3)(x + 5) = 0$
Solution set: $\left\{-2, -\frac{3}{2}, -5\right\}$

b.

Possible rational roots

$\pm 1, \pm \frac{1}{3}$

Test $x = \frac{1}{3}$:

$$\begin{array}{r|rrrr}
\frac{1}{3} & 3 & 2 & 2 & -1 \\
   &   & 1 & 1 & 1 \\
\hline
   & 3 & 3 & 3 & 0
\end{array}$$

Descartes' Rule of Signs

# pos	# neg	# nonreal
1	2	0
1	0	2

$3x^3 + 2x^2 + 2x - 1 = 0$
$\left(x - \frac{1}{3}\right)(3x^2 + 3x + 3) = 0$
$3x^2 + 3x + 3$ does not factor
rationally.
Rational solutions: $\left\{\frac{1}{3}\right\}$

c.

Possible rat. roots

$\pm 1, \pm 2, \pm 3, \pm 4,$
$\pm 6, \pm 9, \pm 12,$
$\pm 18, \pm 36, \pm \frac{1}{2},$
$\pm \frac{3}{2}, \pm \frac{9}{2}, \pm \frac{1}{4},$
$\pm \frac{3}{4}, \pm \frac{9}{4}$

Test $x = 2$:

$$\begin{array}{r|rrrrr}
2 & 4 & 0 & -25 & 0 & 36 \\
  &   & 8 & 16 & -18 & -36 \\
\hline
  & 4 & 8 & -9 & -18 & 0
\end{array}$$

Descartes' Rule of Signs

# pos	# neg	# nonreal
2	2	0
2	0	2
0	2	2
0	0	4

$4x^4 - 25x^2 + 36 = 0$
$(x - 2)(4x^3 + 8x^2 - 9x - 18) = 0$
$(x - 2)(x + 2)(4x^2 - 9) = 0$
$(x - 2)(x + 2)(2x + 3)(2x - 3) = 0$
Solution set: $\left\{2, -2, -\frac{3}{2}, \frac{3}{2}\right\}$

Test $x = -2$:

$$\begin{array}{r|rrrr}
-2 & 4 & 8 & -9 & -18 \\
   &   & -8 & 0 & 18 \\
\hline
   & 4 & 0 & -9 & 0
\end{array}$$

d.

Possible rat. roots

$\pm 1, \pm 2, \pm 4,$
$\pm 8, \pm 16, \pm 32,$
$\pm \frac{1}{2}$

Test $x = 4$:

$$\begin{array}{r|rrrrr}
4 & 2 & -11 & -6 & 64 & 32 \\
  &   & 8 & -12 & -72 & -32 \\
\hline
  & 2 & -3 & -18 & -8 & 0
\end{array}$$

Descartes' Rule of Signs

# pos	# neg	# nonreal
2	2	0
2	0	2
0	2	2
0	0	4

$2x^4 - 11x^3 - 6x^2 + 64x + 32 = 0$
$(x - 4)(2x^3 - 3x^2 - 18x - 8) = 0$
$(x - 4)(x - 4)(2x^2 + 5x + 2) = 0$
$(x - 4)(x - 4)(2x + 1)(x + 2) = 0$
Solution set: $\left\{4, 4, -\frac{1}{2}, -2\right\}$

Test $x = 4$:

$$\begin{array}{r|rrrr}
4 & 2 & -3 & -18 & -8 \\
  &   & 8 & 20 & 8 \\
\hline
  & 2 & 5 & 2 & 0
\end{array}$$

10. a. $P(x) = 5x^3 + 37x^2 + 59x + 18$
$P(-1) = -9; P(0) = 18$

b. $P(x) = 6x^3 - x^2 - 10x - 3$
$P(1) = -8; P(2) = 21$

11. **a.** $P(x) = x^3 - 2x^2 - 9x - 2$. Note that $P(4) = -6$ and $P(5) = 28$.

There is a root between 4 and 5.

STEP	x_l	c	x_r	$P(x_l)$	$P(x_c)$	$P(x_r)$
0	4	4.5	5	-8.0	8.125	21.0
1	4	4.25	4.5	-8.0	0.3906	8.125
2	4	4.125	4.25	-8.0	-2.9668	0.3906
3	4.125	4.1875	4.25	-2.9668	-1.3293	0.3906
4	4.1875	4.21875	4.25	-1.3293	-0.4798	0.3906
5	4.21875	4.234375	4.25	-0.4798	-0.0472	0.3906
6	4.234375	4.2421875	2.25	-0.0472	0.1711	0.3906
7	4.234275	4.23823125	4.2421875			

To the nearest tenth, the solution is $x = 4.2$.

b. $P(x) = 6x^2 - 13x - 5$. Note that $P(2) = -7$ and $P(3) = 10$.

There is a root between 2 and 3.

STEP	x_l	c	x_r	$P(x_l)$	$P(x_c)$	$P(x_r)$
0	2	2.5	3	-7.0	0	10.0

The solution is exactly $x = 2.5$.

12. **a.** Find the x-coordinates of any positive x-intercepts:

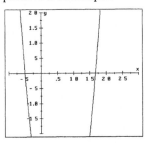

$x = 1.67$

b. Find the x-coordinates of any positive x-intercepts:

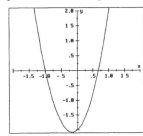

$x = 0.67$

13. Let x = the length of the shortest panel. Then $x + 2$ and $x + 5$ are the lengths of the other panels.

Since each panel has an area of 60, the widths are $\dfrac{60}{x}$, $\dfrac{60}{x + 2}$ and $\dfrac{60}{x + 5}$, respectively.

$$15 = \frac{60}{x} + \frac{60}{x + 2} + \frac{60}{x + 5}$$

$$x(x + 2)(x + 5)(15) = x(x + 2)(x + 5)\left(\frac{60}{x} + \frac{60}{x + 2} + \frac{60}{x + 5}\right)$$

$$15x(x + 2)(x + 5) = 60(x + 2)(x + 5) + 60x(x + 5) + 60x(x + 2)$$

$$15x^3 + 105x^2 + 150x = 60x^2 + 420x + 600 + 60x^2 + 300x + 60x^2 + 120x$$

continued on next page...

213

13. **continued...**

$$15x^3 - 75x^2 - 690x - 600 = 0 \quad \text{Test } x = 10:$$
$$15(x^3 - 5x^2 - 46x - 40) = 0$$
$$x^3 - 5x^2 - 46x - 40 = 0$$
$$(x - 10)(x^2 + 5x + 4) = 0$$
$$(x - 10)(x + 4)(x + 1) = 0$$

$$\begin{array}{r|rrr} 10 & 1 & -5 & -46 & -40 \\ & & 10 & 50 & 40 \\ \hline & 1 & 5 & 4 & 0 \end{array}$$

The positive solution $x = 10$ is the only one that makes sense.

The dimensions are 10 m by 6 m, 12 m by 5 m, and 15 m by 4 m.

14. Let $x =$ the radius. Then $x + 3$ is the height. Graph $y = x^3 + 3x^2 - 6047.89$ and find the x-coordinates of any x-intercepts:

$$\pi r^2 h = V$$
$$r^2 h = \frac{V}{\pi}$$
$$x^2(x + 3) = \frac{19000}{\pi}$$
$$x^3 + 3x^2 = 6047.89$$
$$x^3 + 3x^2 - 6047.89 = 0$$

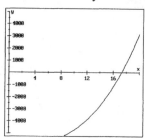

The radius should be about 17.27 feet.

Chapter 5 Test (page 419)

1.
$$\begin{array}{r|rrr} 2 & 3 & 0 & -9 & -5 \\ & & 6 & 12 & 6 \\ \hline & 3 & 6 & 3 & 1 \end{array}$$
$$P(2) = 1$$

2.
$$\begin{array}{r|rrrrrr} -2 & 1 & 0 & 0 & 0 & 0 & 2 \\ & & -2 & 4 & -8 & 16 & -32 \\ \hline & 1 & -2 & 4 & -8 & 16 & -30 \end{array}$$
$$P(-2) = -30$$

3. $(x - 5)(x + 1)(x - 0) = (x^2 - 4x - 5)x = x^3 - 4x^2 - 5x$

4. $(x - i)(x + i)\left(x - \sqrt{3}\right)\left(x + \sqrt{3}\right) = (x^2 - i^2)(x^2 - 3) = (x^2 + 1)(x^2 - 3) = x^4 - 2x^2 - 3$

5.
$$\begin{array}{r|rrrr} 1 & 3 & -2 & 0 & 4 \\ & & 3 & 1 & 1 \\ \hline & 3 & 1 & 1 & 5 \end{array} \quad P(1) = 5$$

6.
$$\begin{array}{r|rrrr} -2 & 3 & -2 & 0 & 4 \\ & & -6 & 16 & -32 \\ \hline & 3 & -8 & 16 & -28 \end{array} \quad P(-2) = -28$$

7.
$$\begin{array}{r|rrrr} -\frac{1}{3} & 3 & -2 & 0 & 4 \\ & & -1 & 1 & -\frac{1}{3} \\ \hline & 3 & -3 & 1 & \frac{11}{3} \end{array} \quad P\left(-\tfrac{1}{3}\right) = \tfrac{11}{3}$$

8.
$$\begin{array}{r|rrrr} i & 3 & -2 & 0 & 4 \\ & & 3i & -3 - 2i & 2 - 3i \\ \hline & 3 & -2 + 3i & -3 - 2i & 6 - 3i \end{array}$$
$$P(i) = 6 - 3i$$

9.
$$\begin{array}{r|rrrr} 2 & 2 & -3 & -4 & -1 \\ & & 4 & 2 & -4 \\ \hline & 2 & 1 & -2 & -5 \end{array}$$ $\quad 2x^3 - 3x^2 - 4x - 1 = (x-2)\left(2x^2 + x - 2\right) - 5$

10.
$$\begin{array}{r|rrrr} -1 & 2 & -3 & -4 & -1 \\ & & -2 & 5 & -1 \\ \hline & 2 & -5 & 1 & -2 \end{array}$$ $\quad 2x^3 - 3x^2 - 4x - 1 = (x+1)\left(2x^2 - 5x + 1\right) - 2$

11.
$$\begin{array}{r|rrr} 5 & 2 & -7 & -15 \\ & & 10 & 15 \\ \hline & 2 & 3 & 0 \end{array}$$ $\quad 2x + 3$

12.
$$\begin{array}{r|rrrr} -2 & 3 & 7 & 2 & 0 \\ & & -6 & -2 & 0 \\ \hline & 3 & 1 & 0 & 0 \end{array}$$ $\quad 3x^2 + x$

13. If i is a root, then $-i$ is a root also:
$$(x-2)(x-i)(x+i) = 0$$
$$(x-2)\left(x^2 - i^2\right) = 0$$
$$(x-2)\left(x^2 + 1\right) = 0$$
$$x^3 - 2x^2 + x - 2 = 0$$

14. If $2 + i$ is a root, then $2 - i$ is a root also:
$$(x-1)[x - (2+i)][x - (2-i)] = 0$$
$$(x-1)\left[x^2 - (2-i)x - (2+i)x + (2+i)(2-i)\right] = 0$$
$$(x-1)\left[x^2 - 2x + ix - 2x - ix + 4 - i^2\right] = 0$$
$$(x-1)\left[x^2 - 4x + 5\right] = 0$$
$$x^3 - 5x^2 + 9x - 5 = 0$$

15. $P(x) = 3x^5 - 2x^4 + 2x^2 - x - 3$: 3 sign variations $\Rightarrow$ 3 or 1 positive roots
$P(-x) = 3(-x)^5 - 2(-x)^4 + 2(-x)^2 - (-x) - 3$
$\qquad = -3x^5 - 2x^4 + 2x^2 + x - 4$: 2 sign variations $\Rightarrow$ 2 or 0 negative roots

# pos	# neg	# nonreal
3	2	0
3	0	2
1	2	2
1	0	4

16. $P(x) = 2x^3 - 5x^2 - 2x - 1$: 1 sign variation $\Rightarrow$ 1 positive root
$P(-x) = 2(-x)^3 - 5(-x)^2 - 2(-x) - 1$
$\qquad = -2x^3 - 5x^2 + 2x - 1$: 2 sign variations $\Rightarrow$ 2 or 0 negative roots

# pos	# neg	# nonreal
1	2	0
1	0	2

17. $P(x) = x^5 - x^4 - 5x^3 + 5x^2 + 4x - 5$

$$
\begin{array}{r|rrrrrr}
3 & 1 & -1 & -5 & 5 & 4 & -5 \\
  &   & 3 & 6 & 3 & 24 & 84 \\
\hline
  & 1 & 2 & 1 & 8 & 28 & 79
\end{array}
$$

Upper bound: 3

$$
\begin{array}{r|rrrrrr}
-3 & 1 & -1 & -5 & 5 & 4 & -5 \\
   &   & -3 & 12 & -21 & 48 & -156 \\
\hline
   & 1 & -4 & 7 & -16 & 52 & -161
\end{array}
$$

Lower bound: -3

18. $P(x) = 2x^3 - 11x^2 + 10x + 3$

$$
\begin{array}{r|rrrr}
6 & 2 & -11 & 10 & 3 \\
  &   & 12 & 6 & 96 \\
\hline
  & 2 & 1   & 16 & 99
\end{array}
$$

Upper bound: 6

$$
\begin{array}{r|rrrr}
-1 & 2 & -11 & 10 & 3 \\
   &   & -2 & 13 & -23 \\
\hline
   & 2 & -13 & 23 & -20
\end{array}
$$

Lower bound: -1

19. Possible rational roots

$\pm 1, \pm 2, \pm 3, \pm 6,$

$\pm \frac{1}{2}, \pm \frac{3}{2}$

Descartes' Rule of Signs

# pos	# neg	# nonreal
1	2	0
1	0	2

$2x^3 + 3x^2 - 11x - 6 = 0$

$(x-2)(2x^2 + 7x + 3) = 0$

$(x-2)(2x+1)(x+3) = 0$

Solution set: $\left\{ 2, -\frac{1}{2}, -3 \right\}$

Test $x = 2$:

$$
\begin{array}{r|rrrr}
2 & 2 & 3 & -11 & -6 \\
  &   & 4 & 14 & 6 \\
\hline
  & 2 & 7 & 3   & 0
\end{array}
$$

20. $P(x) = x^2 - 11$. Note that $P(3) = -2$ and $P(4) = 5$. There is a root between 3 and 4.

STEP	x_l	c	x_r	$P(x_l)$	$P(x_c)$	$P(x_r)$
0	3	3.5	4	-2.0	1.25	5.0
1	3	3.25	3.5	-2.0	-0.4375	1.25
2	3.25	3.375	3.5	-0.4375	0.3906	1.25
3	3.25	3.3125	3.375	-0.4375	-0.0273	0.3906
4	3.3125	3.34375	3.375			

To the nearest tenth, the solution is $x = 3.3$.

Cumulative Review Exercises (page 420)

1. $y = 3^x - 2$

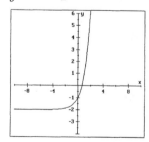

2. $y = 2e^x$

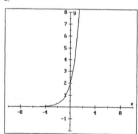

3. $y = \log_3 x$

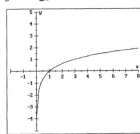

4. $y = \ln(x - 2)$

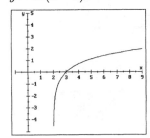

5. $\log_2 64 = 6$ (because $2^6 = 64$)

6. $\log_{1/2} 8 = -3$ $\left(\text{because } \left(\frac{1}{2}\right)^{-3} = 8\right)$

7. $\ln e^3 = 3 \ln e = 3$

8. $2^{\log_2 2} = 2$

9. $\log abc = \log a + \log b + \log c$

10. $\log \dfrac{a^2 b}{c} = \log a^2 b - \log c$
$$= \log a^2 + \log b - \log c$$
$$= 2 \log a + \log b - \log c$$

11. $\log \sqrt{\dfrac{ab}{c^3}} = \log \left(\dfrac{ab}{c^3}\right)^{1/2}$
$$= \tfrac{1}{2} \log \left(\dfrac{ab}{c^3}\right)$$
$$= \tfrac{1}{2}(\log ab - \log c^3)$$
$$= \tfrac{1}{2}(\log a + \log b - 3 \log c)$$

12. $\log \dfrac{\sqrt{ab^2}}{c} = \log \dfrac{(ab^2)^{1/2}}{c}$
$$= \log (ab^2)^{1/2} - \log c$$
$$= \tfrac{1}{2} \log (ab^2) - \log c$$
$$= \tfrac{1}{2}(\log a + \log b^2) - \log c$$
$$= \tfrac{1}{2}(\log a + 2 \log b) - \log c$$
$$= \tfrac{1}{2} \log a + \log b - \log c$$

13. $3 \log a - 3 \log b = \log a^3 - \log b^3 = \log \dfrac{a^3}{b^3}$

14. $\dfrac{1}{2} \log a + 3 \log b - \dfrac{2}{3} \log c = \log a^{1/2} + \log b^3 - \log c^{2/3} = \log \dfrac{a^{1/2} b^3}{c^{2/3}} = \log \dfrac{\sqrt{a}\, b^3}{\sqrt[3]{c^2}}$

15. $3^{x+1} = 8$
$$\log 3^{x+1} = \log 8$$
$$(x + 1) \log 3 = \log 8$$
$$x + 1 = \dfrac{\log 8}{\log 3}$$
$$x = \dfrac{\log 8}{\log 3} - 1$$

16. $3^{x-1} = 3^{2x}$
$$x - 1 = 2x$$
$$-1 = x$$

17. $\log x + \log 2 = 3$
$\log 2x = 3$
$10^3 = 2x$
$1000 = 2x$
$500 = x$

18. $\log(x+1) + \log(x-1) = 1$
$\log(x+1)(x-1) = 1$
$\log(x^2-1) = 1$
$10^1 = x^2 - 1$
$10 = x^2 - 1$
$11 = x^2$
$\pm\sqrt{11} = x$

Only the positive answer, $x = \sqrt{11}$, checks

19.
$$
\begin{array}{r|rrrr}
1 & 4 & 0 & 3 & 2 \\
  &   & 4 & 4 & 7 \\
\hline
  & 4 & 4 & 7 & 9
\end{array}
$$
$P(1) = 9$

20.
$$
\begin{array}{r|rrrr}
-2 & 4 & 0 & 3 & 2 \\
   &   & -8 & 16 & -38 \\
\hline
   & 4 & -8 & 19 & -36
\end{array}
$$
$P(-2) = -36$

21.
$$
\begin{array}{r|rrrr}
\frac{1}{2} & 4 & 0 & 3 & 2 \\
   &   & 2 & 1 & 2 \\
\hline
   & 4 & 2 & 4 & 4
\end{array}
$$
$P(\frac{1}{2}) = 4$

22.
$$
\begin{array}{r|rrrr}
i & 4 & 0 & 3 & 2 \\
  &   & 4i & -4 & -i \\
\hline
  & 4 & 4i & -1 & 2-i
\end{array}
$$
$P(i) = 2 - i$

23.
$$
\begin{array}{r|rrrr}
-1 & 1 & 2 & -1 & -2 \\
   &   & -1 & -1 & 2 \\
\hline
   & 1 & 1 & -2 & 0
\end{array}
$$
factor

24.
$$
\begin{array}{r|rrrr}
2 & 1 & 2 & -1 & -2 \\
  &   & 2 & 8 & 14 \\
\hline
  & 1 & 4 & 7 & 12
\end{array}
$$
not a factor

25.
$$
\begin{array}{r|rrrr}
1 & 1 & 2 & -1 & -2 \\
  &   & 1 & 3 & 2 \\
\hline
  & 1 & 3 & 2 & 0
\end{array}
$$
factor

26.
$$
\begin{array}{r|rrrr}
-2 & 1 & 2 & -1 & -2 \\
   &   & -2 & 0 & 2 \\
\hline
   & 1 & 0 & -1 & 0
\end{array}
$$
factor

27. $x^{12} - 4x^8 + 2x^4 + 12 = 0 \Rightarrow 12$ roots

28. $x^{2000} - 1 = 0 \Rightarrow 2000$ roots

29. $P(x) = x^4 + 2x^3 - 3x^2 + x + 2$: 2 sign variations $\Rightarrow$ 2 or 0 positive roots
$P(-x) = (-x)^4 + 2(-x)^3 - 3(-x)^2 + (-x) + 2$
$= x^4 - 2x^3 - 3x^2 - x + 2$: 2 sign variations $\Rightarrow$ 2 or 0 negative roots

# pos	# neg	# nonreal
2	2	0
2	0	2
0	2	2
0	0	4

30. $P(x) = x^4 - 3x^3 - 2x^2 - 3x - 5$
1 sign variation $\Rightarrow$ 1 positive root
$P(-x) = (-x)^4 - 3(-x)^3 - 2(-x)^2 - 3(-x) - 5$
$= x^4 + 3x^3 - 2x^2 + 3x - 3$
3 sign variations $\Rightarrow$ 3 or 1 negative roots

# pos	# neg	# nonreal
1	3	0
1	1	2

31. Possible rational roots

$\pm 1, \pm 3, \pm 9$

Descartes' Rule of Signs

# pos	# neg	# nonreal
1	2	0
1	0	2

$x^3 + x^2 - 9x - 9 = 0$
$(x+1)(x^2-9) = 0$
$(x+1)(x+3)(x-3) = 0$
Solution set: $\{-1, -3, 3\}$

Test $x = -1$:

$\begin{array}{r|rrrr} -1 & 1 & 1 & -9 & -9 \\ & & -1 & 0 & 9 \\ \hline & 1 & 0 & -9 & 0 \end{array}$

32. Possible rational roots

$\pm 1, \pm 2$

Descartes' Rule of Signs

# pos	# neg	# nonreal
2	1	0
0	1	2

$x^3 - 2x^2 - x + 2 = 0$
$(x+1)(x^2-3x+2) = 0$
$(x+1)(x-1)(x-2) = 0$
Solution set: $\{-1, 1, 2\}$

Test $x = -1$:

$\begin{array}{r|rrrr} -1 & 1 & -2 & -1 & 2 \\ & & -1 & 3 & -2 \\ \hline & 1 & -3 & 2 & 0 \end{array}$

Exercise 6.1 (page 432)

1. system

3. consistent

5. independent

7. consistent

9. dependent

11. is

13. $\begin{cases} y = -3x + 5 \\ x - 2y = -3 \end{cases}$

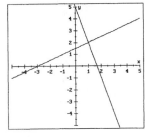

solution: $(1, 2)$

15. $\begin{cases} 3x + 2y = 2 \\ -2x + 3y = 16 \end{cases}$

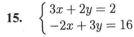
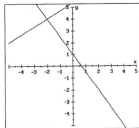

solution: $(-2, 4)$

17. $\begin{cases} y = -5.7x + 7.8 \\ y = 37.2 - 19.1x \end{cases}$

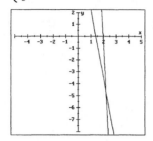

solution: $(2.2, -4.7)$

19. $\begin{cases} y = \dfrac{5.5 - 2.7x}{3.5} \\ 5.3x - 9.2y = 6.0 \end{cases}$

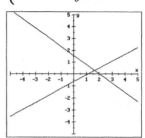

solution: $(1.7, 0.3)$

21. $\begin{cases} (1) & y = x - 1 \\ (2) & y = 2x \end{cases}$

Substitute $y = x - 1$ from (1) into (2):

$y = 2x$

$x - 1 = 2x$

$-1 = x$

Substitute and solve for y:

$y = 2x$

$y = 2(-1) = -2$

$\boxed{x = -1, y = -2}$

23. $\begin{cases} (1) & 2x + 3y = 0 \\ (2) & y = 3x - 11 \end{cases}$

Substitute $y = 3x - 11$ from (2) into (1):

$2x + 3y = 0$

$2x + 3(3x - 11) = 0$

$2x + 9x - 33 = 0$

$11x = 33$

$x = 3$

Substitute and solve for y:

$y = 3x - 11$

$y = 3(3) - 11 = -2$

$\boxed{x = 3, y = -2}$

25. $\begin{cases} (1) & 4x + 3y = 3 \\ (2) & 2x - 6y = -1 \end{cases}$

Substitute $y = \dfrac{3 - 4x}{3}$ from (1) into (2):

$2x - 6y = -1$

$2x - 6 \cdot \dfrac{3 - 4x}{3} = -1$

$2x - 2(3 - 4x) = -1$

$2x - 6 + 8x = -1$

$10x = 5$

$x = \frac{1}{2}$

Substitute and solve for y:

$4x + 3y = 3$

$4\left(\frac{1}{2}\right) + 3y = 3$

$2 + 3y = 3$

$3y = 1$

$y = \frac{1}{3}$

Solution:

$\boxed{x = \frac{1}{2}, y = \frac{1}{3}}$

27. $\begin{cases} (1) & x + 3y = 1 \\ (2) & 2x + 6y = 3 \end{cases}$

Substitute $x = 1 - 3y$ from (1) into (2):

$$2x + 6y = 3$$
$$2(1 - 3y) + 6y = 3$$
$$2 - 6y + 6y = 3$$
$$2 \neq 3 \Rightarrow \boxed{\text{Inconsistent system} \Rightarrow \text{No solution}}$$

29. $\begin{aligned} 5x - 3y &= 12 \Rightarrow \times (-1) \\ 2x - 3y &= 3 \end{aligned}$ $\quad \begin{aligned} -5x + 3y &= -12 \\ 2x - 3y &= 3 \\ \hline -3x &= -9 \\ x &= 3 \end{aligned}$ $\quad \begin{aligned} 2x - 3y &= 3 \\ 2(3) - 3y &= 3 \\ 6 - 3y &= 3 \\ -3y &= -3 \\ y &= 1 \end{aligned}$ $\quad \begin{aligned} &\text{Solution:} \\ &\boxed{x = 3, y = 1} \end{aligned}$

31. $\begin{aligned} x - 7y &= -11 \Rightarrow \times (-8) \\ 8x + 2y &= 28 \end{aligned}$ $\quad \begin{aligned} -8x + 56y &= 88 \\ 8x + 2y &= 28 \\ \hline 58y &= 116 \\ y &= 2 \end{aligned}$ $\quad \begin{aligned} x - 7y &= -11 \\ x - 7(2) &= -11 \\ x - 14 &= -11 \\ x &= 3 \end{aligned}$ $\quad \begin{aligned} &\text{Solution:} \\ &\boxed{x = 3, y = 2} \end{aligned}$

33. $\begin{aligned} 3(x - y) &= y - 9 \Rightarrow 3x - 3y = y - 9 \Rightarrow 3x - 4y = -9 \Rightarrow \times (5) \\ 5(x + y) &= -15 \Rightarrow 5x + 5y = -15 \Rightarrow 5x + 5y = -15 \Rightarrow \times (4) \end{aligned}$ $\quad \begin{aligned} 15x - 20y &= -45 \\ 20x + 20y &= -60 \\ \hline 35x &= -105 \\ x &= -3 \end{aligned}$

$$\begin{aligned} 5x + 5y &= -15 \\ 5(-3) + 5y &= -15 \\ -15 + 5y &= -15 \\ 5y &= 0 \\ y &= 0 \end{aligned}$$

Solution:
$$\boxed{x = -3, y = 0}$$

35. $\begin{aligned} 2 &= \dfrac{1}{x + y} \Rightarrow 2(x + y) = 1 \Rightarrow 2x + 2y = 1 \\ 2 &= \dfrac{3}{x - y} \Rightarrow 2(x - y) = 3 \Rightarrow 2x - 2y = 3 \\ \hline & 4x = 4 \\ & x = 1 \end{aligned}$

$$\begin{aligned} 2x + 2y &= 1 \\ 2(1) + 2y &= 1 \\ 2 + 2y &= 1 \\ 2y &= -1 \\ y &= -\tfrac{1}{2} \end{aligned}$$

Solution:
$$\boxed{x = 1, y = -\tfrac{1}{2}}$$

37. $\begin{aligned} y + 2x &= 5 \Rightarrow 2x + y = 5 \\ 0.5y &= 2.5 - x \Rightarrow x + 0.5y = 2.5 \Rightarrow \times (-2) \end{aligned}$ $\quad \begin{aligned} 2x + y &= 5 \\ -2x - y &= -5 \\ \hline 0 &= 0 \end{aligned}$

$\boxed{\text{Dependent Equations}}$
$$2x + y = 5$$
$$y = -2x + 5$$
$$\boxed{\text{Gen. sol.: } (x, -2x + 5)}$$

39. $\begin{aligned} x + 2(x - y) &= 2 \Rightarrow x + 2x - 2y = 2 \Rightarrow 3x - 2y = 2 \\ 3(y - x) - y &= 5 \Rightarrow 3y - 3x - y = 5 \Rightarrow -3x + 2y = 5 \\ \hline & 0 \neq 7 \end{aligned}$

$\boxed{\text{No Solution}}$
$\boxed{\text{Inconsistent system}}$

41. $\dfrac{3}{2}x + \dfrac{1}{3}y = 2 \Rightarrow \times (6) \quad 9x + 2y = 12 \Rightarrow$

$\dfrac{2}{3}x + \dfrac{1}{9}y = 1 \Rightarrow \times (9) \quad 6x + y = 9 \Rightarrow \times (-2)$

$$9x + 2y = 12$$
$$-12x - 2y = -18$$
$$\overline{}$$
$$-3x \quad\;\;\; = -6$$
$$x \quad\;\;\; = 2$$

$6x + y = 9$
$6(2) + y = 9$
$12 + y = 9$
$y = -3$

Solution:

$$\boxed{x = 2,\, y = -3}$$

43. $\dfrac{x-y}{5} + \dfrac{x+y}{2} = 6 \Rightarrow \times (10) \quad 2(x-y) + 5(x+y) = 60 \Rightarrow 2x - 2y + 5x + 5y = 60$

$\dfrac{x-y}{2} - \dfrac{x+y}{4} = 3 \Rightarrow \times (8) \quad 4(x-y) - 2(x+y) = 24 \Rightarrow 4x - 4y - 2x - 2y = 24$

$7x + 3y = 60 \Rightarrow \times (2) \quad 14x + 6y = 120$
$2x - 6y = 24 \qquad\qquad\quad 2x - 6y = 24$

$$\overline{}$$
$$16x \quad\;\; = 144$$
$$x \quad\;\; = 9$$

$7x + 3y = 60$
$7(9) + 3y = 60$
$63 + 3y = 60$
$3y = -3$
$y = -1$

Solution:

$$\boxed{x = 9,\, y = -1}$$

45.

(1) $x + y + z = 3$
(2) $2x + y + z = 4$
(3) $3x + y - z = 5$

Add (1) and (3):

(1) $\quad x + \;\, y + z = 3$
(3) $\quad 3x + \;\, y - z = 5$
(4) $\quad \overline{4x + 2y \quad\;\; = 8}$

Add equations (2) and (3):

(2) $\quad 2x + \;\, y + z = 4$
(3) $\quad 3x + \;\, y - z = 5$
(5) $\quad \overline{5x + 2y \quad\;\; = 9}$

Solve the system of two equations and two unknowns formed by equations (4) and (5):

$4x + 2y = 8 \Rightarrow \times (-1) \quad -4x - 2y = -8$
$5x + 2y = 9 \Rightarrow \qquad\qquad\;\; 5x + 2y = 9$

$$\overline{}$$
$$x \quad\;\; = 1$$

$4x + 2y = 8$
$4(1) + 2y = 8$
$2y = 4$
$y = 2$

$x + y + z = 3$
$1 + 2 + z = 3$
$3 + z = 3$
$z = 0$

Solution:

$$\boxed{\begin{aligned} x &= 1 \\ y &= 2 \\ z &= 0 \end{aligned}}$$

47.

(1) $x - y + z = 0$
(2) $x + y + 2z = -1$
(3) $-x - y + z = 0$

Add (1) and (2):

(1) $\quad x - y + \;\, z = \;\;\, 0$
(2) $\quad x + y + 2z = -1$
(4) $\quad \overline{2x \quad\;\;\, + 3z = -1}$

Add equations (2) and (3):

(2) $\quad x + y + 2z = -1$
(3) $\quad -x - y + \;\, z = \;\;\; 0$
(5) $\quad \overline{3z = -1}$

$$z = -\tfrac{1}{3}$$

Solve the system of two equations and two unknowns formed by equations (4) and (5):

$2x + 3z = -1$
$2x + 3\left(-\tfrac{1}{3}\right) = -1$
$2x = 0$
$x = 0$

$x - y + z = 0$
$0 - y + \left(-\tfrac{1}{3}\right) = 0$
$-y = \tfrac{1}{3}$
$y = -\tfrac{1}{3}$

Solution:

$$\boxed{\begin{aligned} x &= 0 \\ y &= -\tfrac{1}{3} \\ z &= -\tfrac{1}{3} \end{aligned}}$$

49. (1) $2x + y = 4$ Add (2) and (3):   (2) $x \quad - z = 2$
     (2) $x - z = 2$ (3) $\quad y + z = 1$
     (3) $y + z = 1$ (4) $\overline{x + y \quad = 3}$

Solve the system of two equations and two unknowns formed by equations (1) and (4):

$2x + y = 4 \Rightarrow \times(-1)$ $-2x - y = -4$ $x + y = 3$ $y + z = 1$ Solution:
$\underline{x + y = 3 \Rightarrow} \qquad \underline{x + y = \quad 3} \quad 1 + y = 3 \quad 2 + z = 1$

$$\begin{array}{rl} -x & = -1 \\ x & = 1 \end{array} \qquad y = 2 \qquad z = -1$$

$$\boxed{\begin{array}{l} x = 1 \\ y = 2 \\ z = -1 \end{array}}$$

51. (1) $x + y + z = 6$ Add (1) and $-(2)$:   Add equations (1) and $-(3)$:
     (2) $2x + y + 3z = 17$ (1) $\quad x + y + \quad z = \quad 6$ (1) $\quad x + y + \quad z = \quad 6$
     (3) $x + y + 2z = 11$ $-(2)$ $\underline{-2x - y - 3z = -17}$ $-(3)$ $\underline{-x - y - 2z = -11}$
                          (4) $\quad -x \quad - 2z = -11$ (5) $\quad\quad\quad\quad - z = \quad -5$
                                                        $z = \quad 5$

Solve the system of two equations and two unknowns formed by equations (4) and (5):

$-x - 2z = -11$ $x + y + z = 6$ Solution:
$-x - 2(5) = -11$ $1 + y + 5 = 6$
$\quad -x = -1$ $y = 0$
$\quad\quad x = 1$

$$\boxed{\begin{array}{l} x = 1 \\ y = 0 \\ z = 5 \end{array}}$$

53. (1) $x + y + z = 3$ Add $-2 \cdot (1)$ and (3):   No solution; inconsistent system
     (2) $\quad x + z = 2$ $-2 \cdot (1)$ $\quad -2x - 2y - 2z = -6$
     (3) $2x + 2y + 2z = 3$ (3) $\quad \underline{2x + 2y + 2z = \quad 3}$
                       (4) $\quad\quad\quad\quad\quad 0 \neq -3$

55. (1) $x + 2y - z = 2$ Add (1) and (3):
     (2) $\quad 2x - y = -1$ (1) $\quad x + 2y - z = 2$
     (3) $3x + y + z = 1$ (3) $\quad \underline{3x + y + z = 1}$
                       (4) $\quad 4x + 3y \quad = 3$

Solve the system of two equations and two unknowns formed by equations (2) and (4):

$2x - y = -1 \Rightarrow \times(3)$ $6x - 3y = -3$ $2x - y = -1$ $3x + y + z = 1$ Solution:
$\underline{4x + 3y = \quad 3 \Rightarrow} \qquad \underline{4x + 3y = \quad 3} \quad 2(0) - y = -1 \quad 3(0) + 1 + z = 1$

$$\begin{array}{rl} 10x & = 0 \\ x & = 0 \end{array} \qquad \begin{array}{rl} -y & = -1 \\ y & = 1 \end{array} \qquad z = 0$$

$$\boxed{\begin{array}{l} x = 0 \\ y = 1 \\ z = 0 \end{array}}$$

57.　(1) $3x + 4y + 2z = 4$　　Add (1) and $2 \cdot$ (2):　　　　Add equations $2 \cdot$ (1) and (3):

(2)　$6x - 2y + z = 4$

(3) $3x - 8y - 6z = -3$

$$
\begin{array}{ll}
(1) & 3x + 4y + 2z = 4 \\
2 \cdot (2) & 12x - 4y + 2z = 8 \\
\hline
(4) & 15x \quad\quad + 4z = 12
\end{array}
$$

$$
\begin{array}{ll}
2 \cdot (1) & 6x + 8y + 4z = 8 \\
(3) & 3x - 8y - 6z = -3 \\
\hline
(5) & 9x \quad\quad - 2z = 5
\end{array}
$$

Solve the system of two equations and two unknowns formed by equations (4) and (5):

$$
\begin{aligned}
15x + 4z = 12 \Rightarrow & \quad 15x + 4z = 12 \\
9x - 2z = 5 \Rightarrow \times (2) & \quad \underline{18x - 4z = 10} \\
& \quad 33x \quad\quad = 22 \\
& \quad\quad x = \tfrac{2}{3}
\end{aligned}
$$

$$
\begin{aligned}
15x + 4z &= 12 \\
15\left(\tfrac{2}{3}\right) + 4z &= 12 \\
4z &= 2 \\
z &= \tfrac{1}{2}
\end{aligned}
$$

$$
\begin{aligned}
3x + 4y + 2z &= 4 \\
3\left(\tfrac{2}{3}\right) + 4y + 2\left(\tfrac{1}{2}\right) &= 4 \\
4y &= 1 \\
y &= \tfrac{1}{4}
\end{aligned}
$$

Solution: $x = \tfrac{2}{3},\ y = \tfrac{1}{4},\ z = \tfrac{1}{2}$

59.　(1) $2x - y - z = 0$　　Add (1) and $-2 \cdot$ (2):　　　Add equations (1) and $-2 \cdot$ (3):

(2) $x - 2y - z = -1$

(3) $x - y - 2z = -1$

$$
\begin{array}{ll}
(1) & 2x - y - z = 0 \\
-2 \cdot (2) & -2x + 4y + 2z = 2 \\
\hline
(4) & 3y + z = 2
\end{array}
$$

$$
\begin{array}{ll}
(1) & 2x - y - z = 0 \\
(3) & -2x + 2y + 4z = 2 \\
\hline
(5) & y + 3z = 2
\end{array}
$$

Solve the system of two equations and two unknowns formed by equations (4) and (5):

$$
\begin{aligned}
3y + z = 2 \Rightarrow & \quad 3y + z = 2 \\
y + 3z = 2 \Rightarrow \times (-3) & \quad \underline{-3y - 9z = -6} \\
& \quad -8z = -4 \\
& \quad\quad z = \tfrac{1}{2}
\end{aligned}
$$

$$
\begin{aligned}
y + 3z &= 2 \\
y + 3\left(\tfrac{1}{2}\right) &= 2 \\
y &= \tfrac{1}{2}
\end{aligned}
$$

$$
\begin{aligned}
x - y - 2z &= -1 \\
x - \tfrac{1}{2} - 2\left(\tfrac{1}{2}\right) &= -1 \\
x &= \tfrac{1}{2}
\end{aligned}
$$

Solution: $x = \tfrac{1}{2},\ y = \tfrac{1}{2},\ z = \tfrac{1}{2}$

61.　(1) $(x+y) + (y+z) + (z+x) = 6 \Rightarrow 2x + 2y + 2z = 6$　　Add (1) and $-2 \cdot$ (3):

(2) $(x-y) + (y-z) + (z-x) = 0 \Rightarrow \quad\quad\quad\quad 0 = 0$

(3) $\quad\quad\quad\quad\quad\quad x + y + 2z = 4 \Rightarrow \quad x + y + 2z = 4$

$$
\begin{array}{ll}
(1) & 2x + 2y + 2z = 6 \\
-2 \cdot (3) & -2x - 2y - 4z = -8 \\
\hline
(4) & -2z = -2 \\
& z = 1
\end{array}
$$

Since (2) is always true, the equations are dependent. z must equal 1. Then, from (3), $x + y = 2$.

Let $y = $ any real #. Then $x = 2 - y$. Solution: $(2 - y, y, 1)$

63.　Let $x = $ acres of corn and let $y = $ acres of soybeans. Then $\begin{cases} (1) & x + y = 350 \\ (2) & x = y + 100 \end{cases}$

Substitute $x = y + 100$ from (2) into (1):

$$
\begin{aligned}
x + y &= 350 \\
y + 100 + y &= 350 \\
2y &= 250 \\
y &= 125
\end{aligned}
$$

Substitute and solve for x:

$$
\begin{aligned}
x &= y + 100 \\
&= 125 + 100 = 225
\end{aligned}
$$

The farmer should plant 225 acres of corn and 125 acres of soybeans.

65. Let $w =$ width and let $l =$ length. Then $\begin{cases} (1) & 2w + 2l = 1900 \\ (2) & w = l - 250 \end{cases}$

Substitute $w = l - 250$ from (2) into (1):   Substitute and solve for w:

$$2w + 2l = 1900$$
$$2(l - 250) + 2l = 1900$$
$$2l - 500 + 2l = 1900$$
$$4l = 2400$$
$$l = 600$$

$$w = l - 250$$
$$= 600 - 250 = 350$$
$$\text{Area} = lw = (600 \text{ cm})(350 \text{ cm})$$

| The area is 210,000 cm². |

67. Let $x =$ grams of 9% alloy. Let $y =$ grams of 84% alloy. Then the following system applies: (note: 34% of 60 is $0.34 \times 60 = 20.4$)

$$\begin{array}{ll} x + y = 60 & \Rightarrow \times (-9) \\ 0.09x + 0.84y = 20.4 & \Rightarrow \times (100) \end{array} \quad \begin{array}{rl} -9x - 9y = & -540 \\ 9x + 84y = & 2040 \\ \hline 75y = & 1500 \\ y = & 20 \end{array}$$

| She must use 40 grams of the 9% and 20 grams of the 84% alloy. |

69.

$$\begin{array}{ll} 448x = 112y & \Rightarrow 448x - 112y = 0 \Rightarrow \times(-1) \\ 448(x + 1) = 192(y - 1) & \Rightarrow 448x - 192y = -640 \Rightarrow \end{array} \quad \begin{array}{rl} -448x + 112y = & 0 \\ 448x - 192y = & -640 \\ \hline -80y = & -640 \\ y = & 8 \end{array}$$

$$448x = 112y$$
$$448x = 112(8)$$
$$x = 2$$

| The lever has a length of 10 feet. |

71. $E(x) = 43.53x + 742.72$

$R(x) = 89.95x$

$$E(x) = R(x)$$
$$43.53x + 742.72 = 89.95x$$
$$742.72 = 46.42x$$
$$16 = x \Rightarrow \boxed{\text{Daily production should be 16 pairs.}}$$

73. Let $x =$ hours at fast food restaurant, $y =$ hours at gas station and $z =$ janitorial hours.

(1) $x = 15$ Substitute $x = 15$ into (2) and (3):

(2) $x + y + z = 30$ (2) $15 + y + z = 30$

(3) $5.7x + 6.3y + 10z = 198.50$ (3) $5.7(15) + 6.3y + 10z = 198.5$

Solve the system of two equations and two unknowns formed by equations (2) and (3):

$$\begin{array}{ll} y + z = 15 & \Rightarrow \times(-10) \\ 6.3y + 10z = 113 & \Rightarrow \times(-3) \end{array} \quad \begin{array}{rl} -10y - 10z = & -150 \\ 6.3y + 10z = & 113 \\ \hline -3.7y = & -37 \\ y = & 10 \end{array} \quad \begin{array}{l} y + z = 15 \\ 10 + z = 15 \\ z = 5 \end{array}$$

| He spends 15 hours cooking, 10 hours at the gas station and 5 hours doing janitorial work. |

75. Let $x = \#$ between 0-14, $y = \#$ between 15-49 and $z = \#$ 50 or over.

(1) $x + y + z = 3$ Add (1) and $-(2)$:

(2) $\quad x + y = 2.61$ (1) $\quad x + y + z = 3$

(3) $\quad y + z = 1.95$ $-(2)$ $\quad \underline{-x - y \qquad = -2.61}$

$\qquad\qquad\qquad\qquad\qquad\qquad z = 0.39$

Substitute $z = 0.39$ into (3):   Substitute $y = 1.56$ into (2):   There are 1.05 million between

$\quad y + z = 1.95$ $x + y = 2.61$ 0-14, 1.56 million between 15-49

$\quad y + 0.39 = 1.95$ $x + 1.56 = 2.61$ and 0.39 million over 50.

$\qquad\quad y = 1.56$ $x = 1.05$

77. Let $x =$ the smallest angle, $y =$ the middle angle and $z =$ the largest angle.

(1) $x + y + z = 180$ Substitute (3) into (1):   Substitute (3) and (2):

(2) $\qquad z = x + y + 20$ $x + y + 3x + 10 = 180$ $3x + 10 = x + y + 20$

(3) $\qquad z = 3x + 10$ $4x + y = 170$ (4) $2x - y = 10$ (5)

Solve the system of two equations and two unknowns formed by equations (4) and (5):

$\quad 4x + y = 170$ $4x + y = 170$ $z = 3x + 10$ Solution: | The angles have measures

$\quad \underline{2x - y = 10}$ $4(30) + y = 170$ $z = 3(30) + 10$ of 30°, 50° and 100°.

$\qquad 6x = 180$ $120 + y = 170$ $z = 90 + 10$

$\qquad\quad x = 30$ $y = 50$ $z = 100$

79. **Answers may vary.**   **81.** **Answers may vary.**   **83.** **Answers may vary.**

85. $y = 3^x$; points: $(0, 1), (1, 3)$ **87.** $\quad 3 = \log_2 x$

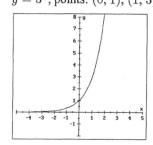

$\qquad\qquad\qquad\qquad\qquad\qquad\qquad\qquad\qquad\qquad 2^3 = x$

$\qquad\qquad\qquad\qquad\qquad\qquad\qquad\qquad\qquad\qquad\quad 8 = x$

89. $\log \dfrac{x}{y^2 z} = \log x - \log \left(y^2 z\right) = \log x - \left(\log y^2 + \log z\right) = \log x - 2\log y - \log z$

91. $\log x + 3\log y - \dfrac{1}{2}\log z = \log x + \log y^3 + \log z^{-1/2} = \log x + \log y^3 + \log \dfrac{1}{\sqrt{z}} = \log \dfrac{xy^3}{\sqrt{z}}$

Exercise 6.2   (page 445)

1. matrix   **3.** coefficient   **5.** equation

7. row equivalent   **9.** interchanged   **11.** adding, multiple

13. $\begin{cases} (1) & x + y = 7 \\ (2) & x - 2y = -1 \end{cases} \Rightarrow \begin{cases} (1) & x + y = 7 \\ (2) & -3y = -8 \end{cases} \Rightarrow \begin{cases} (1) & x + y = 7 \\ (2) & y = \dfrac{8}{3} \end{cases}$

$\qquad\qquad\qquad -(1) + (2) \Rightarrow (2) \qquad\qquad -\frac{1}{3}(2) \Rightarrow (2)$

From (2): $y = \dfrac{8}{3}$ $\qquad$ From (1): $x + y = 7$ $\qquad$ Solution:

$\qquad\qquad\qquad\qquad\qquad\qquad x + \dfrac{8}{3} = 7 \qquad \boxed{x = \frac{13}{3},\, y = \frac{8}{3}}$

$\qquad\qquad\qquad\qquad\qquad\qquad\qquad x = \dfrac{13}{3}$

15. $\begin{cases} (1) & x - y = 1 \\ (2) & 2x - y = 8 \end{cases} \Rightarrow \begin{cases} (1) & x - y = 1 \\ (2) & y = 6 \end{cases}$

$\qquad\qquad\qquad\qquad -2(1) + (2) \Rightarrow (2)$

From (2): $y = 6$ $\quad$ From (1): $x - y = 1$ $\quad$ Solution:

$\qquad\qquad\qquad\qquad\qquad\qquad x - 6 = 1 \qquad \boxed{x = 7,\, y = 6}$

$\qquad\qquad\qquad\qquad\qquad\qquad\quad x = 7$

17. $\begin{cases} (1) & x + 2y - z = 2 \\ (2) & x - 3y + 2z = 1 \\ (3) & x + y - 3z = -6 \end{cases} \Rightarrow \begin{cases} (1) & x + 2y - z = 2 \\ (2) & -5y + 3z = -1 \\ (3) & -y - 2z = -8 \end{cases} \Rightarrow \begin{cases} (1) & x + 2y - z = 2 \\ (2) & -5y + 3z = -1 \\ (3) & 13z = 39 \end{cases} \Rightarrow$

$\qquad\qquad\qquad\qquad\qquad\qquad -(1) + (2) \Rightarrow (2) \qquad\qquad\qquad -5(3) + (2) \Rightarrow (3)$

$\qquad\qquad\qquad\qquad\qquad\qquad -(1) + (3) \Rightarrow (3)$

$\begin{cases} (1) & x + 2y - z = 2 \\ (2) & y - \frac{3}{5}z = \frac{1}{5} \\ (3) & z = 3 \end{cases}$ $\quad$ From (3): $z = 3$ $\qquad\qquad$ From (1): $\quad x + 2y - z = 2$

$\quad -\frac{1}{5}(2) \Rightarrow (2)$ $\qquad$ From (2): $y - \dfrac{3}{5}z = \dfrac{1}{5}$ $\qquad\qquad x + 2(2) - (3) = 2$

$\quad \frac{1}{13}(3) \Rightarrow (3)$ $\qquad\qquad\qquad\qquad y - \dfrac{3}{5}(3) = \dfrac{1}{5}$ $\qquad\qquad\qquad\qquad x = 1$

$\qquad\qquad\qquad\qquad\qquad\qquad\qquad y = 2$ $\qquad$ Solution: $\boxed{x = 1,\, y = 2,\, z = 3}$

19. $\begin{cases} (1) & x - y - z = -3 \\ (2) & 5x + y = 6 \\ (3) & y + z = 4 \end{cases} \Rightarrow \begin{cases} (1) & x - y - z = -3 \\ (2) & 6y + 5z = 21 \\ (3) & y + z = 4 \end{cases} \Rightarrow \begin{cases} (1) & x - y - z = -3 \\ (2) & y + \frac{5}{6}z = \frac{7}{2} \\ (3) & -z = -3 \end{cases} \Rightarrow$

$\qquad\qquad\qquad\qquad\qquad\qquad -5(1) + (2) \Rightarrow (2) \qquad\qquad\qquad \frac{1}{6}(2) \Rightarrow (2)$

$\qquad\qquad\qquad\qquad\qquad\qquad\qquad\qquad\qquad\qquad\qquad -6(3) + (2) \Rightarrow (3)$

$\begin{cases} (1) & x - y - z = -3 \\ (2) & y + \frac{5}{6}z = \frac{7}{2} \\ (3) & z = 3 \end{cases} \Rightarrow$ $\quad$ From (3): $z = 3$ $\qquad$ From (1): $\quad x - y - z = -3$

$\quad -(3) \Rightarrow (3)$ $\qquad\qquad$ From (2): $\quad y + \dfrac{5}{6}z = \dfrac{7}{2}$ $\qquad\qquad x - (1) - (3) = -3$

$\qquad\qquad\qquad\qquad\qquad\qquad\qquad\qquad y + \dfrac{5}{6}(3) = \dfrac{7}{2}$ $\qquad\qquad\qquad\qquad x = 1$

$\qquad\qquad\qquad\qquad\qquad\qquad\qquad\qquad\qquad y = 1$ $\qquad$ Solution: $\boxed{x = 1,\, y = 1,\, z = 3}$

21. row echelon form $\qquad\qquad\qquad\qquad\qquad$ **23.** reduced row echelon form

25. $\begin{bmatrix} 2 & 1 & | & 3 \\ 1 & -3 & | & 5 \end{bmatrix} \Rightarrow \begin{bmatrix} 1 & -3 & | & 5 \\ 2 & 1 & | & 3 \end{bmatrix} \Rightarrow \begin{bmatrix} 1 & -3 & | & 5 \\ 0 & 7 & | & -7 \end{bmatrix} \Rightarrow \begin{bmatrix} 1 & -3 & | & 5 \\ 0 & 1 & | & -1 \end{bmatrix}$

$ R_1 \Leftrightarrow R_2 \qquad -2R_1 + R_2 \Rightarrow R_2 \qquad \frac{1}{7}R_2 \Rightarrow R_2$

From R_2: $y = -1$ $\quad$ From R_1: $\quad x - 3y = 5$ $\quad$ Solution: $\boxed{x = 2, y = -1}$

$ x - 3(-1) = 5$

$ x + 3 = 5$

$ x = 2$

27. $\begin{bmatrix} 1 & -7 & | & -2 \\ 5 & -2 & | & -10 \end{bmatrix} \Rightarrow \begin{bmatrix} 1 & -7 & | & -2 \\ 0 & 33 & | & 0 \end{bmatrix} \Rightarrow \begin{bmatrix} 1 & -7 & | & -2 \\ 0 & 1 & | & 0 \end{bmatrix}$

$ -5R_1 + R_2 \Rightarrow R_2 \qquad \frac{1}{33}R_2 \Rightarrow R_2$

From R_2: $y = 0$ $\quad$ From R_1: $\quad x - 7y = -2$ $\quad$ Solution: $\boxed{x = -2, y = 0}$

$ x - 7(0) = -2$

$ x = -2$

29. $\begin{bmatrix} 2 & -1 & | & 5 \\ 1 & 3 & | & 6 \end{bmatrix} \Rightarrow \begin{bmatrix} 1 & 3 & | & 6 \\ 2 & -1 & | & 5 \end{bmatrix} \Rightarrow \begin{bmatrix} 1 & 3 & | & 6 \\ 0 & -7 & | & -7 \end{bmatrix} \Rightarrow \begin{bmatrix} 1 & 3 & | & 6 \\ 0 & 1 & | & 1 \end{bmatrix}$

$ R_1 \Leftrightarrow R_2 \qquad -2R_1 + R_2 \Rightarrow R_2 \qquad -\frac{1}{7}R_2 \Rightarrow R_2$

From R_2: $y = 1$ $\quad$ From R_1: $\quad x + 3y = 6$ $\quad$ Solution: $\boxed{x = 3, y = 1}$

$ x + 3(1) = 6$

$ x + 3 = 6$

$ x = 3$

31. $\begin{bmatrix} 1 & -2 & | & 3 \\ -2 & 4 & | & 6 \end{bmatrix} \Rightarrow \begin{bmatrix} 1 & -2 & | & 3 \\ 0 & 0 & | & 12 \end{bmatrix} \Rightarrow$ From R_2, $0x + 0y = 12$. This is impossible.

$ 2R_1 + R_2 \Rightarrow R_2 \qquad \boxed{\text{No solution} \Rightarrow \text{inconsistent system}}$

33. $\begin{bmatrix} 1 & -1 & 1 & | & 3 \\ 2 & -1 & 1 & | & 4 \\ 1 & 2 & -1 & | & -1 \end{bmatrix} \Rightarrow \begin{bmatrix} 1 & -1 & 1 & | & 3 \\ 0 & 1 & -1 & | & -2 \\ 0 & 3 & -2 & | & -4 \end{bmatrix} \Rightarrow \begin{bmatrix} 1 & -1 & 1 & | & 3 \\ 0 & 1 & -1 & | & -2 \\ 0 & 0 & 1 & | & 2 \end{bmatrix} \Rightarrow$

$ -2R_1 + R_2 \Rightarrow R_2 \qquad -3R_2 + R_3 \Rightarrow R_3$

$ -R_1 + R_3 \Rightarrow R_3$

From R_3: $z = 2$ $\qquad\qquad$ From R_1: $\qquad x - y + z = 3$

From R_2: $\quad y - z = -2$ $\qquad\qquad\qquad\quad x - (0) + (2) = 3$

$ y - (2) = -2 \qquad\qquad\qquad\qquad\qquad x = 1$

$ y = 0 \qquad$ Solution: $\boxed{x = 1, y = 0, z = 2}$

35. $\begin{bmatrix} 1 & 1 & -1 & | & -1 \\ 3 & 1 & 0 & | & 4 \\ 0 & 1 & -2 & | & -4 \end{bmatrix} \Rightarrow \begin{bmatrix} 1 & 1 & -1 & | & -1 \\ 0 & -2 & 3 & | & 7 \\ 0 & 1 & -2 & | & -4 \end{bmatrix} \Rightarrow \begin{bmatrix} 1 & 1 & -1 & | & -1 \\ 0 & 1 & -\frac{3}{2} & | & -\frac{7}{2} \\ 0 & 0 & -1 & | & -1 \end{bmatrix} \Rightarrow \begin{bmatrix} 1 & 1 & -1 & | & -1 \\ 0 & 1 & -\frac{3}{2} & | & -\frac{7}{2} \\ 0 & 0 & 1 & | & 1 \end{bmatrix}$

$\qquad\qquad\qquad\qquad -3R_1 + R_2 \Rightarrow R_2 \qquad\qquad -\frac{1}{2}R_2 \Rightarrow R_2 \qquad\qquad -R_3 \Rightarrow R_3$

$\qquad\qquad\qquad\qquad\qquad\qquad\qquad\qquad\qquad R_2 + 2R_3 \Rightarrow R_3$

From (3): $z = 1$ $\qquad\qquad$ From (1): $\qquad x + y - z = -1$ $\qquad$ Solution:

From (2): $\quad y - \dfrac{3}{2}z = -\dfrac{7}{2}$ $\qquad\qquad x + (-2) - (1) = -1$ $\qquad$ $\boxed{x = 2,\, y = -2,\, z = 1}$

$\qquad\qquad y - \dfrac{3}{2}(1) = -\dfrac{7}{2}$ $\qquad\qquad\qquad x = 2$

$\qquad\qquad\qquad\quad y = -2$

37. $\begin{bmatrix} 1 & -1 & 1 & | & 2 \\ 2 & 1 & 1 & | & 5 \\ 3 & 0 & -4 & | & -5 \end{bmatrix} \Rightarrow \begin{bmatrix} 1 & -1 & 1 & | & 2 \\ 0 & 3 & -1 & | & 1 \\ 0 & 3 & -7 & | & -11 \end{bmatrix} \Rightarrow \begin{bmatrix} 1 & -1 & 1 & | & 2 \\ 0 & 1 & -\frac{1}{3} & | & \frac{1}{3} \\ 0 & 0 & 6 & | & 12 \end{bmatrix} \Rightarrow$

$\qquad\qquad\qquad\qquad -2R_1 + R_2 \Rightarrow R_2 \qquad\qquad \frac{1}{3}R_2 \Rightarrow R_2$

$\qquad\qquad\qquad\qquad -3R_1 + R_3 \Rightarrow R_3 \qquad\qquad -R_3 + R_2 \Rightarrow R_3$

$\begin{bmatrix} 1 & -1 & 1 & | & 2 \\ 0 & 1 & -\frac{1}{3} & | & \frac{1}{3} \\ 0 & 0 & 1 & | & 2 \end{bmatrix}$ $\quad$ From (3): $z = 2$ $\qquad$ From (1): $\qquad x - y + z = 2$

$\qquad \frac{1}{6}R_3 \Rightarrow R_3$ $\qquad$ From (2): $\quad y - \dfrac{1}{3}z = \dfrac{1}{3}$ $\qquad\qquad x - (1) + (2) = 2$

$\qquad\qquad\qquad\qquad\qquad\qquad\qquad\quad y - \dfrac{1}{3}(2) = \dfrac{1}{3}$ $\qquad\qquad\qquad\qquad x = 1$

$\qquad\qquad\qquad\qquad\qquad\qquad\qquad\qquad\qquad\qquad y = 1$ $\qquad$ Solution:

$\qquad\qquad\qquad\qquad\qquad\qquad\qquad\qquad\qquad\qquad\qquad\qquad \boxed{x = 1,\, y = 1,\, z = 2}$

39. $\begin{bmatrix} 1 & 1 & 2 & | & 4 \\ -1 & -1 & -3 & | & -5 \\ 2 & 1 & 1 & | & 2 \end{bmatrix} \Rightarrow \begin{bmatrix} 1 & 1 & 2 & | & 4 \\ 0 & 0 & -1 & | & -1 \\ 0 & -1 & -3 & | & -6 \end{bmatrix} \Rightarrow \begin{bmatrix} 1 & 1 & 2 & | & 4 \\ 0 & -1 & -3 & | & -6 \\ 0 & 0 & -1 & | & -1 \end{bmatrix} \Rightarrow \begin{bmatrix} 1 & 1 & 2 & | & 4 \\ 0 & 1 & 3 & | & 6 \\ 0 & 0 & 1 & | & 1 \end{bmatrix}$

$\qquad\qquad\qquad\qquad R_1 + R_2 \Rightarrow R_2 \qquad\qquad R_2 \Leftrightarrow R_3 \qquad\qquad -R_2 \Rightarrow R_2$

$\qquad\qquad\qquad\qquad -2R_1 + R_3 \Rightarrow R_3 \qquad\qquad\qquad\qquad\qquad\qquad -R_3 \Rightarrow R_3$

From (3): $z = 1$ $\qquad\qquad$ From (1): $\qquad x + y + 2z = 4$ $\qquad$ Solution:

From (2): $\quad y + 3z = 6$ $\qquad\qquad x + (3) + 2(1) = 4$ $\qquad$ $\boxed{x = -1,\, y = 3,\, z = 1}$

$\qquad\quad y + 3(1) = 6$ $\qquad\qquad\qquad x = -1$

$\qquad\qquad\quad y = 3$

41. $\begin{bmatrix} 1 & -2 & | & 7 \\ 0 & 1 & | & 3 \end{bmatrix} \Rightarrow \begin{bmatrix} 1 & 0 & | & 13 \\ 0 & 1 & | & 3 \end{bmatrix} \Rightarrow$ Solution: $\boxed{x = 13,\, y = 3}$

$\qquad\qquad\qquad\qquad 2R_2 + R_1 \Rightarrow R_1$

43. $\begin{bmatrix} 1 & 2 & -1 & | & 3 \\ 0 & 1 & 3 & | & 1 \\ 0 & 0 & 1 & | & -2 \end{bmatrix} \Rightarrow \begin{bmatrix} 1 & 0 & -7 & | & 1 \\ 0 & 1 & 3 & | & 1 \\ 0 & 0 & 1 & | & -2 \end{bmatrix} \Rightarrow \begin{bmatrix} 1 & 0 & 0 & | & -13 \\ 0 & 1 & 0 & | & 7 \\ 0 & 0 & 1 & | & -2 \end{bmatrix}$ Solution:

$\qquad\qquad\qquad\qquad\qquad -2R_2 + R_1 \Rightarrow R_1 \qquad 7R_3 + R_1 \Rightarrow R_1$ $\qquad$ $\boxed{x = -13,\, y = 7,\, z = -2}$

$\qquad\qquad\qquad\qquad\qquad\qquad\qquad\qquad\qquad -3R_3 + R_2 \Rightarrow R_2$

45. $\begin{bmatrix} 1 & -1 & | & 7 \\ 1 & 1 & | & 13 \end{bmatrix} \Rightarrow \begin{bmatrix} 1 & -1 & | & 7 \\ 0 & 2 & | & 6 \end{bmatrix} \Rightarrow \begin{bmatrix} 1 & -1 & | & 7 \\ 0 & 1 & | & 3 \end{bmatrix} \Rightarrow \begin{bmatrix} 1 & 0 & | & 10 \\ 0 & 1 & | & 3 \end{bmatrix}$ Solution:

$ -R_1 + R_2 \Rightarrow R_2 \qquad \frac{1}{2}R_2 \Rightarrow R_2 \qquad R_2 + R_1 \Rightarrow R_1 \quad \boxed{x = 10, y = 3}$

47. $\begin{bmatrix} 1 & -\frac{1}{2} & | & 0 \\ 1 & 2 & | & 0 \end{bmatrix} \Rightarrow \begin{bmatrix} 1 & -\frac{1}{2} & | & 0 \\ 0 & \frac{5}{2} & | & 0 \end{bmatrix} \Rightarrow \begin{bmatrix} 1 & -\frac{1}{2} & | & 0 \\ 0 & 1 & | & 0 \end{bmatrix} \Rightarrow \begin{bmatrix} 1 & 0 & | & 0 \\ 0 & 1 & | & 0 \end{bmatrix}$ Solution:

$ -R_1 + R_2 \Rightarrow R_2 \qquad \frac{2}{5}R_2 \Rightarrow R_2 \qquad \frac{1}{2}R_2 + R_1 \Rightarrow R_1 \quad \boxed{x = 0, y = 0}$

49. $\begin{bmatrix} 1 & 1 & 2 & | & 0 \\ 1 & 1 & 1 & | & 2 \\ 1 & 0 & 1 & | & 1 \end{bmatrix} \Rightarrow \begin{bmatrix} 1 & 1 & 2 & | & 0 \\ 0 & 0 & -1 & | & 2 \\ 0 & -1 & -1 & | & 1 \end{bmatrix} \Rightarrow \begin{bmatrix} 1 & 1 & 2 & | & 0 \\ 0 & -1 & -1 & | & 1 \\ 0 & 0 & -1 & | & 2 \end{bmatrix} \Rightarrow \begin{bmatrix} 1 & 1 & 2 & | & 0 \\ 0 & 1 & 1 & | & -1 \\ 0 & 0 & 1 & | & -2 \end{bmatrix}$

$ -R_1 + R_2 \Rightarrow R_2 \qquad\quad R_2 \Leftrightarrow R_3 \qquad\qquad -R_2 \Rightarrow R_2$
$ -R_1 + R_3 \Rightarrow R_3 \qquad\qquad\qquad\qquad\qquad -R_3 \Rightarrow R_3$

$\begin{bmatrix} 1 & 0 & 1 & | & 1 \\ 0 & 1 & 1 & | & -1 \\ 0 & 0 & 1 & | & -2 \end{bmatrix} \Rightarrow \begin{bmatrix} 1 & 0 & 0 & | & 3 \\ 0 & 1 & 0 & | & 1 \\ 0 & 0 & 1 & | & -2 \end{bmatrix}$ Solution: $\boxed{x = 3, y = 1, z = -2}$

$-R_2 + R_1 \Rightarrow R_1 \qquad -R_3 + R_1 \Rightarrow R_1$
$ -R_3 + R_2 \Rightarrow R_2$

51. $\begin{bmatrix} 2 & 1 & -2 & | & 1 \\ -1 & 1 & -3 & | & 0 \\ 4 & 3 & 0 & | & 4 \end{bmatrix} \Rightarrow \begin{bmatrix} 2 & 1 & -2 & | & 1 \\ 0 & 3 & -8 & | & 1 \\ 0 & 1 & 4 & | & 2 \end{bmatrix} \Rightarrow \begin{bmatrix} -6 & 0 & -2 & | & -2 \\ 0 & 3 & -8 & | & 1 \\ 0 & 0 & -20 & | & -5 \end{bmatrix} \Rightarrow \begin{bmatrix} -6 & 0 & -2 & | & -2 \\ 0 & 3 & -8 & | & 1 \\ 0 & 0 & 1 & | & \frac{1}{4} \end{bmatrix}$

$ R_1 + 2R_2 \Rightarrow R_2 \qquad -3R_1 + R_2 \Rightarrow R_1 \qquad -\frac{1}{20}R_3 \Rightarrow R_3$
$ -2R_1 + R_3 \Rightarrow R_3 \qquad -3R_3 + R_2 \Rightarrow R_3$

$\begin{bmatrix} -6 & 0 & 0 & | & -\frac{3}{2} \\ 0 & 3 & 0 & | & 3 \\ 0 & 0 & 1 & | & \frac{1}{4} \end{bmatrix} \Rightarrow \begin{bmatrix} 1 & 0 & 0 & | & \frac{1}{4} \\ 0 & 1 & 0 & | & 1 \\ 0 & 0 & 1 & | & \frac{1}{4} \end{bmatrix}$ Solution: $\boxed{x = \frac{1}{4}, y = 1, z = \frac{1}{4}}$

$2R_3 + R_1 \Rightarrow R_1 \qquad -\frac{1}{6}R_1 \Rightarrow R_1$
$8R_3 + R_2 \Rightarrow R_2 \qquad \frac{1}{3}R_2 \Rightarrow R_2$

53. $\begin{bmatrix} 2 & -2 & 3 & 1 & | & 2 \\ 1 & 1 & 1 & 1 & | & 5 \\ -1 & 2 & -3 & 2 & | & 2 \\ 1 & 1 & 2 & -1 & | & 4 \end{bmatrix} \Rightarrow \begin{bmatrix} 2 & -2 & 3 & 1 & | & 2 \\ 0 & -4 & 1 & -1 & | & -8 \\ 0 & 2 & -3 & 5 & | & 6 \\ 0 & -4 & -1 & 3 & | & -6 \end{bmatrix} \Rightarrow \begin{bmatrix} -4 & 0 & -5 & -3 & | & -12 \\ 0 & -4 & 1 & -1 & | & -8 \\ 0 & 0 & -5 & 9 & | & 4 \\ 0 & 0 & 2 & -4 & | & -2 \end{bmatrix}$

$ -2R_2 + R_1 \Rightarrow R_2 \qquad\qquad -2R_1 + R_2 \Rightarrow R_1$
$ 2R_3 + R_1 \Rightarrow R_3 \qquad\qquad\quad 2R_3 + R_2 \Rightarrow R_3$
$ -2R_4 + R_1 \Rightarrow R_4 \qquad\qquad -R_4 + R_2 \Rightarrow R_4$

continued on next page...

53. **continued...**

$$\begin{bmatrix} -4 & 0 & -5 & -3 & | & -12 \\ 0 & -4 & 1 & -1 & | & -8 \\ 0 & 0 & -5 & 9 & | & 4 \\ 0 & 0 & 1 & -2 & | & -1 \end{bmatrix} \Rightarrow \begin{bmatrix} 4 & 0 & 0 & 12 & | & 16 \\ 0 & -20 & 0 & 4 & | & -36 \\ 0 & 0 & -5 & 9 & | & 4 \\ 0 & 0 & 0 & -1 & | & -1 \end{bmatrix} \Rightarrow \begin{bmatrix} 1 & 0 & 0 & 3 & | & 4 \\ 0 & -5 & 0 & 1 & | & -9 \\ 0 & 0 & -5 & 9 & | & 4 \\ 0 & 0 & 0 & 1 & | & 1 \end{bmatrix}$$

$$\frac{1}{2}R_4 \Rightarrow R_4 \qquad \begin{matrix} -R_1 + R_3 \Rightarrow R_1 \\ 5R_2 + R_3 \Rightarrow R_2 \\ 5R_4 + R_3 \Rightarrow R_4 \end{matrix} \qquad \begin{matrix} \frac{1}{4}R_1 \Rightarrow R_1 \\ \frac{1}{4}R_2 \Rightarrow R_2 \\ -R_4 \Rightarrow R_4 \end{matrix}$$

$$\begin{bmatrix} 1 & 0 & 0 & 0 & | & 1 \\ 0 & -5 & 0 & 0 & | & -10 \\ 0 & 0 & -5 & 0 & | & -5 \\ 0 & 0 & 0 & 1 & | & 1 \end{bmatrix} \Rightarrow \begin{bmatrix} 1 & 0 & 0 & 0 & | & 1 \\ 0 & 1 & 0 & 0 & | & 2 \\ 0 & 0 & 1 & 0 & | & 1 \\ 0 & 0 & 0 & 1 & | & 1 \end{bmatrix}$$ Solution:

$$\boxed{x = 1, y = 2, z = 1, t = 1}$$

$$\begin{matrix} -3R_4 + R_1 \Rightarrow R_1 \\ -R_4 + R_2 \Rightarrow R_2 \\ -9R_4 + R_3 \Rightarrow R_3 \end{matrix} \qquad \begin{matrix} -\frac{1}{5}R_2 \Rightarrow R_2 \\ -\frac{1}{5}R_3 \Rightarrow R_3 \end{matrix}$$

55.

$$\begin{bmatrix} 1 & 1 & 0 & 1 & | & 4 \\ 1 & 0 & 1 & 1 & | & 2 \\ 2 & 2 & 1 & 2 & | & 8 \\ 1 & -1 & 1 & -1 & | & -2 \end{bmatrix} \Rightarrow \begin{bmatrix} 1 & 1 & 0 & 1 & | & 4 \\ 0 & 1 & -1 & 0 & | & 2 \\ 0 & 0 & 1 & 0 & | & 0 \\ 0 & 2 & -1 & 2 & | & 6 \end{bmatrix} \Rightarrow \begin{bmatrix} 1 & 0 & 1 & 1 & | & 2 \\ 0 & 1 & -1 & 0 & | & 2 \\ 0 & 0 & 1 & 0 & | & 0 \\ 0 & 0 & 1 & 2 & | & 2 \end{bmatrix} \Rightarrow$$

$$\begin{matrix} -R_2 + R_1 \Rightarrow R_2 \\ -2R_1 + R_3 \Rightarrow R_3 \\ -R_4 + R_1 \Rightarrow R_4 \end{matrix} \qquad \begin{matrix} -R_2 + R_1 \Rightarrow R_1 \\ -2R_2 + R_4 \Rightarrow R_4 \end{matrix}$$

$$\begin{bmatrix} 1 & 0 & 0 & 1 & | & 2 \\ 0 & 1 & 0 & 0 & | & 2 \\ 0 & 0 & 1 & 0 & | & 0 \\ 0 & 0 & 0 & 2 & | & 2 \end{bmatrix} \Rightarrow \begin{bmatrix} 1 & 0 & 0 & 0 & | & 1 \\ 0 & 1 & 0 & 0 & | & 2 \\ 0 & 0 & 1 & 0 & | & 0 \\ 0 & 0 & 0 & 1 & | & 1 \end{bmatrix}$$ Solution: $\boxed{x = 1, y = 2, z = 0, t = 1}$

$$\begin{matrix} -R_3 + R_1 \Rightarrow R_1 \\ R_3 + R_2 \Rightarrow R_2 \\ -R_3 + R_4 \Rightarrow R_4 \end{matrix} \qquad \begin{matrix} -\frac{1}{2}R_4 + R_1 \Rightarrow R_1 \\ \frac{1}{2}R_4 \Rightarrow R_4 \end{matrix}$$

57.

$$\begin{bmatrix} \frac{1}{3} & \frac{3}{4} & -\frac{2}{3} & | & -2 \\ 1 & \frac{1}{2} & \frac{1}{3} & | & 1 \\ \frac{1}{6} & -\frac{1}{8} & -1 & | & 0 \end{bmatrix} \Rightarrow \begin{bmatrix} 1 & \frac{9}{4} & -2 & | & -6 \\ 6 & 3 & 2 & | & 6 \\ 4 & -3 & -24 & | & 0 \end{bmatrix} \Rightarrow \begin{bmatrix} 1 & \frac{9}{4} & -2 & | & -6 \\ 0 & -\frac{21}{2} & 14 & | & 42 \\ 0 & -12 & -16 & | & 24 \end{bmatrix} \Rightarrow$$

$$\begin{matrix} 3R_1 \Rightarrow R_1 \\ 6R_2 \Rightarrow R_2 \\ 24R_3 \Rightarrow R_3 \end{matrix} \qquad \begin{matrix} -6R_1 + R_2 \Rightarrow R_2 \\ -4R_1 + R_3 \Rightarrow R_3 \end{matrix}$$

$$\begin{bmatrix} 1 & \frac{9}{4} & -2 & | & -6 \\ 0 & 1 & -\frac{4}{3} & | & -4 \\ 0 & 3 & 4 & | & -6 \end{bmatrix} \Rightarrow \begin{bmatrix} 1 & 0 & 1 & | & 3 \\ 0 & 1 & -\frac{4}{3} & | & -4 \\ 0 & 0 & 8 & | & 6 \end{bmatrix} \Rightarrow \begin{bmatrix} 1 & 0 & 0 & | & \frac{9}{4} \\ 0 & 1 & 0 & | & -3 \\ 0 & 0 & 1 & | & \frac{3}{4} \end{bmatrix}$$ Solution:

$$\boxed{x = \frac{9}{4}, y = -3, z = \frac{3}{4}}$$

$$\begin{matrix} -\frac{2}{21}R_2 \Rightarrow R_2 \\ -\frac{1}{4}R_3 \Rightarrow R_3 \end{matrix} \qquad \begin{matrix} -\frac{9}{4}R_2 + R_1 \Rightarrow R_1 \\ -3R_2 + R_3 \Rightarrow R_3 \end{matrix} \qquad \begin{matrix} -\frac{1}{8}R_3 + R_1 \Rightarrow R_1 \\ \frac{1}{6}R_3 + R_2 \Rightarrow R_2 \\ \frac{1}{8}R_3 \Rightarrow R_3 \end{matrix}$$

59.
$$\begin{bmatrix} \frac{1}{2} & \frac{1}{4} & -1 & \bigm| & 2 \\ \frac{2}{3} & \frac{1}{4} & \frac{1}{2} & \bigm| & \frac{3}{2} \\ \frac{2}{3} & 0 & 1 & \bigm| & -\frac{1}{3} \end{bmatrix} \Rightarrow \begin{bmatrix} 1 & \frac{1}{2} & -2 & \bigm| & 4 \\ 8 & 3 & 6 & \bigm| & 18 \\ 2 & 0 & 3 & \bigm| & -1 \end{bmatrix} \Rightarrow \begin{bmatrix} 1 & \frac{1}{2} & -2 & \bigm| & 4 \\ 0 & -1 & 22 & \bigm| & -14 \\ 0 & -1 & 7 & \bigm| & -9 \end{bmatrix} \Rightarrow$$

$$2R_1 \Rightarrow R_1, 12R_2 \Rightarrow R_2 \qquad -8R_1 + R_2 \Rightarrow R_2$$
$$3R_3 \Rightarrow R_3 \qquad\qquad -2R_1 + R_3 \Rightarrow R_3$$

$$\begin{bmatrix} 1 & \frac{1}{2} & -2 & \bigm| & 4 \\ 0 & 1 & -22 & \bigm| & 14 \\ 0 & -1 & 7 & \bigm| & -9 \end{bmatrix} \Rightarrow \begin{bmatrix} 1 & 0 & 9 & \bigm| & -3 \\ 0 & 1 & -22 & \bigm| & 14 \\ 0 & 0 & -15 & \bigm| & 5 \end{bmatrix} \Rightarrow \begin{bmatrix} 1 & 0 & 9 & \bigm| & -3 \\ 0 & 1 & -22 & \bigm| & 14 \\ 0 & 0 & 1 & \bigm| & -\frac{1}{3} \end{bmatrix} \Rightarrow$$

$$-R_2 \Rightarrow R_2 \qquad -\frac{1}{2}R_2 + R_1 \Rightarrow R_1 \qquad -\frac{1}{15}R_3 \Rightarrow R_3$$
$$R_2 + R_3 \Rightarrow R_3$$

$$\begin{bmatrix} 1 & 0 & 0 & \bigm| & 0 \\ 0 & 1 & 0 & \bigm| & \frac{20}{3} \\ 0 & 0 & 1 & \bigm| & -\frac{1}{3} \end{bmatrix}$$

Solution: $\boxed{x = 0, \; y = \frac{20}{3}, \; z = -\frac{1}{3}}$

$$-9R_3 + R_1 \Rightarrow R_1$$
$$22R_3 + R_2 \Rightarrow R_2$$

61.
$$\begin{bmatrix} 1 & 1 & \bigm| & -2 \\ 3 & -1 & \bigm| & 6 \\ 2 & 2 & \bigm| & -4 \\ 1 & -1 & \bigm| & 4 \end{bmatrix} \Rightarrow \begin{bmatrix} 1 & 1 & \bigm| & -2 \\ 0 & -4 & \bigm| & 12 \\ 0 & 0 & \bigm| & 0 \\ 0 & -2 & \bigm| & 6 \end{bmatrix} \Rightarrow \begin{bmatrix} 1 & 1 & \bigm| & -2 \\ 0 & 1 & \bigm| & -3 \\ 0 & -2 & \bigm| & 6 \\ 0 & 0 & \bigm| & 0 \end{bmatrix} \Rightarrow \begin{bmatrix} 1 & 0 & \bigm| & 1 \\ 0 & 1 & \bigm| & -3 \\ 0 & 0 & \bigm| & 0 \\ 0 & 0 & \bigm| & 0 \end{bmatrix}$$

$$-3R_1 + R_2 \Rightarrow R_2 \qquad -\frac{1}{4}R_2 \Rightarrow R_2 \qquad -R_2 + R_1 \Rightarrow R_1$$
$$-2R_1 + R_3 \Rightarrow R_3 \qquad R_3 \Leftrightarrow R_4 \qquad 2R_2 + R_3 \Rightarrow R_3$$
$$-R_1 + R_4 \Rightarrow R_4$$

Solution:
$x = 1$
$y = -3$

63.
$$\begin{bmatrix} 1 & 2 & 1 & \bigm| & 4 \\ 3 & -1 & -1 & \bigm| & 2 \end{bmatrix} \Rightarrow \begin{bmatrix} 1 & 2 & 1 & \bigm| & 4 \\ 0 & -7 & -4 & \bigm| & -10 \end{bmatrix} \Rightarrow \begin{bmatrix} 1 & 2 & 1 & \bigm| & 4 \\ 0 & 1 & \frac{4}{7} & \bigm| & \frac{10}{7} \end{bmatrix} \Rightarrow \begin{bmatrix} 1 & 0 & -\frac{1}{7} & \bigm| & \frac{8}{7} \\ 0 & 1 & \frac{4}{7} & \bigm| & \frac{10}{7} \end{bmatrix}$$

$$-3R_1 + R_2 \Rightarrow R_2 \qquad -\frac{1}{7}R_2 \Rightarrow R_2 \qquad -2R_2 + R_1 \Rightarrow R_1$$

From R_1: $\; x - \frac{1}{7}z = \frac{8}{7}$ $\qquad$ From R_2: $\; y + \frac{4}{7}z = \frac{10}{7}$

$\qquad\qquad\quad x = \frac{8}{7} + \frac{1}{7}z$ $\qquad\qquad\qquad\quad y = \frac{10}{7} - \frac{4}{7}z$

Solution:
$x = \frac{8}{7} + \frac{1}{7}z, \; y = \frac{10}{7} - \frac{4}{7}z$
$z = $ any real number

65.
$$\begin{bmatrix} 1 & 1 & 0 & 0 & \bigm| & 1 \\ 1 & 0 & 1 & 0 & \bigm| & 0 \\ 0 & 1 & 0 & 1 & \bigm| & 0 \end{bmatrix} \Rightarrow \begin{bmatrix} 1 & 1 & 0 & 0 & \bigm| & 1 \\ 0 & 1 & -1 & 0 & \bigm| & 1 \\ 0 & 1 & 0 & 1 & \bigm| & 0 \end{bmatrix} \Rightarrow \begin{bmatrix} 1 & 0 & 1 & 0 & \bigm| & 0 \\ 0 & 1 & -1 & 0 & \bigm| & 1 \\ 0 & 0 & 1 & 1 & \bigm| & -1 \end{bmatrix} \Rightarrow$$

$$-R_2 + R_1 \Rightarrow R_2 \qquad\qquad -R_2 + R_1 \Rightarrow R_1$$
$$\qquad\qquad\qquad\qquad\qquad -R_2 + R_3 \Rightarrow R_3$$

$$\begin{bmatrix} 1 & 0 & 0 & -1 & \bigm| & 1 \\ 0 & 1 & 0 & 1 & \bigm| & 0 \\ 0 & 0 & 1 & 1 & \bigm| & -1 \end{bmatrix} \Rightarrow$$

$$-R_3 + R_1 \Rightarrow R_1$$
$$R_3 + R_2 \Rightarrow R_2$$

From R_1: $\qquad$ From R_2: $\qquad$ From R_3:

$w - z = 1 \qquad x + z = 0 \qquad y + z = -1$

$\quad w = 1 + z \qquad\quad x = -z \qquad\quad y = -1 - z$

Solution:
$w = 1 + z,$
$x = -z,$
$y = -1 - z,$
$z = $ any real #

67. $\begin{bmatrix} 1 & 1 & | & 3 \\ 2 & 1 & | & 1 \\ 3 & 2 & | & 2 \end{bmatrix} \Rightarrow \begin{bmatrix} 1 & 1 & | & 3 \\ 0 & -1 & | & -5 \\ 0 & -1 & | & -7 \end{bmatrix} \Rightarrow \begin{bmatrix} 1 & 1 & | & 3 \\ 0 & -1 & | & -5 \\ 0 & 0 & | & -2 \end{bmatrix}$ R_3 indicates that $0x + 0y = -2$.

$\qquad\qquad\quad -2R_1 + R_2 \Rightarrow R_2 \qquad\quad -R_2 + R_3 \Rightarrow R_3$ This is impossible. The system is

$\qquad\qquad\quad -3R_1 + R_3 \Rightarrow R_3$ inconsistent. $\Rightarrow$ no solution

69. Let p = speed with no wind. Let w = the speed of the wind. Then the following system applies:

$\begin{cases} p + w = 300 \\ p - w = 220 \end{cases}$ $\begin{bmatrix} 1 & 1 & | & 300 \\ 1 & -1 & | & 220 \end{bmatrix} \Rightarrow \begin{bmatrix} 1 & 1 & | & 300 \\ 0 & 2 & | & 80 \end{bmatrix} \Rightarrow \begin{bmatrix} 1 & 1 & | & 300 \\ 0 & 1 & | & 40 \end{bmatrix} \Rightarrow \begin{bmatrix} 1 & 0 & | & 260 \\ 0 & 1 & | & 40 \end{bmatrix}$

$\qquad\qquad\qquad\qquad\qquad\quad -R_2 + R_1 \Rightarrow R_2 \qquad\quad \frac{1}{2}R_2 \Rightarrow R_2 \qquad -R_2 + R_1 \Rightarrow R_1$

The plane has a speed of 260 miles per hour with no wind, so it could travel 1300 miles in 5 hours.

71. Let d = width of a dictionary, a = width of an atlas and t = width of a thesaurus.

$\begin{cases} 3d + 5a + t = 35 \\ 6d + 2t = 35 \\ 2d + 4a + 3t = 35 \end{cases}$ $\begin{bmatrix} 3 & 5 & 1 & | & 35 \\ 6 & 0 & 2 & | & 35 \\ 2 & 4 & 3 & | & 35 \end{bmatrix} \Rightarrow \begin{bmatrix} 3 & 5 & 1 & | & 35 \\ 0 & -10 & 0 & | & -35 \\ 0 & \frac{2}{3} & \frac{7}{3} & | & \frac{35}{3} \end{bmatrix} \Rightarrow \begin{bmatrix} 3 & 5 & 1 & | & 35 \\ 0 & 1 & 0 & | & 3.5 \\ 0 & 2 & 7 & | & 35 \end{bmatrix}$

$\qquad\qquad\qquad\qquad\qquad\qquad\qquad -2R_1 + R_2 \Rightarrow R_2 \qquad\quad -\frac{1}{10}R_2 \Rightarrow R_2$

$\qquad\qquad\qquad\qquad\qquad\qquad\qquad -\frac{2}{3}R_1 + R_3 \Rightarrow R_3 \qquad\quad 3R_3 \Rightarrow R_3$

$\Rightarrow \begin{bmatrix} 3 & 0 & 1 & | & 17.5 \\ 0 & 1 & 0 & | & 3.5 \\ 0 & 0 & 7 & | & 28 \end{bmatrix} \Rightarrow \begin{bmatrix} 3 & 0 & 0 & | & 13.5 \\ 0 & 1 & 0 & | & 3.5 \\ 0 & 0 & 1 & | & 4 \end{bmatrix} \Rightarrow \begin{bmatrix} 1 & 0 & 0 & | & 4.5 \\ 0 & 1 & 0 & | & 3.5 \\ 0 & 0 & 1 & | & 4 \end{bmatrix}$ Dictionaries are 4.5 in.

$\quad -5R_2 + R_1 \Rightarrow R_1 \qquad -\frac{1}{7}R_3 + R_1 \Rightarrow R_1 \qquad \frac{1}{3}R_1 \Rightarrow R_1$ wide. Atlases are 3.5 in.

$\quad -2R_2 + R_3 \Rightarrow R_3 \qquad \frac{1}{7}R_3 \Rightarrow R_3$ wide. Thesauruses are

$\qquad\qquad\qquad\qquad\qquad\qquad\qquad\qquad\qquad\qquad\qquad$ 4 in. wide.

73. **Answers may vary.**   **75.** **Answers may vary.**

77. $\begin{bmatrix} 1 & 1 & 1 & | & 14 \\ 2 & 3 & -2 & | & -7 \\ 1 & -5 & 1 & | & 8 \end{bmatrix} \Rightarrow \begin{bmatrix} 1 & 1 & 1 & | & 14 \\ 0 & 1 & -4 & | & -35 \\ 0 & -6 & 0 & | & -6 \end{bmatrix} \Rightarrow \begin{bmatrix} 1 & 1 & 1 & | & 14 \\ 0 & -6 & 0 & | & -6 \\ 0 & 1 & -4 & | & -35 \end{bmatrix} \Rightarrow$

$\qquad\qquad\qquad\qquad\qquad -2R_1 + R_2 \Rightarrow R_2 \qquad\qquad R_2 \Leftrightarrow R_3$

$\qquad\qquad\qquad\qquad\qquad -R_1 + R_3 \Rightarrow R_3$

$\begin{bmatrix} 1 & 0 & 1 & | & 13 \\ 0 & -6 & 0 & | & -6 \\ 0 & 0 & -4 & | & -36 \end{bmatrix} \Rightarrow \begin{bmatrix} 1 & 0 & 0 & | & 4 \\ 0 & 1 & 0 & | & 1 \\ 0 & 0 & 1 & | & 9 \end{bmatrix}$ $x^2 = 4 \Rightarrow \boxed{x = \pm 2}$

$\quad \frac{1}{6}R_2 + R_1 \Rightarrow R_1 \qquad \frac{1}{4}R_3 + R_1 \Rightarrow R_1$ $y^2 = 1 \Rightarrow \boxed{y = \pm 1}$

$\quad \frac{1}{6}R_2 + R_3 \Rightarrow R_3 \qquad -\frac{1}{6}R_2 \Rightarrow R_2$ $z^2 = 9 \Rightarrow \boxed{z = \pm 3}$

$\qquad\qquad\qquad\qquad\qquad -\frac{1}{4}R_3 \Rightarrow R_3$

79. $y = mx + b$ **81.** equal   **83.** $y = mx + b$ **85.** $x = 2$

$\qquad\qquad\qquad\qquad\qquad\qquad\qquad\qquad\qquad\qquad\qquad y = 2x + 7$

Exercise 6.3 (page 456)

1. i, j **3.** every element   **5.** additive identity

7. $x = 2, y = 5$

9. $x + y = 3$
$3 + x = 4 \Rightarrow x = 1$
$5y = 10 \Rightarrow y = 2$

11. $5A = 5\begin{bmatrix} 3 & -3 \\ 0 & -2 \end{bmatrix} = \begin{bmatrix} 15 & -15 \\ 0 & -10 \end{bmatrix}$

13. $5A = 5\begin{bmatrix} 5 & 15 & -2 \\ -2 & -5 & 1 \end{bmatrix}$
$= \begin{bmatrix} 25 & 75 & -10 \\ -10 & -25 & 5 \end{bmatrix}$

15. $A + B = \begin{bmatrix} 2 & 1 & -1 \\ -3 & 2 & 5 \end{bmatrix} + \begin{bmatrix} -3 & 1 & 2 \\ -3 & -2 & -5 \end{bmatrix} = \begin{bmatrix} -1 & 2 & 1 \\ -6 & 0 & 0 \end{bmatrix}$

17. $A - B = \begin{bmatrix} -3 & 2 & -2 \\ -1 & 4 & -5 \end{bmatrix} - \begin{bmatrix} 3 & -3 & -2 \\ -2 & 5 & -5 \end{bmatrix} = \begin{bmatrix} -6 & 5 & 0 \\ 1 & -1 & 0 \end{bmatrix}$

19. $5A + 3B = 5\begin{bmatrix} 3 & 1 & -2 \\ -4 & 3 & -2 \end{bmatrix} + 3\begin{bmatrix} 1 & -2 & 2 \\ -5 & -5 & 3 \end{bmatrix} = \begin{bmatrix} 15 & 5 & -10 \\ -20 & 15 & -10 \end{bmatrix} + \begin{bmatrix} 3 & -6 & 6 \\ -15 & -15 & 9 \end{bmatrix}$
$= \begin{bmatrix} 18 & -1 & -4 \\ -35 & 0 & -1 \end{bmatrix}$

21. additive inverse of $A = \begin{bmatrix} -5 & 2 & -7 \\ 5 & 0 & -3 \\ 2 & -3 & 5 \end{bmatrix}$

23. $\begin{bmatrix} 2 & 3 \\ 3 & -2 \end{bmatrix}_{2\times2} \begin{bmatrix} 1 & 2 \\ 0 & -2 \end{bmatrix}_{2\times2} = \begin{bmatrix} (2)(1)+(3)(0) & (2)(2)+(3)(-2) \\ (3)(1)+(-2)(0) & (3)(2)+(-2)(-2) \end{bmatrix}_{2\times2} = \begin{bmatrix} 2 & -2 \\ 3 & 10 \end{bmatrix}$

25. $\begin{bmatrix} -4 & -2 \\ 21 & 0 \end{bmatrix}_{2\times2} \begin{bmatrix} -5 & 6 \\ 21 & -1 \end{bmatrix}_{2\times2} = \begin{bmatrix} (-4)(-5)+(-2)(21) & (-4)(6)+(-2)(-1) \\ (21)(-5)+(0)(21) & (21)(6)+(0)(-1) \end{bmatrix}_{2\times2}$
$= \begin{bmatrix} -22 & -22 \\ -105 & 126 \end{bmatrix}$

27. $\begin{bmatrix} 2 & 1 & 3 \\ 1 & 2 & -1 \\ 0 & 1 & 0 \end{bmatrix}_{3\times3} \begin{bmatrix} 1 & 2 & 3 \\ 2 & -2 & 1 \\ 0 & 0 & 1 \end{bmatrix}_{3\times3}$
$= \begin{bmatrix} (2)(1)+(1)(2)+(3)(0) & (2)(2)+(1)(-2)+(3)(0) & (2)(3)+(1)(1)+(3)(1) \\ (1)(1)+(2)(2)+(-1)(0) & (1)(2)+(2)(-2)+(-1)(0) & (1)(3)+(2)(1)+(-1)(1) \\ (0)(1)+(1)(2)+(0)(0) & (0)(2)+(1)(-2)+(0)(0) & (0)(3)+(1)(1)+(0)(1) \end{bmatrix}_{3\times3}$
$= \begin{bmatrix} 4 & 2 & 10 \\ 5 & -2 & 4 \\ 2 & -2 & 1 \end{bmatrix}$

29. $\begin{bmatrix} 1 & -2 & -3 \\ 2 & 0 & 1 \end{bmatrix}_{2\times3} \begin{bmatrix} 4 \\ -5 \\ -6 \end{bmatrix}_{3\times1} = \begin{bmatrix} (1)(4)+(-2)(-5)+(-3)(-6) \\ (2)(4)+(0)(-5)+(1)(-6) \end{bmatrix}_{2\times1} = \begin{bmatrix} 32 \\ 2 \end{bmatrix}$

31. $[1 \quad 2 \quad 3]_{1\times 3} \begin{bmatrix} 4 & 5 & 6 \\ 7 & 8 & 9 \end{bmatrix}_{2\times 3} \Rightarrow$ Not possible

33. $\begin{bmatrix} 2 & 5 \\ -3 & 1 \\ 0 & -2 \\ 1 & -5 \end{bmatrix}_{4\times 2} \begin{bmatrix} 3 & -2 & 4 \\ -2 & -3 & 1 \end{bmatrix}_{2\times 3}$

$= \begin{bmatrix} (2)(3)+(5)(-2) & (2)(-2)+(5)(-3) & (2)(4)+(5)(1) \\ (-3)(3)+(1)(-2) & (-3)(-2)+(1)(-3) & (-3)(4)+(1)(1) \\ (0)(3)+(-2)(-2) & (0)(-2)+(-2)(-3) & (0)(4)+(-2)(1) \\ (1)(3)+(-5)(-2) & (1)(-2)+(-5)(-3) & (1)(4)+(-5)(1) \end{bmatrix}_{4\times 3} = \begin{bmatrix} -4 & -19 & 13 \\ -11 & 3 & -11 \\ 4 & 6 & -2 \\ 13 & 13 & -1 \end{bmatrix}$

35. $AB = \begin{bmatrix} 2.3 & -1.7 & 3.1 \\ -2 & 3.5 & 1 \\ -8 & 4.7 & 9.1 \end{bmatrix} \begin{bmatrix} -2.5 \\ 5.2 \\ -7 \end{bmatrix} = \begin{bmatrix} -36.29 \\ 16.2 \\ -19.26 \end{bmatrix}$

37. $A^2 = \begin{bmatrix} 2.3 & -1.7 & 3.1 \\ -2 & 3.5 & 1 \\ -8 & 4.7 & 9.1 \end{bmatrix} \begin{bmatrix} 2.3 & -1.7 & 3.1 \\ -2 & 3.5 & 1 \\ -8 & 4.7 & 9.1 \end{bmatrix} = \begin{bmatrix} -16.11 & 4.71 & 33.64 \\ -19.6 & 20.35 & 6.4 \\ -100.6 & 72.82 & 62.71 \end{bmatrix}$

39. $A(B+C) = \begin{bmatrix} 2 & 3 \\ 1 & 3 \end{bmatrix} \left(\begin{bmatrix} 2 & 1 & -5 \\ 1 & 1 & 2 \end{bmatrix} + \begin{bmatrix} -2 & -1 & 6 \\ 0 & -1 & -1 \end{bmatrix} \right) = \begin{bmatrix} 2 & 3 \\ 1 & 3 \end{bmatrix} \begin{bmatrix} 0 & 0 & 1 \\ 1 & 0 & 1 \end{bmatrix} = \begin{bmatrix} 3 & 0 & 5 \\ 3 & 0 & 4 \end{bmatrix}$

$AB + AC = \begin{bmatrix} 2 & 3 \\ 1 & 3 \end{bmatrix} \begin{bmatrix} 2 & 1 & -5 \\ 1 & 1 & 2 \end{bmatrix} + \begin{bmatrix} 2 & 3 \\ 1 & 3 \end{bmatrix} \begin{bmatrix} -2 & -1 & 6 \\ 0 & -1 & -1 \end{bmatrix}$

$= \begin{bmatrix} 7 & 5 & -4 \\ 5 & 4 & 1 \end{bmatrix} + \begin{bmatrix} -4 & -5 & 9 \\ -2 & -4 & 3 \end{bmatrix} = \begin{bmatrix} 3 & 0 & 5 \\ 3 & 0 & 4 \end{bmatrix}$

41. $3(AB) = 3 \left(\begin{bmatrix} 2 & 3 \\ 1 & 3 \end{bmatrix} \begin{bmatrix} 2 & 1 & -5 \\ 1 & 1 & 2 \end{bmatrix} \right) = 3 \begin{bmatrix} 7 & 5 & -4 \\ 5 & 4 & 1 \end{bmatrix} = \begin{bmatrix} 21 & 15 & -12 \\ 15 & 12 & 3 \end{bmatrix}$

$(3A)B = \left(3 \begin{bmatrix} 2 & 3 \\ 1 & 3 \end{bmatrix} \right) \begin{bmatrix} 2 & 1 & -5 \\ 1 & 1 & 2 \end{bmatrix} = \begin{bmatrix} 6 & 9 \\ 3 & 9 \end{bmatrix} \begin{bmatrix} 2 & 1 & -5 \\ 1 & 1 & 2 \end{bmatrix} = \begin{bmatrix} 21 & 15 & -12 \\ 15 & 12 & 3 \end{bmatrix}$

43. $A - BC = \begin{bmatrix} 1 & 3 \\ 2 & 5 \end{bmatrix} - \begin{bmatrix} -1 \\ 3 \end{bmatrix} [3 \quad 2] = \begin{bmatrix} 1 & 3 \\ 2 & 5 \end{bmatrix} - \begin{bmatrix} -3 & -2 \\ 9 & 6 \end{bmatrix} = \begin{bmatrix} 4 & 5 \\ -7 & -1 \end{bmatrix}$

45. $CB - AB = [3 \quad 2] \begin{bmatrix} -1 \\ 3 \end{bmatrix} - \begin{bmatrix} 1 & 3 \\ 2 & 5 \end{bmatrix} \begin{bmatrix} -1 \\ 3 \end{bmatrix} = [3] - \begin{bmatrix} 8 \\ 13 \end{bmatrix} \Rightarrow$ not possible

47. $ABC = \begin{bmatrix} 1 & 3 \\ 2 & 5 \end{bmatrix} \begin{bmatrix} -1 \\ 3 \end{bmatrix} [3 \quad 2] = \begin{bmatrix} 8 \\ 13 \end{bmatrix} [3 \quad 2] = \begin{bmatrix} 24 & 16 \\ 39 & 26 \end{bmatrix}$

49. $A^2 B = \begin{bmatrix} 1 & 3 \\ 2 & 5 \end{bmatrix} \begin{bmatrix} 1 & 3 \\ 2 & 5 \end{bmatrix} \begin{bmatrix} -1 \\ 3 \end{bmatrix} = \begin{bmatrix} 7 & 18 \\ 12 & 31 \end{bmatrix} \begin{bmatrix} -1 \\ 3 \end{bmatrix} = \begin{bmatrix} 47 \\ 81 \end{bmatrix}$

51. $Q = \begin{bmatrix} 217 & 23 & 319 \\ 347 & 24 & 340 \\ 3 & 97 & 750 \end{bmatrix}, P = \begin{bmatrix} 0.75 \\ 1.00 \\ 1.25 \end{bmatrix}$

$QP = \begin{bmatrix} 217 & 23 & 319 \\ 347 & 24 & 340 \\ 3 & 97 & 750 \end{bmatrix} \begin{bmatrix} 0.75 \\ 1.00 \\ 1.25 \end{bmatrix} = \begin{bmatrix} 584.50 \\ 709.25 \\ 1036.75 \end{bmatrix}$ $\begin{array}{l} \text{\$ spent by adult males} \\ \text{\$ spent by adult females} \\ \text{\$ spent by children} \end{array}$

53. $A^2 = \begin{bmatrix} 0 & 1 & 1 \\ 1 & 0 & 0 \\ 0 & 1 & 0 \end{bmatrix}^2 = \begin{bmatrix} 0 & 1 & 1 \\ 1 & 0 & 0 \\ 0 & 1 & 0 \end{bmatrix}\begin{bmatrix} 0 & 1 & 1 \\ 1 & 0 & 0 \\ 0 & 1 & 0 \end{bmatrix} = \begin{bmatrix} 1 & 1 & 0 \\ 0 & 1 & 1 \\ 1 & 0 & 0 \end{bmatrix}$

55. $A^2 = \begin{bmatrix} 0 & 2 & 1 & 0 \\ 2 & 0 & 1 & 0 \\ 1 & 1 & 0 & 2 \\ 0 & 0 & 2 & 0 \end{bmatrix}^2 = \begin{bmatrix} 0 & 2 & 1 & 0 \\ 2 & 0 & 1 & 0 \\ 1 & 1 & 0 & 2 \\ 0 & 0 & 2 & 0 \end{bmatrix}\begin{bmatrix} 0 & 2 & 1 & 0 \\ 2 & 0 & 1 & 0 \\ 1 & 1 & 0 & 2 \\ 0 & 0 & 2 & 0 \end{bmatrix} = \begin{bmatrix} 5 & 1 & 2 & 2 \\ 1 & 5 & 2 & 2 \\ 2 & 2 & 6 & 0 \\ 2 & 2 & 0 & 4 \end{bmatrix}$

A^2 represents the number of ways two cities can be linked through one intermediary.

57. Let $A = \begin{bmatrix} 1 & 1 \\ 1 & 1 \end{bmatrix}$ and $B = \begin{bmatrix} 1 & 0 \\ 0 & 0 \end{bmatrix}$.

$(AB)^2 = \left(\begin{bmatrix} 1 & 1 \\ 1 & 1 \end{bmatrix}\begin{bmatrix} 1 & 0 \\ 0 & 0 \end{bmatrix}\right)^2 = \begin{bmatrix} 1 & 0 \\ 1 & 0 \end{bmatrix}^2 = \begin{bmatrix} 1 & 0 \\ 1 & 0 \end{bmatrix}\begin{bmatrix} 1 & 0 \\ 1 & 0 \end{bmatrix} = \begin{bmatrix} 1 & 0 \\ 1 & 0 \end{bmatrix}$

$A^2B^2 = \begin{bmatrix} 1 & 1 \\ 1 & 1 \end{bmatrix}^2\begin{bmatrix} 1 & 0 \\ 0 & 0 \end{bmatrix}^2 = \begin{bmatrix} 2 & 2 \\ 2 & 2 \end{bmatrix}\begin{bmatrix} 1 & 0 \\ 0 & 0 \end{bmatrix} = \begin{bmatrix} 2 & 0 \\ 2 & 0 \end{bmatrix}. \quad (AB)^2 \neq A^2B^2$

59. Let $A = \begin{bmatrix} 1 & 2 \\ 1 & 2 \end{bmatrix}$ and $B = \begin{bmatrix} 2 & 2 \\ -1 & -1 \end{bmatrix}$. $AB = \begin{bmatrix} 1 & 2 \\ 1 & 2 \end{bmatrix}\begin{bmatrix} 2 & 2 \\ -1 & -1 \end{bmatrix} = \begin{bmatrix} 0 & 0 \\ 0 & 0 \end{bmatrix}$

61. $(3x + 2)(2x - 3) - (2 - x) = 6x^2 - 9x + 4x - 6 - 2 + x = 6x^2 - 4x - 8$

63. $\dfrac{1 + \frac{1}{x}}{1 - \frac{1}{x}} = \dfrac{x\left(1 + \frac{1}{x}\right)}{x\left(1 - \frac{1}{x}\right)} = \dfrac{x + 1}{x - 1}$

65. $s = \dfrac{n(a + l)}{2}$

$2s = n(a + l)$

$\dfrac{2s}{n} = a + l$

$\dfrac{2s}{n} - l = a$

Exercise 6.4 (page 466)

1. $AB = BA = I$

3. $[I \mid A^{-1}]$

5. $\begin{bmatrix} 3 & -4 & | & 1 & 0 \\ -2 & 3 & | & 0 & 1 \end{bmatrix} \Rightarrow \begin{bmatrix} 1 & -\frac{4}{3} & | & \frac{1}{3} & 0 \\ -2 & 3 & | & 0 & 1 \end{bmatrix} \Rightarrow \begin{bmatrix} 1 & -\frac{4}{3} & | & \frac{1}{3} & 0 \\ 0 & \frac{1}{3} & | & \frac{2}{3} & 1 \end{bmatrix} \Rightarrow$

$\qquad\qquad\qquad \frac{1}{3}R_1 \Rightarrow R_1 \qquad\qquad 2R_1 + R_2 \Rightarrow R_2$

$\begin{bmatrix} 1 & 0 & | & 3 & 4 \\ 0 & \frac{1}{3} & | & \frac{2}{3} & 1 \end{bmatrix} \Rightarrow \begin{bmatrix} 1 & 0 & | & 3 & 4 \\ 0 & 1 & | & 2 & 3 \end{bmatrix} \Rightarrow \quad \text{Inverse:} \begin{bmatrix} 3 & 4 \\ 2 & 3 \end{bmatrix}$

$4R_2 + R_1 \Rightarrow R_1 \qquad\quad 3R_2 \Rightarrow R_2$

7. $\begin{bmatrix} 3 & 7 & | & 1 & 0 \\ 2 & 5 & | & 0 & 1 \end{bmatrix} \Rightarrow \begin{bmatrix} 1 & \frac{7}{3} & | & \frac{1}{3} & 0 \\ 2 & 5 & | & 0 & 1 \end{bmatrix} \Rightarrow \begin{bmatrix} 1 & \frac{7}{3} & | & \frac{1}{3} & 0 \\ 0 & \frac{1}{3} & | & -\frac{2}{3} & 1 \end{bmatrix} \Rightarrow$

$\qquad\qquad\qquad \frac{1}{3}R_1 \Rightarrow R_1 \qquad -2R_1 + R_2 \Rightarrow R_2$

$\begin{bmatrix} 1 & 0 & | & 5 & -7 \\ 0 & \frac{1}{3} & | & -\frac{2}{3} & 1 \end{bmatrix} \Rightarrow \begin{bmatrix} 1 & 0 & | & 5 & -7 \\ 0 & 1 & | & -2 & 3 \end{bmatrix} \Rightarrow \quad \text{Inverse:} \begin{bmatrix} 5 & -7 \\ -2 & 3 \end{bmatrix}$

$-7R_2 + R_1 \Rightarrow R_1 \qquad\qquad 3R_2 \Rightarrow R_2$

9. $\begin{bmatrix} 1 & 0 & 3 & | & 1 & 0 & 0 \\ -1 & 1 & 3 & | & 0 & 1 & 0 \\ -2 & 1 & 1 & | & 0 & 0 & 1 \end{bmatrix} \Rightarrow \begin{bmatrix} 1 & 0 & 3 & | & 1 & 0 & 0 \\ 0 & 1 & 6 & | & 1 & 1 & 0 \\ 0 & 1 & 7 & | & 2 & 0 & 1 \end{bmatrix} \Rightarrow \begin{bmatrix} 1 & 0 & 3 & | & 1 & 0 & 0 \\ 0 & 1 & 6 & | & 1 & 1 & 0 \\ 0 & 0 & 1 & | & 1 & -1 & 1 \end{bmatrix} \Rightarrow$

$\qquad\qquad\qquad\qquad\qquad\qquad R_1 + R_2 \Rightarrow R_2 \qquad\qquad -R_2 + R_3 \Rightarrow R_3$

$\qquad\qquad\qquad\qquad\qquad\qquad 2R_1 + R_3 \Rightarrow R_3$

$\begin{bmatrix} 1 & 0 & 0 & | & -2 & 3 & -3 \\ 0 & 1 & 0 & | & -5 & 7 & -6 \\ 0 & 0 & 1 & | & 1 & -1 & 1 \end{bmatrix} \Rightarrow \text{Inverse:} \begin{bmatrix} -2 & 3 & -3 \\ -5 & 7 & -6 \\ 1 & -1 & 1 \end{bmatrix}$

$\qquad -3R_3 + R_1 \Rightarrow R_1$

$\qquad -6R_3 + R_2 \Rightarrow R_2$

11. $\begin{bmatrix} 3 & 2 & 1 & | & 1 & 0 & 0 \\ 1 & 1 & -1 & | & 0 & 1 & 0 \\ 4 & 3 & 1 & | & 0 & 0 & 1 \end{bmatrix} \Rightarrow \begin{bmatrix} 1 & 1 & -1 & | & 0 & 1 & 0 \\ 3 & 2 & 1 & | & 1 & 0 & 0 \\ 4 & 3 & 1 & | & 0 & 0 & 1 \end{bmatrix} \Rightarrow \begin{bmatrix} 1 & 1 & -1 & | & 0 & 1 & 0 \\ 0 & -1 & 4 & | & 1 & -3 & 0 \\ 0 & -1 & 5 & | & 0 & -4 & 1 \end{bmatrix} \Rightarrow$

$\qquad\qquad\qquad\qquad\qquad\qquad R_1 \Leftrightarrow R_2 \qquad\qquad -3R_1 + R_2 \Rightarrow R_2$

$\qquad\qquad\qquad\qquad\qquad\qquad\qquad\qquad\qquad -4R_1 + R_3 \Rightarrow R_3$

$\begin{bmatrix} 1 & 1 & -1 & | & 0 & 1 & 0 \\ 0 & 1 & -4 & | & -1 & 3 & 0 \\ 0 & -1 & 5 & | & 0 & -4 & 1 \end{bmatrix} \Rightarrow \begin{bmatrix} 1 & 0 & 3 & | & 1 & -2 & 0 \\ 0 & 1 & -4 & | & -1 & 3 & 0 \\ 0 & 0 & 1 & | & -1 & -1 & 1 \end{bmatrix} \Rightarrow$

$\qquad -R_2 \Rightarrow R_2 \qquad\qquad\qquad -R_2 + R_1 \Rightarrow R_1$

$\qquad\qquad\qquad\qquad\qquad\qquad R_2 + R_3 \Rightarrow R_3$

$\begin{bmatrix} 1 & 0 & 0 & | & 4 & 1 & -3 \\ 0 & 1 & 0 & | & -5 & -1 & 4 \\ 0 & 0 & 1 & | & -1 & -1 & 1 \end{bmatrix} \Rightarrow \text{Inverse} = \begin{bmatrix} 4 & 1 & -3 \\ -5 & -1 & 4 \\ -1 & -1 & 1 \end{bmatrix}$

$\qquad -3R_3 + R_1 \Rightarrow R_1$

$\qquad\quad 4R_3 + R_2 \Rightarrow R_2$

13. $\begin{bmatrix} 1 & 3 & 5 & | & 1 & 0 & 0 \\ 0 & 1 & 6 & | & 0 & 1 & 0 \\ 1 & 4 & 11 & | & 0 & 0 & 1 \end{bmatrix} \Rightarrow \begin{bmatrix} 1 & 3 & 5 & | & 1 & 0 & 0 \\ 0 & 1 & 6 & | & 0 & 1 & 0 \\ 0 & 1 & 6 & | & -1 & 0 & 1 \end{bmatrix} \Rightarrow \begin{bmatrix} 1 & 3 & 5 & | & 1 & 0 & 0 \\ 0 & 1 & 6 & | & 0 & 1 & 0 \\ 0 & 0 & 0 & | & -1 & -1 & 1 \end{bmatrix}$

$$-R_1 + R_3 \Rightarrow R_3 \qquad\qquad -R_2 + R_3 \Rightarrow R_3$$

Since the original matrix cannot be changed into the identity, there is no inverse matrix.

15. $\begin{bmatrix} 1 & 2 & 3 & | & 1 & 0 & 0 \\ 0 & 1 & 2 & | & 0 & 1 & 0 \\ 0 & 0 & 1 & | & 0 & 0 & 1 \end{bmatrix} \Rightarrow \begin{bmatrix} 1 & 0 & -1 & | & 1 & -2 & 0 \\ 0 & 1 & 2 & | & 0 & 1 & 0 \\ 0 & 0 & 1 & | & 0 & 0 & 1 \end{bmatrix} \Rightarrow \begin{bmatrix} 1 & 0 & 0 & | & 1 & -2 & 1 \\ 0 & 1 & 0 & | & 0 & 1 & -2 \\ 0 & 0 & 1 & | & 0 & 0 & 2 \end{bmatrix} \Rightarrow$

$$-2R_2 + R_1 \Rightarrow R_1 \qquad\qquad R_3 + R_1 \Rightarrow R_1$$
$$-2R_3 + R_2 \Rightarrow R_2$$

$$\text{Inverse} = \begin{bmatrix} 1 & -2 & 1 \\ 0 & 1 & -2 \\ 0 & 0 & 1 \end{bmatrix}$$

17. $\begin{bmatrix} 1 & 6 & 4 & | & 1 & 0 & 0 \\ 1 & -2 & -5 & | & 0 & 1 & 0 \\ 2 & 4 & -1 & | & 0 & 0 & 1 \end{bmatrix} \Rightarrow \begin{bmatrix} 1 & 6 & 4 & | & 1 & 0 & 0 \\ 0 & -8 & -9 & | & -1 & 1 & 0 \\ 0 & -8 & -9 & | & -2 & 0 & 1 \end{bmatrix} \Rightarrow \begin{bmatrix} 1 & 6 & 4 & | & 1 & 0 & 0 \\ 0 & -8 & -9 & | & -1 & 1 & 0 \\ 0 & 0 & 0 & | & -1 & -1 & 1 \end{bmatrix}$

$$-R_1 + R_2 \Rightarrow R_2 \qquad\qquad -R_2 + R_3 \Rightarrow R_3$$
$$-2R_1 + R_3 \Rightarrow R_3$$

Since the original matrix cannot be changed into the identity, there is no inverse matrix.

19. $\text{Inverse} = \begin{bmatrix} 1 & -2 & 1 & 0 \\ 0 & 1 & -2 & 1 \\ 0 & 0 & 1 & -2 \\ 0 & 0 & 0 & 1 \end{bmatrix}$

21. $\text{Inverse} = \begin{bmatrix} 8 & -2 & -6 \\ -5 & 2 & 4 \\ 2 & 0 & -2 \end{bmatrix}$

23. $\text{Inverse} = \begin{bmatrix} -2.5 & 5 & 3 & 5.5 \\ 5.5 & -8 & -6 & -9.5 \\ -1 & 3 & 1 & 3 \\ -5.5 & 9 & 6 & 10.5 \end{bmatrix}$

25. $\begin{bmatrix} 3 & -4 \\ -2 & 3 \end{bmatrix} \begin{bmatrix} x \\ y \end{bmatrix} = \begin{bmatrix} 1 \\ 5 \end{bmatrix}$

$\begin{bmatrix} x \\ y \end{bmatrix} = \begin{bmatrix} 3 & -4 \\ -2 & 3 \end{bmatrix}^{-1} \begin{bmatrix} 1 \\ 5 \end{bmatrix}$

$\begin{bmatrix} x \\ y \end{bmatrix} = \begin{bmatrix} 3 & 4 \\ 2 & 3 \end{bmatrix} \begin{bmatrix} 1 \\ 5 \end{bmatrix}$

$\begin{bmatrix} x \\ y \end{bmatrix} = \begin{bmatrix} 23 \\ 17 \end{bmatrix}$

27. $\begin{bmatrix} 3 & -4 \\ -2 & 3 \end{bmatrix} \begin{bmatrix} x \\ y \end{bmatrix} = \begin{bmatrix} 0 \\ 0 \end{bmatrix}$

$\begin{bmatrix} x \\ y \end{bmatrix} = \begin{bmatrix} 3 & -4 \\ -2 & 3 \end{bmatrix}^{-1} \begin{bmatrix} 0 \\ 0 \end{bmatrix}$

$\begin{bmatrix} x \\ y \end{bmatrix} = \begin{bmatrix} 3 & 4 \\ 2 & 3 \end{bmatrix} \begin{bmatrix} 0 \\ 0 \end{bmatrix}$

$\begin{bmatrix} x \\ y \end{bmatrix} = \begin{bmatrix} 0 \\ 0 \end{bmatrix}$

29.
$$\begin{bmatrix} 2 & 1 & -1 \\ 2 & 2 & -1 \\ -1 & -1 & 1 \end{bmatrix} \begin{bmatrix} x \\ y \\ z \end{bmatrix} = \begin{bmatrix} 2 \\ 4 \\ -1 \end{bmatrix}$$

$$\begin{bmatrix} x \\ y \\ z \end{bmatrix} = \begin{bmatrix} 2 & 1 & -1 \\ 2 & 2 & -1 \\ -1 & -1 & 1 \end{bmatrix}^{-1} \begin{bmatrix} 2 \\ 4 \\ -1 \end{bmatrix}$$

$$\begin{bmatrix} x \\ y \\ z \end{bmatrix} = \begin{bmatrix} 1 & 0 & 1 \\ -1 & 1 & 0 \\ 0 & 1 & 2 \end{bmatrix} \begin{bmatrix} 2 \\ 4 \\ -1 \end{bmatrix} = \begin{bmatrix} 1 \\ 2 \\ 2 \end{bmatrix}$$

31.
$$\begin{bmatrix} -2 & 1 & -3 \\ 2 & 3 & 0 \\ 1 & 0 & 1 \end{bmatrix} \begin{bmatrix} x \\ y \\ z \end{bmatrix} = \begin{bmatrix} 2 \\ -3 \\ 5 \end{bmatrix}$$

$$\begin{bmatrix} x \\ y \\ z \end{bmatrix} = \begin{bmatrix} -2 & 1 & -3 \\ 2 & 3 & 0 \\ 1 & 0 & 1 \end{bmatrix}^{-1} \begin{bmatrix} 2 \\ -3 \\ 5 \end{bmatrix}$$

$$\begin{bmatrix} x \\ y \\ z \end{bmatrix} = \begin{bmatrix} 3 & -1 & 9 \\ -2 & 1 & -6 \\ -3 & 1 & -8 \end{bmatrix} \begin{bmatrix} 2 \\ -3 \\ 5 \end{bmatrix} = \begin{bmatrix} 54 \\ -37 \\ -49 \end{bmatrix}$$

33.
$$\begin{bmatrix} 5 & 3 \\ -7 & 5 \end{bmatrix} \begin{bmatrix} x \\ y \end{bmatrix} = \begin{bmatrix} 13 \\ -9 \end{bmatrix}$$

$$\begin{bmatrix} x \\ y \end{bmatrix} = \begin{bmatrix} 5 & 3 \\ -7 & 5 \end{bmatrix}^{-1} \begin{bmatrix} 13 \\ -9 \end{bmatrix}$$

$$\begin{bmatrix} x \\ y \end{bmatrix} = \begin{bmatrix} 2 \\ 1 \end{bmatrix}$$

35.
$$\begin{bmatrix} 5 & 2 & 3 \\ 2 & 0 & 5 \\ 3 & 0 & 1 \end{bmatrix} \begin{bmatrix} x \\ y \\ z \end{bmatrix} = \begin{bmatrix} 12 \\ 7 \\ 4 \end{bmatrix}$$

$$\begin{bmatrix} x \\ y \\ z \end{bmatrix} = \begin{bmatrix} 5 & 2 & 3 \\ 2 & 0 & 5 \\ 3 & 0 & 1 \end{bmatrix}^{-1} \begin{bmatrix} 12 \\ 7 \\ 4 \end{bmatrix}$$

$$\begin{bmatrix} x \\ y \\ z \end{bmatrix} = \begin{bmatrix} 1 \\ 2 \\ 1 \end{bmatrix}$$

37.
$$\begin{bmatrix} 23 & 27 \\ 21 & 22 \end{bmatrix} \begin{bmatrix} x \\ y \end{bmatrix} = \begin{bmatrix} 127 \\ 108 \end{bmatrix}$$

$$\begin{bmatrix} x \\ y \end{bmatrix} = \begin{bmatrix} 23 & 27 \\ 21 & 22 \end{bmatrix}^{-1} \begin{bmatrix} 127 \\ 108 \end{bmatrix}$$

$$\begin{bmatrix} x \\ y \end{bmatrix} = \begin{bmatrix} 2 \\ 3 \end{bmatrix}$$

2 of model A and 3 of model B can be made.

39. **Answers may vary.**

41. $A^2 = \begin{bmatrix} -1 & -1 \\ 1 & 1 \end{bmatrix} \begin{bmatrix} -1 & -1 \\ 1 & 1 \end{bmatrix} = \begin{bmatrix} 0 & 0 \\ 0 & 0 \end{bmatrix}$

43. $\begin{bmatrix} 1 & 0 & 0 \\ -2 & -3 & -2 \\ 3 & 6 & 1 \end{bmatrix} \begin{bmatrix} x \\ y \\ z \end{bmatrix} = \begin{bmatrix} 0 \\ 0 \\ 0 \end{bmatrix}$

$\begin{bmatrix} x \\ y \\ z \end{bmatrix} = \begin{bmatrix} 1 & 0 & 0 \\ -2 & -3 & -2 \\ 3 & 6 & 1 \end{bmatrix}^{-1} \begin{bmatrix} 0 \\ 0 \\ 0 \end{bmatrix} = \begin{bmatrix} 0 \\ 0 \\ 0 \end{bmatrix}$

45.
$$AB = AC$$
$$A^{-1}AB = A^{-1}AC$$
$$IB = IC$$
$$B = C$$

47.
$$(I - B)(I + B) = I^2 + IB - BI - B^2$$
$$= I + B - B - B^2$$
$$= I - B^2$$
$$= I - \mathbf{0} = I$$
Thus, $I - B$ and $I + B$ are inverses.

49. $y = \dfrac{3x - 5}{x^2 - 4} = \dfrac{3x - 5}{(x + 2)(x - 2)}$

domain: $(-\infty, -2) \cup (-2, 2) \cup (2, \infty)$

51. $y = \dfrac{3x - 5}{\sqrt{x^2 + 4}}$

domain $= (-\infty, \infty)$

53. $y = x^2$; range $= [0, \infty)$

55. $y = \log x$; range $= (-\infty, \infty)$

Exercise 6.5 (page 477)

1. $|A|,\ \det A$ **3.** 0 **5.** 0

7. $\begin{vmatrix} 2 & 1 \\ -2 & 3 \end{vmatrix} = (2)(3) - (1)(-2)$
$= 6 - (-2) = 8$

9. $\begin{vmatrix} 2 & -3 \\ -3 & 5 \end{vmatrix} = (2)(5) - (-3)(-3)$
$= 10 - 9 = 1$

11. $M_{21} = \begin{vmatrix} -2 & 3 \\ 8 & 9 \end{vmatrix}$

13. $M_{33} = \begin{vmatrix} 1 & -2 \\ 4 & 5 \end{vmatrix}$

15. $C_{21} = -\begin{vmatrix} -2 & 3 \\ 8 & 9 \end{vmatrix}$

17. $C_{33} = \begin{vmatrix} 1 & -2 \\ 4 & 5 \end{vmatrix}$

19. $\begin{vmatrix} 2 & -3 & 5 \\ -2 & 1 & 3 \\ 1 & 3 & -2 \end{vmatrix} = 2\begin{vmatrix} 1 & 3 \\ 3 & -2 \end{vmatrix} - (-3)\begin{vmatrix} -2 & 3 \\ 1 & -2 \end{vmatrix} + 5\begin{vmatrix} -2 & 1 \\ 1 & 3 \end{vmatrix}$

$= 2(-11) + 3(1) + 5(-7) = -22 + 3 - 35 = -54$

21. $\begin{vmatrix} 1 & -1 & 2 \\ 2 & 1 & 3 \\ 1 & 1 & -1 \end{vmatrix} = 1\begin{vmatrix} 1 & 3 \\ 1 & -1 \end{vmatrix} - (-1)\begin{vmatrix} 2 & 3 \\ 1 & -1 \end{vmatrix} + 2\begin{vmatrix} 2 & 1 \\ 1 & 1 \end{vmatrix}$

$= 1(-4) + 1(-5) + 2(1) = -4 - 5 + 2 = -7$

23. $\begin{vmatrix} 2 & 1 & -1 \\ 1 & 3 & 5 \\ 2 & -5 & 3 \end{vmatrix} = 2\begin{vmatrix} 3 & 5 \\ -5 & 3 \end{vmatrix} - 1\begin{vmatrix} 1 & 5 \\ 2 & 3 \end{vmatrix} + (-1)\begin{vmatrix} 1 & 3 \\ 2 & -5 \end{vmatrix}$

$$= 2(34) - 1(-7) - 1(-11) = 68 + 7 + 11 = 86$$

25. $\begin{vmatrix} 0 & 1 & -3 \\ -3 & 5 & 2 \\ 2 & -5 & 3 \end{vmatrix} = 0\begin{vmatrix} 5 & 2 \\ -5 & 3 \end{vmatrix} - 1\begin{vmatrix} -3 & 2 \\ 2 & 3 \end{vmatrix} + (-3)\begin{vmatrix} -3 & 5 \\ 2 & -5 \end{vmatrix}$

$$= 0 - 1(-13) - 3(5) = 0 + 13 - 15 = -2$$

27. $\begin{vmatrix} 0 & 0 & 1 & 0 \\ -2 & 1 & 0 & 1 \\ 1 & 0 & 1 & 2 \\ 2 & 0 & 1 & 2 \end{vmatrix} = 0(***) - 0(***) + 1\begin{vmatrix} -2 & 1 & 1 \\ 1 & 0 & 2 \\ 2 & 0 & 2 \end{vmatrix} - 0(***)$

$$= 1\left(-2\begin{vmatrix} 0 & 2 \\ 0 & 2 \end{vmatrix} - 1\begin{vmatrix} 1 & 2 \\ 2 & 2 \end{vmatrix} - 1\begin{vmatrix} 1 & 0 \\ 2 & 0 \end{vmatrix}\right) = -2(0) - 1(-2) - 1(0) = 2$$

29. $\begin{vmatrix} 1 & 2 & 1 & 3 \\ -2 & 1 & -3 & 1 \\ -1 & 0 & 1 & -2 \\ 2 & -1 & -1 & 3 \end{vmatrix} = \begin{vmatrix} 1 & 2 & 1 & 3 \\ 0 & 0 & -4 & 4 \\ 0 & 2 & 2 & 1 \\ 0 & -5 & -3 & -3 \end{vmatrix} \begin{array}{l} \\ R_2 + R_4 \\ R_1 + R_3 \\ -2R_1 + R_4 \end{array}$

$$= 1\begin{vmatrix} 0 & -4 & 4 \\ 2 & 2 & 1 \\ -5 & -3 & -3 \end{vmatrix} \quad \text{(expand along first column)}$$

$$= 1\left(0\begin{vmatrix} 2 & 1 \\ -3 & -3 \end{vmatrix} - (-4)\begin{vmatrix} 2 & 1 \\ -5 & -3 \end{vmatrix} + 4\begin{vmatrix} 2 & 2 \\ -5 & -3 \end{vmatrix}\right)$$

$$= 0 + 4(-1) + 4(4) = 12$$

31. R_1 and R_2 have been switched. This multiplies the determinant by -1. TRUE

33. R_1 and R_2 have both been multiplied by -1. This multiplies the determinant by -1 twice. FALSE

35. R_1 and R_2 have been switched. This multiplies the determinant by -1. However, R_3 has been multiplied by -1, which also multiplies the determinant by -1. Thus, the determinant remains equal to 3.

37. R_1 has been added to R_3. This leaves the determinant equal to 3.

39. $x = \dfrac{\begin{vmatrix} 7 & 2 \\ -4 & -3 \end{vmatrix}}{\begin{vmatrix} 3 & 2 \\ 2 & -3 \end{vmatrix}} = \dfrac{-13}{-13} = 1 \quad y = \dfrac{\begin{vmatrix} 3 & 7 \\ 2 & -4 \end{vmatrix}}{\begin{vmatrix} 3 & 2 \\ 2 & -3 \end{vmatrix}} = \dfrac{-26}{-13} = 2$

41. $x = \dfrac{\begin{vmatrix} 3 & -1 \\ 9 & -7 \end{vmatrix}}{\begin{vmatrix} 1 & -1 \\ 3 & -7 \end{vmatrix}} = \dfrac{-12}{-4} = 3 \qquad y = \dfrac{\begin{vmatrix} 1 & 3 \\ 3 & 9 \end{vmatrix}}{\begin{vmatrix} 1 & -1 \\ 3 & -7 \end{vmatrix}} = \dfrac{0}{-4} = 0$

43. $x = \dfrac{\begin{vmatrix} 2 & 2 & 1 \\ 2 & -1 & 1 \\ 4 & 1 & 3 \end{vmatrix}}{\begin{vmatrix} 1 & 2 & 1 \\ 1 & -1 & 1 \\ 1 & 1 & 3 \end{vmatrix}} = \dfrac{-6}{-6} = 1 \qquad y = \dfrac{\begin{vmatrix} 1 & 2 & 1 \\ 1 & 2 & 1 \\ 1 & 4 & 3 \end{vmatrix}}{\begin{vmatrix} 1 & 2 & 1 \\ 1 & -1 & 1 \\ 1 & 1 & 3 \end{vmatrix}} = \dfrac{0}{-6} = 0 \qquad z = \dfrac{\begin{vmatrix} 1 & 2 & 2 \\ 1 & -1 & 2 \\ 1 & 1 & 4 \end{vmatrix}}{\begin{vmatrix} 1 & 2 & 1 \\ 1 & -1 & 1 \\ 1 & 1 & 3 \end{vmatrix}} = \dfrac{-6}{-6} = 1$

45. $x = \dfrac{\begin{vmatrix} 5 & -1 & 1 \\ 10 & -3 & 2 \\ 0 & 3 & 1 \end{vmatrix}}{\begin{vmatrix} 2 & -1 & 1 \\ 3 & -3 & 2 \\ 1 & 3 & 1 \end{vmatrix}} = \dfrac{-5}{-5} = 1 \qquad y = \dfrac{\begin{vmatrix} 2 & 5 & 1 \\ 3 & 10 & 2 \\ 1 & 0 & 1 \end{vmatrix}}{\begin{vmatrix} 2 & -1 & 1 \\ 3 & -3 & 2 \\ 1 & 3 & 1 \end{vmatrix}} = \dfrac{5}{-5} = -1 \qquad z = \dfrac{\begin{vmatrix} 2 & -1 & 5 \\ 3 & -3 & 10 \\ 1 & 3 & 0 \end{vmatrix}}{\begin{vmatrix} 2 & -1 & 1 \\ 3 & -3 & 2 \\ 1 & 3 & 1 \end{vmatrix}} = \dfrac{-10}{-5} = 2$

47. Rewrite system:

$$3x + 2y + 3z = 66$$
$$2x + 6y - z = 36$$
$$3x + 1y + 6z = 96$$

$x = \dfrac{\begin{vmatrix} 66 & 2 & 3 \\ 36 & 6 & -1 \\ 96 & 1 & 6 \end{vmatrix}}{\begin{vmatrix} 3 & 2 & 3 \\ 2 & 6 & -1 \\ 3 & 1 & 6 \end{vmatrix}} = \dfrac{198}{33} = 6$

$y = \dfrac{\begin{vmatrix} 3 & 66 & 3 \\ 2 & 36 & -1 \\ 3 & 96 & 6 \end{vmatrix}}{\begin{vmatrix} 3 & 2 & 3 \\ 2 & 6 & -1 \\ 3 & 1 & 6 \end{vmatrix}} = \dfrac{198}{33} = 6 \qquad z = \dfrac{\begin{vmatrix} 3 & 2 & 66 \\ 2 & 6 & 36 \\ 3 & 1 & 96 \end{vmatrix}}{\begin{vmatrix} 3 & 2 & 3 \\ 2 & 6 & -1 \\ 3 & 1 & 6 \end{vmatrix}} = \dfrac{396}{33} = 12$

49. $p = \dfrac{\begin{vmatrix} 0 & -1 & 3 & -1 \\ -1 & 1 & 0 & -1 \\ 2 & 0 & -1 & 0 \\ 7 & -2 & 0 & 3 \end{vmatrix}}{\begin{vmatrix} 2 & -1 & 3 & -1 \\ 1 & 1 & 0 & -1 \\ 3 & 0 & -1 & 0 \\ 1 & -2 & 0 & 3 \end{vmatrix}} = \dfrac{-15}{-18} = \dfrac{5}{6} \qquad q = \dfrac{\begin{vmatrix} 2 & 0 & 3 & -1 \\ 1 & -1 & 0 & -1 \\ 3 & 2 & -1 & 0 \\ 1 & 7 & 0 & 3 \end{vmatrix}}{\begin{vmatrix} 2 & -1 & 3 & -1 \\ 1 & 1 & 0 & -1 \\ 3 & 0 & -1 & 0 \\ 1 & -2 & 0 & 3 \end{vmatrix}} = \dfrac{-12}{-18} = \dfrac{2}{3}$

continued on next page...

49. continued...

$$r = \frac{\begin{vmatrix} 2 & -1 & 0 & -1 \\ 1 & 1 & -1 & -1 \\ 3 & 0 & 2 & 0 \\ 1 & -2 & 7 & 3 \end{vmatrix}}{\begin{vmatrix} 2 & -1 & 3 & -1 \\ 1 & 1 & 0 & -1 \\ 3 & 0 & -1 & 0 \\ 1 & -2 & 0 & 3 \end{vmatrix}} = \frac{-9}{-18} = \frac{1}{2} \qquad s = \frac{\begin{vmatrix} 2 & -1 & 3 & 0 \\ 1 & 1 & 0 & -1 \\ 3 & 0 & -1 & 2 \\ 1 & -2 & 0 & 7 \end{vmatrix}}{\begin{vmatrix} 2 & -1 & 3 & -1 \\ 1 & 1 & 0 & -1 \\ 3 & 0 & -1 & 0 \\ 1 & -2 & 0 & 3 \end{vmatrix}} = \frac{-45}{-18} = \frac{5}{2}$$

51.
$$\begin{vmatrix} x & y & 1 \\ 0 & 0 & 1 \\ 4 & 6 & 1 \end{vmatrix} = 0$$

$$x \begin{vmatrix} 0 & 1 \\ 6 & 1 \end{vmatrix} - y \begin{vmatrix} 0 & 1 \\ 4 & 1 \end{vmatrix} + 1 \begin{vmatrix} 0 & 0 \\ 4 & 6 \end{vmatrix} = 0$$
$$x(-6) - y(-4) + 1(0) = 0$$
$$-6x + 4y = 0$$
$$3x - 2y = 0$$

53.
$$\begin{vmatrix} x & y & 1 \\ -2 & 3 & 1 \\ 5 & -3 & 1 \end{vmatrix} = 0$$

$$x \begin{vmatrix} 3 & 1 \\ -3 & 1 \end{vmatrix} - y \begin{vmatrix} -2 & 1 \\ 5 & 1 \end{vmatrix} + 1 \begin{vmatrix} -2 & 3 \\ 5 & -3 \end{vmatrix} = 0$$
$$x(6) - y(-7) + 1(-9) = 0$$
$$6x + 7y - 9 = 0$$
$$6x + 7y = 9$$

55.
$$\pm \frac{1}{2} \begin{vmatrix} 0 & 0 & 1 \\ 12 & 0 & 1 \\ 12 & 5 & 1 \end{vmatrix} = \pm \frac{1}{2}(60)$$
$$= 30 \text{ square units}$$

57.
$$\pm \frac{1}{2} \begin{vmatrix} 2 & 3 & 1 \\ 10 & 8 & 1 \\ 0 & 20 & 1 \end{vmatrix} = \pm \frac{1}{2}(146)$$
$$= 73 \text{ square units}$$

59.
$$\begin{vmatrix} a & b \\ c & d \end{vmatrix} = ad - bc$$
$$\begin{vmatrix} b & a \\ d & c \end{vmatrix} = bc - ad = -(ad - bc)$$

61.
$$\begin{vmatrix} a & b \\ c & d \end{vmatrix} = ad - bc$$
$$\begin{vmatrix} a & b + ka \\ c & d + kc \end{vmatrix} = a(d + kc) - (b + ka)c$$
$$= ad + akc - bc - akc$$
$$= ad - bc$$

63.
$$\begin{vmatrix} 3 & x \\ 1 & 2 \end{vmatrix} = \begin{vmatrix} 2 & -1 \\ x & -5 \end{vmatrix}$$
$$6 - x = -10 + x$$
$$-2x = -16$$
$$x = 8$$

65.
$$\begin{vmatrix} 3 & x & 1 \\ x & 0 & -2 \\ 4 & 0 & 1 \end{vmatrix} = \begin{vmatrix} 2 & x \\ x & 4 \end{vmatrix}$$
$$-x \begin{vmatrix} x & -2 \\ 4 & 1 \end{vmatrix} = 8 - x^2$$
$$-x(x + 8) = 8 - x^2$$
$$-x^2 - 8x = -x^2 + 8$$
$$x = -1$$

67. Let $x = \$$ invested in HiTech, $y = \$$ invested in SaveTel, and $z = \$$ invested in OilCo.

$$\begin{cases} x + y + z = 20{,}000 \\ y + z = 3x \\ 0.10x + 0.05y + 0.06z = 0.066(20{,}000) \end{cases} \qquad \begin{array}{l} x + y + z = 20000 \\ -3x + y + z = 0 \\ 10x + 5y + 6z = 132000 \end{array}$$

$$x = \frac{\begin{vmatrix} 20000 & 1 & 1 \\ 0 & 1 & 1 \\ 132000 & 5 & 6 \end{vmatrix}}{\begin{vmatrix} 1 & 1 & 1 \\ -3 & 1 & 1 \\ 10 & 5 & 6 \end{vmatrix}} = \frac{20000}{4} = 5000, \quad y = \frac{\begin{vmatrix} 1 & 20000 & 1 \\ -3 & 0 & 1 \\ 10 & 132000 & 6 \end{vmatrix}}{\begin{vmatrix} 1 & 1 & 1 \\ -3 & 1 & 1 \\ 10 & 5 & 6 \end{vmatrix}} = \frac{32000}{4} = 8000$$

$$z = \frac{\begin{vmatrix} 1 & 1 & 20000 \\ -3 & 1 & 0 \\ 10 & 5 & 132000 \end{vmatrix}}{\begin{vmatrix} 1 & 1 & 1 \\ -3 & 1 & 1 \\ 10 & 5 & 6 \end{vmatrix}} = \frac{28000}{4} = 7000 \Rightarrow$$ He should invest \$5000 in HiTech, \$8000 in SaveTel, and \$7000 in OilCo.

69. $\begin{vmatrix} 1 & 3 & 4 \\ 0 & 5 & 2 \\ 0 & 0 & 2 \end{vmatrix} = 10$

$1 \cdot 5 \cdot 2 = 10$

71. $\begin{vmatrix} 1 & 2 & 4 & 3 \\ 0 & 2 & 2 & 1 \\ 0 & 0 & 3 & 2 \\ 0 & 0 & 0 & 4 \end{vmatrix} = 24$

$1 \cdot 2 \cdot 3 \cdot 4 = 24$

73. Answers may vary.

75. Answers may vary.

77. yes

79. $\begin{vmatrix} 2.3 & 5.7 & 6.1 \\ 3.4 & 6.2 & 8.3 \\ 5.8 & 8.2 & 9.2 \end{vmatrix} = 21.468$

81. $x^2 + 3x - 4 = (x + 4)(x - 1)$

83. $9x^3 - x = x(9x^2 - 1) = x(3x + 1)(3x - 1)$

85. $\dfrac{1}{x - 2} + \dfrac{2}{2x - 1} = \dfrac{1(2x - 1)}{(x - 2)(2x - 1)} + \dfrac{2(x - 2)}{(2x - 1)(x - 2)} = \dfrac{2x - 1 + 2x - 4}{(x - 2)(2x - 1)} = \dfrac{4x - 5}{(x - 2)(2x - 1)}$

87. $\dfrac{2}{x^2 + 1} + \dfrac{1}{x} = \dfrac{2x}{x(x^2 + 1)} + \dfrac{x^2 + 1}{x(x^2 + 1)} = \dfrac{x^2 + 2x + 1}{x(x^2 + 1)}$

Exercise 6.6 (page 486)

1. first-degree, second-degree

3.
$$\frac{3x-1}{x(x-1)} = \frac{A}{x} + \frac{B}{x-1}$$
$$\frac{3x-1}{x(x-1)} = \frac{A(x-1)}{x(x-1)} + \frac{Bx}{x(x-1)}$$
$$\frac{3x-1}{x(x-1)} = \frac{Ax - A + Bx}{x(x-1)}$$
$$\frac{3x-1}{x(x-1)} = \frac{(A+B)x - A}{x(x-1)}$$
$$\begin{cases} A+B= 3 \\ -A = -1 \end{cases} \Rightarrow A=1, B=2$$
$$\frac{3x-1}{x(x-1)} = \frac{1}{x} + \frac{2}{x-1}$$

5.
$$\frac{2x-15}{x(x-3)} = \frac{A}{x} + \frac{B}{x-3}$$
$$\frac{2x-15}{x(x-3)} = \frac{A(x-3)}{x(x-3)} + \frac{Bx}{x(x-3)}$$
$$\frac{2x-15}{x(x-3)} = \frac{Ax - 3A + Bx}{x(x-3)}$$
$$\frac{2x-15}{x(x-3)} = \frac{(A+B)x - 3A}{x(x-3)}$$
$$\begin{cases} A+B= 2 \\ -3A = -15 \end{cases} \Rightarrow A=5, B=-3$$
$$\frac{2x-15}{x(x-3)} = \frac{5}{x} - \frac{3}{x-3}$$

7.
$$\frac{3x+1}{(x+1)(x-1)} = \frac{A}{x+1} + \frac{B}{x-1}$$
$$\frac{3x+1}{(x+1)(x-1)} = \frac{A(x-1)}{(x+1)(x-1)} + \frac{B(x+1)}{(x+1)(x-1)}$$
$$\frac{3x+1}{(x+1)(x-1)} = \frac{Ax - A + Bx + B}{(x+1)(x-1)}$$
$$\frac{3x+1}{(x+1)(x-1)} = \frac{(A+B)x + (-A+B)}{(x+1)(x-1)}$$

$$\begin{cases} A+B=3 \\ -A+B=1 \end{cases} \Rightarrow A=1, B=2$$
$$\frac{3x+1}{(x+1)(x-1)} = \frac{1}{x+1} + \frac{2}{x-1}$$

9.
$$\frac{-2x+11}{(x+2)(x-3)} = \frac{A}{x+2} + \frac{B}{x-3}$$
$$\frac{-2x+11}{(x+2)(x-3)} = \frac{A(x-3)}{(x+2)(x-3)} + \frac{B(x+2)}{(x+2)(x-3)}$$
$$\frac{-2x+11}{(x+2)(x-3)} = \frac{Ax - 3A + Bx + 2B}{(x+2)(x-3)}$$
$$\frac{-2x+11}{(x+2)(x-3)} = \frac{(A+B)x + (-3A+2B)}{(x+2)(x-3)}$$

$$\frac{-2x+11}{x^2-x-6} = \frac{-2x+11}{(x+2)(x-3)}$$
$$\begin{cases} A+ B=-2 \\ -3A+2B= 11 \end{cases} \Rightarrow A=-3, B=1$$
$$\frac{-2x+11}{(x+2)(x-3)} = \frac{-3}{x+2} + \frac{1}{x-3}$$

11.
$$\frac{3x-23}{(x+3)(x-1)} = \frac{A}{x+3} + \frac{B}{x-1}$$
$$\frac{3x-23}{(x+3)(x-1)} = \frac{A(x-1)}{(x+3)(x-1)} + \frac{B(x+3)}{(x+3)(x-1)}$$
$$\frac{3x-23}{(x+3)(x-1)} = \frac{Ax - A + Bx + 3B}{(x+3)(x-1)}$$
$$\frac{3x-23}{(x+3)(x-1)} = \frac{(A+B)x + (-A+3B)}{(x+3)(x-1)}$$

$$\frac{3x-23}{x^2+2x-3} = \frac{3x-23}{(x+3)(x-1)}$$
$$\begin{cases} A+ B= 3 \\ -A+3B= -23 \end{cases} \Rightarrow A=8, B=-5$$
$$\frac{3x-23}{(x+3)(x-1)} = \frac{8}{x+3} - \frac{5}{x-1}$$

13. $$\frac{9x-31}{2x^2-13x+15} = \frac{9x-31}{(2x-3)(x-5)} = \frac{A}{2x-3} + \frac{B}{x-5}$$

$$\frac{9x-31}{(2x-3)(x-5)} = \frac{A(x-5)}{(2x-3)(x-5)} + \frac{B(2x-3)}{(2x-3)(x-5)}$$

$$\frac{9x-31}{(2x-3)(x-5)} = \frac{Ax-5A+2Bx-3B}{(2x-3)(x-5)}$$

$$\frac{9x-31}{(2x-3)(x-5)} = \frac{(A+2B)x+(-5A-3B)}{(2x-3)(x-5)}$$

$$\begin{cases} A+2B = 9 \\ -5A-3B = -31 \end{cases} \Rightarrow A=5,\, B=2 \qquad \frac{9x-31}{(2x-3)(x-5)} = \frac{5}{2x-3} + \frac{2}{x-5}$$

15. $$\frac{4x^2+4x-2}{x(x^2-1)} = \frac{4x^2+4x-2}{x(x+1)(x-1)} = \frac{A}{x} + \frac{B}{x+1} + \frac{C}{x-1}$$

$$\frac{4x^2+4x-2}{x(x+1)(x-1)} = \frac{A(x+1)(x-1)}{x(x+1)(x-1)} + \frac{Bx(x-1)}{x(x+1)(x-1)} + \frac{Cx(x+1)}{x(x+1)(x-1)}$$

$$\frac{4x^2+4x-2}{x(x+1)(x-1)} = \frac{Ax^2-A+Bx^2-Bx+Cx^2+Cx}{x(x+1)(x-1)}$$

$$\frac{4x^2+4x-2}{x(x+1)(x-1)} = \frac{(A+B+C)x^2+(-B+C)x+(-A)}{x(x+1)(x-1)}$$

$$\begin{cases} A+B+C = 4 \\ -B+C = 4 \\ -A = -2 \end{cases} \begin{matrix} A=2 \\ \Rightarrow B=-1 \\ C=3 \end{matrix} \qquad \frac{4x^2+4x-2}{x(x+1)(x-1)} = \frac{2}{x} - \frac{1}{x+1} + \frac{3}{x-1}$$

17. $$\frac{x^2+x+3}{x(x^2+3)} = \frac{A}{x} + \frac{Bx+C}{x^2+3}$$

$$\frac{x^2+x+3}{x(x^2+3)} = \frac{A(x^2+3)}{x(x^2+3)} + \frac{(Bx+C)x}{x(x^2+3)}$$

$$\frac{x^2+x+3}{x(x^2+3)} = \frac{Ax^2+3A+Bx^2+Cx}{x(x^2+3)}$$

$$\frac{x^2+x+3}{x(x^2+3)} = \frac{(A+B)x^2+Cx+3A}{x(x^2+3)}$$

$$\begin{cases} A+B = 1 \\ C = 1 \\ 3A = 3 \end{cases} \begin{matrix} A=1 \\ \Rightarrow B=0 \\ C=1 \end{matrix} \qquad \frac{x^2+x+3}{x(x^2+3)} = \frac{1}{x} + \frac{1}{x^2+3}$$

9.
$$\frac{3x^2 + 8x + 11}{(x+1)(x^2 + 2x + 3)} = \frac{A}{x+1} + \frac{Bx + C}{x^2 + 2x + 3}$$

$$\frac{3x^2 + 8x + 11}{(x+1)(x^2 + 2x + 3)} = \frac{A(x^2 + 2x + 3)}{(x+1)(x^2 + 2x + 3)} + \frac{(Bx + C)(x+1)}{(x+1)(x^2 + 2x + 3)}$$

$$\frac{3x^2 + 8x + 11}{(x+1)(x^2 + 2x + 3)} = \frac{Ax^2 + 2Ax + 3A + Bx^2 + Bx + Cx + C}{(x+1)(x^2 + 2x + 3)}$$

$$\frac{3x^2 + 8x + 11}{(x+1)(x^2 + 2x + 3)} = \frac{(A+B)x^2 + (2A + B + C)x + (3A + C)}{(x+1)(x^2 + 2x + 3)}$$

$$\begin{cases} A + B & = 3 \\ 2A + B + C = 8 \\ 3A \quad + C = 11 \end{cases} \Rightarrow \begin{matrix} A = 3 \\ B = 0 \\ C = 2 \end{matrix} \qquad \frac{3x^2 + 8x + 11}{(x+1)(x^2 + 2x + 3)} = \frac{3}{x+1} + \frac{2}{x^2 + 2x + 3}$$

11.
$$\frac{5x^2 + 9x + 3}{x(x+1)^2} = \frac{A}{x} + \frac{B}{x+1} + \frac{C}{(x+1)^2}$$

$$= \frac{A(x+1)^2}{x(x+1)^2} + \frac{Bx(x+1)}{x(x+1)^2} + \frac{Cx}{x(x+1)^2}$$

$$= \frac{Ax^2 + 2Ax + A + Bx^2 + Bx + Cx}{x(x+1)^2}$$

$$= \frac{(A+B)x^2 + (2A + B + C)x + A}{x(x+1)^2}$$

$$\begin{cases} A + B & = 5 \\ 2A + B + C = 9 \\ A \quad\quad = 3 \end{cases} \Rightarrow \begin{matrix} A = 3 \\ B = 2 \\ C = 1 \end{matrix} \qquad \frac{5x^2 + 9x + 3}{x(x+1)^2} = \frac{3}{x} + \frac{2}{x+1} + \frac{1}{(x+1)^2}$$

13.
$$\frac{-2x^2 + x - 2}{x^2(x-1)} = \frac{A}{x} + \frac{B}{x^2} + \frac{C}{x-1}$$

$$= \frac{Ax(x-1)}{x^2(x-1)} + \frac{B(x-1)}{x^2(x-1)} + \frac{Cx^2}{x^2(x-1)}$$

$$= \frac{Ax^2 - Ax + Bx - B + Cx^2}{x^2(x-1)}$$

$$= \frac{(A+C)x^2 + (-A+B)x + (-B)}{x^2(x-1)}$$

$$\begin{cases} A \quad\quad + C = -2 \\ -A + B \quad = 1 \\ \quad - B \quad = -2 \end{cases} \Rightarrow \begin{matrix} A = 1 \\ B = 2 \\ C = -3 \end{matrix} \qquad \frac{-2x^2 + x - 2}{x^2(x-1)} = \frac{1}{x} + \frac{2}{x^2} - \frac{3}{x-1}$$

25.
$$\frac{3x^2 - 13x + 18}{x(x-3)^2} = \frac{A}{x} + \frac{B}{x-3} + \frac{C}{(x-3)^2}$$

$$= \frac{A(x-3)^2}{x(x-3)^2} + \frac{Bx(x-3)}{x(x-3)^2} + \frac{Cx}{x(x-3)^2}$$

$$= \frac{Ax^2 - 6Ax + 9A + Bx^2 - 3Bx + Cx}{x(x-3)^2}$$

$$= \frac{(A+B)x^2 + (-6A - 3B + C)x + 9A}{x(x-3)^2}$$

$$\begin{cases} A + B & = 3 \\ -6A - 3B + C & = -13 \\ 9A & = 18 \end{cases} \Rightarrow \begin{matrix} A = 2 \\ B = 1 \\ C = 2 \end{matrix} \qquad \frac{3x^2 - 13x + 18}{x(x-3)^2} = \frac{2}{x} + \frac{1}{x-3} + \frac{2}{(x-3)^2}$$

27.
$$\frac{x^2 - 2x - 3}{(x-1)^3} = \frac{A}{x-1} + \frac{B}{(x-1)^2} + \frac{C}{(x-1)^3}$$

$$= \frac{A(x-1)^2}{(x-1)^3} + \frac{B(x-1)}{(x-1)^3} + \frac{C}{(x-1)^3}$$

$$= \frac{Ax^2 - 2Ax + A + Bx - B + C}{(x-1)^3}$$

$$= \frac{Ax^2 + (-2A + B)x + (A - B + C)}{(x-1)^3}$$

$$\begin{cases} A & = 1 \\ -2A + B & = -2 \\ A - B + C & = -3 \end{cases} \Rightarrow \begin{matrix} A = 1 \\ B = 0 \\ C = -4 \end{matrix} \qquad \frac{x^2 - 2x - 3}{(x-1)^3} = \frac{1}{x-1} - \frac{4}{(x-1)^3}$$

29.
$$\frac{x^3 + 4x^2 + 2x + 1}{x^2(x^2 + x + 1)} = \frac{A}{x} + \frac{B}{x^2} + \frac{Cx + D}{x^2 + x + 1}$$

$$= \frac{Ax(x^2 + x + 1)}{x^2(x^2 + x + 1)} + \frac{B(x^2 + x + 1)}{x^2(x^2 + x + 1)} + \frac{(Cx + D)x^2}{x^2(x^2 + x + 1)}$$

$$= \frac{Ax^3 + Ax^2 + Ax + Bx^2 + Bx + B + Cx^3 + Dx^2}{x^2(x^2 + x + 1)}$$

$$= \frac{(A+C)x^3 + (A + B + D)x^2 + (A + B)x + B}{x^2(x^2 + x + 1)}$$

$$\begin{cases} A & + C & = 1 \\ A + B & + D & = 4 \\ A + B & & = 2 \\ B & & = 1 \end{cases} \Rightarrow \begin{matrix} A = 1 \\ B = 1 \\ C = 0 \\ D = 2 \end{matrix} \qquad \frac{x^3 + 4x^2 + 2x + 1}{x^2(x^2 + x + 1)} = \frac{1}{x} + \frac{1}{x^2} + \frac{2}{x^2 + x + 1}$$

31.
$$\frac{4x^3 + 5x^2 + 3x + 4}{x^2(x^2 + 1)} = \frac{A}{x} + \frac{B}{x^2} + \frac{Cx + D}{x^2 + 1}$$
$$= \frac{Ax(x^2 + 1)}{x^2(x^2 + 1)} + \frac{B(x^2 + 1)}{x^2(x^2 + 1)} + \frac{(Cx + D)x^2}{x^2(x^2 + 1)}$$
$$= \frac{Ax^3 + Ax + Bx^2 + B + Cx^3 + Dx^2}{x^2(x^2 + 1)}$$
$$= \frac{(A + C)x^3 + (B + D)x^2 + Ax + B}{x^2(x^2 + 1)}$$

$$\begin{cases} A \quad + C \quad = 4 \\ \quad B \quad + D = 5 \\ A \quad\quad = 3 \\ \quad B \quad\quad = 4 \end{cases} \Rightarrow \begin{matrix} A = 3 \\ B = 4 \\ C = 1 \\ D = 1 \end{matrix} \qquad \frac{4x^3 + 5x^2 + 3x + 4}{x^2(x^2 + 1)} = \frac{3}{x} + \frac{4}{x^2} + \frac{x + 1}{x^2 + 1}$$

33.
$$\frac{-x^2 - 3x - 5}{(x + 1)(x^2 + 2)} = \frac{A}{x + 1} + \frac{Bx + C}{x^2 + 2}$$
$$= \frac{A(x^2 + 2)}{(x + 1)(x^2 + 2)} + \frac{(Bx + C)(x + 1)}{(x + 1)(x^2 + 2)}$$
$$= \frac{Ax^2 + 2A + Bx^2 + Bx + Cx + C}{(x + 1)(x^2 + 2)}$$
$$= \frac{(A + B)x^2 + (B + C)x + (2A + C)}{(x + 1)(x^2 + 2)}$$

$$\begin{cases} A + B \quad\quad = -1 \\ \quad B + C = -3 \\ 2A \quad + C = -5 \end{cases} \Rightarrow \begin{matrix} A = -1 \\ B = 0 \\ C = -3 \end{matrix} \qquad \frac{-x^2 - 3x - 5}{(x + 1)(x^2 + 2)} = \frac{-1}{x + 1} - \frac{3}{x^2 + 2}$$

35.
$$\frac{x^3 + 4x^2 + 3x + 6}{(x^2 + 2)(x^2 + x + 2)} = \frac{Ax + B}{x^2 + 2} + \frac{Cx + D}{x^2 + x + 2}$$
$$= \frac{(Ax + B)(x^2 + x + 2)}{(x^2 + 2)(x^2 + x + 2)} + \frac{(Cx + D)(x^2 + 2)}{(x^2 + 2)(x^2 + x + 2)}$$
$$= \frac{Ax^3 + Ax^2 + 2Ax + Bx^2 + Bx + 2B + Cx^3 + 2Cx + Dx^2 + 2D}{(x^2 + 2)(x^2 + x + 2)}$$
$$= \frac{(A + C)x^3 + (A + B + D)x^2 + (2A + B + 2C)x + (2B + 2D)}{(x^2 + 2)(x^2 + x + 2)}$$

$$\begin{cases} A \quad\quad + C \quad\quad = 1 \\ A + B \quad\quad + D = 4 \\ 2A + B + 2C \quad\quad = 3 \\ \quad 2B \quad\quad + 2D = 6 \end{cases} \Rightarrow \begin{matrix} A = 1 \\ B = 1 \\ C = 0 \\ D = 2 \end{matrix} \qquad \frac{x^3 + 4x^2 + 3x + 6}{(x^2 + 2)(x^2 + x + 2)} = \frac{x + 1}{x^2 + 2} + \frac{2}{x^2 + x + 2}$$

37.
$$\frac{2x^4 + 6x^3 + 20x^2 + 22x + 25}{x(x^2 + 2x + 5)^2}$$

$$= \frac{A}{x} + \frac{Bx + C}{x^2 + 2x + 5} + \frac{Dx + E}{(x^2 + 2x + 5)^2}$$

$$= \frac{A(x^2 + 2x + 5)^2}{x(x^2 + 2x + 5)^2} + \frac{(Bx + C)(x)(x^2 + 2x + 5)}{x(x^2 + 2x + 5)^2} + \frac{(Dx + E)x}{x(x^2 + 2x + 5)^2}$$

$$= \frac{(A + B)x^4 + (4A + 2B + C)x^3 + (14A + 5B + 2C + D)x^2 + (20A + 5C + E)x + (25A)}{x(x^2 + 2x + 5)^2}$$

$$\begin{cases} A + B & = 2 \\ 4A + 2B + C & = 6 \\ 14A + 5B + 2C + D & = 20 \\ 20A \quad + 5C \quad + E = 22 \\ 25A & = 25 \end{cases} \begin{aligned} A &= 1 \\ B &= 1 \\ \Rightarrow C &= 0 \\ D &= 1 \\ E &= 2 \end{aligned}$$

$$\frac{2x^4 + 6x^3 + 20x^2 + 22x + 25}{x(x^2 + 2x + 5)^2} = \frac{1}{x} + \frac{x}{x^2 + 2x + 5} + \frac{x + 2}{(x^2 + 2x + 5)^2}$$

39. Use long division first: $\frac{x^3}{x^2 + 3x + 2} = x - 3 + \frac{7x + 6}{x^2 + 3x + 2} = x - 3 + \frac{7x + 6}{(x + 1)(x + 2)}$

$$\frac{7x + 6}{(x + 1)(x + 2)} = \frac{A}{x + 1} + \frac{B}{x + 2}$$

$$= \frac{A(x + 2)}{(x + 1)(x + 2)} + \frac{B(x + 1)}{(x + 1)(x + 2)}$$

$$= \frac{Ax + 2A + Bx + B}{(x + 1)(x + 2)}$$

$$= \frac{(A + B)x + (2A + B)}{(x + 1)(x + 2)}$$

$$\begin{cases} A + B = 7 \\ 2A + B = 6 \end{cases} \Rightarrow A = -1, B = 8 \quad x - 3 + \frac{7x + 6}{(x + 1)(x + 2)} = x - 3 - \frac{1}{x + 1} + \frac{8}{x + 2}$$

41. Use long division first: $\frac{3x^3 + 3x^2 + 6x + 4}{3x^3 + x^2 + 3x + 1} = 1 + \frac{2x^2 + 3x + 3}{3x^3 + x^2 + 3x + 1} = 1 + \frac{2x^2 + 3x + 3}{(3x + 1)(x^2 + 1)}$

$$\frac{2x^2 + 3x + 3}{(3x + 1)(x^2 + 1)} = \frac{A}{3x + 1} + \frac{Bx + C}{x^2 + 1}$$

$$= \frac{A(x^2 + 1)}{(3x + 1)(x^2 + 1)} + \frac{(Bx + C)(3x + 1)}{(3x + 1)(x^2 + 1)}$$

$$= \frac{Ax^2 + A + 3Bx^2 + Bx + 3Cx + C}{(3x + 1)(x^2 + 1)}$$

$$= \frac{(A + 3B)x^2 + (B + 3C)x + (A + C)}{(3x + 1)(x^2 + 1)}$$

$$\begin{cases} A + 3B & = 2 \\ B + 3C = 3 \\ A \quad + C = 3 \end{cases} \Rightarrow \begin{aligned} A &= 2 \\ B &= 0 \\ C &= 1 \end{aligned} \quad 1 + \frac{2x^2 + 3x + 3}{(3x + 1)(x^2 + 1)} = 1 + \frac{2}{3x + 1} + \frac{1}{x^2 + 1}$$

43. Use long division first: $\dfrac{x^3 + 3x^2 + 2x + 1}{x^3 + x^2 + x} = 1 + \dfrac{2x^2 + x + 1}{x^3 + x^2 + x} = 1 + \dfrac{2x^2 + x + 1}{x(x^2 + x + 1)}$

$$\dfrac{2x^2 + x + 1}{x(x^2 + x + 1)} = \dfrac{A}{x} + \dfrac{Bx + C}{x^2 + x + 1}$$

$$= \dfrac{A(x^2 + x + 1)}{x(x^2 + x + 1)} + \dfrac{(Bx + C)x}{x(x^2 + x + 1)}$$

$$= \dfrac{Ax^2 + Ax + A + Bx^2 + Cx}{x(x^2 + x + 1)}$$

$$= \dfrac{(A + B)x^2 + (A + C)x + (A)}{x(x^2 + x + 1)}$$

$$\begin{cases} A + B &= 2 \\ A &+ C = 1 \\ A & = 1 \end{cases} \begin{matrix} A = 1 \\ B = 1 \\ C = 0 \end{matrix} \quad 1 + \dfrac{2x^2 + x + 1}{x(x^2 + x + 1)} = 1 + \dfrac{1}{x} + \dfrac{x}{x^2 + x + 1}$$

45. Use long division first: $\dfrac{2x^4 + 2x^3 + 3x^2 - 1}{(x^2 - x)(x^2 + 1)} = 2 + \dfrac{4x^3 + x^2 + 2x - 1}{x(x - 1)(x^2 + 1)}$

$$\dfrac{4x^3 + x^2 + 2x - 1}{x(x - 1)(x^2 + 1)} = \dfrac{A}{x} + \dfrac{B}{x - 1} + \dfrac{Cx + D}{x^2 + 1}$$

$$= \dfrac{A(x - 1)(x^2 + 1)}{x(x - 1)(x^2 + 1)} + \dfrac{Bx(x^2 + 1)}{x(x - 1)(x^2 + 1)} + \dfrac{(Cx + D)(x)(x - 1)}{x(x - 1)(x^2 + 1)}$$

$$= \dfrac{Ax^3 - Ax^2 + Ax - A + Bx^3 + Bx + Cx^3 - Cx^2 + Dx^2 - Dx}{x(x - 1)(x^2 + 1)}$$

$$= \dfrac{(A + B + C)x^3 + (-A - C + D)x^2 + (A + B - D)x + (-A)}{x(x - 1)(x^2 + 1)}$$

$$\begin{cases} A + B + C & = 4 \\ -A & - C + D = 1 \\ A + B & - D = 2 \\ -A & = -1 \end{cases} \Rightarrow \begin{matrix} A = 1 \\ B = 3 \\ C = 0 \\ D = 2 \end{matrix} \quad 2 + \dfrac{4x^3 + x^2 + 2x - 1}{x(x - 1)(x^2 + 1)} = 2 + \dfrac{1}{x} + \dfrac{3}{x - 1} + \dfrac{2}{x^2 + 1}$$

47. $x^3 + 1 = (x + 1)(x^2 - x + 1) \Rightarrow$ not prime   **49.** $\sqrt{8a^3b} = \sqrt{4a^2}\sqrt{2ab} = 2|a|\sqrt{2ab}$

51. $\sqrt{18x^5} + x^2\sqrt{50x} - 5x^2\sqrt{2x} = \sqrt{9x^4}\sqrt{2x} + x^2\sqrt{25}\sqrt{2x} - 5x^2\sqrt{2x}$

$$= 3x^2\sqrt{2x} + 5x^2\sqrt{2x} - 5x^2\sqrt{2x} = 3x^2\sqrt{2x}$$

53.
$$\sqrt{x - 5} = x - 7$$
$$\left(\sqrt{x - 5}\right)^2 = (x - 7)^2$$
$$x - 5 = x^2 - 14x + 49$$
$$0 = x^2 - 15x + 54$$
$$0 = (x - 9)(x - 6)$$
$$x = 9 \quad \text{or} \quad x = 6$$

$x = 6$ does not check and is extraneous.

Exercise 6.7 (page 492)

1. half-plane, boundary

3. is not

5. $2x + 3y < 12$

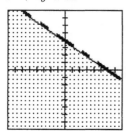

7. $x < 3$

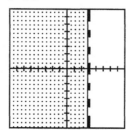

9. $4x - y > 4$

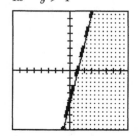

11. $y > 2x$

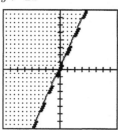

13. $y \leq \dfrac{1}{2}x + 1$

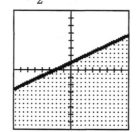

15. $2y \geq 3x - 2$

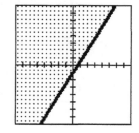

17. $\begin{cases} y < 3 \\ x \geq 2 \end{cases}$

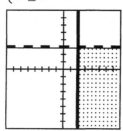

19. $\begin{cases} y \geq 1 \\ x < 2 \end{cases}$

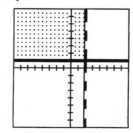

21. $\begin{cases} y \leq x - 2 \\ y \geq 2x + 1 \end{cases}$

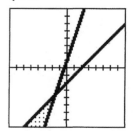

23. $\begin{cases} x + y < 2 \\ x + y \le 1 \end{cases}$

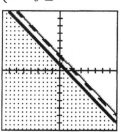

25. $\begin{cases} x + 2y < 3 \\ 2x - 4y < 8 \end{cases}$

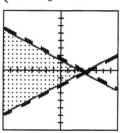

27. $\begin{cases} 2x - 3y \ge 6 \\ 3x + 2y < 6 \end{cases}$

29. $\begin{cases} 2x - y \le 0 \\ x + 2y \le 10 \\ y \ge 0 \end{cases}$

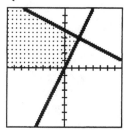

31. $\begin{cases} x - 2y \ge 0 \\ x - y \le 2 \\ x \ge 0 \end{cases}$

33. $\begin{cases} x + y \le 4 \\ x - y \le 4 \\ x \ge 0 \\ y \ge 0 \end{cases}$

35. $\begin{cases} 3x - 2y \le 6 \\ x + 2y \le 10 \\ x \ge 0 \\ y \ge 0 \end{cases}$

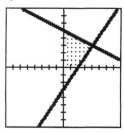

37. **Answers may vary.**

39. **Answers may vary.**

41. **Answers may vary.**

43. one, one

45. 0

Exercise 6.8 (page 501)

1. constraints

3. objective

5.

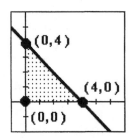

Point	$P = 2x + 3y$
$(0, 0)$	$= 2(0) + 3(0) = 0$
$(0, 4)$	$= 2(0) + 3(4) = 12$
$(4, 0)$	$= 2(4) + 3(0) = 8$

Max: $P = 12$ at $(0, 4)$

7.

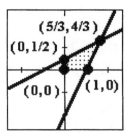

Point	$P = y + \frac{1}{2}x$
$(0, 0)$	$= 0 + \frac{1}{2}(0) = 0$
$(0, \frac{1}{2})$	$= \frac{1}{2} + \frac{1}{2}(0) = \frac{1}{2}$
$(\frac{5}{3}, \frac{4}{3})$	$= \frac{4}{3} + \frac{1}{2}(\frac{5}{3}) = \frac{13}{6}$
$(1, 0)$	$= 0 + \frac{1}{2}(1) = \frac{1}{2}$

Max: $P = \frac{13}{6}$ at $(\frac{5}{3}, \frac{4}{3})$

9.

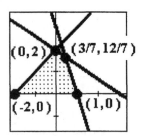

Point	$P = 2x + y$
$(-2, 0)$	$= 2(-2) + 0 = -4$
$(1, 0)$	$= 2(1) + 0 = 2$
$(\frac{3}{7}, \frac{12}{7})$	$= 2(\frac{3}{7}) + \frac{12}{7} = \frac{18}{7}$
$(0, 2)$	$= 2(0) + 2 = 2$

Max: $P = \frac{18}{7}$ at $(\frac{3}{7}, \frac{12}{7})$

11.

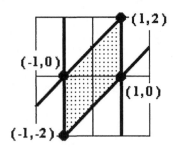

Point	$P = 3x - 2y$
$(1, 0)$	$= 3(1) - 2(0) = 3$
$(1, 2)$	$= 3(1) - 2(2) = -1$
$(-1, 0)$	$= 3(-1) - 2(0) = -3$
$(-1, -2)$	$= 3(-1) - 2(-2) = 1$

Max: $P = 3$ at $(1, 0)$

13.

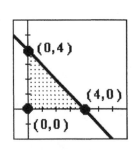

Point	$P = 5x + 12y$
$(0, 0)$	$= 5(0) + 12(0) = 0$
$(0, 4)$	$= 5(0) + 12(4) = 48$
$(4, 0)$	$= 5(4) + 12(0) = 20$

Min: $P = 0$ at $(0, 0)$

15.

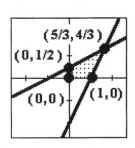

Point	$P = 3y + x$
$(0, 0)$	$= 3(0) + 0 = 0$
$(0, \frac{1}{2})$	$= 3(\frac{1}{2}) + 0 = \frac{3}{2}$
$(\frac{5}{3}, \frac{4}{3})$	$= 3(\frac{4}{3}) + \frac{5}{3} = \frac{17}{3}$
$(1, 0)$	$= 3(0) + 1 = 1$

Min: $P = 0$ at $(0, 0)$

17.

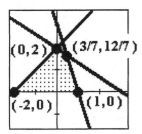

Point	$P = 6x + 2y$
$(-2, 0)$	$= 6(-2) + 2(0) = -12$
$(1, 0)$	$= 6(1) + 2(0) = 6$
$(\frac{3}{7}, \frac{12}{7})$	$= 6(\frac{3}{7}) + 2(\frac{12}{7}) = 6$
$(0, 2)$	$= 6(0) + 2(2) = 4$

Min: $P = -12$ at $(-2, 0)$

19.

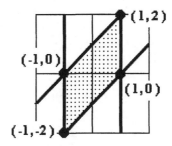

Point	$P = 2x - 2y$
$(1, 0)$	$= 2(1) - 2(0) = 2$
$(1, 2)$	$= 2(1) - 2(2) = -2$
$(-1, 0)$	$= 2(-1) - 2(0) = -2$
$(-1, -2)$	$= 2(-1) - 2(-2) = 2$

Min: $P = -2$ on the edge joining $(1, 2)$ and $(-1, 0)$

21.   Let $x =$ # tables and $y =$ # chairs.

Maximize $P = 100x + 80y$

subject to $\begin{cases} 2x + 3y \leq 42 \\ 6x + 2y \leq 42 \\ x \geq 0, y \geq 0 \end{cases}$

Point	$P = 100x + 80y$
$(7, 0)$	$= 100(7) + 80(0) = 700$
$(3, 12)$	$= 100(3) + 80(12) = 1260$
$(0, 14)$	$= 100(0) + 80(14) = 1120$
$(-1, -2)$	$= 2(-1) - 2(-2) = 2$

They should make 3 tables and 12 chairs, for a maximum profit of $1260.

23. Let $x = $ # IBM and $y = $ # Macintosh.

Maximize $P = 50x + 40y$

subject to $\begin{cases} x + y \le 60 \\ 20 \le x \le 30 \\ 30 \le y \le 50 \end{cases}$

Point	$P = 50x + 40y$
$(20, 30)$	$= 50(20) + 40(30) = 2200$
$(30, 30)$	$= 50(30) + 40(30) = 2700$
$(20, 40)$	$= 50(20) + 40(40) = 2600$

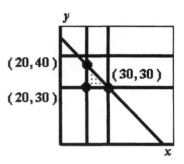

She should stock 30 IBM and 30 Macintosh computers, for a maximum commission of $2700.

25. Let $x = $ # VCRs and $y = $ # TVs.

Maximize $P = 40x + 32y$

subject to $\begin{cases} 3x + 4y \le 180 \\ 2x + 3y \le 120 \\ 2x + y \le 60 \\ x \ge 0, y \ge 0 \end{cases}$

Point	$P = 40x + 32y$
$(0, 0)$	$= 40(0) + 32(0) = 0$
$(0, 40)$	$= 40(0) + 32(40) = 1280$
$(15, 30)$	$= 40(15) + 32(30) = 1560$
$(30, 0)$	$= 40(30) + 32(0) = 1200$

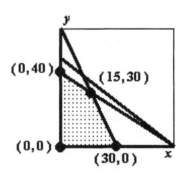

15 VCRs and 30 TVs should be made, for a maximum profit of $1560.

27. Let $x = $ \$ in stocks and $y = $ \$ in bonds.

Maximize $P = 0.09x + 0.07y$

subject to $\begin{cases} x + y \le 200000 \\ x \ge 100000 \\ y \ge 50000 \end{cases}$

Point	$P = 0.09x + 0.07y$
$(100000, 50000)$	$= 12500$
$(150000, 50000)$	$= 17000$
$(100000, 100000)$	$= 16000$

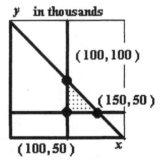

She should invest $150,000 in stocks and $50,000 in bonds, for a maximum return of $17,000.

29. Let $x = $ # buses and $y = $ # trucks.

Minimize $P = 350x + 200y$

subject to $\begin{cases} 40x + 10y \geq 100 \\ 3x + 6y \geq 18 \\ x \geq 0, y \geq 0 \end{cases}$

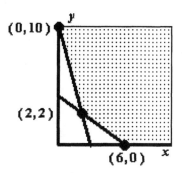

Point	$P = 350x + 200y$
$(0, 10)$	$= 350(0) + 200(10) = 2000$
$(2, 2)$	$= 350(2) + 200(2) = 1100$
$(6, 0)$	$= 350(6) + 200(0) = 2100$

2 buses and 2 trucks should be rented, for a minimum cost of $1100.

31. Answers may vary.

33. $\begin{bmatrix} 1 & 2 & 3 \\ 1 & -2 & 3 \\ 0 & 2 & -3 \\ 2 & 0 & 6 \end{bmatrix} \Rightarrow \begin{bmatrix} 1 & 2 & 3 \\ 0 & -4 & 0 \\ 0 & 2 & -3 \\ 0 & -4 & 0 \end{bmatrix} \Rightarrow \begin{bmatrix} 1 & 2 & 3 \\ 0 & 1 & 0 \\ 0 & 2 & -3 \\ 0 & 0 & 0 \end{bmatrix} \Rightarrow \begin{bmatrix} 1 & 0 & 3 \\ 0 & 1 & 0 \\ 0 & 0 & -3 \\ 0 & 0 & 0 \end{bmatrix} \Rightarrow \begin{bmatrix} 1 & 0 & 0 \\ 0 & 1 & 0 \\ 0 & 0 & 1 \\ 0 & 0 & 0 \end{bmatrix}$

$-R_1 + R_2 \Rightarrow R_2$ $-\frac{1}{4}R_2 \Rightarrow R_2$ $-2R_2 + R_1 \Rightarrow R_1$ $R_3 + R_1 \Rightarrow R_1$

$-2R_1 + R_4 \Rightarrow R_4$ $-R_4 + R_2 \Rightarrow R_4$ $-2R_2 + R_3 \Rightarrow R_3$ $-\frac{1}{3}R_3 \Rightarrow R_3$

35. $\left(\frac{1}{3} - 3y, y, -\frac{1}{3}\right)$

Chapter 6 Summary (page 505)

1. a. $\begin{cases} 2x - y = -1 \\ x + y = 7 \end{cases}$

b. $\begin{cases} 5x + 2y = 1 \\ 2x - y = -5 \end{cases}$

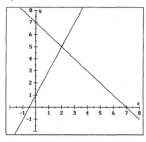

solution: $(2, 5)$

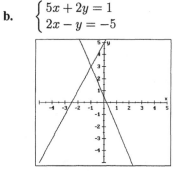

solution: $(-1, 3)$

c. $\begin{cases} y = 5x + 7 \\ x = y - 7 \end{cases}$

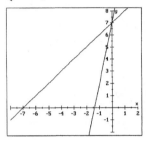

solution: $(0, 7)$

2. a. $\begin{cases} (1) \quad 2y + x = 0 \\ (2) \quad x = y + 3 \end{cases}$

Substitute $x = y + 3$ into (1):

$2y + x = 0$

$2y + (y + 3) = 0$

$3y + 3 = 0$

$3y = -3$

$y = -1$

Substitute and solve for x:

$x = y + 3$

$x = -1 + 3 = 2$

$\boxed{x = 2, y = -1}$

b. $\begin{cases} (1) \quad 2x + y = -3 \\ (2) \quad x - y = 3 \end{cases}$

Substitute $x = y + 3$ into (1):

$2x + y = -3$

$2(y + 3) + y = -3$

$2y + 6 + y = -3$

$3y = -9$

$y = -3$

Substitute and solve for x:

$x = y + 3$

$x = -3 + 3 = 0$

$\boxed{x = 0, y = -3}$

c. $\begin{cases} (1) \quad \dfrac{x+y}{2} + \dfrac{x-y}{3} = 1 \\ (2) \quad y = 3x - 2 \end{cases}$

Substitute $y = 3x - 2$ into (1):

$\dfrac{x + 3x - 2}{2} + \dfrac{x - (3x - 2)}{3} = 1$

$\dfrac{4x - 2}{2} + \dfrac{-2x + 2}{3} = 1$

$3(4x - 2) + 2(-2x + 2) = 6$

$12x - 6 - 4x + 4 = 6$

$8x = 8$

$x = 1$

Substitute and solve for y:

$y = 3x - 2$

$y = 3(1) - 2 = 1$

$\boxed{x = 1, y = 1}$

3. a.

$\begin{aligned} x + 5y &= 7 \\ 3x + y &= -7 \Rightarrow \times (-5) \end{aligned}$

$\begin{aligned} x + 5y &= 7 \\ -15x - 5y &= 35 \\ \hline -14x &= 42 \\ x &= -3 \end{aligned}$

$\begin{aligned} x + 5y &= 7 \\ -3 + 5y &= 7 \\ 5y &= 10 \\ y &= 2 \end{aligned}$

Solution:

$\boxed{x = -3, y = 2}$

b. $\quad \begin{aligned} 2x + 3y &= 11 \Rightarrow \times (3) \\ 3x - 7y &= -41 \Rightarrow \times (-2) \end{aligned}$ $\quad \begin{aligned} 6x + 9y &= 33 \\ -6x + 14y &= 82 \\ \hline 23y &= 115 \\ y &= 5 \end{aligned}$ $\quad \begin{aligned} 2x + 3y &= 11 \\ 2x + 3(5) &= 11 \\ 2x &= -4 \\ x &= -2 \end{aligned}$ $\quad$ Solution:

$\boxed{x = -2, y = 5}$

c. $\quad \begin{aligned} 2(x + y) - x &= 0 \Rightarrow \\ 3(x + y) + 2y &= 1 \Rightarrow \end{aligned}$ $\begin{aligned} x + 2y &= 0 \Rightarrow \\ 3x + 5y &= 1 \Rightarrow \end{aligned}$ $\begin{aligned} -3x - 6y &= 0 \\ 3x + 5y &= 1 \\ \hline -y &= 1 \\ y &= -1 \end{aligned}$ $\quad \begin{aligned} x + 2y &= 0 \\ x + 2(-1) &= 0 \\ x &= 2 \end{aligned}$ $\quad$ Solution:

$\boxed{\begin{aligned} x &= 2 \\ y &= -1 \end{aligned}}$

4. **a.** $\quad$ (1) $3x + 2y - z = 2$ $\quad$ Add (1) and $-$(2): $\qquad$ Add equations (1) and $-$(3):

$\quad$ (2) $\quad x + y - z = 0$ $\qquad$ (1) $\quad 3x + 2y - z = 2$ $\qquad$ (1) $\quad 3x + 2y - z = 2$

$\quad$ (3) $2x + 3y - z = 1$ $\qquad$ $-$(2) $\dfrac{-x - y + z = 0}{}$ $\qquad$ $-$(3) $\dfrac{-2x - 3y + z = -1}{}$

$\qquad\qquad\qquad\qquad\qquad$ (4) $\quad 2x + y \; = 2$ $\qquad$ (5) $\qquad x - y \; = 1$

Solve the system of two equations and two unknowns formed by equations (4) and (5):

$\begin{aligned} 2x + y &= 2 \\ x - y &= 1 \\ \hline 3x &= 3 \\ x &= 1 \end{aligned}$ $\qquad \begin{aligned} 2x + y &= 2 \\ 2(1) + y &= 2 \\ y &= 0 \end{aligned}$ $\qquad \begin{aligned} x + y - z &= 0 \\ 1 + 0 - z &= 0 \\ -z &= -1 \\ z &= 1 \end{aligned}$ $\qquad$ Solution:

$\boxed{x = 1, y = 0, z = 1}$

b. $\quad$ (1) $\quad 5x - y + z = 3$ $\quad$ Add (1) and (2): $\qquad$ Add equations (1) and (3):

$\quad$ (2) $3x + y + 2z = 2$ $\qquad$ (1) $\quad 5x - y + z = 3$ $\qquad$ (1) $\quad 5x - y + z = 3$

$\quad$ (3) $\qquad x + y = 2$ $\qquad$ (2) $\quad 3x + y + 2z = 2$ $\qquad$ (3) $\dfrac{x + y \; = 2}{}$

$\qquad\qquad\qquad\qquad\qquad$ (4) $\dfrac{8x + 3z = 5}{}$ $\qquad$ (5) $\quad 6x + z = 5$

Solve the system of two equations and two unknowns formed by equations (4) and (5):

$\begin{aligned} 8x + 3z &= 5 \\ 6x + z &= 5 \Rightarrow \times (-3) \end{aligned}$ $\qquad \begin{aligned} 8x + 3z &= 5 \\ -18x - 3z &= -15 \\ \hline -10x &= -10 \\ x &= 1 \end{aligned}$ $\qquad \begin{aligned} 6x + z &= 5 \\ 6(1) + z &= 5 \\ z &= -1 \end{aligned}$ $\qquad \begin{aligned} x + y &= 2 \\ 1 + y &= 2 \\ y &= 1 \end{aligned}$

Solution: $\boxed{x = 1, y = 1, z = -1}$

c. $\quad$ (1) $2x - y + z = 1$ $\quad$ Add (1) and $-$(2): $\qquad$ Add equations (1) and $-$(3):

$\quad$ (2) $x - y + 2z = 3$ $\qquad$ (1) $\quad 2x - y + z = 1$ $\qquad$ (1) $\quad 2x - y + z = 1$

$\quad$ (3) $\quad x - y + z = 1$ $\qquad$ $-$(2) $\dfrac{-x + y - 2z = -3}{}$ $\qquad$ $-$(3) $\dfrac{-x + y - z = -1}{}$

$\qquad\qquad\qquad\qquad\qquad$ (4) $\quad x - z = -2$ $\qquad$ (5) $\qquad x \; = 0$

Solve the system of two equations and two unknowns formed by equations (4) and (5):

$\begin{aligned} x - z &= -2 \\ 0 - z &= -2 \\ z &= 2 \end{aligned}$ $\qquad \begin{aligned} x - y + z &= 1 \\ 0 - y + 2 &= 1 \\ -y &= -1 \\ y &= 1 \end{aligned}$ $\qquad$ Solution: $\boxed{x = 0, y = 1, z = 2}$

5. Let $x =$ cost of fake fur and let $y =$ cost of leather. Then $\begin{cases} (1) & 25x + 15y = 9300 \\ (2) & 10x + 30y = 12600 \end{cases}$

$$25x + 15y = 9300 \Rightarrow \times (-2) \quad \begin{array}{r} -50x + 30y = -18600 \\ 10x + 30y = 12600 \\ \hline -40x = -6000 \\ x = 150 \end{array} \quad \begin{array}{r} 25x + 15y = 9300 \\ 25(150) + 15y = 9300 \\ 15y = 5550 \\ y = 370 \end{array}$$

$$10x + 30y = 12600$$

The fake fur coats cost \$150 while the leather coats cost \$370. The cost will be \$10,400.

6. Let $x = \#$ adult tickets, $y = \#$ senior tickets and $z = \#$ children tickets.

(1) $x + y + z = 1800$ Add $-4(1)$ and (2): Add equations (1) and $-(3)$:

(2) $5x + 4y + 2.5z = 7425$ $\quad -4(1) \quad -4x - 4y - 4z = -7200 \quad (1) \quad x + y + z = 1800$

(3) $y + z = 900$ $\quad (2) \quad \underline{5x + 4y + 2.5z = 7425} \quad -(3) \quad \underline{-y - z = -900}$

$\quad\quad\quad\quad\quad\quad\quad\quad\quad\quad (4) \quad x - 1.5z = 225 \quad (5) \quad x = 900$

Solve the system of two equations and two unknowns formed by equations (4) and (5):

$$\begin{array}{ll} x - 1.5z = 225 & y + z = 900 \\ 900 - 1.5z = 225 & y + 450 = 900 \\ -1.5z = -675 & y = 450 \\ z = 450 \end{array}$$

There were 900 adult tickets, 450 senior tickets, and 450 children's tickets sold.

7. a. $\begin{bmatrix} 2 & 5 & | & 7 \\ 3 & -1 & | & 2 \end{bmatrix} \Rightarrow \begin{bmatrix} 1 & -6 & | & -5 \\ 3 & -1 & | & 2 \end{bmatrix} \Rightarrow \begin{bmatrix} 1 & -6 & | & -5 \\ 0 & 17 & | & 17 \end{bmatrix} \Rightarrow \begin{bmatrix} 1 & -6 & | & -5 \\ 0 & 1 & | & 1 \end{bmatrix} \Rightarrow$

$ -R_1 + R_2 \Rightarrow R_1 \quad\quad -3R_1 + R_2 \Rightarrow R_2 \quad\quad \frac{1}{17}R_2 \Rightarrow R_2$

$\begin{bmatrix} 1 & 0 & | & 1 \\ 0 & 1 & | & 1 \end{bmatrix}$ Solution: $\boxed{x = 1, y = 1}$

$6R_2 + R_1 \Rightarrow R_1$

b. $\begin{bmatrix} 1 & 3 & -1 & | & 8 \\ 2 & 1 & -2 & | & 11 \\ 1 & -1 & 5 & | & -8 \end{bmatrix} \Rightarrow \begin{bmatrix} 1 & 3 & -1 & | & 8 \\ 0 & -5 & 0 & | & -5 \\ 0 & -4 & 6 & | & -16 \end{bmatrix} \Rightarrow \begin{bmatrix} 1 & 3 & -1 & | & 8 \\ 0 & 1 & 0 & | & 1 \\ 0 & -4 & 6 & | & -16 \end{bmatrix} \Rightarrow$

$ -2R_1 + R_2 \Rightarrow R_2 \quad\quad\quad -\frac{1}{5}R_2 \Rightarrow R_2$

$ -R_1 + R_3 \Rightarrow R_3$

$\begin{bmatrix} 1 & 0 & -1 & | & 5 \\ 0 & 1 & 0 & | & 1 \\ 0 & 0 & 6 & | & -12 \end{bmatrix} \Rightarrow \begin{bmatrix} 1 & 0 & 0 & | & 3 \\ 0 & 1 & 0 & | & 1 \\ 0 & 0 & 1 & | & -2 \end{bmatrix}$ Solution: $\boxed{x = 3, y = 1, z = -2}$

$ -3R_2 + R_1 \Rightarrow R_1 \quad\quad \frac{1}{6}R_3 + R_1 \Rightarrow R_1$

$ 4R_2 + R_3 \Rightarrow R_3 \quad\quad\quad \frac{1}{6}R_3 \Rightarrow R_3$

c.
$$\begin{bmatrix} 1 & 3 & 1 & | & 3 \\ 2 & -1 & 1 & | & -11 \\ 3 & 2 & 3 & | & 2 \end{bmatrix} \Rightarrow \begin{bmatrix} 1 & 3 & 1 & | & 3 \\ 0 & -7 & -1 & | & -17 \\ 0 & -7 & 0 & | & -7 \end{bmatrix} \Rightarrow \begin{bmatrix} 1 & 3 & 1 & | & 3 \\ 0 & 1 & 0 & | & 1 \\ 0 & -7 & -1 & | & -17 \end{bmatrix} \Rightarrow$$

$$\begin{matrix} & -2R_1 + R_2 \Rightarrow R_2 & & R_2 \Leftrightarrow -\frac{1}{7}R_3 \\ & -3R_1 + R_3 \Rightarrow R_3 & & \end{matrix}$$

$$\begin{bmatrix} 1 & 0 & 1 & | & 0 \\ 0 & 1 & 0 & | & 1 \\ 0 & 0 & -1 & | & -10 \end{bmatrix} \Rightarrow \begin{bmatrix} 1 & 0 & 0 & | & -10 \\ 0 & 1 & 0 & | & 1 \\ 0 & 0 & 1 & | & 10 \end{bmatrix} \quad \text{Solution:} \quad \boxed{x = -10, y = 1, z = 10}$$

$$\begin{matrix} -3R_2 + R_1 \Rightarrow R_1 & & R_3 + R_1 \Rightarrow R_1 \\ 7R_2 + R_3 \Rightarrow R_3 & & -R_3 \Rightarrow R_3 \end{matrix}$$

d.
$$\begin{bmatrix} 1 & 1 & 1 & | & 4 \\ 3 & -2 & -2 & | & -3 \\ 4 & -1 & -1 & | & 0 \end{bmatrix} \Rightarrow \begin{bmatrix} 1 & 1 & 1 & | & 4 \\ 0 & -5 & -5 & | & -15 \\ 0 & -5 & -5 & | & -16 \end{bmatrix} \Rightarrow \begin{bmatrix} 1 & 1 & 1 & | & 4 \\ 0 & -5 & -5 & | & -15 \\ 0 & 0 & 0 & | & -1 \end{bmatrix}$$

$$\begin{matrix} & -3R_1 + R_2 \Rightarrow R_2 & & -R_2 + R_3 \Rightarrow R_3 \\ & -4R_1 + R_3 \Rightarrow R_3 & & \end{matrix}$$

The last row indicates $0x + 0y + 0z = -1$. This is impossible. $\Rightarrow$ no solution

8. $\quad -4 = x, x = -4, 0 = x + 4, x + 7 = y \Rightarrow \boxed{x = -4, y = 3}$

9. **a.**
$$\begin{bmatrix} 3 & 2 & 1 \\ 3 & 2 & 1 \end{bmatrix} + \begin{bmatrix} -2 & 1 & 3 \\ 1 & -2 & 1 \end{bmatrix} = \begin{bmatrix} 1 & 3 & 4 \\ 4 & 0 & 2 \end{bmatrix}$$

b.
$$\begin{bmatrix} 2 & 3 & 5 \\ 1 & -2 & 4 \\ 2 & 1 & -2 \end{bmatrix} - \begin{bmatrix} 0 & -2 & 1 \\ 3 & 4 & -2 \\ 6 & -4 & 1 \end{bmatrix} = \begin{bmatrix} 2 & 5 & 4 \\ -2 & -6 & 6 \\ -4 & 5 & -3 \end{bmatrix}$$

c.
$$\begin{bmatrix} 1 & -2 \\ -3 & 1 \end{bmatrix}\begin{bmatrix} 2 & 3 \\ -1 & 2 \end{bmatrix} = \begin{bmatrix} 4 & -1 \\ -7 & -7 \end{bmatrix}$$

d.
$$\begin{bmatrix} -2 & 3 & 5 \\ 1 & -2 & -3 \end{bmatrix}\begin{bmatrix} 2 & 1 \\ -1 & 2 \\ -2 & 3 \end{bmatrix} = \begin{bmatrix} -17 & 19 \\ 10 & -12 \end{bmatrix}$$

e.
$$\begin{bmatrix} 1 & -3 & 2 \end{bmatrix}\begin{bmatrix} 2 \\ 1 \\ 3 \end{bmatrix} = \begin{bmatrix} 5 \end{bmatrix}$$

f.
$$\begin{bmatrix} 1 \\ 2 \\ 1 \\ 5 \end{bmatrix}\begin{bmatrix} 2 & -1 & 1 & 3 \end{bmatrix} = \begin{bmatrix} 2 & -1 & 1 & 3 \\ 4 & -2 & 2 & 6 \\ 2 & -1 & 1 & 3 \\ 10 & -5 & 5 & 15 \end{bmatrix}$$

g.
$$\begin{bmatrix} 1 & -5 & 3 \\ 2 & 1 & -1 \end{bmatrix}\begin{bmatrix} 2 \\ -2 \\ 3 \end{bmatrix}\begin{bmatrix} 1 & -1 \\ -1 & 3 \end{bmatrix}\begin{bmatrix} 1 \\ -2 \end{bmatrix} = \begin{bmatrix} 21 \\ -1 \end{bmatrix}\begin{bmatrix} 1 & -1 \\ -1 & 3 \end{bmatrix}\begin{bmatrix} 1 \\ -2 \end{bmatrix} \Rightarrow \text{undefined}$$

h. $\begin{bmatrix} 1 & -3 & 2 \end{bmatrix} \begin{bmatrix} 2 \\ 1 \\ -5 \end{bmatrix} + \begin{bmatrix} 1 & -3 \end{bmatrix} \begin{bmatrix} 2 \\ 5 \end{bmatrix} = [-11] + [-13] = [-24]$

i. $\left(\begin{bmatrix} 1 & -3 \\ 3 & 1 \end{bmatrix} + \begin{bmatrix} -1 & 3 \\ 1 & 1 \end{bmatrix} \right) \begin{bmatrix} 1 \\ -5 \end{bmatrix} = \begin{bmatrix} 0 & 0 \\ 4 & 2 \end{bmatrix} \begin{bmatrix} 1 \\ -5 \end{bmatrix} = \begin{bmatrix} 0 \\ -6 \end{bmatrix}$

10. a. $\begin{bmatrix} 2 & 3 & | & 1 & 0 \\ 3 & 5 & | & 0 & 1 \end{bmatrix} \Rightarrow \begin{bmatrix} 1 & \frac{3}{2} & | & \frac{1}{2} & 0 \\ 3 & 5 & | & 0 & 1 \end{bmatrix} \Rightarrow \begin{bmatrix} 1 & \frac{3}{2} & | & \frac{1}{2} & 0 \\ 0 & \frac{1}{2} & | & -\frac{3}{2} & 1 \end{bmatrix} \Rightarrow$

$\qquad \frac{1}{2}R_1 \Rightarrow R_1 \qquad\qquad -3R_1 + R_2 \Rightarrow R_2$

$\begin{bmatrix} 1 & 0 & | & 5 & -3 \\ 0 & \frac{1}{2} & | & -\frac{3}{2} & 1 \end{bmatrix} \Rightarrow \begin{bmatrix} 1 & 0 & | & 5 & -3 \\ 0 & 1 & | & -3 & 2 \end{bmatrix} \Rightarrow \quad \text{Inverse:} \begin{bmatrix} 5 & -3 \\ -3 & 2 \end{bmatrix}$

$\quad -3R_2 + R_1 \Rightarrow R_1 \qquad\qquad 2R_2 \Rightarrow R_2$

b. $\begin{bmatrix} 1 & 0 & 0 & | & 1 & 0 & 0 \\ 2 & 0 & -2 & | & 0 & 1 & 0 \\ 1 & 2 & 2 & | & 0 & 0 & 1 \end{bmatrix} \Rightarrow \begin{bmatrix} 1 & 0 & 0 & | & 1 & 0 & 0 \\ 1 & 2 & 2 & | & 0 & 0 & 1 \\ 2 & 0 & -2 & | & 0 & 1 & 0 \end{bmatrix} \Rightarrow \begin{bmatrix} 1 & 0 & 0 & | & 1 & 0 & 0 \\ 0 & 2 & 2 & | & -1 & 0 & 1 \\ 0 & 0 & -2 & | & -2 & 1 & 0 \end{bmatrix} \Rightarrow$

$\qquad\qquad\qquad R_2 \Leftrightarrow R_3 \qquad\qquad\qquad\qquad -R_1 + R_2 \Rightarrow R_2$

$\qquad\qquad\qquad\qquad\qquad\qquad\qquad\qquad\qquad\qquad -2R_1 + R_3 \Rightarrow R_3$

$\begin{bmatrix} 1 & 0 & 0 & | & 1 & 0 & 0 \\ 0 & 2 & 0 & | & -3 & 1 & 1 \\ 0 & 0 & -2 & | & -2 & 1 & 0 \end{bmatrix} \Rightarrow \begin{bmatrix} 1 & 0 & 0 & | & 1 & 0 & 0 \\ 0 & 1 & 0 & | & -\frac{3}{2} & \frac{1}{2} & \frac{1}{2} \\ 0 & 0 & 1 & | & 1 & -\frac{1}{2} & 0 \end{bmatrix} : \text{Inverse} = \begin{bmatrix} 1 & 0 & 0 \\ -\frac{3}{2} & \frac{1}{2} & \frac{1}{2} \\ 1 & -\frac{1}{2} & 0 \end{bmatrix}$

$\qquad R_2 + R_3 \Rightarrow R_3 \qquad\qquad\qquad \frac{1}{2}R_2 \Rightarrow R_2$

$\qquad\qquad\qquad\qquad\qquad\qquad -\frac{1}{2}R_3 \Rightarrow R_3$

c. $\begin{bmatrix} 1 & 0 & 8 & | & 1 & 0 & 0 \\ 3 & 7 & 6 & | & 0 & 1 & 0 \\ 1 & 2 & 3 & | & 0 & 0 & 1 \end{bmatrix} \Rightarrow \begin{bmatrix} 1 & 0 & 8 & | & 1 & 0 & 0 \\ 0 & 7 & -18 & | & -3 & 1 & 0 \\ 0 & 2 & -5 & | & -1 & 0 & 1 \end{bmatrix} \Rightarrow$

$\qquad\qquad\qquad\qquad\qquad -3R_1 + R_2 \Rightarrow R_2$

$\qquad\qquad\qquad\qquad\qquad -R_1 + R_3 \Rightarrow R_3$

$\begin{bmatrix} 1 & 0 & 8 & | & 1 & 0 & 0 \\ 0 & 1 & -\frac{18}{7} & | & -\frac{3}{7} & \frac{1}{7} & 0 \\ 0 & 2 & -5 & | & -1 & 0 & 1 \end{bmatrix} \Rightarrow \begin{bmatrix} 1 & 0 & 8 & | & 1 & 0 & 0 \\ 0 & 1 & -\frac{18}{7} & | & -\frac{3}{7} & \frac{1}{7} & 0 \\ 0 & 0 & \frac{1}{7} & | & -\frac{1}{7} & -\frac{2}{7} & 1 \end{bmatrix} \Rightarrow$

$\qquad \frac{1}{7}R_2 \Rightarrow R_2 \qquad\qquad\qquad\qquad -2R_2 + R_3 \Rightarrow R_3$

$\begin{bmatrix} 1 & 0 & 0 & | & 9 & 16 & -56 \\ 0 & 1 & 0 & | & -3 & -5 & 18 \\ 0 & 0 & 1 & | & -1 & -2 & 7 \end{bmatrix} \Rightarrow \quad \text{Inverse} = \begin{bmatrix} 9 & 16 & -56 \\ -3 & -5 & 18 \\ -1 & -2 & 7 \end{bmatrix}$

$\qquad -56R_3 + R_1 \Rightarrow R_1$

$\qquad\quad 18R_3 + R_2 \Rightarrow R_2$

$\qquad\qquad 7R_3 \Rightarrow R_3$

d.
$$\begin{bmatrix} 4 & 4 & 1 & | & 1 & 0 & 0 \\ 1 & 1 & 1 & | & 0 & 1 & 0 \\ -1 & -1 & 0 & | & 0 & 0 & 1 \end{bmatrix} \Rightarrow \begin{bmatrix} 1 & 1 & 1 & | & 0 & 1 & 0 \\ 4 & 4 & 1 & | & 1 & 0 & 0 \\ -1 & -1 & 0 & | & 0 & 0 & 1 \end{bmatrix} \Rightarrow$$
$$R_1 \Leftrightarrow R_2$$

$$\begin{bmatrix} 1 & 1 & 1 & | & 0 & 1 & 0 \\ 0 & 0 & -3 & | & 1 & -4 & 0 \\ 0 & 0 & 1 & | & 0 & 1 & 1 \end{bmatrix} \Rightarrow \begin{bmatrix} 1 & 1 & 1 & | & 0 & 1 & 0 \\ 0 & 0 & 0 & | & 1 & -1 & 3 \\ 0 & 0 & 1 & | & 0 & 1 & 1 \end{bmatrix} : \text{No inverse}$$
$$-4R_1 + R_2 \Rightarrow R_2 \qquad\qquad 3R_3 + R_2 \Rightarrow R_2$$
$$R_1 + R_3 \Rightarrow R_3$$

11. a.
$$\begin{bmatrix} 4 & -1 & 2 \\ 1 & 1 & 2 \\ 1 & 0 & 1 \end{bmatrix} \begin{bmatrix} x \\ y \\ z \end{bmatrix} = \begin{bmatrix} 0 \\ 1 \\ 0 \end{bmatrix}$$

$$\begin{bmatrix} x \\ y \\ z \end{bmatrix} = \begin{bmatrix} 4 & -1 & 2 \\ 1 & 1 & 2 \\ 1 & 0 & 1 \end{bmatrix}^{-1} \begin{bmatrix} 0 \\ 1 \\ 0 \end{bmatrix}$$

$$\begin{bmatrix} x \\ y \\ z \end{bmatrix} = \begin{bmatrix} 1 & 1 & -4 \\ 1 & 2 & -6 \\ -1 & -1 & 5 \end{bmatrix} \begin{bmatrix} 0 \\ 1 \\ 0 \end{bmatrix} = \begin{bmatrix} 1 \\ 2 \\ -1 \end{bmatrix}$$

b.
$$\begin{bmatrix} 1 & 3 & 1 & 3 \\ 1 & 4 & 1 & 3 \\ 0 & 1 & 1 & 0 \\ 1 & 2 & -1 & 2 \end{bmatrix} \begin{bmatrix} w \\ x \\ y \\ z \end{bmatrix} = \begin{bmatrix} 1 \\ 2 \\ 1 \\ 1 \end{bmatrix}$$

$$\begin{bmatrix} w \\ x \\ y \\ z \end{bmatrix} = \begin{bmatrix} 1 & 3 & 1 & 3 \\ 1 & 4 & 1 & 3 \\ 0 & 1 & 1 & 0 \\ 1 & 2 & -1 & 2 \end{bmatrix}^{-1} \begin{bmatrix} 1 \\ 2 \\ 1 \\ 1 \end{bmatrix}$$

$$\begin{bmatrix} w \\ x \\ y \\ z \end{bmatrix} = \begin{bmatrix} 3 & -5 & 5 & 3 \\ -1 & 1 & 0 & 0 \\ 1 & -1 & 1 & 0 \\ 0 & 1 & -2 & -1 \end{bmatrix} \begin{bmatrix} 1 \\ 2 \\ 1 \\ 1 \end{bmatrix} = \begin{bmatrix} 1 \\ 1 \\ 0 \\ -1 \end{bmatrix}$$

12 a.
$$\begin{vmatrix} 3 & -2 \\ 1 & -3 \end{vmatrix} = (3)(-3) - (-2)(1) = -9 + 2 = -7$$

b.
$$\begin{vmatrix} 1 & -2 & 3 \\ 2 & -1 & 3 \\ 1 & -1 & 0 \end{vmatrix} = 1 \begin{vmatrix} -1 & 3 \\ -1 & 0 \end{vmatrix} - (-2) \begin{vmatrix} 2 & 3 \\ 1 & 0 \end{vmatrix} + 3 \begin{vmatrix} 2 & -1 \\ 1 & -1 \end{vmatrix}$$
$$= 1(3) + 2(-3) + 3(-1) = 3 - 6 - 3 = -6$$

c.
$$\begin{vmatrix} 1 & 3 & -1 \\ 1 & 2 & 1 \\ 1 & 0 & 2 \end{vmatrix} = 1\begin{vmatrix} 2 & 1 \\ 0 & 2 \end{vmatrix} - 3\begin{vmatrix} 1 & 1 \\ 1 & 2 \end{vmatrix} + (-1)\begin{vmatrix} 1 & 2 \\ 1 & 0 \end{vmatrix}$$
$$= 1(4) - 3(1) - 1(-2) = 4 - 3 + 2 = 3$$

d. Expand along 3rd row...
$$\begin{vmatrix} 1 & 2 & 3 & 4 \\ -1 & 3 & -3 & 2 \\ 0 & 0 & 0 & -1 \\ 3 & 3 & 4 & 3 \end{vmatrix} = 0(***) - 0(***) + 0(***) - (-1)\begin{vmatrix} 1 & 2 & 3 \\ -1 & 3 & -3 \\ 3 & 3 & 4 \end{vmatrix}$$
$$= 1\left(1\begin{vmatrix} 3 & -3 \\ 3 & 4 \end{vmatrix} - 2\begin{vmatrix} -1 & -3 \\ 3 & 4 \end{vmatrix} + 3\begin{vmatrix} -1 & 3 \\ 3 & 3 \end{vmatrix} \right)$$
$$= 1(21) - 2(5) + 3(-12) = -25$$

13. a.
$$x = \frac{\begin{vmatrix} -5 & 3 \\ -4 & 1 \end{vmatrix}}{\begin{vmatrix} 1 & 3 \\ -2 & 1 \end{vmatrix}} = \frac{7}{7} = 1 \quad y = \frac{\begin{vmatrix} 1 & -5 \\ -2 & -4 \end{vmatrix}}{\begin{vmatrix} 1 & 3 \\ -2 & 1 \end{vmatrix}} = \frac{-14}{7} = -2$$

b.
$$x = \frac{\begin{vmatrix} -1 & -1 & 1 \\ -4 & -1 & 3 \\ -1 & -3 & 1 \end{vmatrix}}{\begin{vmatrix} 1 & -1 & 1 \\ 2 & -1 & 3 \\ 1 & -3 & 1 \end{vmatrix}} = \frac{2}{2} = 1 \quad y = \frac{\begin{vmatrix} 1 & -1 & 1 \\ 2 & -4 & 3 \\ 1 & -1 & 1 \end{vmatrix}}{\begin{vmatrix} 1 & -1 & 1 \\ 2 & -1 & 3 \\ 1 & -3 & 1 \end{vmatrix}} = \frac{0}{2} = 0$$

$$z = \frac{\begin{vmatrix} 1 & -1 & -1 \\ 2 & -1 & -4 \\ 1 & -3 & -1 \end{vmatrix}}{\begin{vmatrix} 1 & -1 & 1 \\ 2 & -1 & 3 \\ 1 & -3 & 1 \end{vmatrix}} = \frac{-4}{2} = -2$$

c.
$$x = \frac{\begin{vmatrix} 7 & -3 & 1 \\ -9 & 1 & -3 \\ 3 & 1 & 1 \end{vmatrix}}{\begin{vmatrix} 1 & -3 & 1 \\ 1 & 1 & -3 \\ 1 & 1 & 1 \end{vmatrix}} = \frac{16}{16} = 1 \quad y = \frac{\begin{vmatrix} 1 & 7 & 1 \\ 1 & -9 & -3 \\ 1 & 3 & 1 \end{vmatrix}}{\begin{vmatrix} 1 & -3 & 1 \\ 1 & 1 & -3 \\ 1 & 1 & 1 \end{vmatrix}} = \frac{-16}{16} = -1$$

continued on next page...

13. **c.** continued...

$$z = \dfrac{\begin{vmatrix} 1 & -3 & 7 \\ 1 & 1 & -9 \\ 1 & 1 & 3 \end{vmatrix}}{\begin{vmatrix} 1 & -3 & 1 \\ 1 & 1 & -3 \\ 1 & 1 & 1 \end{vmatrix}} = \dfrac{48}{16} = 3$$

d.

$$w = \dfrac{\begin{vmatrix} 4 & 1 & -1 & 1 \\ 4 & 1 & 0 & 1 \\ 0 & 1 & 2 & 1 \\ 2 & 0 & 1 & 1 \end{vmatrix}}{\begin{vmatrix} 1 & 1 & -1 & 1 \\ 2 & 1 & 0 & 1 \\ 0 & 1 & 2 & 1 \\ 1 & 0 & 1 & 1 \end{vmatrix}} = \dfrac{-4}{-4} = 1 \qquad x = \dfrac{\begin{vmatrix} 1 & 4 & -1 & 1 \\ 2 & 4 & 0 & 1 \\ 0 & 0 & 2 & 1 \\ 1 & 2 & 1 & 1 \end{vmatrix}}{\begin{vmatrix} 1 & 1 & -1 & 1 \\ 2 & 1 & 0 & 1 \\ 0 & 1 & 2 & 1 \\ 1 & 0 & 1 & 1 \end{vmatrix}} = \dfrac{0}{-4} = 0$$

$$y = \dfrac{\begin{vmatrix} 1 & 1 & 4 & 1 \\ 2 & 1 & 4 & 1 \\ 0 & 1 & 0 & 1 \\ 1 & 0 & 2 & 1 \end{vmatrix}}{\begin{vmatrix} 1 & 1 & -1 & 1 \\ 2 & 1 & 0 & 1 \\ 0 & 1 & 2 & 1 \\ 1 & 0 & 1 & 1 \end{vmatrix}} = \dfrac{4}{-4} = -1 \qquad z = \dfrac{\begin{vmatrix} 1 & 1 & -1 & 4 \\ 2 & 1 & 0 & 4 \\ 0 & 1 & 2 & 0 \\ 1 & 0 & 1 & 2 \end{vmatrix}}{\begin{vmatrix} 1 & 1 & -1 & 1 \\ 2 & 1 & 0 & 1 \\ 0 & 1 & 2 & 1 \\ 1 & 0 & 1 & 1 \end{vmatrix}} = \dfrac{-8}{-4} = 2$$

14. **a.** $\begin{vmatrix} 3a & 3b & 3c \\ d & e & f \\ g & h & i \end{vmatrix} = 3 \begin{vmatrix} a & b & c \\ d & e & f \\ g & h & i \end{vmatrix} = 21$ \qquad **b.** $\begin{vmatrix} a & b & c \\ d+g & e+h & f+i \\ g & h & i \end{vmatrix} = 7$

15. **a.**
$$\dfrac{7x+3}{x(x+1)} = \dfrac{A}{x} + \dfrac{B}{x+1}$$
$$= \dfrac{A(x+1)}{x(x+1)} + \dfrac{Bx}{x(x+1)}$$
$$= \dfrac{Ax + A + Bx}{x(x+1)}$$
$$= \dfrac{(A+B)x + A}{x(x+1)}$$

$$\begin{cases} A + B = 7 \\ A \qquad\; = 3 \end{cases} \Rightarrow \begin{matrix} A = 3 \\ B = 4 \end{matrix} \qquad \dfrac{7x+3}{x(x+1)} = \dfrac{3}{x} + \dfrac{4}{x+1}$$

b.
$$\frac{4x^3 + x^2 + 3x + 2}{x^2(x^2+1)} = \frac{A}{x} + \frac{B}{x^2} + \frac{Cx+D}{x^2+1}$$

$$= \frac{Ax(x^2+1)}{x^2(x^2+1)} + \frac{B(x^2+1)}{x^2(x^2+1)} + \frac{(Cx+D)x^2}{x^2(x^2+1)}$$

$$= \frac{Ax^3 + Ax + Bx^2 + B + Cx^3 + Dx^2}{x^2(x^2+1)}$$

$$= \frac{(A+C)x^3 + (B+D)x^2 + Ax + B}{x^2(x^2+1)}$$

$$\begin{cases} A & +C & =4 \\ & B & +D=1 \\ A & & =3 \\ & B & =2 \end{cases} \Rightarrow \begin{aligned} A&=3 \\ B&=2 \\ C&=1 \\ D&=-1 \end{aligned} \qquad \frac{4x^3+x^2+3x+2}{x^2(x^2+1)} = \frac{3}{x} + \frac{2}{x^2} + \frac{x-1}{x^2+1}$$

c.
$$\frac{x^2+5}{x(x^2+x+5)} = \frac{A}{x} + \frac{Bx+C}{x^2+x+5}$$

$$= \frac{A(x^2+x+5)}{x(x^2+x+5)} + \frac{(Bx+C)x}{x(x^2+x+5)}$$

$$= \frac{Ax^2 + Ax + 5A + Bx^2 + Cx}{x(x^2+x+5)}$$

$$= \frac{(A+B)x^2 + (A+C)x + (5A)}{x(x^2+x+5)}$$

$$\begin{cases} A+B & =1 \\ A & +C=0 \\ 5A & =5 \end{cases} \Rightarrow \begin{aligned} A&=1 \\ B&=0 \\ C&=-1 \end{aligned} \qquad \frac{x^2+5}{x(x^2+x+5)} = \frac{1}{x} - \frac{1}{x^2+x+5}$$

d.
$$\frac{x^2+1}{(x+1)^3} = \frac{A}{x+1} + \frac{B}{(x+1)^2} + \frac{C}{(x+1)^3}$$

$$= \frac{A(x+1)^2}{(x+1)^3} + \frac{B(x+1)}{(x+1)^3} + \frac{C}{(x+1)^3}$$

$$= \frac{Ax^2 + 2Ax + A + Bx + B + C}{(x+1)^3}$$

$$= \frac{Ax^2 + (2A+B)x + (A+B+C)}{(x+1)^3}$$

$$\begin{cases} A & =1 \\ 2A+B & =0 \\ A+B+C & =1 \end{cases} \Rightarrow \begin{aligned} A&=1 \\ B&=-2 \\ C&=2 \end{aligned} \qquad \frac{x^2+1}{(x+1)^3} = \frac{1}{x+1} - \frac{2}{(x+1)^2} + \frac{2}{(x+1)^3}$$

16. **a.**
$$\begin{cases} 3x + 2y \le 6 \\ x - y > 3 \end{cases}$$

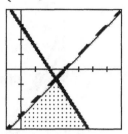

b.
$$\begin{cases} y \le x^2 + 1 \\ y \ge x^2 - 1 \end{cases}$$

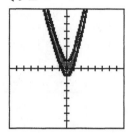

17. **a.**

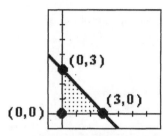

Point	$P = 2x + y$
$(0, 0)$	$= 2(0) + 0 = 0$
$(0, 3)$	$= 2(0) + 3 = 3$
$(3, 0)$	$= 2(3) + 0 = 6$

Max: $P = 6$ at $(3, 0)$

b.

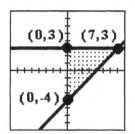

Point	$P = 2x - 3y$
$(0, 3)$	$= 2(0) - 3(3) = -9$
$(7, 3)$	$= 2(7) - 3(3) = 5$
$(0, -4)$	$= 2(0) - 3(-4) = 12$

Max: $P = 12$ at $(0, -4)$

c.

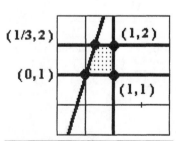

Point	$P = 3x - y$
$(0, 1)$	$= 3(0) - 1 = -1$
$(1, 1)$	$= 3(1) - 1 = 2$
$(1, 2)$	$= 3(1) - 2 = 1$
$\left(\frac{1}{3}, 2\right)$	$= 3\left(\frac{1}{3}\right) - 2 = -1$

Max: $P = 2$ at $(1, 1)$

d.

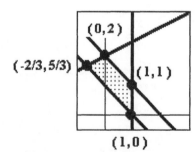

Point	$P = y - 2x$
$(0, 2)$	$= 2 - 2(0) = 2$
$(1, 1)$	$= 1 - 2(1) = -1$
$(1, 0)$	$= 0 - 2(1) = -2$
$\left(-\frac{2}{3}, \frac{5}{3}\right)$	$= \frac{5}{3} - 2\left(-\frac{2}{3}\right) = 3$

Max: $P = 3$ at $\left(-\frac{2}{3}, \frac{5}{3}\right)$

18. Let $x =$ bags of Fertilizer x and $y =$ bags of Fertilizer y.
Maximize $P = 6x + 5y$

subject to $\begin{cases} 6x + 10y \leq 20000 \\ 8x + 6y \leq 16400 \\ 6x + 4y \leq 12000 \\ x \geq 0, y \geq 0 \end{cases}$

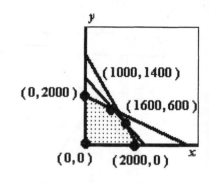

Point	$P = 6x + 5y$
$(0, 0)$	$= 0$
$(0, 2000)$	$= 10000$
$(1000, 1400)$	$= 13000$
$(1600, 600)$	$= 6(1600) + 5(600) = 12600$
$(2000, 0)$	$= 6(2000) + 5(0) = 12000$

1000 bags of x and 1400 bags of y should be made, for a maximum profit of $12,600.

Chapter 6 Test (page 510)

1. $\begin{cases} x - 3y = -5 \\ 2x - y = 0 \end{cases}$

2. $\begin{cases} x = 2y + 5 \\ y = 2x - 4 \end{cases}$

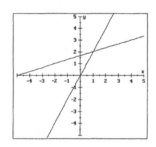

solution:
$(1, 2)$

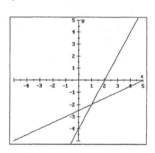

solution:
$(1, -2)$

3. $\begin{array}{l} 3x + y = 0 \Rightarrow \times (5) \\ 2x - 5y = 17 \end{array}$ $\quad \begin{array}{l} 15x + 5y = 0 \\ 2x - 5y = 17 \\ \hline 17x = 17 \\ x = 1 \end{array}$ $\quad \begin{array}{l} 3x + y = 0 \\ 3(1) + y = 0 \\ y = -3 \end{array}$ $\quad$ Solution:
$\boxed{x = 1, y = -3}$

4. $\dfrac{x+y}{2} + x = 7 \Rightarrow 3x + y = 14 \Rightarrow 3x + y = 14$

$\dfrac{x - y}{2} - y = -6 \Rightarrow x - 3y = -12 \Rightarrow -3x + 9y = 36$

$\qquad\qquad\qquad\qquad\qquad\qquad\qquad\qquad \overline{10y = 50}$
$\qquad\qquad\qquad\qquad\qquad\qquad\qquad\qquad\qquad y = 5$

$x - 3y = -12$
$x - 3(5) = -12$
$x = 3$

Solution:
$\boxed{x = 3}$
$\boxed{y = 5}$

5. Let $x =$ liters of 20% solution and $y =$ liters of 45% solution. The following system applies:

$x + y = 10 \Rightarrow \times (-2) \quad -2x - 2y = -20$
$0.2x + 0.45y = 3 \Rightarrow \times (10) \quad \underline{2x + 4.5y = 30}$
$\qquad\qquad\qquad\qquad\qquad\qquad\qquad 2.5y = 10$
$\qquad\qquad\qquad\qquad\qquad\qquad\qquad\quad y = 4$

She must use 4 liters of the 45%
and 6 liters of the 20% solution.

6. Let $x = \#$ from Ace, $y = \#$ from Hi-Fi and $z = \#$ from CD World.

(1) $\qquad x + y + z = 175$ Add (1) and (2):

(2) $\qquad -x - y + z = 25$

(3) $170x + 165y + 160z = 28500$

(1) $\quad x + y + \; z = 175$

(2) $\quad \underline{-x - y + \; z = \; 25}$

(4) $\qquad\qquad\qquad 2z = 200$

$\qquad\qquad\qquad\qquad z = 100$

Add equations 170(2) and (3):

170(2) $\quad -170x - 170y + 170z = \; 4250$

(3) $\qquad \underline{170x + 165y + 160z = 28500}$

(5) $\qquad\qquad - \; 5y + 330z = 32750$

Solve the system of two equations and two unknowns formed by equations (4) and (5):

$-5y + 330z = 32750$ $x + y + z = 175$ Ace buys 25 players per month. Hi-Fi buys

$-5y + 330(100) = 32750$ $x + 50 + 100 = 175$ 50 players per month. CD World buys 100

$-5y = -250$ $y = 25$ players per month.

$y = 50$

7. $\begin{bmatrix} 3 & -2 & | & 4 \\ 2 & 3 & | & 7 \end{bmatrix} \Rightarrow \underset{-R_2 + R_1 \Rightarrow R_1}{\begin{bmatrix} 1 & -5 & | & -3 \\ 2 & 3 & | & 7 \end{bmatrix}} \Rightarrow \underset{-2R_1 + R_2 \Rightarrow R_2}{\begin{bmatrix} 1 & -5 & | & -3 \\ 0 & 13 & | & 13 \end{bmatrix}} \Rightarrow \underset{\frac{1}{13}R_2 \Rightarrow R_2}{\begin{bmatrix} 1 & -5 & | & -3 \\ 0 & 1 & | & 1 \end{bmatrix}}$

From R_2: $y = 1$ From R_1: $x - 5y = -3$ Solution: $\boxed{x = 2, y = 1}$

$x - 5(1) = -3$

$x = 2$

8. $\begin{bmatrix} 1 & 3 & -1 & | & 6 \\ 2 & -1 & -2 & | & -2 \\ 1 & 2 & 1 & | & 6 \end{bmatrix} \Rightarrow \underset{\substack{-2R_1 + R_2 \Rightarrow R_2 \\ -R_3 + R_1 \Rightarrow R_3}}{\begin{bmatrix} 1 & 3 & -1 & | & 6 \\ 0 & -7 & 0 & | & -14 \\ 0 & 1 & -2 & | & 0 \end{bmatrix}} \Rightarrow \underset{\substack{-\frac{1}{7}R_2 \Rightarrow R_2 \\ 7R_3 + R_2 \Rightarrow R_3}}{\begin{bmatrix} 1 & 3 & -1 & | & 6 \\ 0 & 1 & 0 & | & 2 \\ 0 & 0 & -14 & | & -14 \end{bmatrix}} \Rightarrow$

$\underset{-\frac{1}{14}R_3 \Rightarrow R_3}{\begin{bmatrix} 1 & 3 & -1 & | & 6 \\ 0 & 1 & 0 & | & 2 \\ 0 & 0 & 1 & | & 1 \end{bmatrix}}$ From (3): $z = 1$ From (1): $\quad x + 3y - z = 6$

                         From (2): $y = 2$ $x + 3(2) - (1) = 6$

$\qquad\qquad\qquad\qquad\qquad\qquad\qquad\qquad\qquad\qquad\qquad\qquad x = 1$

Solution: $\boxed{x = 1, y = 2, z = 1}$

9. $\begin{bmatrix} 1 & 2 & 3 & | & -5 \\ 3 & 1 & -2 & | & 7 \\ 0 & 1 & -1 & | & 2 \end{bmatrix} \Rightarrow \underset{-3R_1 + R_2 \Rightarrow R_2}{\begin{bmatrix} 1 & 2 & 3 & | & -5 \\ 0 & -5 & -11 & | & 22 \\ 0 & 1 & -1 & | & 2 \end{bmatrix}} \Rightarrow \underset{R_2 \Leftrightarrow R_3}{\begin{bmatrix} 1 & 2 & 3 & | & -5 \\ 0 & 1 & -1 & | & 2 \\ 0 & -5 & -11 & | & 22 \end{bmatrix}} \Rightarrow$

$\underset{\substack{-2R_2 + R_1 \Rightarrow R_1 \\ 5R_2 + R_3 \Rightarrow R_3}}{\begin{bmatrix} 1 & 0 & 5 & | & -9 \\ 0 & 1 & -1 & | & 2 \\ 0 & 0 & -16 & | & 32 \end{bmatrix}} \Rightarrow \underset{-\frac{1}{16}R_3 \Rightarrow R_3}{\begin{bmatrix} 1 & 0 & 5 & | & -9 \\ 0 & 1 & -1 & | & 2 \\ 0 & 0 & 1 & | & -2 \end{bmatrix}} \Rightarrow \underset{\substack{-5R_3 + R_1 \Rightarrow R_1 \\ R_2 + R_3 \Rightarrow R_2}}{\begin{bmatrix} 1 & 0 & 0 & | & 1 \\ 0 & 1 & 0 & | & 0 \\ 0 & 0 & 1 & | & -2 \end{bmatrix}}$ Solution: $\boxed{\begin{aligned} x &= 1 \\ y &= 0 \\ z &= -2 \end{aligned}}$

10. $\begin{bmatrix} 1 & 2 & 1 & | & 0 \\ 3 & -2 & -2 & | & 7 \\ 4 & 0 & -1 & | & 7 \end{bmatrix} \Rightarrow \begin{bmatrix} 1 & 2 & 1 & | & 0 \\ 0 & -8 & -5 & | & 7 \\ 0 & -8 & -5 & | & 7 \end{bmatrix} \Rightarrow \begin{bmatrix} 1 & 2 & 1 & | & 0 \\ 0 & 1 & \frac{5}{8} & | & -\frac{7}{8} \\ 0 & 0 & 0 & | & 0 \end{bmatrix} \Rightarrow$

$ -3R_1 + R_2 \Rightarrow R_2 -R_2 + R_3 \Rightarrow R_3$

$ -4R_1 + R_3 \Rightarrow R_3 -\frac{1}{8}R_2 \Rightarrow R_2$

$\begin{bmatrix} 1 & 0 & -\frac{1}{4} & | & \frac{7}{4} \\ 0 & 1 & \frac{5}{8} & | & -\frac{7}{8} \\ 0 & 0 & 0 & | & 0 \end{bmatrix} \Rightarrow$ Solution: $\boxed{\begin{array}{l} x = \frac{7}{4} + \frac{1}{4}z \\ y = -\frac{7}{8} - \frac{5}{8}z \\ z = \text{any real number} \end{array}}$ Note: This answer is equivalent to the answer provided in the textbook.

$ -2R_2 + R_1 \Rightarrow R_1$

11. $3\begin{bmatrix} 2 & -3 & 5 \\ 0 & 3 & -1 \end{bmatrix} - 5\begin{bmatrix} -2 & 1 & -1 \\ 0 & 3 & 2 \end{bmatrix} = \begin{bmatrix} 6 & -9 & 15 \\ 0 & 9 & -3 \end{bmatrix} - \begin{bmatrix} -10 & 5 & -5 \\ 0 & 15 & 10 \end{bmatrix}$

$ = \begin{bmatrix} 16 & -14 & 20 \\ 0 & -6 & -13 \end{bmatrix}$

12. $[1 \quad 2 \quad 3]\begin{bmatrix} 2 & -2 \\ -2 & 2 \\ 1 & 0 \end{bmatrix}\begin{bmatrix} 3 \\ -2 \end{bmatrix} = [1 \quad 2]\begin{bmatrix} 3 \\ -2 \end{bmatrix} = [-1]$

13. $\begin{bmatrix} 5 & 19 & | & 1 & 0 \\ 2 & 7 & | & 0 & 1 \end{bmatrix} \Rightarrow \begin{bmatrix} 1 & \frac{19}{5} & | & \frac{1}{5} & 0 \\ 2 & 7 & | & 0 & 1 \end{bmatrix} \Rightarrow \begin{bmatrix} 1 & \frac{19}{5} & | & \frac{1}{5} & 0 \\ 0 & -\frac{3}{5} & | & -\frac{2}{5} & 1 \end{bmatrix} \Rightarrow$

$ \frac{1}{5}R_1 \Rightarrow R_1 -2R_1 + R_2 \Rightarrow R_2$

$\begin{bmatrix} 1 & \frac{19}{5} & | & \frac{1}{5} & 0 \\ 0 & 1 & | & \frac{2}{3} & -\frac{5}{3} \end{bmatrix} \Rightarrow \begin{bmatrix} 1 & 0 & | & -\frac{7}{3} & \frac{19}{3} \\ 0 & 1 & | & \frac{2}{3} & -\frac{5}{3} \end{bmatrix} \Rightarrow$ Inverse: $\begin{bmatrix} -\frac{7}{3} & \frac{19}{3} \\ \frac{2}{3} & -\frac{5}{3} \end{bmatrix}$

$ -\frac{5}{3}R_2 \Rightarrow R_1 -\frac{19}{5}R_2 + R_1 \Rightarrow R_1$

14. $\begin{bmatrix} -1 & 3 & -2 & | & 1 & 0 & 0 \\ 4 & 1 & 4 & | & 0 & 1 & 0 \\ 0 & 3 & -1 & | & 0 & 0 & 1 \end{bmatrix} \Rightarrow \begin{bmatrix} 1 & -3 & 2 & | & -1 & 0 & 0 \\ 0 & 13 & -4 & | & 4 & 1 & 0 \\ 0 & 3 & -1 & | & 0 & 0 & 1 \end{bmatrix} \Rightarrow$

$ 4R_1 + R_2 \Rightarrow R_2$

$ -R_1 \Rightarrow R_1$

$\begin{bmatrix} 1 & -3 & 2 & | & -1 & 0 & 0 \\ 0 & 1 & 0 & | & 4 & 1 & -4 \\ 0 & 3 & -1 & | & 0 & 0 & 1 \end{bmatrix} \Rightarrow \begin{bmatrix} 1 & 0 & 2 & | & 11 & 3 & -12 \\ 0 & 1 & 0 & | & 4 & 1 & -4 \\ 0 & 0 & -1 & | & -12 & -3 & 13 \end{bmatrix} \Rightarrow$

$ -4R_3 + R_2 \Rightarrow R_2 3R_2 + R_1 \Rightarrow R_1$

$ -3R_2 + R_3 \Rightarrow R_3$

$\begin{bmatrix} 1 & 0 & 0 & | & -13 & -3 & 14 \\ 0 & 1 & 0 & | & 4 & 1 & -4 \\ 0 & 0 & 1 & | & 12 & 3 & -13 \end{bmatrix}$ Inverse: $\begin{bmatrix} -13 & -3 & 14 \\ 4 & 1 & -4 \\ 12 & 3 & -13 \end{bmatrix}$

$ 2R_3 + R_1 \Rightarrow R_1$

$ -R_3 \Rightarrow R_3$

15. $\begin{bmatrix} 5 & 19 \\ 2 & 7 \end{bmatrix} \begin{bmatrix} x \\ y \end{bmatrix} = \begin{bmatrix} 3 \\ 2 \end{bmatrix}$

$\begin{bmatrix} x \\ y \end{bmatrix} = \begin{bmatrix} 5 & 19 \\ 2 & 7 \end{bmatrix}^{-1} \begin{bmatrix} 3 \\ 2 \end{bmatrix} = \begin{bmatrix} -\frac{7}{3} & \frac{19}{3} \\ \frac{2}{3} & -\frac{5}{3} \end{bmatrix} \begin{bmatrix} 3 \\ 2 \end{bmatrix} = \begin{bmatrix} \frac{17}{3} \\ -\frac{4}{3} \end{bmatrix}$

16. $\begin{bmatrix} -1 & 3 & -2 \\ 4 & 1 & 4 \\ 0 & 3 & -1 \end{bmatrix} \begin{bmatrix} x \\ y \\ z \end{bmatrix} = \begin{bmatrix} 1 \\ 3 \\ -1 \end{bmatrix}$

$\begin{bmatrix} x \\ y \\ z \end{bmatrix} = \begin{bmatrix} -1 & 3 & -2 \\ 4 & 1 & 4 \\ 0 & 3 & -1 \end{bmatrix}^{-1} \begin{bmatrix} 1 \\ 3 \\ -1 \end{bmatrix}$

$\begin{bmatrix} x \\ y \\ z \end{bmatrix} = \begin{bmatrix} -13 & -3 & 14 \\ 4 & 1 & -4 \\ 12 & 3 & -13 \end{bmatrix} \begin{bmatrix} 1 \\ 3 \\ -1 \end{bmatrix} = \begin{bmatrix} -36 \\ 11 \\ 34 \end{bmatrix}$

17. $\begin{vmatrix} 3 & -5 \\ -3 & 1 \end{vmatrix} = (3)(1) - (-5)(-3) = 3 - 15 = -12$

18. $\begin{vmatrix} 3 & 5 & -1 \\ -2 & 3 & -2 \\ 1 & 5 & -3 \end{vmatrix} = 3 \begin{vmatrix} 3 & -2 \\ 5 & -3 \end{vmatrix} - 5 \begin{vmatrix} -2 & -2 \\ 1 & -3 \end{vmatrix} + (-1) \begin{vmatrix} -2 & 3 \\ 1 & 5 \end{vmatrix}$

$= 3(1) - 5(8) - 1(-13) = 3 - 40 + 13 = -24$

19. $y = \dfrac{\begin{vmatrix} 3 & 3 \\ -3 & 2 \end{vmatrix}}{\begin{vmatrix} 3 & -5 \\ -3 & 1 \end{vmatrix}} = \dfrac{15}{-12} = -\dfrac{5}{4}$

20. $y = \dfrac{\begin{vmatrix} 3 & 2 & -1 \\ -2 & 1 & -2 \\ 1 & 0 & -3 \end{vmatrix}}{\begin{vmatrix} 3 & 5 & -1 \\ -2 & 3 & -2 \\ 1 & 5 & -3 \end{vmatrix}} = \dfrac{-24}{-24} = 1$

21. $\dfrac{5x}{(2x-3)(x+1)} = \dfrac{A}{2x-3} + \dfrac{B}{x+1}$

$= \dfrac{A(x+1)}{(2x-3)(x+1)} + \dfrac{B(2x-3)}{(2x-3)(x+1)}$

$= \dfrac{Ax + A + 2Bx - 3B}{(2x-3)(x+1)}$

$= \dfrac{(A+2B)x + (A-3B)}{(2x-3)(x+1)}$

$\begin{cases} A + 2B = 5 \\ A - 3B = 0 \end{cases} \Rightarrow \begin{matrix} A = 3 \\ B = 1 \end{matrix} \quad \dfrac{5x}{(2x-3)(x+1)} = \dfrac{3}{2x-3} + \dfrac{1}{x+1}$

22. $\dfrac{3x^2 + x + 2}{x(x^2 + 2)} = \dfrac{A}{x} + \dfrac{Bx + C}{x^2 + 2}$

$$= \dfrac{A(x^2 + 2)}{x(x^2 + 2)} + \dfrac{(Bx + C)x}{x(x^2 + 2)}$$

$$= \dfrac{Ax^2 + 2A + Bx^2 + Cx}{x(x^2 + 2)}$$

$$= \dfrac{(A + B)x^2 + Cx + 2A}{x(x^2 + 2)}$$

$\begin{cases} A + B & = 3 \\ & C = 1 \\ 2A & = 2 \end{cases} \Rightarrow \begin{matrix} A = 1 \\ B = 2 \\ C = 1 \end{matrix}$ $\quad \dfrac{3x^2 + x + 2}{x(x^2 + 2)} = \dfrac{1}{x} + \dfrac{2x + 1}{x^2 + 2}$

23. $\begin{cases} x - 3y \ge 3 \\ x + 3y \le 3 \end{cases}$

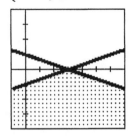

24. $\begin{cases} 3x + 4y \le 12 \\ 3x + 4y \ge 6 \\ x \ge 0, y \ge 0 \end{cases}$

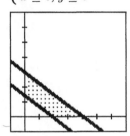

25.

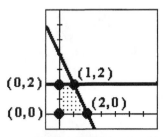

Point	$P = 3x + 2y$
$(0, 2)$	$= 3(0) + 2(2) = 4$
$(1, 2)$	$= 3(1) + 2(2) = 7$
$(2, 0)$	$= 3(2) + 2(0) = 6$
$(0, 0)$	$= 3(0) + 2(0) = 0$

Max: $P = 7$ at $(1, 2)$

26.

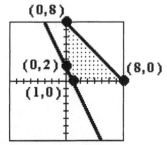

Point	$P = y - x$
$(0, 2)$	$= 2 - 0 = 2$
$(0, 8)$	$= 8 - 0 = 8$
$(8, 0)$	$= 0 - 8 = -8$
$(1, 0)$	$= 0 - 1 = -1$

Min: $P = -8$ at $(8, 0)$

Exercise 7.1 (page 523)

1. $(2, -5), 3$ **3.** $(0, 0), \sqrt{5}$ **5.** to the left **7.** down

9. directrix, focus

11. $(x - h)^2 + (y - k)^2 = r^2$
$(x - 0)^2 + (y - 0)^2 = 7^2$
$x^2 + y^2 = 49$

13. $r = \sqrt{(3 - 2)^2 + (2 - (-2))^2} = \sqrt{17}$
$(x - h)^2 + (y - k)^2 = r^2$
$(x - 2)^2 + (y - (-2))^2 = \left(\sqrt{17}\right)^2$
$(x - 2)^2 + (y + 2)^2 = 17$

15.
$$\begin{array}{l} 3x + \ y = 1 \Rightarrow \times \ (3) \\ \underline{-2x - 3y = 4} \end{array} \qquad \begin{array}{r} 9x + 3y = 3 \\ \underline{-2x - 3y = 4} \\ 7x \quad\ = 7 \\ x \quad\ = 1 \end{array}$$
$$3x + y = 1 \qquad \text{Center:}$$
$$3(1) + y = 1 \qquad (1, -2)$$
$$y = -2$$
$$(x - h)^2 + (y - k)^2 = r^2$$
$$(x - 1)^2 + (y - (-2))^2 = 6^2$$
$$(x - 1)^2 + (y + 2)^2 = 36$$

17.
$$x^2 + y^2 = 4$$
$$(x - 0)^2 + (y - 0)^2 = 2^2$$
$$C(0, 0), r = 2$$

19.
$$3x^2 + 3y^2 - 12x - 6y = 12$$
$$x^2 - 4x + y^2 - 2y = 4$$
$$x^2 - 4x + 4 + y^2 - 2y + 1 = 4 + 4 + 1$$
$$(x - 2)^2 + (y - 1)^2 = 3^2$$
$$C(2, 1), r = 3$$

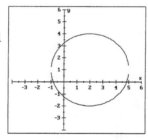

21. Vertical (up), $p = 3$
$$(x - h)^2 = 4p(y - k)$$
$$(x - 0)^2 = 4(3)(y - 0)$$
$$x^2 = 12y$$

23. Horizontal (left), $p = 3$
$$(y - k)^2 = -4p(x - h)$$
$$(y - 0)^2 = -4(3)(x - 0)$$
$$y^2 = -12x$$

25. Vertical (down), $p = 3$

$$(x - h)^2 = -4p(y - k)$$
$$(x - 3)^2 = -4(3)(y - 5)$$
$$(x - 3)^2 = -12(y - 5)$$

27. Vertical (down), $p = 7$

$$(x - h)^2 = -4p(y - k)$$
$$(x - 3)^2 = -4(7)(y - 5)$$
$$(x - 3)^2 = -28(y - 5)$$

29. $(x - 2)^2 = 4p(y - 2)$ **OR** $(y - 2)^2 = 4p(x - 2)$

$$(0 - 2)^2 = 4p(0 - 2) \qquad\qquad (0 - 2)^2 = 4p(0 - 2)$$
$$4 = -8p \qquad\qquad\qquad\qquad 4 = -8p$$
$$-\frac{1}{2} = p \qquad\qquad\qquad\qquad -\frac{1}{2} = p$$
$$-2 = 4p \qquad\qquad\qquad\qquad -2 = 4p$$
$$(x - 2)^2 = -2(y - 2) \qquad\qquad (y - 2)^2 = -2(x - 2)$$

31. $(x - (-4))^2 = 4p(y - 6)$ **OR** $(y - 6)^2 = 4p(x - (-4))$

$$(0 + 4)^2 = 4p(3 - 6) \qquad\qquad (3 - 6)^2 = 4p(0 + 4)$$
$$16 = -12p \qquad\qquad\qquad\qquad 9 = 16p$$
$$-\frac{4}{3} = p \qquad\qquad\qquad\qquad \frac{9}{16} = p$$
$$-\frac{16}{3} = 4p \qquad\qquad\qquad\qquad \frac{9}{4} = 4p$$
$$(x + 4)^2 = -\tfrac{16}{3}(y - 6) \qquad\qquad (y - 6)^2 = \tfrac{9}{4}(x + 4)$$

33. $(x - 6)^2 = 4p(y - 8)$ **OR** $(y - 8)^2 = 4p(x - 6)$

$$(5 - 6)^2 = 4p(10 - 8) \qquad\qquad (10 - 8)^2 = 4p(5 - 6)$$
$$1 = 8p \qquad\qquad\qquad\qquad 4 = -4p$$
$$\frac{1}{8} = p \qquad\qquad\qquad\qquad -1 = p$$
$$\frac{1}{2} = 4p \qquad\qquad\qquad\qquad -4 = 4p$$
$$(x - 6)^2 = \tfrac{1}{2}(y - 8) \qquad\qquad (y - 8)^2 = -4(x - 6)$$

Check to see which equation is satisfied by $(5, 6)$ as well. Answer: $(y - 8)^2 = -4(x - 6)$

35. $(x - 3)^2 = 4p(y - 1)$ **OR** $(y - 1)^2 = 4p(x - 3)$

$$(4 - 3)^2 = 4p(3 - 1) \qquad\qquad (3 - 1)^2 = 4p(4 - 3)$$
$$1 = 8p \qquad\qquad\qquad\qquad 4 = 4p$$
$$\frac{1}{8} = p \qquad\qquad\qquad\qquad (y - 1)^2 = 4(x - 3)$$
$$\frac{1}{2} = 4p$$
$$(x - 3)^2 = \tfrac{1}{2}(y - 1)$$

Check to see which equation is satisfied by $(2, 3)$ as well. Answer: $(x - 3)^2 = \tfrac{1}{2}(y - 1)$

37.
$$y = x^2 + 4x + 5$$
$$y - 5 = x^2 + 4x$$
$$y - 5 + 4 = x^2 + 4x + 4$$
$$y - 1 = (x + 2)^2$$

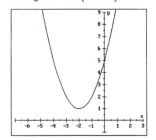

39.
$$y^2 + 4x - 6y = -1$$
$$y^2 - 6y = -4x - 1$$
$$y^2 - 6y + 9 = -4x - 1 + 9$$
$$(y - 3)^2 = -4x + 8$$
$$(y - 3)^2 = -4(x - 2)$$

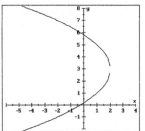

41.
$$y^2 + 2x - 2y = 5$$
$$y^2 - 2y = -2x + 5$$
$$y^2 - 2y + 1 = -2x + 5 + 1$$
$$(y - 1)^2 = -2x + 6$$
$$(y - 1)^2 = -2(x - 3)$$

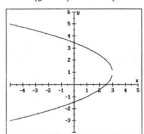

43.
$$x^2 - 6y + 22 = -4x$$
$$x^2 + 4x = 6y - 22$$
$$x^2 + 4x + 4 = 6y - 22 + 4$$
$$(x + 2)^2 = 6y - 18$$
$$(x + 2)^2 = 6(y - 3)$$

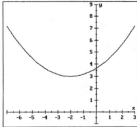

45.
$$4x^2 - 4x + 32y = 47$$
$$x^2 - x = -8y + \frac{47}{4}$$
$$x^2 - x + \frac{1}{4} = -8y + \frac{47}{4} + \frac{1}{4}$$
$$\left(x - \frac{1}{2}\right)^2 = -8y + 12$$
$$\left(x - \frac{1}{2}\right)^2 = -8\left(y - \frac{3}{2}\right)$$

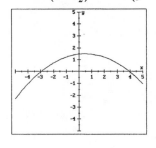

47. Check the coordinates:
$$x^2 + y^2 = 50^2 + 70^2$$
$$= 2500 + 4900 = 7400$$
$7400 < 8100$, so the city can receive.

49. Graph both circles: $\begin{cases} x^2 + y^2 = 1600 \\ x^2 + (y - 35)^2 = 625 \end{cases}$

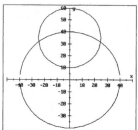

The point farthest from the transmitter $(0, 0)$ is the point $(0, 60)$. The greatest distance is 60 miles.

51. $C(4, 0), r = 4$
$$(x - h)^2 + (y - k)^2 = r^2$$
$$(x - 4)^2 + (y - 0)^2 = 4^2$$
$$(x - 4)^2 + y^2 = 16$$

53. Find the distance to the focus:
$$4p = 8 \Rightarrow p = 2$$
It will be hottest 2 feet from the vertex.

55. The vertex is $(0, 0)$, while $(15, -10)$ is on the curve (vertical parabola):
$$(x - h)^2 = 4p(y - k)$$
$$(15 - 0)^2 = 4p(-10 - 0)$$
$$225 = -40p$$
$$-\tfrac{45}{2} = 4p \Rightarrow x^2 = -\tfrac{45}{2}y$$

57. $\boxed{\text{Income}} = \boxed{\text{Price}} \cdot \boxed{\text{\# rented}}$
$$y = -45\left(\tfrac{n}{32} - \tfrac{1}{2}\right)n$$
$$y = -\tfrac{45}{32}n^2 + \tfrac{45}{2}n$$
$$-\tfrac{32}{45}y = n^2 - 16n$$
$$-\tfrac{32}{45}y + 64 = n^2 - 16n + 64$$
$$-\tfrac{32}{45}y + 64 = (n - 8)^2$$
She should build 8 cabins.

59. Place the vertex at $(0, 0)$, with the focus at $(1, 0) \Rightarrow p = 1, 4p = 4$.
$$(y - k)^2 = 4p(x - h)$$
$$y^2 = 4x$$
Let $x = 10$: $y^2 = 4x$
$$y^2 = 4(10) \Rightarrow y = \pm\sqrt{40}$$
The width $= 2\sqrt{40} \approx 12.6$ cm.

61. The vertex is $(0, 0)$, while $(315, -630)$ is on the curve (vertical parabola):
$$(x - h)^2 = 4p(y - k)$$
$$(315 - 0)^2 = 4p(-630 - 0)$$
$$99{,}225 = -2520p$$
$$-\tfrac{19845}{504} = p \Rightarrow 4p = -\tfrac{19845}{126}$$

Let $y = -430$:
$$x^2 = -\tfrac{19845}{126}y$$
$$x^2 = -\tfrac{19845}{126}(-430)$$
$$x = \pm\sqrt{-\tfrac{19845}{126}(-430)}$$
$$x \approx \pm 260$$
The width is about 520 feet.

63.
$$(y - 2)^2 = 8(x - 1)$$
$$y^2 - 4y + 4 = 8x - 8$$
$$y^2 - 8x - 4y + 12 = 0$$
$$0x^2 + 0xy + y^2 - 8x - 4y + 12 = 0$$

65.

$$\begin{array}{ll}
(x-h)^2 + (y-k)^2 = r^2 & (x-h)^2 + (y-k)^2 = r^2 \\
(0-h)^2 + (8-k)^2 = r^2 & (5-h)^2 + (3-k)^2 = r^2 \\
h^2 + 64 - 16k + k^2 = r^2 & 25 - 10h + h^2 + 9 - 6k + k^2 = r^2 \\
h^2 + k^2 - r^2 = 16k - 64 & h^2 + k^2 - r^2 = 10h + 6k - 34
\end{array}$$

$$\begin{array}{c}
(x-h)^2 + (y-k)^2 = r^2 \\
(4-h)^2 + (6-k)^2 = r^2 \\
16 - 8h + h^2 + 36 - 12k + k^2 = r^2 \\
h^2 + k^2 - r^2 = 8h + 12k - 52
\end{array}$$

$$\begin{cases} 16k - 64 = 10h + 6k - 34 & \Rightarrow & 10k - 10h = 30 \\ 16k - 64 = 8h + 12k - 52 & \Rightarrow & 4k - 8h = 12 \end{cases} \Rightarrow k = 3, h = 0$$

Substitute into one of the above equations to get $r = 5$.   Circle: $x^2 + (y-3)^2 = 25$

67.

$$\begin{array}{lll}
y = ax^2 + bx + c & y = ax^2 + bx + c & y = ax^2 + bx + c \\
8 = a(1)^2 + b(1) + c & -1 = a(-2)^2 + b(-2) + c & 15 = a(2)^2 + b(2) + c \\
8 = a + b + c & -1 = 4a - 2b + c & 15 = 4a + 2b + c
\end{array}$$

$$\begin{cases} a + b + c = 8 \\ 4a - 2b + c = -1 \\ 4a + 2b + c = 15 \end{cases} \Rightarrow a = 1, b = 4, c = 3 \Rightarrow y = x^2 + 4x + 3$$

69.   The stone hits the ground when $s = 0$:   Find s when $t = 8 - x$:

$$0 = -16t^2 + 128t$$
$$0 = -16t(t - 8)$$

It hits the ground after 8 seconds.
Find s when $t = x$:

$$s = -16x^2 + 128x$$

$$\begin{aligned}
s &= -16(8-x)^2 + 128(8-x) \\
&= -16(64 - 16x + x^2) + 1024 - 128x \\
&= -1024 + 256x - 16x^2 + 1024 - 128x \\
&= -16x^2 + 128x
\end{aligned}$$

71. $x^2 + 4x + \boxed{4}$

73. $x^2 - 7x + \boxed{\dfrac{49}{4}}$

75.
$$\begin{aligned}
x^2 + 4x &= 5 \\
x^2 + 4x - 5 &= 0 \\
(x+5)(x-1) &= 0 \\
x = -5 \text{ or } x &= 1
\end{aligned}$$

77.
$$\begin{aligned}
x^2 - 7x - 18 &= 0 \\
(x+2)(x-9) &= 0 \\
x = -2 \text{ or } x &= 9
\end{aligned}$$

Exercise 7.2   (page 535)

1.   sum, constant

3.   vertices

5. $(a, 0), (-a, 0)$

7. $2a = 26 \Rightarrow$ String: 26 inches long
$2b = 10 \Rightarrow b = 5$
$b^2 = a^2 - c^2$
$5^2 = 13^2 - c^2 \Rightarrow c = 12$
Thumbtacks: $2c = 24$ inches apart

9. $c = 3, a = 5$; horizontal
$b^2 = a^2 - c^2$
$\quad = 25 - 9 = 16$
$\dfrac{x^2}{25} + \dfrac{y^2}{16} = 1$

11. $c = 1, b = \frac{4}{3}$; vertical
$a^2 = b^2 + c^2 = \dfrac{16}{9} + 1 = \dfrac{25}{9}$
$\dfrac{x^2}{16/9} + \dfrac{y^2}{25/9} = 1$
$\dfrac{9x^2}{16} + \dfrac{9y^2}{25} = 1$

13. $c = 3, a = 4$; vertical
$b^2 = a^2 - c^2 = 16 - 9 = 7$
$\dfrac{x^2}{7} + \dfrac{y^2}{16} = 1$

15. vertical; $\dfrac{(x-3)^2}{4} + \dfrac{(y-4)^2}{9} = 1$

17. horizontal; $\dfrac{(x-3)^2}{9} + \dfrac{(y-4)^2}{4} = 1$

19. Center: $(3, 4), b = 4, c = 5$, horizontal
$a^2 = b^2 + c^2 = 16 + 25 = 41$
$\dfrac{(x-3)^2}{41} + \dfrac{(y-4)^2}{16} = 1$

21. Center: $(0, 4), c = 4, a = 6$, horizontal
$b^2 = a^2 - c^2 = 36 - 16 = 20$
$\dfrac{x^2}{36} + \dfrac{(y-4)^2}{20} = 1$

23. Center: $(0, 0), c = 6, a = 10$, horizontal
$b^2 = a^2 - c^2 = 100 - 36 = 64$
$\dfrac{x^2}{100} + \dfrac{y^2}{64} = 1$

25. $\dfrac{x^2}{25} + \dfrac{y^2}{49} = 1$
Center: $(0, 0), a = 7, b = 5$, vertical

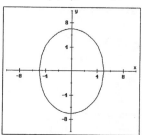

27. $\dfrac{x^2}{16} + \dfrac{(y+2)^2}{36} = 1$
Center: $(0, -2), a = 6, b = 4$, vertical

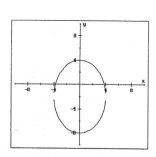

29.
$$x^2 + 4y^2 - 4x + 8y + 4 = 0$$
$$x^2 - 4x + 4(y^2 + 2y) = -4$$
$$x^2 - 4x + 4 + 4(y^2 + 2y + 1) = -4 + 4 + 4$$
$$(x-2)^2 + 4(y+1)^2 = 4$$
$$\frac{(x-2)^2}{4} + \frac{(y+1)^2}{1} = 1$$
Center: $(2, -1)$, $a = 2$, $b = 1$, horizontal

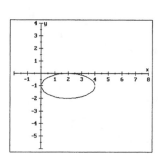

31.
$$16x^2 + 25y^2 - 160x - 200y + 400 = 0$$
$$16(x^2 - 10x) + 25(y^2 - 8y) = -400$$
$$16(x^2 - 10x + 25) + 25(y^2 - 8y + 16) = -400 + 400 + 400$$
$$16(x-5)^2 + 25(y-4)^2 = 400$$
$$\frac{(x-5)^2}{25} + \frac{(y-4)^2}{16} = 1$$
Center: $(5, 4)$, $a = 5$, $b = 4$, horizontal

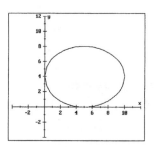

33. The farthest distance $= a + c$: $\quad a = \dfrac{378000}{2} = 189000$

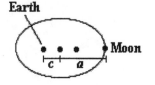

$$\frac{c}{a} = \frac{11}{200}$$
$$c = \frac{11a}{200} = 10395$$
distance $= a + c = 199{,}395$ miles

35. $a = 50$, $b = 30$

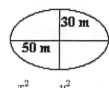

$$\frac{x^2}{2500} + \frac{y^2}{900} = 1$$

$$c^2 = a^2 - b^2$$
$$= 2500 - 900$$
$$= 1600$$
$$c = 40$$

$$\frac{40^2}{2500} + \frac{y^2}{900} = 1$$
$$\frac{y^2}{900} = \frac{900}{2500}$$
$$\frac{y}{30} = \pm\frac{30}{50}$$
$$y = \pm 18$$
The focal width is 36 meters.

37. $a = 24$, $b = 12 \qquad \dfrac{x^2}{144} + \dfrac{y^2}{576} = 1$

$$\frac{x^2}{144} + \frac{y^2}{576} = 1 \qquad \frac{x^2}{144} + \frac{(-12)^2}{576} = 1$$

$$\frac{x^2}{144} = \pm\frac{432}{576} \Rightarrow x \approx \pm 10.4 \Rightarrow \text{The width is about 20.8 inches.}$$

39. $FB = \sqrt{(c-0)^2 + (0-b)^2} = \sqrt{c^2 + b^2} = \sqrt{a^2} = |a| = a \ (a \geq 0)$

41. The equation of the ellipse is $\dfrac{x^2}{a^2} + \dfrac{y^2}{b^2} = 1$. Thus, $y = \pm\sqrt{\dfrac{b^4}{a^2}} = \pm\dfrac{b^2}{a}$.

Let $x = c$ and solve for y^2:

The coordinates of A' and A are

$\dfrac{x^2}{a^2} + \dfrac{y^2}{b^2} = 1$

$\dfrac{c^2}{a^2} + \dfrac{y^2}{b^2} = 1$

$\left(c, \dfrac{b^2}{a}\right)$ and $\left(c, -\dfrac{b^2}{a}\right)$. Therefore, the focal width is $\dfrac{2b^2}{a}$.

$$y^2 = b^2\left(1 - \dfrac{c^2}{a^2}\right)$$

$$= b^2\left(1 - \dfrac{a^2 - b^2}{a^2}\right)$$

$$= b^2\left(1 - 1 + \dfrac{b^2}{a^2}\right) = \dfrac{b^4}{a^2}$$

43. Let the origin be at the midpoint of the line segment between the two thumbtacks and let the x-axis be parallel to that segment. Then $2a = 6$, so $a = 3$. Also, $2c = 2$, so $c = 1$. Find b:

$b^2 = a^2 - c^2 = 3^2 - 1^2 = 8$. The equation is $\dfrac{x^2}{a^2} + \dfrac{y^2}{b^2} = 1$, or $\dfrac{x^2}{9} + \dfrac{y^2}{8} = 1$.

45. **Answers may vary.**

47. $AB = \begin{bmatrix} 3 & -1 & 2 \\ 0 & 2 & -1 \\ 3 & 1 & 1 \end{bmatrix} \begin{bmatrix} 1 & 2 \\ 2 & 0 \\ -1 & 1 \end{bmatrix} = \begin{bmatrix} -1 & 8 \\ 5 & -1 \\ 4 & 7 \end{bmatrix}$

49. $5B - 2C = 5\begin{bmatrix} 1 & 2 \\ 2 & 0 \\ -1 & 1 \end{bmatrix} - 2\begin{bmatrix} 0 & 1 \\ -1 & 1 \\ -2 & 0 \end{bmatrix} = \begin{bmatrix} 5 & 10 \\ 10 & 0 \\ -5 & 5 \end{bmatrix} - \begin{bmatrix} 0 & 2 \\ -2 & 2 \\ -4 & 0 \end{bmatrix} = \begin{bmatrix} 5 & 8 \\ 12 & -2 \\ -1 & 5 \end{bmatrix}$

51. $\begin{bmatrix} -1 & -2 \\ -2 & 0 \\ 1 & -1 \end{bmatrix}$

Exercise 7.3 (page 544)

1. difference, constant

3. $(a, 0), (-a, 0)$

5. transverse axis

7. $a = 5, c = 7$; horizontal
$b^2 = c^2 - a^2 = 49 - 25 = 24$
$\dfrac{x^2}{25} - \dfrac{y^2}{24} = 1$

9. $a = 2, b = 3$; horizontal
$\dfrac{(x-2)^2}{4} - \dfrac{(y-4)^2}{9} = 1$

11. $a = 3$; vertical

$$\frac{(y-3)^2}{9} - \frac{(x-5)^2}{b^2} = 1$$

$$\frac{(8-3)^2}{9} - \frac{(1-5)^2}{b^2} = 1$$

$$\frac{25}{9} - \frac{16}{b^2} = 1$$

$$-\frac{16}{b^2} = -\frac{16}{9}$$

$$b^2 = 9$$

$$\frac{(y-3)^2}{9} - \frac{(x-5)^2}{9} = 1$$

13. Center: $(0, 0)$, $a = 3$, $c = 5$; vertical

$$b^2 = c^2 - a^2$$

$$= 25 - 9 = 16$$

$$\frac{y^2}{9} - \frac{x^2}{16} = 1$$

15. $\dfrac{(x-1)^2}{4} - \dfrac{(y+3)^2}{16} = 1$ **OR** $\dfrac{(y+3)^2}{4} - \dfrac{(x-1)^2}{16} = 1$

17.

$$\frac{x^2}{a^2} - \frac{y^2}{b^2} = 1$$

$$\frac{4^2}{a^2} - \frac{2^2}{b^2} = 1$$

$$\frac{16}{a^2} - \frac{4}{b^2} = 1$$

$$\frac{16}{a^2} = 1 + \frac{4}{b^2}$$

$$\frac{64}{a^2} = 4 + \frac{16}{b^2}$$

$$\frac{x^2}{a^2} - \frac{y^2}{b^2} = 1$$

$$\frac{8^2}{a^2} - \frac{(-6)^2}{b^2} = 1$$

$$\frac{64}{a^2} - \frac{36}{b^2} = 1$$

$$\frac{64}{a^2} - \frac{36}{b^2} = 1$$

$$4 + \frac{16}{b^2} - \frac{36}{b^2} = 1$$

$$3 = \frac{20}{b^2}$$

$$b^2 = \frac{20}{3}$$

$$a^2 = 10$$

$$\frac{x^2}{10} - \frac{3y^2}{20} = 1$$

19. $4(x-1)^2 - 9(y+2)^2 = 36$

$$\frac{4(x-1)^2}{36} - \frac{9(y+2)^2}{36} = \frac{36}{36}$$

$$\frac{(x-1)^2}{9} - \frac{(y+2)^2}{4} = 1$$

$a = 3, b = 2$

Area $= (2a)(2b) = (6)(4) = 24$

21.

$$x^2 + 6x - y^2 + 2y = -11$$

$$x^2 + 6x - (y^2 - 2y) = -11$$

$$x^2 + 6x + 9 - (y^2 - 2y + 1) = -11 + 9 - 1$$

$$(x+3)^2 - (y-1)^2 = -3$$

$$\frac{(x+3)^2}{-3} - \frac{(y-1)^2}{-3} = 1$$

$$\frac{(y-1)^2}{3} - \frac{(x+3)^2}{3} = 1$$

$a = \sqrt{3}, b = \sqrt{3}$

Area $= (2a)(2b) = \left(2\sqrt{3}\right)\left(2\sqrt{3}\right) = 12$

23. $(2a)(2b) = 36$
$4(2b) = 36$
$b = \dfrac{9}{2}$
$\dfrac{(x+2)^2}{4} - \dfrac{4(y+4)^2}{81} = 1$
OR
$\dfrac{(y+4)^2}{4} - \dfrac{4(x+2)^2}{81} = 1$

25. Center: $(0, 0)$, $a = 6$, $b = \dfrac{5}{4}$
$\dfrac{x^2}{6^2} - \dfrac{y^2}{\left(\frac{5}{4}\right)^2} = 1$
$\dfrac{x^2}{36} - \dfrac{16y^2}{25} = 1$

27. $\dfrac{x^2}{9} - \dfrac{y^2}{4} = 1$
Center: $(0, 0)$, $a = 3$, $b = 2$, horizontal

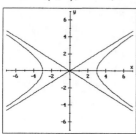

29. $4x^2 - 3y^2 = 36$
$\dfrac{4x^2}{36} - \dfrac{3y^2}{36} = \dfrac{36}{36}$
$\dfrac{x^2}{9} - \dfrac{y^2}{12} = 1$
Center: $(0, 0)$, $a = 3$, $b = 2\sqrt{3}$, horizontal

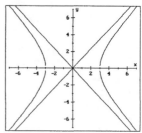

31. $y^2 - x^2 = 1$
$\dfrac{y^2}{1} - \dfrac{x^2}{1} = 1$
Center: $(0, 0)$, $a = 1$, $b = 1$, vertical

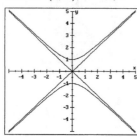

33.

$$4x^2 - 2y^2 + 8x - 8y = 8$$
$$4(x^2 + 2x) - 2(y^2 + 4y) = 8$$
$$4(x^2 + 2x + 1) - 2(y^2 + 4y + 4) = 8 + 4 - 8$$
$$4(x + 1)^2 - 2(y + 2)^2 = 4$$
$$\frac{4(x + 1)^2}{4} - \frac{2(y + 2)^2}{4} = \frac{4}{4}$$
$$\frac{(x + 1)^2}{1} - \frac{(y + 2)^2}{2} = 1$$

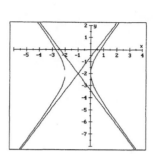

Center: $(-1, -2)$, $a = 1$, $b = \sqrt{2}$, horizontal

35.

$$y^2 - 4x^2 + 6y + 32x = 59$$
$$y^2 + 6y - 4(x^2 - 8x) = 59$$
$$y^2 + 6y + 9 - 4(x^2 - 8x + 16) = 59 + 9 - 64$$
$$(y + 3)^2 - 4(x - 4)^2 = 4$$
$$\frac{(y + 3)^2}{4} - \frac{(x - 4)^2}{1} = 1$$

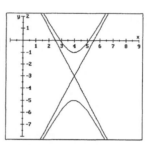

Center: $(4, -3)$, $a = 2$, $b = 1$, vertical

37. $-xy = 6$

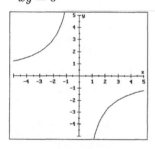

39. Foci: $(-2, 1)$, $(8, 1)$
Center: $(3, 1)$, $c = 5$
$2a = 6 \Rightarrow a = 3$, $b = 4$

$$\frac{(x - 3)^2}{9} - \frac{(y - 1)^2}{16} = 1$$

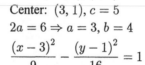

41. The distance between the point (x, y) and the line $x = -2$ is the difference between the y-coordinates, or $y - (-2) = y + 2$.

$$\sqrt{(x - 0)^2 + (y - 3)^2} = \frac{3}{2}(y + 2)$$
$$x^2 + (y - 3)^2 = \frac{9}{4}(y + 2)^2$$
$$4x^2 + 4(y - 3)^2 = 9(y + 2)^2$$
$$4x^2 + 4(y^2 - 6y + 9) = 9(y^2 + 4y + 4)$$
$$4x^2 - 5y^2 - 60y = 0$$

43. $a = 100{,}000{,}000$
$c = 200{,}000{,}000$

$$\frac{x^2}{100{,}000{,}000^2} + \frac{y^2}{200{,}000{,}000^2} = 1$$

45. $2a = 24 \Rightarrow a = 12$
$c = 13 \Rightarrow b = 5$

$$\frac{x^2}{144} - \frac{y^2}{25} = 1$$

47. Answers may vary.

49. Answers may vary.

51. $f(x) = 3x - 2$
$y = 3x - 2$
$x = 3y - 2$
$x + 2 = 3y$
$\dfrac{x + 2}{3} = y \Rightarrow f^{-1}(x) = \dfrac{x + 2}{3}$

53. $f(x) = \dfrac{5x}{x + 2}$

$y = \dfrac{5x}{x + 2}$

$x = \dfrac{5y}{y + 2}$

$x(y + 2) = 5y$
$xy + 2x = 5y$
$2x = 5y - xy$
$2x = y(5 - x)$
$\dfrac{2x}{5 - x} = y \Rightarrow f^{-1}(x) = \dfrac{2x}{5 - x}$

55. $f(g(x)) = f\big((x+1)^2\big) = \big((x+1)^2\big)^2 + 1 = (x+1)^4 + 1$

57. $f(f(x)) = f(x^2 + 1) = (x^2 + 1)^2 + 1$

Exercise 7.4 (page 550)

1. graphs

3. $\begin{cases} 8x^2 + 32y^2 = 256 \\ x = 2y \end{cases}$

5. $\begin{cases} x^2 + y^2 = 90 \\ y = x^2 \end{cases}$

7. $\begin{cases} x^2 + y^2 = 25 \\ 12x^2 + 64y^2 = 768 \end{cases}$

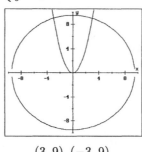

$(4, 2), (-4, 2)$

$(3, 9), (-3, 9)$

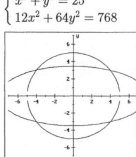

$(-4, 3), (4, 3)$
$(-4, -3), (4, -3)$

9. $\begin{cases} x^2 - 13 = -y^2 \\ y = 2x - 4 \end{cases}$ **11.** $\begin{cases} x^2 - 6x - y = -5 \\ x^2 - 6x + y = -5 \end{cases}$ **13.** $\begin{cases} y = x + 1 \\ y = x^2 + x \end{cases}$

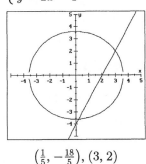

$\left(\frac{1}{5}, -\frac{18}{5}\right), (3, 2)$ $(1, 0), (5, 0)$ $(1, 2), (-1, 0)$

15. $\begin{cases} 6x^2 + 9y^2 = 10 \Rightarrow y = \pm\sqrt{\frac{10 - 6x^2}{9}} \\ 3y - 2x = 0 \quad\;\; \Rightarrow y = \frac{2}{3}x \end{cases}$

17. $\qquad 5x + 3y = 15 \Rightarrow y = \frac{15 - 5x}{3}$

$$25x^2 + 9y^2 = 225$$

$$25x^2 + 9\left(\frac{15 - 5x}{3}\right)^2 = 225$$

$$25x^2 + (15 - 5x)^2 = 225$$

$$25x^2 + 225 - 150x + 25x^2 = 225$$

$$50x^2 - 150x = 0$$

$$50x(x - 3) = 0$$

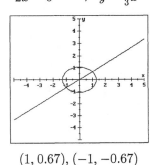

$(1, 0.67), (-1, -0.67)$

$x = 0$	$x = 3$
$y = \dfrac{15 - 5(0)}{3} = 5$	$y = \dfrac{15 - 5(3)}{3} = 0$
$(0, 5)$	$(3, 0)$

19. $x + y = 2 \Rightarrow y = 2 - x$

$$x^2 + y^2 = 2$$
$$x^2 + (2 - x)^2 = 2$$
$$x^2 + 4 - 4x + x^2 = 2$$
$$2x^2 - 4x + 2 = 0$$
$$2(x - 1)(x - 1) = 0$$

$x = 1$

$y = 2 - 1 = 1$

$(1, 1)$

21. $x + y = 3 \Rightarrow y = 3 - x$

$$x^2 + y^2 = 5$$
$$x^2 + (3 - x)^2 = 5$$
$$x^2 + 9 - 6x + x^2 = 5$$
$$2x^2 - 6x + 4 = 0$$
$$2(x - 1)(x - 2) = 0$$

$x = 1$	$x = 2$
$y = 3 - 1 = 2$	$y = 3 - 2 = 1$
$(1, 2)$	$(2, 1)$

23. $y = x^2 - 1 \Rightarrow x^2 = y + 1$

$$x^2 + y^2 = 13$$
$$y + 1 + y^2 = 13$$
$$y^2 + y - 12 = 0$$
$$(y - 3)(y + 4) = 0$$

$y = 3$	$y = -4$
$3 + 1 = x^2 \Rightarrow x = \pm 2$	$-4 + 1 = x^2$
$(2, 3), (-2, 3)$	no real solutions

25.
$$y = x^2 \Rightarrow x^2 = y$$
$$x^2 + y^2 = 30$$
$$y + y^2 = 30$$
$$y^2 + y - 30 = 0$$
$$(y - 5)(y + 6) = 0$$

$$\underline{\quad y = 5 \quad} \qquad \underline{\quad y = -6 \quad}$$
$$5 = x^2 \Rightarrow x = \pm\sqrt{5} \qquad -6 = x^2$$
$$\left(\sqrt{5}, 5\right), \left(-\sqrt{5}, 5\right) \qquad \text{no real solutions}$$

27.
$$\begin{array}{rr} x^2 + y^2 = & 13 \\ \underline{x^2 - y^2 = \quad 5} \\ 2x^2 \quad = & 18 \\ x^2 \quad = & 9 \\ x \quad = & \pm 3 \end{array} \qquad \begin{array}{r} x^2 + y^2 = 13 \\ 9 + y^2 = 13 \\ y^2 = 4 \\ y = \pm 2 \end{array}$$
$$(3, 2), (-3, 2), (3, -2), (-3, -2)$$

29.
$$\begin{array}{rr} x^2 + y^2 = & 20 \\ \underline{x^2 - y^2 = -12} \\ 2x^2 \quad = & 8 \\ x^2 \quad = & 4 \\ x \quad = & \pm 2 \end{array} \qquad \begin{array}{r} x^2 + y^2 = 20 \\ 4 + y^2 = 20 \\ y^2 = 16 \\ y = \pm 4 \end{array}$$
$$(2, 4), (-2, 4), (2, -4), (-2, -4)$$

31.
$$y = x^2 - 10 \Rightarrow x^2 = y + 10$$
$$y^2 = 40 - x^2$$
$$y^2 = 40 - y - 10$$
$$y^2 + y - 30 = 0$$
$$(y - 5)(y + 6) = 0$$

$$\underline{\quad y = 5 \quad} \qquad \underline{\quad y = -6 \quad}$$
$$x^2 = 5 + 10 \Rightarrow x = \pm\sqrt{15} \qquad x^2 = -6 + 10 \Rightarrow x = \pm 2$$
$$\left(\sqrt{15}, 5\right), \left(-\sqrt{15}, 5\right) \qquad (2, -6), (-2, -6)$$

33.
$$y = x^2 - 4 \Rightarrow x^2 = y + 4$$
$$x^2 - y^2 = -16$$
$$y + 4 - y^2 = -16$$
$$y^2 - y - 20 = 0$$
$$(y - 5)(y + 4) = 0$$

$$\underline{\quad y = 5 \quad} \qquad \underline{\quad y = -4 \quad}$$
$$x^2 = 5 + 4 \Rightarrow x = \pm 3 \qquad x^2 = -4 + 4 \Rightarrow x = 0$$
$$(3, 5), (-3, 5) \qquad (0, -4)$$

35.
$$\begin{array}{rr} x^2 - \quad y^2 = & -5 \\ \underline{3x^2 + 2y^2 = \quad 30} \\ \\ 2x^2 - 2y^2 = & -10 \\ \underline{3x^2 + 2y^2 = \quad 30} \\ 5x^2 \quad = & 20 \\ x^2 \quad = & 4 \\ x \quad = & \pm 2 \end{array} \qquad \begin{array}{r} 3x^2 + 2y^2 = 30 \\ 3(4) + 2y^2 = 30 \\ 2y^2 = 18 \\ y^2 = 9 \\ y = \pm 3 \end{array}$$
$$(2, 3), (-2, 3), (2, -3), (-2, -3)$$

37.
$$\begin{array}{r} \frac{1}{x} + \frac{2}{y} = 1 \\ \underline{\frac{2}{x} - \frac{1}{y} = \frac{1}{3}} \\ \frac{5}{x} \quad = \frac{5}{3} \\ x \quad = 3 \end{array} \qquad \begin{array}{r} \frac{1}{x} + \frac{2}{y} = 1 \\ \underline{\frac{4}{x} - \frac{2}{y} = \frac{2}{3}} \\ \frac{5}{x} \quad = \frac{5}{3} \\ x \quad = 3 \\ (3, 3) \end{array} \qquad \begin{array}{r} \frac{1}{x} + \frac{2}{y} = 1 \\ \frac{1}{3} + \frac{2}{y} = 1 \\ \frac{2}{y} = \frac{2}{3} \\ y = 3 \end{array}$$

39.

$$3y^2 = xy$$
$$3y^2 - xy = 0$$
$$y(3y - x) = 0$$
$$y = 0 \text{ or } x = 3y$$

$y = 0$
$2x^2 + xy - 84 = 0$

$$2x^2 + x(0) - 84 = 0$$
$$2x^2 = 84$$
$$x^2 = 42$$
$$x = \pm\sqrt{42}$$
$$\left(\sqrt{42}, 0\right), \left(-\sqrt{42}, 0\right)$$

$x = 3y$
$2x^2 + xy - 84 = 0$

$$2(3y)^2 + (3y)y - 84 = 0$$
$$18y^2 + 3y^2 - 84 = 0$$
$$21y^2 = 84$$
$$y^2 = 4 \Rightarrow y = \pm 2$$
$$(6, 2), (-6, -2)$$

41.

$$xy = \tfrac{1}{6} \Rightarrow y = \tfrac{1}{6x}$$
$$y + x = 5xy$$
$$\frac{1}{6x} + x = \frac{5x}{6x}$$
$$1 + 6x^2 = 5x$$
$$6x^2 - 5x + 1 = 0$$
$$(2x - 1)(3x - 1) = 0$$

$x = \tfrac{1}{2}$	$x = \tfrac{1}{3}$
$y = \dfrac{1}{6(1/2)} = \dfrac{1}{3}$	$y = \dfrac{1}{6(1/3)} = \dfrac{1}{2}$
$\left(\tfrac{1}{2}, \tfrac{1}{3}\right)$	$\left(\tfrac{1}{3}, \tfrac{1}{2}\right)$

43. Let $x = $ width and $y = $ length.

$$\begin{cases} xy = 63 \\ 2x + 2y = 32 \end{cases}$$
$$xy = 63 \Rightarrow y = \tfrac{63}{x}$$

$$2x + 2y = 32$$
$$2x + 2\left(\frac{63}{x}\right) = 32$$
$$2x^2 + 126 = 32x$$
$$2x^2 - 32x + 126 = 0$$
$$2(x - 9)(x - 7) = 0$$

$x = 9$	$x = 7$
$y = \tfrac{63}{9} = 7$	$y = \tfrac{63}{7} = 9$

The dimensions are 9 cm by 7 cm.

45. Let $x = $ Carol's principal.

$$\boxed{\text{John's rate}} = \boxed{\text{Carol's rate}} + 0.015$$
$$\frac{94.50}{x + 150} = \frac{67.50}{x} + 0.015$$

	I	P	r
Carol	67.50	x	$\frac{67.50}{x}$
John	94.50	$x + 150$	$\frac{94.50}{x+150}$

$$94.5x = 67.5(x + 150) + 0.015x(x + 150)$$
$$0.015x^2 - 24.75x + 10{,}125 = 0$$
$$x^2 - 1650x + 675{,}000 = 0$$
$$(x - 750)(x - 900) = 0$$

$$x - 750 = 0 \quad \textbf{or} \quad x - 900 = 0$$
$$x = 750 \qquad\qquad x = 900 \Rightarrow$$

Carol invested either $750 at 9% or she invested $900 at 7.5% interest.

47. $\begin{cases} x = 2y \\ (x - 120)^2 + y^2 = 100^2 \end{cases}$

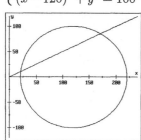

The x-coordinate of the point where the line crosses the circle closest to the origin has the approximate coordinates $(20.5, 10.25)$. The distance to the origin is about 23 miles.

49-53. **Answers may vary.**

55. $y = \dfrac{3x + 1}{x - 1}$

Vertical: $x = 1$
Horizontal: $y = 3$
Slant: none

57. $y = \dfrac{3x + 1}{x^2 - 1}$
Vertical: $x = 1, x = -1$
Horizontal: $y = 0$
Slant: none

59. $y = \dfrac{3x^2 + 5}{5x^2 + 3}$
y-axis

61. $y = \dfrac{3x^3}{5x^2 + 3}$
origin

Chapter 7 Summary (page 554)

1. **a.** $(x - h)^2 + (y - k)^2 = r^2$
$(x - 0)^2 + (y - 0)^2 = 4^2$
$x^2 + y^2 = 16$

b. $r = \sqrt{(6 - 0)^2 + (8 - 0)^2} = 10$
$(x - h)^2 + (y - k)^2 = r^2$
$(x - 0)^2 + (y - 0)^2 = 10^2$
$x^2 + y^2 = 100$

c. $(x - h)^2 + (y - k)^2 = r^2$
$(x - 3)^2 + (y - (-2))^2 = 5^2$
$(x - 3)^2 + (y + 2)^2 = 25$

d. $r = \sqrt{(-2 - 1)^2 + (4 - 0)^2} = 5$
$(x - h)^2 + (y - k)^2 = r^2$
$(x - (-2))^2 + (y - 4)^2 = 5^2$
$(x + 2)^2 + (y - 4)^2 = 25$

e. $C\left(\dfrac{-2+12}{2}, \dfrac{4+16}{2}\right) = C(5, 10)$

$r = \sqrt{(12-5)^2 + (16-10)^2} = \sqrt{85}$

$(x-h)^2 + (y-k)^2 = r^2$

$(x-5)^2 + (y-10)^2 = \left(\sqrt{85}\right)^2$

$(x-5)^2 + (y-10)^2 = 85$

f. $C\left(\dfrac{-3+7}{2}, \dfrac{-6+10}{2}\right) = C(2, 2)$

$r = \sqrt{(7-2)^2 + (10-2)^2} = \sqrt{89}$

$(x-h)^2 + (y-k)^2 = r^2$

$(x-2)^2 + (y-2)^2 = \left(\sqrt{89}\right)^2$

$(x-2)^2 + (y-2)^2 = 89$

2. a.

$x^2 + y^2 - 6x + 4y = 3$

$x^2 - 6x + y^2 + 4y = 3$

$x^2 - 6x + 9 + y^2 + 4y + 4 = 3 + 9 + 4$

$(x-3)^2 + (y+2)^2 = 16$

b.

$x^2 + 4x + y^2 - 10y = -13$

$x^2 + 4x + 4 + y^2 - 10y + 25 = -13 + 4 + 25$

$(x+2)^2 + (y-5)^2 = 16$

3. a.   Horizontal

$(y-0)^2 = 4p(x-0)$

$(4-0)^2 = 4p(-8-0)$

$16 = -32p$

$-2 = 4p$

$y^2 = -2x$

b.   Vertical

$(x-0)^2 = 4p(y-0)$

$(-8-0)^2 = 4p(4-0)$

$64 = 16p$

$16 = 4p$

$x^2 = 16y$

4.   Vertical

$(x+2)^2 = 4p(y-3)$

$(-4+2)^2 = 4p(-8-3)$

$4 = 4p(-11)$

$-\dfrac{4}{11} = 4p$

$(x+2)^2 = -\dfrac{4}{11}(y-3)$

5.   **a.** $x^2 - 4y - 2x + 9 = 0$

$$x^2 - 2x = 4y - 9$$
$$x^2 - 2x + 1 = 4y - 9 + 1$$
$$(x-1)^2 = 4(y-2)$$

b. $y^2 - 6y = 4x - 13$

$$y^2 - 6y + 9 = 4x - 13 + 9$$
$$(y-3)^2 = 4(x-1)$$

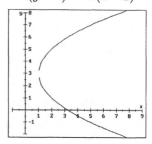

6. $a = 6, b = 4$, horizontal

$$\frac{x^2}{36} + \frac{y^2}{16} = 1$$

7. $a = 5, b = 2$, vertical

$$\frac{y^2}{25} + \frac{x^2}{4} = 1$$

8. $a = 4, b = 3$, horizontal $\Rightarrow \dfrac{(x+2)^2}{16} + \dfrac{(y-3)^2}{9} = 1$

9.

$$4x^2 + y^2 - 16x + 2y = -13$$
$$4\left(x^2 - 4x\right) + y^2 + 2y = -13$$
$$4\left(x^2 - 4x + 4\right) + y^2 + 2y + 1 = -13 + 16 + 1$$
$$4(x-2)^2 + (y+1)^2 = 4$$
$$\frac{(x-2)^2}{1} + \frac{(y+1)^2}{4} = 1$$

Center: $(2, -1)$, $a = 2, b = 1$, vertical

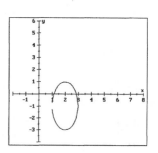

10. $a = 2, c = 4$; horizontal

$$b^2 = c^2 - a^2 = 16 - 4 = 12$$
$$\frac{x^2}{4} - \frac{y^2}{12} = 1$$

11. $a = 3, c = 5$; vertical

$$b^2 = c^2 - a^2 = 25 - 9 = 16$$
$$\frac{y^2}{9} - \frac{x^2}{16} = 1$$

12. $C(0, 3), a = 3, c = 5$; horizontal

$$b^2 = c^2 - a^2 = 25 - 9 = 16$$
$$\frac{x^2}{9} - \frac{(y-3)^2}{16} = 1$$

13. $C(3, 0), a = 3, c = 5$; vertical

$$b^2 = c^2 - a^2 = 25 - 9 = 16$$
$$\frac{y^2}{9} - \frac{(x-3)^2}{16} = 1$$

14. $y = \pm\dfrac{b}{a}x \Rightarrow y = \pm\dfrac{4}{5}x$

15.

$$9x^2 - 4y^2 - 16y - 18x = 43$$
$$9(x^2 - 2x) - 4(y^2 + 4y) = 43$$
$$9(x^2 - 2x + 1) - 4(y^2 + 4y + 4) = 43 + 9 - 16$$
$$9(x - 1)^2 - 4(y + 2)^2 = 36$$
$$\frac{(x - 1)^2}{4} - \frac{(y + 2)^2}{9} = 1$$

Center: $(1, -2)$, $a = 2$, $b = 3$, horizontal

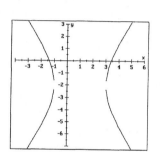

16. $4xy = 1$

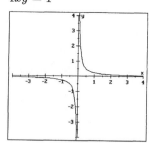

17. $\begin{cases} x^2 + y^2 = 16 \\ y = x + 4 \end{cases}$

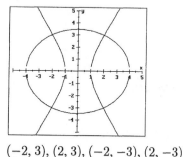

$(-4, 0), (0, 4)$

18. $\begin{cases} 3x^2 + y^2 = 52 \\ x^2 - y^2 = 12 \end{cases}$

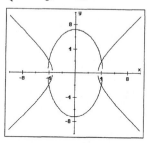

$(-4, 2), (4, 2), (-4, -2), (4, -2)$

19. $\begin{cases} \dfrac{x^2}{16} + \dfrac{y^2}{12} = 1 \\ x^2 - \dfrac{y^2}{3} = 1 \end{cases}$

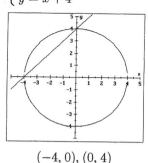

$(-2, 3), (2, 3), (-2, -3), (2, -3)$

20.

$$\begin{aligned} 3x^2 + y^2 &= 52 \\ \underline{x^2 - y^2 = 12} \\ 4x^2 &= 64 \\ x^2 &= 16 \\ x &= \pm 4 \end{aligned}$$

$$\begin{aligned} 3x^2 + y^2 &= 52 \\ 3(16) + y^2 &= 52 \\ y^2 &= 4 \\ y &= \pm 2 \end{aligned}$$

$(4, 2), (-4, 2), (4, -2), (-4, -2)$

21. $-\sqrt{3}y + 4\sqrt{3} = 3x \Rightarrow x = \dfrac{-\sqrt{3}y + 4\sqrt{3}}{3}$

$x^2 + y^2 = 16$

$\left(\dfrac{-\sqrt{3}y + 4\sqrt{3}}{3}\right)^2 + y^2 = 16$

$3y^2 - 24y + 48 + 9y^2 = 144$

$12y^2 - 24y - 96 = 0$

$12(y - 4)(y + 2) = 0$

$\underline{\hspace{4cm} y = 4 \hspace{4cm}}$

$x = \dfrac{-\sqrt{3}(4) + 4\sqrt{3}}{3} = 0 \Rightarrow (0, 4)$

$\underline{\hspace{4cm} y = -2 \hspace{4cm}}$

$x = \dfrac{-\sqrt{3}(-2) + 4\sqrt{3}}{3} = 2\sqrt{3} \Rightarrow \left(2\sqrt{3}, -2\right)$

22. $\dfrac{x^2}{16} + \dfrac{y^2}{12} = 1 \qquad 5y^2 = 45 \qquad 3x^2 - y^2 = 3$

$\underline{x^2 - \dfrac{y^2}{3} = 1} \qquad y^2 = 9 \qquad 3x^2 - 9 = 3$

$\qquad\qquad\qquad\qquad\quad y = \pm 3 \qquad 3x^2 = 12$

$3x^2 + 4y^2 = 48 \qquad\qquad\qquad\qquad x^2 = 4$

$\underline{3x^2 - y^2 = 3} \qquad\qquad x = \pm 2 \Rightarrow (2, 3), (-2, 3), (2, -3), (-2, -3)$

$\qquad\quad 5y^2 = 45$

Chapter 7 Test (page 558)

1. $(x - h)^2 + (y - k)^2 = r^2$

$(x - 2)^2 + (y - 3)^2 = 3^2$

$(x - 2)^2 + (y - 3)^2 = 9$

2. $C\left(\dfrac{-2 + 6}{2}, \dfrac{-2 + 8}{2}\right) = C(2, 3)$

$r = \sqrt{(6 - 2)^2 + (8 - 3)^2} = \sqrt{41}$

$(x - h)^2 + (y - k)^2 = r^2$

$(x - 2)^2 + (y - 3)^2 = \left(\sqrt{41}\right)^2$

$(x - 2)^2 + (y - 3)^2 = 41$

3. $r = \sqrt{(7 - 2)^2 + (7 - (-5))^2} = 13$

$(x - h)^2 + (y - k)^2 = r^2$

$(x - 2)^2 + (y - (-5))^2 = 13^2$

$(x - 2)^2 + (y + 5)^2 = 169$

4. $x^2 + y^2 - 4x + 6y + 4 = 0$

$x^2 - 4x + y^2 + 6y = -4$

$x^2 - 4x + 4 + y^2 + 6y + 9 = -4 + 4 + 9$

$(x - 2)^2 + (y + 3)^2 = 9$

$C(2, -3), r = 3$

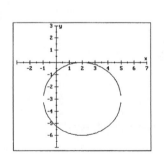

5. Vertical (up), $p = 4$

$(x - h)^2 = 4p(y - k)$

$(x - 3)^2 = 4(4)(y - 2)$

$(x - 3)^2 = 16(y - 2)$

6. Horizontal

$(y + 6)^2 = 4p(x - 4)$

$(-4 + 6)^2 = 4p(3 - 4)$

$4 = -4p$

$-4 = 4p$

$(y + 6)^2 = -4(x - 4)$

7. $(x - 2)^2 = 4p(y + 3)$ **OR** $(y + 3)^2 = 4p(x - 2)$

$(0 - 2)^2 = 4p(0 + 3)$ $\qquad$ $(0 + 3)^2 = 4p(0 - 2)$

$4 = 4p(3)$ $\qquad\qquad$ $9 = 4p(-2)$

$\dfrac{4}{3} = 4p$ $\qquad\qquad$ $-\dfrac{9}{2} = 4p$

$(x - 2)^2 = \frac{4}{3}(y + 3)$ $\qquad$ $(y + 3)^2 = -\frac{9}{2}(x - 2)$

8. $x^2 - 6x - 8y = 7$

$x^2 - 6x = 8y + 7$

$x^2 - 6x + 9 = 8y + 7 + 9$

$(x - 3)^2 = 8(y + 2)$

Vertex: $(3, -2)$, vertical

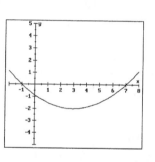

9. $a = 10, c = 6$, horizontal

$b^2 = a^2 - c^2$

$= 100 - 36 = 64$

$\dfrac{x^2}{100} + \dfrac{y^2}{64} = 1$

10. $b = 12, c = 5$, horizontal

$a^2 = b^2 + c^2$

$= 144 + 25 = 169$

$\dfrac{x^2}{169} + \dfrac{y^2}{144} = 1$

11. $a = 6, b = 2$, vertical

$\dfrac{x^2}{4} + \dfrac{y^2}{36} = 1$

12. $9x^2 + 4y^2 - 18x - 16y - 11 = 0$

$9(x^2 - 2x) + 4(y^2 - 4y) = 11$

$9(x^2 - 2x + 1) + 4(y^2 - 4y + 4) = 11 + 9 + 16$

$9(x - 1)^2 + 4(y - 2)^2 = 36$

$\dfrac{(x - 1)^2}{4} + \dfrac{(y - 2)^2}{9} = 1$

Center: $(1, 2)$, $a = 3, b = 2$, vertical

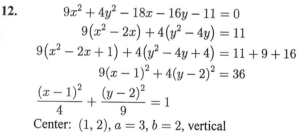

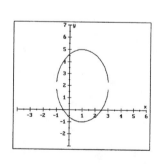

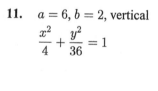

13. $a = 5, c = 13$; horizontal
$b^2 = c^2 - a^2$
$\qquad = 169 - 25 = 144$
$$\frac{x^2}{25} - \frac{y^2}{144} = 1$$

14. $C(0, 0), a = 6, c = \frac{13}{2}$
horizontal
$b^2 = c^2 - a^2$
$\qquad = \frac{169}{4} - 36 = \frac{25}{4}$
$$\frac{x^2}{36} - \frac{4y^2}{25} = 1$$

15. $a = 8, c = 10$; horizontal
$b^2 = c^2 - a^2$
$\qquad = 100 - 64 = 36$
$$\frac{(x-2)^2}{64} - \frac{(y+1)^2}{36} = 1$$

16.
$$x^2 - 4y^2 + 16y = 8$$
$$x^2 - 4\left(y^2 - 4y\right) = 8$$
$$x^2 - 4\left(y^2 - 4y + 4\right) = 8 - 16$$
$$x^2 - 4(y - 2)^2 = -8$$
$$\frac{(y-2)^2}{2} - \frac{x^2}{8} = 1$$
Center: $(0, 2)$, $a = \sqrt{2}, b = \sqrt{8}$, vertical

17. $y = x^2 - 3 \Rightarrow x^2 = y + 3$
$\qquad x^2 + y^2 = 23$
$\qquad y + 3 + y^2 = 23$
$\qquad y^2 + y - 20 = 0$
$\qquad (y - 4)(y + 5) = 0$

$\dfrac{y = 4}{4 + 3 = x^2 \Rightarrow x = \pm\sqrt{7}}$
$$\left(\sqrt{7}, 4\right), \left(-\sqrt{7}, 4\right)$$

$\dfrac{y = -5}{-5 + 3 = x^2}$
no real solutions

18. $x^2 + y^2 = 27 \Rightarrow x^2 = 27 - y^2$
$\qquad 2x^2 - 3y^2 = 9$
$\qquad 2\left(27 - y^2\right) - 3y^2 = 9$
$\qquad 54 - 2y^2 - 3y^2 = 9$
$\qquad\qquad 45 = 5y^2$
$\qquad\qquad 9 = y^2$

$\dfrac{y = 3}{x^2 = 27 - 3^2 \Rightarrow x = \pm 3\sqrt{2}}$
$$\left(3\sqrt{2}, 3\right), \left(-3\sqrt{2}, 3\right)$$

$\dfrac{y = -3}{x^2 = 27 - (-3)^2 \Rightarrow x = \pm 3\sqrt{2}}$
$$\left(3\sqrt{2}, -3\right), \left(-3\sqrt{2}, -3\right)$$

19. $y^2 - 4y - 6x - 14 = 0$
$$y^2 - 4y = 6x + 14$$
$$y^2 - 4x + 4 = 6x + 14 + 4$$
$$(y - 2)^2 = 6(x + 3) \Rightarrow \text{Parabola}$$

20.
$$2x^2 + 3y^2 - 4x + 12y + 8 = 0$$
$$2\left(x^2 - 2x\right) + 3\left(y^2 + 4y\right) = -8$$
$$2\left(x^2 - 2x + 1\right) + 3\left(y^2 + 4y + 4\right) = -8 + 2 + 12$$
$$2(x - 1)^2 + 3(y + 2)^2 = 6$$
$$\frac{(x-1)^2}{3} + \frac{(y+2)^2}{2} = 1 \Rightarrow \text{ellipse}$$

Cumulative Review Exercises (page 560)

1. $64^{2/3} = \left(64^{1/3}\right)^2 = 4^2 = 16$

2. $8^{-1/3} = \dfrac{1}{8^{1/3}} = \dfrac{1}{2}$

3. $\dfrac{y^{2/3}y^{5/3}}{y^{1/3}} = \dfrac{y^{7/3}}{y^{1/3}} = y^{6/3} = y^2$

4. $\dfrac{x^{5/3}x^{1/2}}{x^{3/4}} = \dfrac{x^{13/6}}{x^{3/4}} = x^{34/24} = x^{17/12}$

5. $\left(x^{2/3} - x^{1/3}\right)\left(x^{2/3} + x^{1/3}\right) = x^{4/3} + x^{3/3} - x^{3/3} - x^{2/3} = x^{4/3} - x^{2/3}$

6. $\left(x^{-1/2} + x^{1/2}\right)^2 = \left(x^{-1/2} + x^{1/2}\right)\left(x^{-1/2} + x^{1/2}\right) = x^{-2/2} + x^0 + x^0 + x^{2/2} = \dfrac{1}{x} + 2 + x$

7. $\sqrt[3]{-27x^3} = \sqrt[3]{(-3x)^3} = -3x$

8. $\sqrt{48t^3} = \sqrt{16t^2}\sqrt{3t} = 4t\sqrt{3t}$

9. $\sqrt[3]{\dfrac{128x^4}{2x}} = \sqrt[3]{64x^3} = 4x$

10. $\sqrt{x^2 + 6x + 9} = \sqrt{(x+3)^2} = x + 3$

11. $\sqrt{50} - \sqrt{8} + \sqrt{32} = 5\sqrt{2} - 2\sqrt{2} + 4\sqrt{2} = 7\sqrt{2}$

12. $-3\sqrt[4]{32} - 2\sqrt[4]{162} + 5\sqrt[4]{48} = -3 \cdot 2\sqrt[4]{2} - 2 \cdot 3\sqrt[4]{2} + 5 \cdot 2\sqrt[4]{3} = -12\sqrt[4]{2} + 10\sqrt[4]{3}$

13. $3\sqrt{2}\left(2\sqrt{3} - 4\sqrt{12}\right) = 6\sqrt{6} - 12\sqrt{24} = 6\sqrt{6} - 12 \cdot 2\sqrt{6} = -18\sqrt{6}$

14. $\dfrac{5}{\sqrt[3]{x}} = \dfrac{5\sqrt[3]{x^2}}{\sqrt[3]{x}\sqrt[3]{x^2}} = \dfrac{5\sqrt[3]{x^2}}{x}$

15. $\dfrac{\sqrt{x}+2}{\sqrt{x}-1} = \dfrac{\left(\sqrt{x}+2\right)\left(\sqrt{x}+1\right)}{\left(\sqrt{x}-1\right)\left(\sqrt{x}+1\right)}$

$= \dfrac{x + 3\sqrt{x} + 2}{x - 1}$

16. $\sqrt[6]{x^3y^3} = \left(x^3y^3\right)^{1/6} = x^{3/6}y^{3/6} = x^{1/2}y^{1/2} = (xy)^{1/2} = \sqrt{xy}$

17.
$$5\sqrt{x+2} = x + 8$$
$$\left(5\sqrt{x+2}\right)^2 = (x+8)^2$$
$$25(x+2) = x^2 + 16x + 64$$
$$25x + 50 = x^2 + 16x + 64$$
$$0 = x^2 - 9x + 14$$
$$0 = (x-2)(x-7)$$
$$x = 2 \text{ or } x = 7 \quad (\text{both check})$$

18.
$$\sqrt{x} + \sqrt{x+2} = 2$$
$$\sqrt{x+2} = 2 - \sqrt{x}$$
$$\left(\sqrt{x+2}\right)^2 = \left(2 - \sqrt{x}\right)^2$$
$$x + 2 = 4 - 4\sqrt{x} + x$$
$$4\sqrt{x} = 2$$
$$\left(4\sqrt{x}\right)^2 = 2^2$$
$$16x = 4$$
$$x = \dfrac{1}{4}$$

19. $\quad 2x^2 + x - 3 = 0$

$$x^2 + \frac{1}{2}x = \frac{3}{2}$$

$$x^2 + \frac{1}{2}x + \frac{1}{16} = \frac{24}{16} + \frac{1}{16}$$

$$\left(x + \tfrac{1}{4}\right)^2 = \tfrac{25}{16}$$

$$x + \tfrac{1}{4} = \pm\tfrac{5}{4}$$

$$x = -\tfrac{1}{4} \pm \tfrac{5}{4}$$

$$x = 1 \ \text{ or } \ x = -\tfrac{3}{2}$$

20. $\quad 3x^2 + 4x - 1 = 0 \Rightarrow a = 3, b = 4, c = -1$

$$x = \frac{-b \pm \sqrt{b^2 - 4ac}}{2a}$$

$$= \frac{-4 \pm \sqrt{4^2 - 4(3)(-1)}}{2(3)}$$

$$= \frac{-4 \pm \sqrt{16 + 12}}{6}$$

$$= \frac{-4 \pm \sqrt{28}}{6} = \frac{-2 \pm \sqrt{7}}{3}$$

21. $\quad (3 + 5i) + (4 - 3i) = 3 + 5i + 4 - 3i = 7 + 2i$

22. $\quad (7 - 4i) - (12 + 3i) = 7 - 4i - 12 - 3i = -5 - 7i$

23. $\quad (2 - 3i)(2 + 3i) = 4 + 6i - 6i - 9i^2 = 4 - 9(-1) = 4 + 9 = 13 + 0i$

24. $\quad (3 + i)(3 - 3i) = 9 - 9i + 3i - 3i^2 = 9 - 6i - 3(-1) = 9 - 6i + 3 = 12 - 6i$

25. $\quad (3 - 2i) - (4 + i)^2 = 3 - 2i - \left(16 + 8i + i^2\right) = 3 - 2i - (15 + 8i) = 3 - 2i - 15 - 8i$

$$= -12 - 10i$$

26. $\quad \dfrac{5}{3 - i} = \dfrac{5(3 + i)}{(3 - i)(3 + i)} = \dfrac{5(3 + i)}{9 - i^2} = \dfrac{5(3 + i)}{10} = \dfrac{3 + i}{2} = \dfrac{3}{2} + \dfrac{1}{2}i$

27. $\quad |3 + 2i| = \sqrt{3^2 + 2^2} = \sqrt{13}$

28. $\quad |5 - 6i| = \sqrt{5^2 + (-6)^2} = \sqrt{61}$

29. $\quad 2x^2 + 4x = k \Rightarrow 2x^2 + 4x - k = 0$

$\quad a = 2, b = 4, c = -k$: Set $b^2 - 4ac = 0$.

$$b^2 - 4ac = 0$$

$$4^2 - 4(2)(-k) = 0$$

$$16 + 8k = 0$$

$$k = -2$$

30. $\quad y = \dfrac{1}{2}x^2 - x + 1: \ a = \dfrac{1}{2}, b = -1, c = 1$

$$x = -\frac{b}{2a} = -\frac{-1}{2\left(\tfrac{1}{2}\right)} = 1$$

$$y = \frac{1}{2}(1)^2 - 1 + 1 = \frac{1}{2}$$

31. $\quad x^2 - x - 6 > 0$

$\quad (x + 2)(x - 3) > 0$

$\quad$ factors $= 0$: $\ x = -2, x = 3$

$\quad$ intervals: $(-\infty, -2), (-2, 3), (3, \infty)$

interval	test number	value of x^2-x-6
$(-\infty, -2)$	-3	$+6$
$(-2, 3)$	0	-6
$(3, \infty)$	4	$+6$

Solution: $(-\infty, -2) \cup (3, \infty)$

32. $\quad x^2 - x - 6 \leq 0$

$\quad (x + 2)(x - 3) \leq 0$

$\quad$ factors $= 0$: $\ x = -2, x = 3$

$\quad$ intervals: $(-\infty, -2), (-2, 3), (3, \infty)$

interval	test number	value of x^2-x-6
$(-\infty, -2)$	-3	$+6$
$(-2, 3)$	0	-6
$(3, \infty)$	4	$+6$

Solution: $[-2, 3]$

33. $f(-1) = 3(-1)^2 + 2 = 3 + 2 = 5$

34. $(g \circ f)(2) = g(f(2)) = g\big(3(2)^2 + 2\big)$
$$= g(14)$$
$$= 2(14) - 1 = 27$$

35. $(f \circ g)(x) = f(g(x))$
$$= f(2x - 1)$$
$$= 3(2x - 1)^2 + 2$$
$$= 3\big(4x^2 - 4x + 1\big) + 2$$
$$= 12x^2 - 12x + 3 + 2$$
$$= 12x^2 - 12x + 5$$

36. $(g \circ f)(x) = g(f(x))$
$$= g\big(3x^2 + 2\big)$$
$$= 2\big(3x^2 + 2\big) - 1$$
$$= 6x^2 + 4 - 1$$
$$= 6x^2 + 3$$

37. $y = \log_2 x \Rightarrow 2^y = x$

38. $3^b = a \Rightarrow \log_3 a = b$

39. $\log_x 25 = 2 \Rightarrow x^2 = 25 \Rightarrow x = 5$

40. $\log_5 125 = x \Rightarrow 5^x = 125 \Rightarrow x = 3$

41. $\log_3 x = -3 \Rightarrow 3^{-3} = x \Rightarrow x = \dfrac{1}{27}$

42. $\log_5 x = 0 \Rightarrow 5^0 = x \Rightarrow x = 1$

43. $y = \log_2 x$; inverse: $y = 2^x$

44. $\log_{10} 10^x = x$, so $y = x$.

45. $\log 98 = \log(14 \cdot 7) = \log 14 + \log 7 = 1.1461 + 0.8451 = 1.9912$

46. $\log 2 = \log \dfrac{14}{7} = \log 14 - \log 7 = 1.1461 - 0.8451 = 0.3010$

47. $\log 49 = \log 7^2 = 2\log 7 = 2(0.8451) = 1.6902$

48. $\log \dfrac{7}{5} = \log \dfrac{7}{10/2} = \log 7 - \log \dfrac{10}{2} = \log 7 - (\log 10 - \log 2) = \log 7 - \log 10 + \log 2$
$$= 0.8451 - 1 + 0.3010 = 0.1461$$

49.
$$2^{x+2} = 3^x$$
$$\log 2^{x+2} = \log 3^x$$
$$(x+2)\log 2 = x\log 3$$
$$x\log 2 + 2\log 2 = x\log 3$$
$$2\log 2 = x\log 3 - x\log 2$$
$$2\log 2 = x(\log 3 - \log 2)$$
$$\dfrac{2\log 2}{\log 3 - \log 2} = x$$

50. $2\log 5 + \log x - \log 4 = 2$
$$\log 5^2 + \log x - \log 4 = 2$$
$$\log \dfrac{25x}{4} = 2$$
$$10^2 = \dfrac{25x}{4}$$
$$400 = 25x$$
$$16 = x$$

51. $A = A_0 \left(1 + \dfrac{r}{k}\right)^{kt}$
$$= 9000\left(1 + \dfrac{-0.12}{1}\right)^{1(9)}$$
$$\approx \$2848.31$$

52. $\log_6 8 = \dfrac{\log 8}{\log 6} \approx 1.16056$

53. $\begin{cases} 2x + y = 5 \\ x - 2y = 0 \end{cases}$

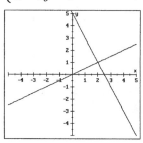

solution: $(2, 1)$

54. $\begin{cases} (1) & 3x + y = 4 \\ (2) & 2x - 3y = -1 \end{cases}$

Substitute $y = -3x + 4$ from (1) into (2):
$$2x - 3y = -1$$
$$2x - 3(-3x + 4) = -1$$
$$2x + 9x - 12 = -1$$
$$11x = 11$$
$$x = 1$$

Substitute and solve for y:
$$y = -3x + 4 = -3(1) + 4 = 1$$
$$\boxed{x = 1, y = 1}$$

55.
$$\begin{array}{l} x + 2y = -2 \\ 2x - y = 6 \end{array} \Rightarrow \times 2$$
$$\begin{array}{l} x + 2y = -2 \\ 4x - 2y = 12 \\ \hline 5x = 10 \\ x = 2 \end{array}$$
$$\begin{array}{l} 2x - y = 6 \\ 2(2) - y = 6 \\ -y = 2 \\ y = -2 \end{array}$$

Solution:
$$\boxed{x = 2, y = -2}$$

56.
$$\dfrac{x}{10} + \dfrac{y}{5} = \dfrac{1}{2} \Rightarrow \times 10 \quad x + 2y = 5$$
$$\dfrac{x}{2} - \dfrac{y}{5} = \dfrac{13}{10} \Rightarrow \times 10 \quad 5x - 2y = 13$$
$$\begin{array}{l} \hline 6x = 18 \\ x = 3 \end{array}$$
$$\begin{array}{l} x + 2y = 5 \\ 3 + 2y = 5 \\ 2y = 2 \\ y = 1 \end{array}$$

Solution:
$$\boxed{x = 3, y = 1}$$

57. $\begin{vmatrix} 3 & -2 \\ 1 & -1 \end{vmatrix} = 3(-1) - (-2)1$
$$= -3 + 2 = -1$$

58. $y = \dfrac{\begin{vmatrix} 4 & -1 \\ 3 & -7 \end{vmatrix}}{\begin{vmatrix} 4 & -3 \\ 3 & 4 \end{vmatrix}} = \dfrac{-25}{25} = -1$

59. $\begin{bmatrix} x \\ y \\ z \end{bmatrix} = \begin{bmatrix} 1 & 1 & 1 \\ 2 & -1 & -1 \\ 1 & -2 & 1 \end{bmatrix}^{-1} \begin{bmatrix} 1 \\ -4 \\ 4 \end{bmatrix} = \begin{bmatrix} \frac{1}{3} & \frac{1}{3} & 0 \\ \frac{1}{3} & 0 & -\frac{1}{3} \\ \frac{1}{3} & -\frac{1}{3} & \frac{1}{3} \end{bmatrix} \begin{bmatrix} 1 \\ -4 \\ 4 \end{bmatrix} = \begin{bmatrix} -1 \\ -1 \\ 3 \end{bmatrix}$

60. $\begin{bmatrix} x \\ y \\ z \end{bmatrix} = \begin{bmatrix} 1 & 2 & 3 \\ 3 & 2 & 1 \\ 2 & 3 & 1 \end{bmatrix}^{-1} \begin{bmatrix} 6 \\ 6 \\ 6 \end{bmatrix} = \begin{bmatrix} -\frac{1}{12} & \frac{7}{12} & -\frac{1}{3} \\ -\frac{1}{12} & -\frac{5}{12} & \frac{2}{3} \\ \frac{5}{12} & \frac{1}{12} & -\frac{1}{3} \end{bmatrix} \begin{bmatrix} 6 \\ 6 \\ 6 \end{bmatrix} = \begin{bmatrix} 1 \\ 1 \\ 1 \end{bmatrix}$

Exercise 8.1 (page 568)

1. power

3. first

5. $7 \cdot 6 \cdot 5 \cdot 4 \cdot 3 \cdot 2 \cdot 1$

7. $(n-1)!$

9. $4! = 4 \cdot 3 \cdot 2 \cdot 1 = 24$

11. $3! \cdot 6! = 6 \cdot 720 = 4320$

13. $6! + 6! = 720 + 720 = 1440$

15. $\dfrac{9!}{12!} = \dfrac{9!}{12 \cdot 11 \cdot 10 \cdot 9!} = \dfrac{1}{1320}$

17. $\dfrac{5! \cdot 7!}{9!} = \dfrac{5! \cdot 7!}{9 \cdot 8 \cdot 7!} = \dfrac{120}{72} = \dfrac{5}{3}$

19. $\dfrac{18!}{6!(18-6)!} = \dfrac{18!}{6! \cdot 12!} = \dfrac{18 \cdot 17 \cdot 16 \cdot 15 \cdot 14 \cdot 13 \cdot 12!}{6! \cdot 12!} = \dfrac{13{,}366{,}080}{720} = 18{,}564$

21. $(a+b)^3 = a^3 + \dfrac{3!}{1!2!}a^2 b + \dfrac{3!}{2!1!}ab^2 + b^3 = a^3 + 3a^2 b + 3ab^2 + b^3$

23. $(a-b)^5 = a^5 + \dfrac{5!}{1!4!}a^4(-b) + \dfrac{5!}{2!3!}a^3(-b)^2 + \dfrac{5!}{3!2!}a^2(-b)^3 + \dfrac{5!}{4!1!}a(-b)^4 + (-b)^5$

$\qquad = a^5 - 5a^4 b + 10a^3 b^2 - 10a^2 b^3 + 5ab^4 - b^5$

25. $(2x+y)^3 = (2x)^3 + \dfrac{3!}{1!2!}(2x)^2 y + \dfrac{3!}{2!1!}(2x)y^2 + y^3 = 8x^3 + 12x^2 y + 6xy^2 + y^3$

27. $(x-2y)^3 = x^3 + \dfrac{3!}{1!2!}x^2(-2y) + \dfrac{3!}{2!1!}x(-2y)^2 + (-2y)^3 = x^3 - 6x^2 y + 12xy^2 - 8y^3$

29. $(2x+3y)^4 = (2x)^4 + \dfrac{4!}{1!3!}(2x)^3(3y) + \dfrac{4!}{2!2!}(2x)^2(3y)^2 + \dfrac{4!}{3!1!}(2x)(3y)^3 + (3y)^4$

$\qquad = 16x^4 + 96x^3 y + 216x^2 y^2 + 216xy^3 + 81y^4$

31. $(x-2y)^4 = x^4 + \dfrac{4!}{1!3!}x^3(-2y) + \dfrac{4!}{2!2!}x^2(-2y)^2 + \dfrac{4!}{3!1!}x(-2y)^3 + (-2y)^4$

$\qquad = x^4 - 8x^3 y + 24x^2 y^2 - 32xy^3 + 16y^4$

33. $(x-3y)^5 = x^5 + \dfrac{5!}{1!4!}x^4(-3y) + \dfrac{5!}{2!3!}x^3(-3y)^2 + \dfrac{5!}{3!2!}x^2(-3y)^3 + \dfrac{5!}{4!1!}x(-3y)^4 + (-3y)^5$

$\qquad = x^5 - 15x^4 y + 90x^3 y^2 - 270x^2 y^3 + 405xy^4 - 243y^5$

35. $\left(\dfrac{x}{2}+y\right)^4 = \left(\dfrac{x}{2}\right)^4 + \dfrac{4!}{1!3!}\left(\dfrac{x}{2}\right)^3 y + \dfrac{4!}{2!2!}\left(\dfrac{x}{2}\right)^2 y^2 + \dfrac{4!}{3!1!}\left(\dfrac{x}{2}\right)y^3 + y^4$

$\qquad = \dfrac{1}{16}x^4 + \dfrac{1}{2}x^3 y + \dfrac{3}{2}x^2 y^2 + 2xy^3 + y^4$

37. The 3rd term will involve b^2.
$$\frac{4!}{2!2!}a^2b^2 = 6a^2b^2$$

39. The 5th term will involve b^4.
$$\frac{7!}{3!4!}a^3b^4 = 35a^3b^4$$

41. The 6th term will involve $(-b)^5$.
$$\frac{5!}{0!5!}a^0(-b)^5 = -b^5$$

43. The 5th term will involve b^4.
$$\frac{17!}{13!4!}a^{13}b^4 = 2380a^{13}b^4$$

45. The 2nd term will involve $\left(-\sqrt{2}\right)^1$.
$$\frac{4!}{3!1!}a^3\left(-\sqrt{2}\right)^1 = -4\sqrt{2}\,a^3$$

47. The 5th term will involve $\left(\sqrt{3b}\right)^4$.
$$\frac{9!}{5!4!}a^5\left(\sqrt{3b}\right)^4 = 1134a^5b^4$$

49. The 3rd term will involve y^2.
$$\frac{4!}{2!2!}\left(\frac{x}{2}\right)^2 y^2 = \frac{3}{2}x^2y^2$$

51. The 10th term will involve $\left(-\frac{s}{2}\right)^9$.
$$\frac{11!}{2!9!}\left(\frac{r}{2}\right)^2\left(-\frac{s}{2}\right)^9 = -\frac{55}{2048}r^2s^9$$

53. The 4th term will involve b^3.
$$\frac{n!}{(n-3)!3!}a^{n-3}b^3$$

55. The rth term will involve b^{r-1}.
$$\frac{n!}{(n-r+1)!(r-1)!}a^{n-r+1}b^{r-1}$$

57. **Answers may vary.**

59. Each term will contain $a^n\left(-\frac{1}{a}\right)^{10-n}$, or $a^n\left(-a^{n-10}\right)$. The constant term will occur when this product is equal to 1, or when the sum of the exponents equals 0.
$$n + n - 10 = 0 \Rightarrow n = 5$$
$$\frac{10!}{5!5!}a^5\left(-\frac{1}{a}\right)^5 = -252$$

61.
$$\frac{n!}{0!(n-0)!} = \frac{n!}{0!n!} = \frac{n!}{1\cdot n!} = 1$$

63. **Answers may vary.**

65. **Answers may vary.**

67. $3x^3y^2z^4 - 6xyz^5 + 15x^2yz^2 = 3xyz^2(x^2yz^2 - 2z^3 + 5x)$

69. $a^4 - b^4 = (a^2 + b^2)(a^2 - b^2) = (a^2 + b^2)(a+b)(a-b)$

71. $\dfrac{\frac{1}{x} + \frac{1}{3}}{\frac{1}{x} - \frac{1}{3}} = \dfrac{3x\left(\frac{1}{x} + \frac{1}{3}\right)}{3x\left(\frac{1}{x} - \frac{1}{3}\right)} = \dfrac{3+x}{3-x}$

Exercise 8.2 (page 575)

1. domain

3. Summation notation

5. 6

7. $5c$

9. $f(1) = 5(1)(1-1) = 0 \quad f(2) = 5(2)(2-1) = 10 \quad f(3) = 5(3)(3-1) = 30$
$f(4) = 5(4)(4-1) = 60 \quad f(5) = 5(5)(5-1) = 100 \quad f(6) = 5(6)(6-1) = 150$

11. $1, 6, 11, 16, \ldots$ Add 5 to get the next term. The next term is 21.

13. $a, a+d, a+2d, a+3d, \ldots$ Add d to get the next term. The next term is $a+4d$.

15. $1, 3, 6, 10, \ldots$ The difference between terms increases by 1 each time. The next term is $10 + 5 = 15$.

17. $1 + 2 + 3 + 4 + 5 = 15$

19. $3 + 3 + 3 + 3 + 3 = 15$

21. $2\left(\frac{1}{3}\right)^1 + 2\left(\frac{1}{3}\right)^2 + 2\left(\frac{1}{3}\right)^3 + 2\left(\frac{1}{3}\right)^4 + 2\left(\frac{1}{3}\right)^5 = \frac{2}{3} + \frac{2}{9} + \frac{2}{27} + \frac{2}{81} + \frac{2}{243} = \frac{242}{243}$

23. $[3(1) - 2] + [3(2) - 2] + [3(3) - 2] + [3(4) - 2] + [3(5) - 2] = 1 + 4 + 7 + 10 + 13 = 35$

25. $a_1 = 3$
$a_2 = 2a_1 + 1 = 2(3) + 1 = 7$
$a_3 = 2a_2 + 1 = 2(7) + 1 = 15$
$a_4 = 2a_3 + 1 = 2(15) + 1 = 31$

27. $a_1 = -4$
$a_2 = \frac{a_1}{2} = \frac{-4}{2} = -2$
$a_3 = \frac{a_2}{2} = \frac{-2}{2} = -1$
$a_4 = \frac{a_3}{2} = \frac{-1}{2} = -\frac{1}{2}$

29. $a_1 = k$
$a_2 = a_1^2 = k^2$
$a_3 = a_2^2 = \left(k^2\right)^2 = k^4$
$a_4 = a_3^2 = \left(k^4\right)^2 = k^8$

31. $a_1 = 8$
$a_2 = \frac{2a_1}{k} = \frac{2(8)}{k} = \frac{16}{k}$
$a_3 = \frac{2a_2}{k} = \frac{2\left(\frac{16}{k}\right)}{k} = \frac{32}{k^2}$
$a_4 = \frac{2a_3}{k} = \frac{2\left(\frac{32}{k^2}\right)}{k} = \frac{64}{k^3}$

33. alternating

35. not alternating

37. $\displaystyle\sum_{k=1}^{5} 2k = 2\sum_{k=1}^{5} k = 2(1 + 2 + 3 + 4 + 5)$
$= 2(15) = 30$

39. $\displaystyle\sum_{k=3}^{4}(-2k^2) = -2\sum_{k=3}^{4} k^2 = -2\left(3^2 + 4^2\right)$
$= -2(25) = -50$

41. $\displaystyle\sum_{k=1}^{5}(3k - 1) = 3\sum_{k=1}^{5} k - \sum_{k=1}^{5} 1 = 3(1 + 2 + 3 + 4 + 5) - 5(1) = 3(15) - 5 = 40$

43. $\displaystyle\sum_{k=1}^{1000} \frac{1}{2} = 1000\left(\frac{1}{2}\right) = 500$

45. $\displaystyle\sum_{x=3}^{4} \frac{1}{x} = \frac{1}{3} + \frac{1}{4} = \frac{4}{12} + \frac{3}{12} = \frac{7}{12}$

47. $\displaystyle\sum_{x=1}^{4}(4x+1)^2 - \sum_{x=1}^{4}(4x-1)^2 = \sum_{x=1}^{4}\left(16x^2+8x+1\right) - \sum_{x=1}^{4}\left(16x^2-8x+1\right)$

$\displaystyle = \sum_{x=1}^{4}16x = 16\sum_{x=1}^{4}x = 16(1+2+3+4) = 16(10) = 160$

49. $\displaystyle\sum_{x=6}^{8}(5x-1)^2 + \sum_{x=6}^{8}(10x-1) = \sum_{x=6}^{8}\left(25x^2-10x+1\right) + \sum_{x=6}^{8}(10x-1)$

$\displaystyle = \sum_{x=6}^{8}\left(25x^2\right) = 25\sum_{x=6}^{8}x^2 = 25\left(6^2+7^2+8^2\right) = 3725$

51. Answers may vary.

53. Answers may vary.

55. $\dfrac{8}{12} = \dfrac{4}{x}$

$8x = 48$

$x = 6$ cm

57. $AB^2 + BC^2 = AC^2$

$24^2 + 10^2 = AC^2$

$676 = AC^2$

26 ft. $= AC$

Exercise 8.3 (page 580)

1. $(n-1)d$ **3.** $l = a + (n-1)d$ **5.** Arithmetic means   **7.** $1, 3, 5, 7, 9, 11$

9. nth term $= a + (n-1)d$

$2 = 5 + (3-1)d$

$2 = 5 + 2d$

$-3 = 2d$

$-\dfrac{3}{2} = d \Rightarrow 5, \dfrac{7}{2}, 2, \dfrac{1}{2}, -1, -\dfrac{5}{2}$

11. nth term $= a + (n-1)d$

$24 = a + (7-1)\dfrac{5}{2}$

$24 = a + 6\left(\dfrac{5}{2}\right)$

$24 = a + 15$

$9 = a \Rightarrow 9, \dfrac{23}{2}, 14, \dfrac{33}{2}, 19, \dfrac{43}{2}$

13. nth term $= a + (n-1)d$

$= 6 + (40-1)8$

$= 6 + 39(8) = 6 + 312 = 318$

15. nth term $= a + (n-1)d$

$28 = -2 + (6-1)d$

$28 = -2 + 5d$

$30 = 5d \Rightarrow d = 6$

17. nth term $= a + (n-1)d$

$= -8 + (55-1)7$

$= -8 + 54(7) = -8 + 378 = 370$

19. 5th term   2nd term

$14 = a + 4d$ $5 = a + d$

Solve the system: $\begin{cases} a + 4d = 14 \\ a + d = 5 \end{cases}$

$a = 2, d = 3$

15th term $= 2 + 14(3) = 44$

21. $a = 10$, 5th term $= 20$

$20 = 10 + 4d$

$10 = 4d$

$\dfrac{5}{2} = d \Rightarrow 10, \dfrac{25}{2}, 15, \dfrac{35}{2}, 20$

23. $a = -7$, 6th term $= \dfrac{2}{3}$

$\dfrac{2}{3} = -7 + 5d$

$\dfrac{23}{3} = 5d$

$\dfrac{23}{15} = d \Rightarrow -7, -\dfrac{82}{15}, -\dfrac{59}{15}, -\dfrac{12}{5}, -\dfrac{13}{15}, \dfrac{2}{3}$

25. $a = 5$, $d = 2$

$l = a + (n-1)d = 5 + 14(2) = 33$

$S = \dfrac{n(a+l)}{2} = \dfrac{15(5+33)}{2} = 285$

27. $a = \dfrac{27}{2}$, $d = \dfrac{3}{2}$

$l = a + (n-1)d = \dfrac{27}{2} + 19\left(\dfrac{3}{2}\right) = 42$

$S = \dfrac{n(a+l)}{2} = \dfrac{20\left(\frac{27}{2} + 42\right)}{2} = 555$

29. $d = \dfrac{1}{2}$, 25th term $= 10$

$10 = a + 24\left(\dfrac{1}{2}\right)$

$10 = a + 12 \Rightarrow a = -2$

$l = a + (n-1)d = -2 + 29\left(\dfrac{1}{2}\right) = \dfrac{25}{2}$

$S = \dfrac{n(a+l)}{2} = \dfrac{30\left(-2 + \frac{25}{2}\right)}{2} = 157\dfrac{1}{2}$

31. $a = 1$, $d = 1$, $n = l = 200$

$S = \dfrac{n(a+l)}{2} = \dfrac{200(1+200)}{2} = 20{,}100$

33. $a = 5000$, $d = -200$, $n = 13$

Note: $n = 13$ occurs at the <u>beginning</u> of the 13th month, right after the 12th payment has been made

13th term $= a + (n-1)d$

$\qquad = 5000 + 12(-200)$

$\qquad = \$2600$

35. $a = \dfrac{1}{2}$, 51st term $= 6\dfrac{3}{4} = \dfrac{27}{4}$

51st term $= a + (n-1)d$

$\dfrac{27}{4} = \dfrac{1}{2} + 50d$

$\dfrac{25}{4} = 50d \Rightarrow d = \dfrac{1}{8}$

The distance increased $\frac{1}{8}$ mile per day.

37. 10th term $= a + (n-1)d$

$\qquad = 16 + 9(32)$

$\qquad = 304$ feet

39. $1 + 2 + 3 + \cdots + 20$

$a = 1$, $d = 1$, $n = l = 20$

$S = \dfrac{n(a+l)}{2} = \dfrac{20(1+20)}{2} = 210$ logs

41. **Answers may vary.**

43. **Answers may vary.**

45. $x + \sqrt{x+3} = 9$

$\sqrt{x+3} = 9 - x$

$\left(\sqrt{x+3}\right)^2 = (9-x)^2$

$x + 3 = 81 - 18x + x^2$

$0 = x^2 - 19x + 78$

$0 = (x-6)(x-13)$

$x = 6 \;\; \text{or} \;\; x = 13$

$x = 13$ does not check and is extraneous.

47. $\dfrac{1}{x} + \dfrac{2}{x^2} + \dfrac{1}{x^3} = 0$

$x^3 \left(\dfrac{1}{x} + \dfrac{2}{x^2} + \dfrac{1}{x^3} \right) = x^3(0)$

$x^2 + 2x + 1 = 0$

$(x+1)(x+1) = 0$

$x = -1$

49. $x^4 - 1 = 0$

$\left(x^2 + 1\right)\left(x^2 - 1\right) = 0$

$x^2 + 1 = 0 \qquad \textbf{OR} \quad x^2 - 1 = 0$

$ x^2 = -1 \qquad\qquad\quad x^2 = 1$

$ x = \pm\sqrt{-1} \qquad\qquad x = \pm\sqrt{1}$

$ x = \pm i \qquad\qquad\quad\;\; x = \pm 1$

Exercise 8.4 (page 587)

1. r^{n-1}

3. $l = ar^{n-1}$

5. Geometric means

7. 10, 20, 40, 80

9. $-2, -6, -18, -54$

11. $3, 3\sqrt{2}, 6, 6\sqrt{2}$

13. nth term $= ar^{n-1}$

$54 = 2r^3$

$27 = r^3$

$3 = r \Rightarrow 2, 6, 18, 54$

15. $a = \dfrac{1}{4}, r = 4$

6th term $= ar^{n-1} = \left(\dfrac{1}{4}\right)4^5 = 256$

17. 2nd term $= ar^1$

$6 = ar^1$

3rd term $= ar^2$

$-18 = ar^2$

$ar^2 = -18$

$ar \cdot r = -18$

$6r = -18$

$r = -3$

$r = -3:\; a = \dfrac{6}{r} = \dfrac{6}{-3} = -2$

4th term $= ar^7 = -2(-3)^4 = -162$

19. 5th term $= ar^4$

$20 = 10r^4$

$2 = r^4$

$\sqrt[4]{2} = r$ (problem specifies positive)

$10, 10\sqrt[4]{2}, 10\sqrt{2}, 10\sqrt[4]{8}, 20$

21. 6th term $= ar^5$

$2048 = 2r^5$

$1024 = r^4$

$4 = r$

$2, 8, 32, 128, 512, 2048$

23. $a = 4, r = 2, n = 5$

$$S = \frac{a - ar^n}{1 - r} = \frac{4 - 4(2)^5}{1 - 2}$$
$$= \frac{-124}{-1} = 124$$

25. $a = 2, r = -3, n = 10$

$$S = \frac{a - ar^n}{1 - r} = \frac{2 - 2(-3)^{10}}{1 - (-3)}$$
$$= \frac{-118{,}096}{4} = -29{,}524$$

27. $a = 3, r = \frac{3}{2}, n = 6$

$$S = \frac{a - ar^n}{1 - r} = \frac{3 - 3\left(\frac{3}{2}\right)^6}{1 - \frac{3}{2}}$$
$$= \frac{-\frac{1995}{64}}{-\frac{1}{2}} = \frac{1995}{32}$$

29. $a = 6, r = \frac{2}{3}$

$$S = \frac{a}{1 - r} = \frac{6}{1 - \frac{2}{3}} = \frac{6}{\frac{1}{3}} = 18$$

31. $a = 12, r = -\frac{1}{2}$

$$S = \frac{a}{1 - r} = \frac{12}{1 - \left(-\frac{1}{2}\right)} = \frac{12}{\frac{3}{2}} = 8$$

33. $a = \frac{5}{10} = \frac{1}{2}, r = \frac{1}{10}$

$$S = \frac{a}{1 - r} = \frac{\frac{1}{2}}{1 - \frac{1}{10}} = \frac{\frac{1}{2}}{\frac{9}{10}} = \frac{5}{9}$$

35. $a = \frac{25}{100} = \frac{1}{4}, r = \frac{1}{100}$

$$S = \frac{a}{1 - r} = \frac{\frac{1}{4}}{1 - \frac{1}{100}} = \frac{\frac{1}{4}}{\frac{99}{100}} = \frac{25}{99}$$

37. $a = 623, r = 1.10$

9th term $= ar^8 = 623(1.1)^8$
$$\approx 1335 \text{ students}$$

\# professors $= \dfrac{1335}{60} \approx 22.25$

23 professors will be needed.

39. $a = 1000, r = 1 + \dfrac{0.0675}{365}, n = 365$

$$ar^{365} = 1000\left(1 + \frac{0.0675}{365}\right)^{365}$$
$$\approx \$1069.82$$

The interest will be \$69.82.

41. $a = c, r = 0.80, n = 5$

$$ar^5 = c(0.80)^5 = 0.32768c, \text{ or about } \tfrac{1}{3}c$$

43. $a = 5 \times 10^9, r = 2, n = 34$

$$ar^{34} = 5 \times 10^9 (2)^{34} \approx 8.6 \times 10^{19}$$

45. $a = 50{,}000; \; r = 1.06, n = 22$

$$ar^{22} = 50{,}000(1.06)^{22}$$
$$\approx \$180{,}176.87$$

47. $a = 1000, r = 1 + \dfrac{0.07}{4}, n = 40$

$$ar^{40} = 1000\left(1 + \frac{0.07}{4}\right)^{40}$$
$$\approx \$2001.60$$

49. $a = 1000, r = 1 + \dfrac{0.07}{365}, n = 3650$

$$ar^{3650} = 1000\left(1 + \frac{0.07}{365}\right)^{3650}$$
$$\approx \$2013.62$$

51. $a = 2000, r = 1 + \dfrac{0.11}{4}, n = 180$

$ar^{180} = 2000\left(1 + \dfrac{0.11}{4}\right)^{180}$

$\approx \$264,094.58$

53. $a = 1, r = 2, n = 64$

$S = \dfrac{a - ar^n}{1 - r} = \dfrac{1 - 1(2)^{64}}{1 - 2}$

$\approx 1.8447 \times 10^{19}$ grains

55. **Answers may vary.**

57. $(3 + 2i) + (2 - 5i) = 3 + 2i + 2 - 5i$

$= 5 - 3i$

59. $(3 + i)(3 - i) = 9 - 3i + 3i - i^2 = 9 - (-1) = 10 + 0i$

61. $\dfrac{2 + 3i}{2 - i} = \dfrac{(2 + 3i)(2 + i)}{(2 - i)(2 + i)} = \dfrac{4 + 2i + 6i + 3i^2}{4 - i^2} = \dfrac{4 + 8i + 3(-1)}{4 - (-1)} = \dfrac{1 + 8i}{5} = \dfrac{1}{5} + \dfrac{8}{5}i$

63. $i^{127} = i^{124}i^3 = (i^4)^{31}i^3 = 1^{31}i^3 = i^3 = -i$

Exercise 8.5 (page 593)

1. two

3. $n = k + 1$

5.

$\dfrac{\quad n = 1 \quad}{}$

$5(1) \overset{?}{=} \dfrac{5(1)(1 + 1)}{2}$

$5 = 5$

$\dfrac{\quad n = 2 \quad}{}$

$5 + 5(2) \overset{?}{=} \dfrac{5(2)(2 + 1)}{2}$

$15 \overset{?}{=} \dfrac{10(3)}{2}$

$15 = 15$

$\dfrac{\quad n = 3 \quad}{}$

$5 + 10 + 5(3) \overset{?}{=} \dfrac{5(3)(3 + 1)}{2}$

$30 \overset{?}{=} \dfrac{15(4)}{2}$

$30 = 30$

$\dfrac{\quad n = 4 \quad}{}$

$5 + 10 + 15 + 5(4) \overset{?}{=} \dfrac{5(4)(4 + 1)}{2}$

$50 \overset{?}{=} \dfrac{20(5)}{2}$

$50 = 50$

7.

$\dfrac{\quad n = 1 \quad}{}$

$3(1) + 4 \overset{?}{=} \dfrac{1(3(1) + 11)}{2}$

$7 \overset{?}{=} \dfrac{1(14)}{2}$

$7 = 7$

$\dfrac{\quad n = 2 \quad}{}$

$7 + 3(2) + 4 \overset{?}{=} \dfrac{2(3(2) + 11)}{2}$

$17 \overset{?}{=} \dfrac{2(17)}{2}$

$17 = 17$

$\dfrac{\quad n = 3 \quad}{}$

$7 + 10 + 3(3) + 4 \overset{?}{=} \dfrac{3(3(3) + 11)}{2}$

$30 \overset{?}{=} \dfrac{3(20)}{2}$

$30 = 30$

$\dfrac{\quad n = 4 \quad}{}$

$7 + 10 + 13 + 3(4) + 4 \overset{?}{=} \dfrac{4(3(4) + 11)}{2}$

$46 \overset{?}{=} \dfrac{4(23)}{2}$

$46 = 46$

9.

Check $n = 1$: $2 \overset{?}{=} 1(1+1)$ True for $n = 1$
$$2 = 2$$

Assume for $n = k$: $2 + 4 + 6 + \cdots + 2k = k(k+1)$

Show for $n = k+1$: $\boxed{2 + 4 + 6 + \cdots + 2k} + 2(k+1) = \boxed{k(k+1)} + 2(k+1)$

$$2 + 4 + 6 + \cdots + 2(k+1) = k^2 + k + 2k + 2$$
$$2 + 4 + 6 + \cdots + 2(k+1) = k^2 + 3k + 2$$
$$2 + 4 + 6 + \cdots + 2(k+1) = (k+1)(k+2)$$

Since this is what results when $n = k+1$ in the formula, we have shown that the formula works for $n = k+1$ if it works for $n = k$.

11.

Check $n = 1$: $4(1) - 1 \overset{?}{=} 1(2(1)+1)$ True for $n = 1$
$$3 = 3$$

Assume for $n = k$ and show for $n = k+1$:
$$3 + 7 + 11 + \cdots + (4k-1) = k(2k+1)$$
$$\boxed{3 + 7 + 11 + \cdots + (4k-1)} + 4(k+1) - 1 = \boxed{k(2k+1)} + 4(k+1) - 1$$
$$3 + 7 + 11 + \cdots + [4(k+1)-1] = 2k^2 + k + 4k + 4 - 1$$
$$3 + 7 + 11 + \cdots + [4(k+1)-1] = 2k^2 + 5k + 3$$
$$3 + 7 + 11 + \cdots + [4(k+1)-1] = (k+1)(2k+3)$$
$$3 + 7 + 11 + \cdots + [4(k+1)-1] = (k+1)(2(k+1)+1)$$

Since this is what results when $n = k+1$ in the formula, we have shown that the formula works for $n = k+1$ if it works for $n = k$.

13.

Check $n = 1$: $14 - 4(1) \overset{?}{=} 12(1) - 2(1)^2$ True for $n = 1$
$$10 = 10$$

Assume for $n = k$ and show for $n = k+1$:
$$10 + 6 + 2 + \cdots + (14 - 4k) = 12k - 2k^2$$
$$\boxed{10 + 6 + 2 + \cdots + (14 - 4k)} + 14 - 4(k+1) = \boxed{12k - 2k^2} + 14 - 4(k+1)$$
$$10 + 6 + 2 + \cdots + (14 - 4(k+1)) = 12k - 2k^2 + 14 - 4k - 4$$
$$10 + 6 + 2 + \cdots + (14 - 4(k+1)) = 12k + 12 - 2k^2 - 4k - 2$$
$$10 + 6 + 2 + \cdots + (14 - 4(k+1)) = 12(k+1) - 2(k^2 + 2k + 1)$$
$$10 + 6 + 2 + \cdots + (14 - 4(k+1)) = 12(k+1) - 2(k+1)^2$$

Since this is what results when $n = k+1$ in the formula, we have shown that the formula works for $n = k+1$ if it works for $n = k$.

15.

Check $n = 1$: $3(1) - 1 \overset{?}{=} \dfrac{1(3(1) + 1)}{2}$ True for $n = 1$

$$2 = 2$$

Assume for $n = k$ and show for $n = k + 1$:

$$2 + 5 + 8 + \cdots + (3k - 1) = \frac{k(3k + 1)}{2}$$

$$\boxed{2 + 5 + 8 + \cdots + (3k - 1)} + 3(k + 1) - 1 = \boxed{\frac{k(3k + 1)}{2}} + 3(k + 1) - 1$$

$$2 + 5 + 8 + \cdots + (3(k + 1) - 1) = \frac{k(3k + 1)}{2} + \frac{2 \cdot (3(k + 1) - 1)}{2}$$

$$2 + 5 + 8 + \cdots + (3(k + 1) - 1) = \frac{3k^2 + k + 6k + 6 - 2}{2}$$

$$2 + 5 + 8 + \cdots + (3(k + 1) - 1) = \frac{3k^2 + 7k + 4}{2}$$

$$2 + 5 + 8 + \cdots + (3(k + 1) - 1) = \frac{(k + 1)(3k + 4)}{2}$$

$$2 + 5 + 8 + \cdots + (3(k + 1) - 1) = \frac{(k + 1)(3(k + 1) + 1)}{2}$$

Since this is what results when $n = k + 1$ in the formula, we have shown that the formula works for $n = k + 1$ if it works for $n = k$.

17.

Check $n = 1$: $1^2 \overset{?}{=} \dfrac{1(1 + 1)(2(1) + 1)}{6}$ True for $n = 1$

$$1 = 1$$

Assume for $n = k$ and show for $n = k + 1$:

$$1^2 + 2^2 + 3^2 + \cdots + k^2 = \frac{k(k + 1)(2k + 1)}{6}$$

$$\boxed{1^2 + 2^2 + 3^2 + \cdots + k^2} + (k + 1)^2 = \boxed{\frac{k(k + 1)(2k + 1)}{6}} + (k + 1)^2$$

$$1^2 + 2^2 + 3^2 + \cdots + (k + 1)^2 = \frac{(2k^2 + k)(k + 1)}{6} + \frac{6(k + 1)(k + 1)}{6}$$

$$1^2 + 2^2 + 3^2 + \cdots + (k + 1)^2 = \frac{(2k^2 + k + 6(k + 1))(k + 1)}{6}$$

$$1^2 + 2^2 + 3^2 + \cdots + (k + 1)^2 = \frac{(2k^2 + 7k + 6)(k + 1)}{6}$$

$$1^2 + 2^2 + 3^2 + \cdots + (k + 1)^2 = \frac{(2k + 3)(k + 2)(k + 1)}{6}$$

$$1^2 + 2^2 + 3^2 + \cdots + (k + 1)^2 = \frac{(k + 1)(k + 2)(2k + 3)}{6}$$

Since this is what results when $n = k + 1$ in the formula, we have shown that the formula works for $n = k + 1$ if it works for $n = k$.

9.

Check $n = 1$: $\quad \dfrac{5}{3}(1) - \dfrac{4}{3} \overset{?}{=} 1\left(\dfrac{5}{6}(1) - \dfrac{1}{2}\right) \quad$ True for $n = 1$

$$\dfrac{1}{3} = \dfrac{1}{3}$$

Assume for $n = k$ and show for $n = k + 1$:

$$\dfrac{1}{3} + 2 + \dfrac{11}{3} + \cdots + \left(\dfrac{5}{3}k - \dfrac{4}{3}\right) = k\left(\dfrac{5}{6}k - \dfrac{1}{2}\right)$$

$$\boxed{\dfrac{1}{3} + 2 + \dfrac{11}{3} + \cdots + \left(\dfrac{5}{3}k - \dfrac{4}{3}\right)} + \left(\dfrac{5}{3}(k+1) - \dfrac{4}{3}\right) = \boxed{k\left(\dfrac{5}{6}k - \dfrac{1}{2}\right)} + \left(\dfrac{5}{3}(k+1) - \dfrac{4}{3}\right)$$

$$\dfrac{1}{3} + 2 + \dfrac{11}{3} + \cdots + \left(\dfrac{5}{3}(k+1) - \dfrac{4}{3}\right) = \dfrac{5}{6}k^2 - \dfrac{1}{2}k + \dfrac{5}{3}k + \dfrac{5}{3} - \dfrac{4}{3}$$

$$\dfrac{1}{3} + 2 + \dfrac{11}{3} + \cdots + \left(\dfrac{5}{3}(k+1) - \dfrac{4}{3}\right) = \dfrac{5}{6}k^2 + \dfrac{7}{6}k + \dfrac{1}{3}$$

$$\dfrac{1}{3} + 2 + \dfrac{11}{3} + \cdots + \left(\dfrac{5}{3}(k+1) - \dfrac{4}{3}\right) = (k+1)\left(\dfrac{5}{6}k + \dfrac{1}{3}\right)$$

$$\dfrac{1}{3} + 2 + \dfrac{11}{3} + \cdots + \left(\dfrac{5}{3}(k+1) - \dfrac{4}{3}\right) = (k+1)\left(\dfrac{5}{6}k + \dfrac{5}{6} - \dfrac{1}{2}\right)$$

$$\dfrac{1}{3} + 2 + \dfrac{11}{3} + \cdots + \left(\dfrac{5}{3}(k+1) - \dfrac{4}{3}\right) = (k+1)\left(\dfrac{5}{6}(k+1) - \dfrac{1}{2}\right)$$

Since this is what results when $n = k + 1$ in the formula, we have shown that the formula works for $n = k + 1$ if it works for $n = k$.

1.

Check $n = 1$: $\quad \left(\dfrac{1}{2}\right)^1 \overset{?}{=} 1 - \left(\dfrac{1}{2}\right)^1 \quad$ True for $n = 1$

$$\dfrac{1}{2} = \dfrac{1}{2}$$

Assume for $n = k$ and show for $n = k + 1$:

$$\dfrac{1}{2} + \dfrac{1}{4} + \dfrac{1}{8} + \cdots + \left(\dfrac{1}{2}\right)^k = 1 - \left(\dfrac{1}{2}\right)^k$$

$$\boxed{\dfrac{1}{2} + \dfrac{1}{4} + \dfrac{1}{8} + \cdots + \left(\dfrac{1}{2}\right)^k} + \left(\dfrac{1}{2}\right)^{k+1} = \boxed{1 - \left(\dfrac{1}{2}\right)^k} + \left(\dfrac{1}{2}\right)^{k+1}$$

$$\dfrac{1}{2} + \dfrac{1}{4} + \dfrac{1}{8} + \cdots + \left(\dfrac{1}{2}\right)^{k+1} = 1 - 2\left(\dfrac{1}{2}\right)\left(\dfrac{1}{2}\right)^k + \left(\dfrac{1}{2}\right)^{k+1}$$

$$\dfrac{1}{2} + \dfrac{1}{4} + \dfrac{1}{8} + \cdots + \left(\dfrac{1}{2}\right)^{k+1} = 1 - 2\left(\dfrac{1}{2}\right)^{k+1} + \left(\dfrac{1}{2}\right)^{k+1}$$

$$\dfrac{1}{2} + \dfrac{1}{4} + \dfrac{1}{8} + \cdots + \left(\dfrac{1}{2}\right)^{k+1} = 1 - \left(\dfrac{1}{2}\right)^{k+1}$$

Since this is what results when $n = k + 1$ in the formula, we have shown that the formula works for $n = k + 1$ if it works for $n = k$.

23.

Check $n = 1$: $\quad 2^{1-1} \overset{?}{=} 2^1 - 1 \quad$ True for $n = 1$
$$1 = 1$$

Assume for $n = k$ and show for $n = k + 1$:
$$2^0 + 2^1 + 2^2 + \cdots + 2^{k-1} = 2^k - 1$$
$$\boxed{2^0 + 2^1 + 2^2 + \cdots + 2^{k-1}} + 2^k = \boxed{2^k - 1} + 2^k$$
$$2^0 + 2^1 + 2^2 + \cdots + 2^k = 2 \cdot 2^k - 1$$
$$2^0 + 2^1 + 2^2 + \cdots + 2^k = 2^{k+1} - 1$$

Since this is what results when $n = k + 1$ in the formula, we have shown that the formula works for $n = k + 1$ if it works for $n = k$.

25.

Check $n = 1$: $\quad x - y$ is a factor of $x^1 - y^1$. $\quad$ True for $n = 1$

Assume for $n = k$ and show for $n = k + 1$:
Thus, we assume that $x^k - y^k = (x - y)(\text{SOMETHING})$.
$$x^{k+1} - y^{k+1} = x^{k+1} - xy^k + xy^k - y^{k+1}$$
$$= x\left(x^k - y^k\right) + y^k(x - y)$$
$$= x(x - y)(\text{SOMETHING}) + y^k(x - y)$$
$$= (x - y)\left[x(\text{SOMETHING}) + y^k\right]$$

We have shown that $x - y$ is a factor of $x^{k+1} - y^{k+1}$ if it is a factor of $x^k - y^k$.

27. The formula is true for $n = 3$, since a triangle has $180° = (3 - 2) \cdot 180°$. Next, assume that a polygon with k sides has an angle sum of $(k - 2) \cdot 180°$. Take a polygon with $k + 1$ sides. Consider two adjacent sides with a common endpoint. Connect the endpoints which are NOT common to both sides. The figure is now a polygon with k sides with a triangle adjacent to it.

$$\boxed{\begin{array}{c}\text{Sum of angles of} \\ (k+1)\text{-sided polygon}\end{array}} = \boxed{\begin{array}{c}\text{Sum of angles of} \\ k\text{-sided polygon}\end{array}} + \boxed{\begin{array}{c}\text{Sum of angles of} \\ \text{triangle}\end{array}}$$
$$= (k - 2) \cdot 180° + 180° = (k - 1) \cdot 180° = [(k + 1) - 2] \cdot 180°$$

Thus, the formula works for $n = k + 1$ if it works for $n = k$.

29. Assume for $n = k$ and show for $n = k + 1$:
$$1 + 2 + 3 + \cdots + k = \tfrac{k}{2}(k + 1) + 1$$
$$\boxed{1 + 2 + 3 + \cdots + k} + (k + 1) = \boxed{\tfrac{k}{2}(k + 1) + 1} + (k + 1)$$
$$1 + 2 + 3 + \cdots + (k + 1) = \tfrac{1}{2}k(k + 1) + (k + 1) + 2$$
$$1 + 2 + 3 + \cdots + (k + 1) = \tfrac{1}{2}(k + 1)(k + 2)$$
$$1 + 2 + 3 + \cdots + (k + 1) = \tfrac{(k+1)}{2}(k + 2)$$

The formula works for $n = k + 1$ if it works for $n = k$. However, the formula does not work for $n = 1$. Thus, the formula does not work for all natural numbers.

31.

Check $n = 1$: $7^1 - 1 = 7 - 1 = 6$ True for $n = 1$

Thus, $7^1 - 1$ is divisible by 6.

Assume for $n = k$ and show for $n = k + 1$:

$7^k - 1$ is divisible by 6, so $7^k - 1 = 6 \cdot x$, where x is some natural number.

Then $7^{k+1} - 1 = 7^k \cdot 7 - 1 = (6x + 1) \cdot 7 - 1 = 42x + 6 = 6(7x + 1)$

Thus, $7^{k+1} - 1$ is divisible by 6.

We have shown that $7^{k+1} - 1$ is divisible by 6 if $7^k - 1$ is divisible by 6.

33.

Check $n = 1$: $1 + r^1 \overset{?}{=} \dfrac{1 - r^2}{1 - r}$ True for $n = 1$

$$1 + r \overset{?}{=} \frac{(1 + r)(1 - r)}{1 - r}$$

$$1 + r = 1 + r$$

Assume for $n = k$ and show for $n = k + 1$:

$$1 + r + r^2 + r^3 + \cdots + r^k = \frac{1 - r^{k+1}}{1 - r}$$

$$\boxed{1 + r + r^2 + r^3 + \cdots + r^k} + r^{k+1} = \boxed{\frac{1 - r^{k+1}}{1 - r}} + r^{k+1}$$

$$1 + r + r^2 + r^3 + \cdots + r^{k+1} = \frac{1 - r^{k+1} + r^{k+1}(1 - r)}{1 - r}$$

$$1 + r + r^2 + r^3 + \cdots + r^{k+1} = \frac{1 - r^{k+1} + r^{k+1} - r^{k+2}}{1 - r}$$

$$1 + r + r^2 + r^3 + \cdots + r^{k+1} = \frac{1 - r^{(k+1)+1}}{1 - r}$$

Since this is what results when $n = k + 1$ in the formula, we have shown that the formula works for $n = k + 1$ if it works for $n = k$.

35.

Check $n = 1$: $a^m a^1 = a^m \cdot a = a^{m+1}$, by definition   True for $n = 1$

Assume for $n = k$ and show for $n = k + 1$:

$a^m a^k = a^{m+k}$

$a^m a^k a = a^{m+k} a$

$a^m a^{k+1} = a^{m+k+1}$

$a^m a^{k+1} = a^{m+(k+1)}$

We have shown that the formula works for $n = k + 1$ if it works for $n = k$.

37. $3x + 4y \leq 12$

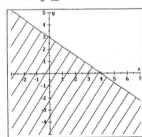

39. $2x + y > 5$

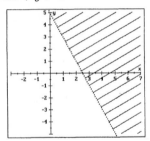

Exercise 8.6 (page 601)

1. $6 \cdot 4 = 24$

3. $\dfrac{n!}{(n-r)!}$

5. 1

7. $\dfrac{n!}{r!(n-r)!}$

9. 1

11. $\dfrac{n!}{a! \cdot b! \cdots}$

13. $P(7, 4) = \dfrac{7!}{(7-4)!} = \dfrac{7!}{3!} = \dfrac{7 \cdot 6 \cdot 5 \cdot 4 \cdot 3!}{3!} = 7 \cdot 6 \cdot 5 \cdot 4 = 840$

15. $C(7, 4) = \dfrac{7!}{4!(7-4)!} = \dfrac{7!}{4!3!} = \dfrac{7 \cdot 6 \cdot 5 \cdot 4!}{4!3!} = \dfrac{7 \cdot 6 \cdot 5}{3 \cdot 2 \cdot 1} = 35$

17. $P(5, 5) = \dfrac{5!}{(5-5)!} = \dfrac{5!}{0!} = \dfrac{5!}{1} = 5 \cdot 4 \cdot 3 \cdot 2 \cdot 1 = 120$

19. $\dbinom{5}{4} = \dfrac{5!}{4!(5-4)!} = \dfrac{5 \cdot 4!}{4!1!} = \dfrac{5}{1} = 5$

21. $\dbinom{5}{0} = \dfrac{5!}{0!(5-0)!} = \dfrac{5!}{0!5!} = \dfrac{1}{1} = 1$

23. $P(5, 4) \cdot C(5, 3) = \dfrac{5!}{(5-4)!} \cdot \dfrac{5!}{3!(5-3)!} = \dfrac{5!}{1!} \cdot \dfrac{5!}{3!2!} = 120 \cdot 10 = 1200$

25. $\dbinom{5}{3}\dbinom{4}{3}\dbinom{3}{3} = \dfrac{5!}{3!(5-3)!} \cdot \dfrac{4!}{3!(4-3)!} \cdot \dfrac{3!}{3!(3-3)!} = 10 \cdot 4 \cdot 1 = 40$

27. $\dbinom{68}{66} = \dfrac{68!}{66!(68-66)!} = \dfrac{68 \cdot 67 \cdot 66!}{66!2!} = \dfrac{68 \cdot 67}{2 \cdot 1} = 2278$

29. $8 \cdot 6 \cdot 3 = 144$

31. $8 \cdot 10 \cdot 10 \cdot 10 \cdot 10 \cdot 10 \cdot 10 = 8{,}000{,}000$

33. Consider the e and the r to be a block that cannot be divided, say x. Then the problem becomes finding the number of ways to rearrange the letters in the word $numbx$. This can be done in 5!, or 120 ways. For each of these possibilities, the e and the r could be reversed, doubling the number of possibilities. The answer is $2 \cdot 120$, or 240 ways.

SECTION 8.6

35. The word must appear as $\square\,L\,U\,\square\,\square$, where one of the Fs must appear in each box. This can be done in 3!, or 6 ways.

37. $8! = 40{,}320$

39. The line will look like this:

$$\frac{5}{M} \quad \frac{4}{M} \quad \frac{3}{M} \quad \frac{2}{M} \quad \frac{1}{M} \quad \frac{5}{W} \quad \frac{4}{W} \quad \frac{3}{W} \quad \frac{2}{W} \quad \frac{1}{W}$$

Then there are $5 \cdot 4 \cdot 3 \cdot 2 \cdot 1 \cdot 5 \cdot 4 \cdot 3 \cdot 2 \cdot 1 = 14{,}400$ arrangements.

41. $P(30, 3) = \dfrac{30!}{27!} = 24{,}360$ **43.** $(8 - 1)! = 7! = 5040$

45. Consider the two people who must sit together as a single person, so that there are 5 "people" who must be arranged in a circle. This can be done in $(5 - 1)! = 4! = 24$ ways. However, the two people who have been seated next to each other could be switched, so that the number of arrangements is doubled. There are $2 \cdot 24 = 48$ possible arrangements.

47. Consider Sally and John as a single person and Martha and Peter as a single person, so that there are 5 "people" who must be arranged in a circle. This can be done in $(5 - 1)! = 4! = 24$ ways. However, each group of 2 people could be switched, so that the number of arrangements will equal $2 \cdot 2 \cdot 24$, or 96.

49. $\dbinom{10}{4} = \dfrac{10!}{4!6!} = 210$ **51.** $4! = 24$

53. $7! = 5040$ **55.** $25 \cdot 24 \cdot 9 \cdot 9 \cdot 8 \cdot 7 = 2{,}721{,}600$

57. $\dbinom{6}{3}\dbinom{8}{3} = 20 \cdot 56 = 1120$ **59.** $\dbinom{12}{4}\dbinom{10}{3} = 495 \cdot 120 = 59{,}400$

61. $\dfrac{17}{H} \cdot \dfrac{16}{W} = 272$ **63.** $\dbinom{8}{2} = 28$ **65.** $\dbinom{10}{5} = 252$

67. $\dbinom{30}{5} = 142{,}506$ **69.** $\dbinom{12}{10} = 66$

71. $C(n, n) = \dfrac{n!}{n!(n - n)!} = \dfrac{n!}{n!0!} = \dfrac{n!}{n!} = 1$

73. $\dbinom{n}{n - r} = \dfrac{n!}{(n - r)!(n - (n - r))!} = \dfrac{n!}{(n - r)!r!} = \dfrac{n!}{r!(n - r)!} = \dbinom{n}{r}$

313

75. **Answers may vary.**

77. $\log_x 16 = 4$
$x^4 = 16 \Rightarrow x = 2$

79. $\log_{\sqrt{7}} 49 = x$
$\left(\sqrt{7}\right)^x = 49 \Rightarrow x = 4$

81. true

83. true

Exercise 8.7 (page 608)

1. experiment

3. $\dfrac{n(E)}{n(S)}$

5. $\{(1, H), (2, H), (3, H), (4, H), (5, H), (6, H), (1, T), (2, T), (3, T), (4, T), (5, T), (6, T)\}$

7. $\{A, B, C, D, E, F, G, H, I, J, K, L, M, N, O, P, Q, R, S, T, U, V, W, X, Y, Z\}$

9. $\dfrac{1}{6}$

11. $\dfrac{4}{6} = \dfrac{2}{3}$

13. $\dfrac{19}{42}$

15. $\dfrac{13}{42}$

17. $\dfrac{3}{8}$

19. $\dfrac{0}{8} = 0$

21. rolls of 4: $\{(1, 3), (2, 2), (3, 1)\}$
Probability $= \dfrac{3}{36} = \dfrac{1}{12}$

23. $\dfrac{\# \text{ aces}}{\# \text{ cards}} \cdot \dfrac{\# \text{ aces}}{\# \text{ cards}} = \dfrac{4}{52} \cdot \dfrac{4}{52} = \dfrac{1}{169}$

25. $\dfrac{\# \text{ red}}{\# \text{ eggs}} = \dfrac{5}{12}$

27. $\dfrac{\substack{\# \text{ ways to get 13} \\ \text{cards of the same suit}}}{\substack{\# \text{ ways to get 13 cards} \\ \text{from the deck of 52}}} = \dfrac{4 \cdot \binom{13}{13}}{\binom{52}{13}} = \dfrac{4}{6.350136 \times 10^{11}} \approx 6.3 \times 10^{-12}$

29. impossible $\Rightarrow 0$

31. $\dfrac{\# \text{ face cards}}{\# \text{ cards in deck}} = \dfrac{12}{52} = \dfrac{3}{13}$

33. $\dfrac{\# \text{ ways to get 5 orange}}{\# \text{ ways to get 5 cubes}} = \dfrac{\binom{5}{5}}{\binom{6}{5}} = \dfrac{1}{6}$

35. rolls of 11: $(1, 4, 6), (1, 5, 5), (1, 6, 4), (2, 3, 6), (2, 4, 5), (2, 5, 4), (2, 6, 3), (3, 2, 6), (3, 3, 5)$
$(3, 4, 4), (3, 5, 3), (3, 6, 2), (4, 1, 6), (4, 2, 5), (4, 3, 4), (4, 4, 3), (4, 5, 2), (4, 6, 1), (5, 1, 5),$
$(5, 2, 4), (5, 3, 3), (5, 4, 2), (5, 5, 1), (6, 1, 4), (6, 2, 3), (6, 3, 2), (6, 4, 1)$
Probability $= \dfrac{27}{216} = \dfrac{1}{8}$

37. $\dfrac{\binom{5}{3}}{2^5} = \dfrac{10}{32} = \dfrac{5}{16}$

39. $SSSS, SSSF, SSFS, SSFF, SFSS, SFSF, SFFS, SFFF,$
$FSSS, FSSF, FSFS, FSFF, FFSS, FFSF, FFFS, FFFF$

41. $\dfrac{1}{4}$

43. $\dfrac{1}{4}$

45. 1

47. $\dfrac{32}{119}$

49. $\dfrac{\binom{8}{4}}{\binom{10}{4}} = \dfrac{70}{210} = \dfrac{1}{3}$

51. $P(A \cap B) = P(A) \cdot P(B|A)$
$ = 0.3(0.6) = 0.18$

53. $P(A \cap B) = P(A) \cdot P(B|A)$
$ = 0.2(0.7) = 0.14$

55. $P(A \cap B) = P(A) \cdot P(B|A)$
$ 0.25 = 0.75 P(B|A)$
$ 0.33 \approx P(B|A)$
$ 33\% \approx P(B|A)$

57. **Answers may vary.**

59. $|x+3| = 7$
$x+3 = 7 \quad \textbf{OR} \quad x + 3 = -7$
$x = 4 x = -10$

61. $|x-3| < 7$
$-7 < x - 3 < 7$
$-4 < x < 10$

$-4 10$

Exercise 8.8 (page 613)

1. compound

3. mutually exclusive

5. $1 - P(A)$

7. $P(A) \cdot P(B)$

9. $\dfrac{26}{52} = \dfrac{1}{2}$

11. $\dfrac{26}{52} + \dfrac{4}{52} - \dfrac{2}{52} = \dfrac{7}{13}$

13. $\dfrac{4}{52} \cdot \dfrac{3}{51} = \dfrac{1}{221}$

15. $\dfrac{13}{52} \cdot \dfrac{25}{51} = \dfrac{25}{204}$

17. $\dfrac{6}{36} + \dfrac{5}{36} = \dfrac{11}{36}$

19. $\dfrac{3}{36} + \dfrac{18}{36} = \dfrac{21}{36} = \dfrac{7}{12}$

21. $\dfrac{7}{16} + \dfrac{3}{16} = \dfrac{10}{16} = \dfrac{5}{8}$

23. $1 - \dfrac{3}{16} = \dfrac{13}{16}$

25. $\dfrac{7}{16} \cdot \dfrac{6}{16} = \dfrac{21}{128}$

27. $\underbrace{\dfrac{9}{16}}_{\text{Not beige}} \cdot \underbrace{\dfrac{8}{15}}_{\text{Not beige}} \cdot \underbrace{\dfrac{7}{14}}_{\text{Not beige}} = \dfrac{504}{3360} = \dfrac{3}{20}$

29. $\dfrac{7}{7} \cdot \dfrac{1}{7} \cdot \dfrac{1}{7} = \dfrac{1}{49}$

31. $\dfrac{365}{365} \cdot \dfrac{364}{365} \cdot \dfrac{363}{365} \cdot \dfrac{362}{365} \cdot \dfrac{361}{365} \approx 0.973$

33. $\dfrac{1}{4}+\dfrac{2}{5}-\dfrac{1}{4}\cdot\dfrac{2}{5}=\dfrac{5+8-2}{20}=\dfrac{11}{20}$ **35.** $\dfrac{1}{2}\cdot\dfrac{1}{3}\cdot\dfrac{3}{4}=\dfrac{1}{8}$

37. $\dfrac{1}{2}\cdot\dfrac{2}{3}\cdot\dfrac{1}{4}+\dfrac{1}{2}\cdot\dfrac{1}{3}\cdot\dfrac{1}{4}+\dfrac{1}{2}\cdot\dfrac{2}{3}\cdot\dfrac{3}{4}=\dfrac{2}{24}+\dfrac{1}{24}+\dfrac{6}{24}=\dfrac{9}{24}=\dfrac{3}{8}$

39. $\dfrac{2}{3}\cdot\dfrac{1}{2}\cdot\dfrac{1}{6}+\dfrac{2}{3}\cdot\dfrac{1}{2}\cdot\dfrac{5}{6}+\dfrac{1}{3}\cdot\dfrac{1}{2}\cdot\dfrac{5}{6}=\dfrac{2}{36}+\dfrac{10}{36}+\dfrac{5}{36}=\dfrac{17}{36}$

41. $P(A)=0.05,\ P(H|A)=0.40\Rightarrow P(A\cap\overline{H})=P(A)\cdot P(\overline{H}|A)=0.05(0.60)=0.03$

43. **Answers may vary.**

45.

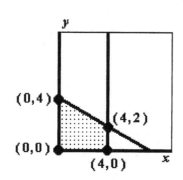

Point	$P=2x+y$
$(0,0)$	$=2(0)+0=0$
$(0,4)$	$=2(0)+4=4$
$(4,2)$	$=2(4)+2=10$
$(4,0)$	$=2(4)+0=8$

Max: $P=10$ at $(4,2)$

Exercise 8.9 (page 617)

1. odds for an event

3. $\dfrac{1}{1+4}=\dfrac{1}{5}$

5. $\dfrac{6}{36}=\dfrac{1}{6}$

7. $\dfrac{\frac{5}{6}}{\frac{1}{6}}=\dfrac{5}{1}=5\text{ to }1$

9. $\dfrac{\frac{1}{2}}{\frac{1}{2}}=\dfrac{1}{1}=1\text{ to }1$

11. $\dfrac{5}{36}$

13. $\dfrac{\frac{31}{36}}{\frac{5}{36}}=\dfrac{31}{5}=31\text{ to }5$

15. $\dfrac{\frac{1}{2}}{\frac{1}{2}}=\dfrac{1}{1}=1\text{ to }1$

17. $\dfrac{\frac{4}{52}}{\frac{48}{52}}=\dfrac{1}{12}=1\text{ to }12$

19. $\dfrac{\frac{12}{52}}{\frac{40}{52}}=\dfrac{3}{10}=3\text{ to }10$

21. $\dfrac{5}{5+2}=\dfrac{5}{7}$

23. 1 to 90

25. $P(4\text{ heads})=\dfrac{1}{2}\cdot\dfrac{1}{2}\cdot\dfrac{1}{2}\cdot\dfrac{1}{2}=\dfrac{1}{16}$

odds against $=\dfrac{\frac{15}{16}}{\frac{1}{16}}=\dfrac{15}{1}=15\text{ to }1$

27. $\dfrac{1}{8+1}=\dfrac{1}{9}$

29. Expected winnings $=\dfrac{4}{52}(5)+\dfrac{4}{52}(4)+\dfrac{44}{52}(0)=\dfrac{36}{52}\approx\$0.69\Rightarrow$ not worth playing

31. $P(3 \text{ boys}) = \dfrac{1}{2} \cdot \dfrac{1}{2} \cdot \dfrac{1}{2} \cdot \dfrac{1}{2} = \dfrac{1}{8} \Rightarrow \text{odds against} = \dfrac{\frac{7}{8}}{\frac{1}{8}} = \dfrac{7}{1} = 7 \text{ to } 1$

33. $P(5 \text{ heads}) = \dfrac{\binom{5}{5}}{2^5} = \dfrac{1}{32},\ P(4 \text{ heads}) = \dfrac{\binom{5}{4}}{2^5} = \dfrac{5}{32},\ P(3 \text{ heads}) = \dfrac{\binom{5}{3}}{2^5} = \dfrac{10}{32}$

Expected winnings $= \dfrac{1}{32}(5) + \dfrac{5}{32}(4) + \dfrac{10}{32}(3) \approx \$1.72 \Leftarrow$ fair price

35. Expected value $= \dfrac{4}{52}(1) + \dfrac{12}{52}(10) + \dfrac{4}{52}(2) + \dfrac{4}{52}(3) + \cdots + \dfrac{4}{52}(10) \approx 6.54$

37. $\dbinom{8}{7} \cdot \left(\dfrac{1}{5}\right)^7 \cdot \dfrac{4}{5} = \dfrac{32}{390{,}625}$ **39.** **Answers may vary.**

41.
$$3x - 2y = 9$$
$$-2y = -3x + 9 \qquad\qquad y - y_1 = m(x - x_1)$$
$$y = \tfrac{3}{2}x - \tfrac{9}{2} \qquad\qquad y - 1 = -\tfrac{2}{3}(x - 1)$$
$$m = \tfrac{3}{2} \qquad\qquad\qquad\quad 3y - 3 = -2(x - 1)$$
$$\text{Use } m = -\tfrac{2}{3}. \qquad\qquad 3y - 3 = -2x + 2$$
$$\boxed{2x + 3y = 5}$$

43. $C(3, 5),\ r = 5 \Rightarrow (x - 3)^2 + (y - 5)^2 = 25$

Chapter 8 Summary (page 620)

1. **a.** $6! = 6 \cdot 5 \cdot 4 \cdot 3 \cdot 2 \cdot 1 = 720$ **b.** $7! \cdot 0! \cdot 1! \cdot 3! = 5040 \cdot 1 \cdot 1 \cdot 6 = 30{,}240$

c. $\dfrac{8!}{7!} = \dfrac{8 \cdot 7!}{7!} = 8$

d. $\dfrac{5! \cdot 7! \cdot 8!}{6! \cdot 9!} = \dfrac{5! \cdot 7 \cdot 6! \cdot 8!}{6! \cdot 9 \cdot 8!} = \dfrac{5! \cdot 7}{9} = \dfrac{280}{3}$

2. **a.** $(x + y)^3 = x^3 + \dfrac{3!}{1!2!}x^2y + \dfrac{3!}{2!1!}xy^2 + y^3 = x^3 + 3x^2y + 3xy^2 + y^3$

b. $(p + q)^4 = p^4 + \dfrac{4!}{1!3!}p^3q + \dfrac{4!}{2!2!}p^2q^2 + \dfrac{4!}{3!1!}pq^3 + q^4 = p^4 + 4p^3q + 6p^2q^2 + 4pq^3 + q^4$

c. $(a - b)^5 = a^5 + \dfrac{5!}{1!4!}a^4(-b) + \dfrac{5!}{2!3!}a^3(-b)^2 + \dfrac{5!}{3!2!}a^2(-b)^3 + \dfrac{5!}{4!1!}a(-b)^4 + (-b)^5$
$$= a^5 - 5a^4b + 10a^3b^2 - 10a^2b^3 + 5ab^4 - b^5$$

d. $(2a - b)^3 = (2a)^3 + \dfrac{3!}{1!2!}(2a)^2(-b) + \dfrac{3!}{2!1!}(2a)(-b)^2 + (-b)^3 = 8a^3 - 12a^2b + 6ab^2 - b^3$

3. **a.** The 4th term will involve b^3.

$$\frac{8!}{5!3!}a^5b^3 = 56a^5b^3$$

b. The 3rd term will involve $(-y)^2$.

$$\frac{5!}{3!2!}(2x)^3(-y)^2 = 80x^3y^2$$

c. The 7th term will involve $(-y)^6$.

$$\frac{9!}{3!6!}x^3(-y)^6 = 84x^3y^6$$

d. The 4th term will involve 7^3.

$$\frac{6!}{3!3!}(4x)^3 7^3 = 439{,}040x^3$$

4. **a.** $4^3 - 1 = 63$

b. $\dfrac{4^2 + 2}{2} = \dfrac{18}{2} = 9$

5. **a.** $a_1 = 5$
$a_2 = 3a_1 + 2 = 3(5) + 2 = 17$
$a_3 = 3a_2 + 2 = 3(17) + 2 = 53$
$a_4 = 3a_3 + 2 = 3(53) + 2 = 161$

b. $a_1 = -2$
$a_2 = 2a_1^2 = 2(-2)^2 = 8$
$a_3 = 2a_2^2 = 2(8)^2 = 128$
$a_4 = 2a_3^2 = 2(128)^2 = 32{,}768$

6. **a.** $\displaystyle\sum_{k=1}^{4} 3k^2 = 3\sum_{k=1}^{4} k^2 = 3\left(1^2 + 2^2 + 3^2 + 4^2\right) = 3(30) = 90$

b. $\displaystyle\sum_{k=1}^{10} 6 = 10(6) = 60$

c. $\displaystyle\sum_{k=5}^{8}\left(k^3 + 3k^2\right) = \sum_{k=5}^{8} k^3 + 3\sum_{k=5}^{8} k^2 = \left(5^3 + 6^3 + 7^3 + 8^3\right) + 3\left(5^2 + 6^2 + 7^2 + 8^2\right)$

$$= 1718$$

d. $\displaystyle\sum_{k=1}^{30}\left(\frac{3}{2}k - 12\right) - \frac{3}{2}\sum_{k=1}^{30} k = \frac{3}{2}\sum_{k=1}^{30} k - \sum_{k=1}^{30} 12 - \frac{3}{2}\sum_{k=1}^{30} k = -\sum_{k=1}^{30} 12 = -360$

7. **a.** $a = 5, d = 4$
29th term $= a + (n-1)d$
$= 5 + (29 - 1)4 = 117$

b. $a = 8, d = 7$
40th term $= a + (n-1)d$
$= 8 + (40 - 1)7 = 281$

c. $a = 6, d = -7$
15th term $= a + (n-1)d$
$= 6 + (15 - 1)(-7) = -92$

d. $a = \frac{1}{2}, d = -2$
35th term $= a + (n-1)d$
$= \frac{1}{2} + (35 - 1)(-2) = -\frac{13\!}{2}$

8. $a = 2$, 5th term $= 8$
$8 = 2 + 4d$
$6 = 4d$
$\dfrac{3}{2} = d \Rightarrow 2, \dfrac{7}{2}, 5, \dfrac{13}{2}, 8$

9. $a = 10$, 7th term $= 100$
$100 = 10 + 6d$
$90 = 6d$
$15 = d \Rightarrow 10, 25, 40, 55, 70, 85, 100$

10. a. $a = 5, d = 4$

$l = a + (n - 1)d = 5 + 39(4) = 161$

$S = \dfrac{n(a + l)}{2} = \dfrac{40(5 + 161)}{2} = 3320$

b. $a = 8, d = 7$

$l = a + (n - 1)d = 8 + 39(7) = 281$

$S = \dfrac{n(a + l)}{2} = \dfrac{40(8 + 281)}{2} = 5780$

c. $a = 6, d = -7$

$l = a + (n - 1)d = 6 + 39(-7) = -267; \quad S = \dfrac{n(a + l)}{2} = \dfrac{40(6 - 267)}{2} = -5220$

d. $a = \frac{1}{2}, d = -2$

$l = a + (n - 1)d = \frac{1}{2} + 39(-2) = -\frac{155}{2}; \quad S = \dfrac{n(a + l)}{2} = \dfrac{40\left(\frac{1}{2} - \frac{155}{2}\right)}{2} = -1540$

11. a. $a = 81, r = \frac{1}{3}$

11th term $= ar^{n-1} = 81\left(\frac{1}{3}\right)^{10} = \frac{1}{729}$

b. $a = 2, r = 3$

9th term $= ar^{n-1} = 2(3)^8 = 13{,}122$

c. $a = 9, r = \frac{1}{2}$

15th term $= ar^{n-1} = 9\left(\frac{1}{2}\right)^{14} = \frac{9}{16{,}384}$

d. $a = 8, r = -\frac{1}{5}$

7th term $= ar^{n-1} = 8\left(-\frac{1}{5}\right)^6 = \frac{8}{15{,}625}$

12. 5th term $= ar^4$

$8 = 2r^4$

$4 = r^4$

$\pm \sqrt[4]{4} = r \Rightarrow r = \pm\sqrt{2}$

Use $r = +\sqrt{2}$:

$2, 2\sqrt{2}, 4, 4\sqrt{2}, 8$

13. 6th term $= ar^5$

$64 = -2r^5$

$-32 = r^5$

$-2 = r$

$-2, 4, -8, 16, -32, 64$

14. 3rd term $= ar^2$

$64 = 4r^2$

$16 = r^2 \Rightarrow r = 4$ (problem specifies positive) $\Rightarrow 4, 16, 64$

15. a. $a = 81, r = \frac{1}{3}, n = 8$

$S = \dfrac{a - ar^n}{1 - r} = \dfrac{81 - 81\left(\frac{1}{3}\right)^8}{1 - \frac{1}{3}}$

$= \dfrac{81 - \frac{1}{81}}{\frac{2}{3}} = \dfrac{3280}{27}$

b. $a = 2, r = 3, n = 8$

$S = \dfrac{a - ar^n}{1 - r} = \dfrac{2 - 2(3)^8}{1 - 3}$

$= \dfrac{-13{,}120}{-2} = 6560$

c. $a = 9, r = \frac{1}{2}, n = 8$

$S = \dfrac{a - ar^n}{1 - r} = \dfrac{9 - 9\left(\frac{1}{2}\right)^8}{1 - \frac{1}{2}}$

$= \dfrac{\frac{2295}{256}}{\frac{1}{2}} = \dfrac{2295}{128}$

d. $a = 8, r = -\frac{1}{5}, n = 8$

$S = \dfrac{a - ar^n}{1 - r} = \dfrac{8 - 8\left(-\frac{1}{5}\right)^8}{1 - \left(-\frac{1}{5}\right)}$

$= \dfrac{\frac{3{,}124{,}992}{390{,}625}}{\frac{6}{5}} = \dfrac{520{,}832}{78{,}125}$

16. $a = \frac{1}{3}, r = 3, n = 8$

$$S = \frac{a - ar^n}{1 - r} = \frac{\frac{1}{3} - \frac{1}{3}(3)^8}{1 - 3}$$

$$= \frac{-\frac{6560}{3}}{-2} = \frac{3280}{3}$$

17. 7th term $= ar^6$

$$= 2\sqrt{2}\left(\sqrt{2}\right)^6$$

$$= 16\sqrt{2}$$

18. a. $a = \frac{1}{3}, r = \frac{1}{2}$

$$S = \frac{a}{1 - r} = \frac{\frac{1}{3}}{1 - \frac{1}{2}} = \frac{\frac{1}{3}}{\frac{1}{2}} = \frac{2}{3}$$

b. $a = \frac{1}{5}, r = -\frac{2}{3}$

$$S = \frac{a}{1 - r} = \frac{\frac{1}{5}}{1 - \left(-\frac{2}{3}\right)} = \frac{\frac{1}{5}}{\frac{5}{3}} = \frac{3}{25}$$

c. $a = \frac{1}{3}, r = \frac{3}{2} > 1 \Rightarrow$ no sum

d. $a = \frac{1}{2}, r = \frac{1}{2}$

$$S = \frac{a}{1 - r} = \frac{\frac{1}{2}}{1 - \frac{1}{2}} = \frac{\frac{1}{2}}{\frac{1}{2}} = 1$$

19. a. $a = \frac{3}{10}, r = \frac{1}{10}$

$$S = \frac{a}{1 - r} = \frac{\frac{3}{10}}{1 - \frac{1}{10}} = \frac{\frac{3}{10}}{\frac{9}{10}} = \frac{1}{3}$$

b. $a = \frac{9}{10}, r = \frac{1}{10}$

$$S = \frac{a}{1 - r} = \frac{\frac{9}{10}}{1 - \frac{1}{10}} = \frac{\frac{9}{10}}{\frac{9}{10}} = 1$$

c. $a = \frac{17}{100}, r = \frac{1}{100}$

$$S = \frac{a}{1 - r} = \frac{\frac{17}{100}}{1 - \frac{1}{100}} = \frac{\frac{17}{100}}{\frac{99}{100}} = \frac{17}{99}$$

d. $a = \frac{45}{100}, r = \frac{1}{100}$

$$S = \frac{a}{1 - r} = \frac{\frac{45}{100}}{1 - \frac{1}{100}} = \frac{\frac{45}{100}}{\frac{99}{100}} = \frac{5}{11}$$

20. $a = 3000, r = 1 + \frac{0.0775}{365}, n = 2190$

$$ar^{365} = 3000\left(1 + \frac{0.0775}{365}\right)^{2190}$$

$$\approx \$4775.81$$

The amount will be \$4775.81.

21. $a = 4000, r = 1.05, n = 10$

$$ar^{10} = 4000(1.05)^{10} \approx 6516 \text{ in 10 years}$$

$$ar^{-5} = 4000(1.05)^{-5} \approx 3134 \text{ 5 years ago}$$

22. $a = 10,000; \ r = 0.90; \ ar^{10} = 10,000(0.90)^{10} \approx \$3486.78$ in 10 years

23.

$$n = 1$$
$$1^3 \overset{?}{=} \frac{1^2(1+1)^2}{4}$$
$$1 = 1$$

$$n = 2$$
$$1^3 + 2^3 \overset{?}{=} \frac{2^2(2+1)^2}{4}$$
$$9 \overset{?}{=} \frac{4(9)}{4}$$
$$9 = 9$$

$$n = 3$$
$$1^3 + 2^3 + 3^3 \overset{?}{=} \frac{3^2(3+1)^2}{4}$$
$$36 \overset{?}{=} \frac{9(16)}{4}$$
$$36 = 36$$

$$n = 4$$
$$1^3 + 2^3 + 3^3 + 4^3 \overset{?}{=} \frac{4^2(4+1)^2}{4}$$
$$100 \overset{?}{=} \frac{16(25)}{4}$$
$$100 = 100$$

24.

Check $n = 1$: $1^3 = \dfrac{1^2(1+1)^2}{4}$ True for $n = 1$

$$1 = 1$$

Assume for $n = k$ and show for $n = k + 1$:

$$1^3 + 2^3 + 3^3 + \cdots + k^3 = \frac{k^2(k+1)^2}{4}$$

$$\boxed{1^3 + 2^3 + 3^3 + \cdots + k^3} + (k+1)^3 = \boxed{\frac{k^2(k+1)^2}{4}} + (k+1)^3$$

$$1^3 + 2^3 + 3^3 + \cdots + (k+1)^3 = \frac{k^2(k+1)^2 + 4(k+1)^3}{4}$$

$$1^2 + 2^2 + 3^2 + \cdots + (k+1)^2 = \frac{(k+1)^2[k^2 + 4(k+1)]}{4}$$

$$1^2 + 2^2 + 3^2 + \cdots + (k+1)^2 = \frac{(k+1)^2(k+2)^2}{4}$$

Since this is what results when $n = k + 1$ in the formula, we have shown that the formula works for $n = k + 1$ if it works for $n = k$.

25.

a. $P(8, 5) = \dfrac{8!}{(8-5)!} = \dfrac{8!}{3!} = 8 \cdot 7 \cdot 6 \cdot 5 \cdot 4 = 6720$

b. $C(7, 4) = \dfrac{7!}{4!(7-4)!} = \dfrac{7!}{4!3!} = \dfrac{7 \cdot 6 \cdot 5 \cdot 4!}{4! \cdot 3 \cdot 2 \cdot 1} = 35$

c. $0! \cdot 1! = 1 \cdot 1 = 1$

d. $P(10, 2) \cdot C(10, 2) = \dfrac{10!}{8!} \cdot \dfrac{10!}{2!8!} = 90 \cdot 45 = 4050$

e. $P(8, 6) \cdot C(8, 6) = \dfrac{8!}{2!} \cdot \dfrac{8!}{6!2!} = 20{,}160 \cdot 28 = 564{,}480$

f. $\quad C(8,5) \cdot C(6,2) = \dfrac{8!}{5!3!} \cdot \dfrac{6!}{2!4!} = 56 \cdot 15 = 840$

g. $\quad C(7,5) \cdot P(4,0) = \dfrac{7!}{5!2!} \cdot \dfrac{4!}{4!} = 21 \cdot 1 = 21$

h. $\quad C(12,10) \cdot C(11,0) = \dfrac{12!}{10!2!} \cdot \dfrac{11!}{0!11!} = 66 \cdot 1 = 66$

i. $\quad \dfrac{P(8,5)}{C(8,5)} = \dfrac{6720}{56} = 120$

j. $\quad \dfrac{C(8,5)}{C(13,5)} = \dfrac{56}{1287}$

k. $\quad \dfrac{C(6,3)}{C(10,3)} = \dfrac{20}{120} = \dfrac{1}{6}$

l. $\quad \dfrac{C(13,5)}{C(52,5)} = \dfrac{1287}{2,598,960} = \dfrac{33}{66,640}$

26. Consider each set of two people who must sit together as a single person, so that there are 8 "people" who must be arranged in a circle. This can be done in $(8-1)! = 7! = 5040$ ways. However, each pair seated next to each other could be switched, so that the number of arrangements is multiplied by 4. There are $4 \cdot 5040 = 20{,}160$ possible arrangements.

27. $\dfrac{9!}{2!2!} = 90{,}720$

28.

29. $\dbinom{4}{3}\dbinom{4}{2} = 4 \cdot 6 = 24$

30. $\dfrac{\binom{4}{3}\binom{4}{2}}{\binom{52}{5}} = \dfrac{24}{2,598,960} = \dfrac{1}{108,290}$

31. $1 - \dfrac{1}{108,290} = \dfrac{108,289}{108,290}$

32. $\dfrac{\binom{4}{4}\binom{4}{4}\binom{4}{4}\binom{4}{1}}{\binom{52}{13}} \approx \dfrac{4}{6.35 \times 10^{11}} \approx 6.3 \times 10^{-}$

33. $\dfrac{\binom{8}{3}\binom{6}{2}}{\binom{14}{5}} = \dfrac{840}{2002} = \dfrac{60}{143}$

34. $\dfrac{13 + 13}{52} = \dfrac{1}{2}$

35. $\dfrac{26 + 4 - 2}{52} = \dfrac{7}{13}$

36. $\dfrac{1}{\binom{52}{5}} = \dfrac{1}{2,598,960}$

37. $\dfrac{4\binom{13}{13}}{\binom{52}{13}} \approx \dfrac{4}{6.35 \times 10^{11}}$

38. Use the tree diagram in #28. $\Rightarrow \dfrac{15}{16}$

CHAPTER 8 SUMMARY

39. $P(2 \text{ or spade}) = P(2) + P(\text{spade}) - P(2 \text{ and spade}) = \frac{4}{52} + \frac{13}{52} - \frac{1}{52} = \frac{16}{52} = \frac{4}{13}$

40. # cured $= 0.83(800) = 664$ people
not cured $= 800 - 664 = 136$

41. $\dfrac{\frac{1}{8}}{\frac{7}{8}} = \dfrac{1}{7} = 1 \text{ to } 7$

42. $P(4 \text{ girls}) = \frac{1}{2} \cdot \frac{1}{2} \cdot \frac{1}{2} \cdot \frac{1}{2} = \frac{1}{16}$

odds in favor $= \dfrac{\frac{1}{16}}{\frac{15}{16}} = \dfrac{1}{15} = 1 \text{ to } 15$

43. $P(\text{neither}) = \frac{1}{6} \cdot \frac{1}{4} = \frac{1}{24}$

odds $= \dfrac{\frac{1}{24}}{\frac{23}{24}} = \dfrac{1}{23} = 1 \text{ to } 23$

44. $P(\text{not graduate}) = \dfrac{10}{10+11} = \dfrac{10}{21} \Rightarrow P(\text{graduate}) = \dfrac{11}{21}$

45. $P(4 \text{ heads}) = \dfrac{\binom{4}{4}}{2^4} = \dfrac{1}{16}, \; P(3 \text{ heads}) = \dfrac{\binom{4}{3}}{2^4} = \dfrac{4}{16}, \; P(2 \text{ heads}) = \dfrac{\binom{4}{2}}{2^4} = \dfrac{6}{16}$

$P(1 \text{ heads}) = \dfrac{\binom{4}{1}}{2^4} = \dfrac{4}{16}, \; P(0 \text{ heads}) = \dfrac{\binom{4}{0}}{2^4} = \dfrac{1}{16}$

Expected value $= 4\left(\dfrac{1}{16}\right) + 3\left(\dfrac{4}{16}\right) + 2\left(\dfrac{6}{16}\right) + 1\left(\dfrac{4}{16}\right) + 0\left(\dfrac{1}{16}\right) = \dfrac{32}{16} = \$2.00$

46. Each addend represents the number of ways of selecting a subset from the set, starting with subsets of size 0, size 1, size 2, and so on until size n. Thus, when added together, the sum represents the total number of subsets (of all size), or 2^n.

Chapter 8 Test (page 624)

1. $3! \cdot 0! \cdot 4! \cdot 1! = 6 \cdot 1 \cdot 24 \cdot 1 = 144$

2. $\dfrac{2! \cdot 4! \cdot 6! \cdot 8!}{3! \cdot 5! \cdot 7!} = 2! \cdot \dfrac{4!}{3!} \cdot \dfrac{6!}{5!} \cdot \dfrac{8!}{7!}$
$= 2 \cdot 4 \cdot 6 \cdot 8 = 384$

3. The 2nd term will involve $(2y)^1$.
$\dfrac{5!}{4!1!} x^4 (2y)^1 = 10x^4 y$

4. The 7th term will involve $(-b)^6$.
$\dfrac{8!}{2!6!} (2a)^2 (-b)^6 = 112a^2 b^6$

5. $\displaystyle\sum_{k=1}^{3}(4k+1) = 4\sum_{k=1}^{3} k + \sum_{k=1}^{3} 1 = 4(1+2+3) + 3(1) = 24 + 3 = 27$

6. $\displaystyle\sum_{k=2}^{4}(3k-21) = 3\sum_{k=2}^{4} k - \sum_{k=2}^{4} 21 = 3(2+3+4) - 3(21) = 27 - 63 = -36$

7. $a = 2, d = 3$
$l = a + (n-1)d = 2 + 9(3) = 29$
$S = \dfrac{n(a+l)}{2} = \dfrac{10(2+29)}{2} = 155$

8. $a = 5, d = -4$
$l = a + (n-1)d = 5 + 9(-4) = -31$
$S = \dfrac{n(a+l)}{2} = \dfrac{10(5-31)}{2} = -130$

9. $a = 4$, 5th term $= 24$

$24 = 4 + 4d$

$20 = 4d$

$5 = d \Rightarrow 4, 9, 14, 19, 24$

10. 4th term $= ar^3$

$-54 = -2r^3$

$27 = r^3$

$3 = r \Rightarrow -2, -6, -18, -54$

11. $a = \frac{1}{4}, r = 2, n = 10$

$S = \dfrac{a - ar^n}{1 - r} = \dfrac{\frac{1}{4} - \frac{1}{4}(2)^{10}}{1 - 2}$

$= \dfrac{-\frac{1023}{4}}{-1}$

$= \dfrac{1023}{4} = 255.75$

12. $a = 6, r = \frac{1}{3}, n = 10$

$S = \dfrac{a - ar^n}{1 - r} = \dfrac{6 - 6\left(\frac{1}{3}\right)^{10}}{1 - \frac{1}{3}}$

$= \dfrac{\frac{354,288}{59,049}}{\frac{2}{3}}$

$= \dfrac{177,144}{19,683} \approx 9$

13. $a = c;\ r = 0.75$

$ar^3 = c(0.75)^3 \approx \$0.42c$ in 3 years

14. $a = c;\ r = 1.10$

$ar^4 = c(1.10)^4 \approx \$1.46c$ in 4 years

15.

Check $n = 1$: $\quad 3 = \dfrac{1}{2}(1)(1 + 5) \quad$ True for $n = 1$

$\qquad\qquad\qquad 3 = 3$

Assume for $n = k$ and show for $n = k + 1$:

$$3 + 4 + 5 + \cdots + (k + 2) = \frac{1}{2}k(k + 5)$$

$$\boxed{3 + 4 + 5 + \cdots + (k + 2)} + ((k + 1) + 2) = \boxed{\frac{1}{2}k(k + 5)} + ((k + 1) + 2)$$

$$3 + 4 + 5 + \cdots + ((k + 1) + 2) = \frac{1}{2}k^2 + \frac{7}{2}k + 3$$

$$3 + 4 + 5 + \cdots + ((k + 1) + 2) = \frac{1}{2}\left(k^2 + 7k + 6\right)$$

$$3 + 4 + 5 + \cdots + ((k + 1) + 2) = \frac{1}{2}(k + 1)(k + 6) = \frac{1}{2}(k + 1)((k + 1) + 5)$$

Since this is what results when $n = k + 1$ in the formula, we have shown that the formula works for $n = k + 1$ if it works for $n = k$.

16. $8 \cdot 10 \cdot 10 \cdot 10 \cdot 10 \cdot 10 = 800,000$

17. $P(7, 2) = \dfrac{7!}{5!} = 42$

18. $P(4, 4) = \dfrac{4!}{0!} = 24$

19. $C(8, 2) = \dfrac{8!}{2!6!} = 28$

20. $C(12, 0) = \dfrac{12!}{0!12!} = 1$

21. $4!4! = 576$

22. $(6 - 1)! = 5! = 120$

23. $\dfrac{5!}{2!} = 60$

24. $\{(H,H,H),(H,H,T),(H,T,H),(H,T,T),(T,H,H),(T,H,T),(T,T,H),(T,T,T)\}$

25. $\dfrac{1}{6}$

26. $\dfrac{4+4}{52} = \dfrac{2}{13}$

27. $\dfrac{\binom{13}{5}}{\binom{52}{5}} = \dfrac{429}{866,320}$

28. $\dfrac{\binom{5}{2}}{2^5} = \dfrac{10}{32} = \dfrac{5}{16}$

29. $0.30 + 0.80 - 0.70 = 0.40$
$= 40\%$

30. $1 - 0.1 = 0.9$

31. $\dfrac{26 + 12 - 6}{52} = \dfrac{8}{13}$

32. $\dfrac{2+18}{36} = \dfrac{5}{9}$

33. $\dfrac{3}{3+1} = \dfrac{3}{4}$

34. Expected winnings $= \frac{1}{6}(4) + \frac{1}{6}(2) + \frac{4}{6}(0) = \$1.00 \Leftarrow$ fair price

Exercise 9.1 (page 635)

1. interest

3. annual

5. future

7. interest

9. periodic rate

11. principal, periodic rate, frequency of compounding, number of years

13. effective

15. $P = 1200, i = 0.08, k = 1, n = 1$
$FV = P(1+i)^{kn}$
$= 1200(1+0.08)^{1\cdot1} \approx \$1296$

17. $P = 1200, i = 0.08, k = 1, n = 5$
$FV = P(1+i)^{kn}$
$= 1200(1+0.08)^{1\cdot5} \approx \$1763.19$

19. $P = 1200, i = 0.03, k = 1, n = 10$
$FV = P(1+i)^{kn}$
$= 1200(1+0.03)^{1\cdot10} \approx \$1612.70$

21. $P = 1200, i = 0.09, k = 1, n = 10$
$FV = P(1+i)^{kn}$
$= 1200(1+0.09)^{1\cdot10} \approx \$2840.84$

23. $P = 1200, i = \frac{0.06}{2}, k = 2, n = 15$
$FV = P(1+i)^{kn}$
$= 1200\left(1 + \frac{0.06}{2}\right)^{2\cdot15} \approx \$2912.71$

25. $P = 1200, i = \frac{0.06}{12}, k = 12, n = 15$
$FV = P(1+i)^{kn}$
$= 1200\left(1 + \frac{0.06}{12}\right)^{12\cdot15} \approx \$2944.91$

27. $k = 4, i = \frac{0.06}{4}$
$R = (1+i)^k - 1$
$= \left(1 + \frac{0.06}{4}\right)^4 - 1$
$\approx 0.0614 = 6.14\%$

29. $k = 2, i = \frac{0.095}{2}$
$R = (1+i)^k - 1$
$= \left(1 + \frac{0.095}{2}\right)^2 - 1$
$\approx 0.0973 = 9.73\%$

31. $FV = 20,000; i = \frac{0.06}{2}, k = 2, n = 6$

$PV = FV(1 + i)^{-kn}$

$\quad = 20,000\left(1 + \frac{0.06}{2}\right)^{-2 \cdot 6}$

$\quad \approx \$14,027.60$

33. $FV = 20,000; i = \frac{0.09}{12}, k = 12, n = 6$

$PV = FV(1 + i)^{-kn}$

$\quad = 20,000\left(1 + \frac{0.09}{12}\right)^{-12 \cdot 6}$

$\quad \approx \$11,678.47$

35. $P = 7000, i = \frac{0.06}{4}, k = 4, n = 18$

$FV = P(1 + i)^{kn}$

$\quad = 7000\left(1 + \frac{0.06}{4}\right)^{4 \cdot 18} \approx \$20,448.11$

37. $P = 147,500; i = \frac{0.075}{2}, k = 2, n = 12$

$FV = P(1 + i)^{kn}$

$\quad = 147,500\left(1 + \frac{0.075}{2}\right)^{2 \cdot 12} \approx \$356,867.1$

39. $P = 137,000; i = \frac{0.11}{1}, k = 1, n = 4$

$FV = P(1 + i)^{kn}$

$\quad = 137,000\left(1 + \frac{0.11}{1}\right)^{1 \cdot 4} \approx \$207,976$

41. $P = 4.3, i = \frac{0.072}{1}, k = 1, n = 10$

$FV = P(1 + i)^{kn}$

$\quad = 4.3\left(1 + \frac{0.072}{1}\right)^{1 \cdot 10} \approx \$8.62 \text{ million ft}^3$

43.

NOW account	Money Market

$k = 4, i = \frac{0.072}{4}$ $\qquad$ $k = 12, i = \frac{0.069}{12}$

$R = (1 + i)^k - 1$ $\qquad$ $R = (1 + i)^k - 1$

$\quad = \left(1 + \frac{0.072}{4}\right)^4 - 1$ $\qquad$ $= \left(1 + \frac{0.069}{12}\right)^{12} - 1$

$\quad \approx 0.0740 = 7.40\%$ $\qquad$ $\approx 0.0712 = 7.12\%$

45. $P = 1230, i = \frac{0.12}{360}, k = 360, n = 0.25$

$FV = P(1 + i)^{kn}$

$\quad = 1230\left(1 + \frac{0.12}{360}\right)^{360(0.25)} \approx \$1267.45$

47. $FV = 4200; i = \frac{0.0575}{12}, k = 12, n = 1.5$

$PV = FV(1 + i)^{-kn}$

$\quad = 4200\left(1 + \frac{0.0575}{12}\right)^{-12(1.5)}$

$\quad \approx \$3853.73$

49.

First ten years	Next four years

$P = 12,000; \; i = \frac{0.075}{12}, k = 12, n = 10$ $\qquad$ $P = 37,344.78; \; i = \frac{0.075}{12}, k = 12, n = 4$

$FV = P(1 + i)^{kn}$ $\qquad$ $FV = P(1 + i)^{kn}$

$\quad = 12,000\left(1 + \frac{0.075}{12}\right)^{12(10)} \approx \$25,344.78$ $\qquad$ $= 37,344.78\left(1 + \frac{0.075}{12}\right)^{12(4)} \approx \$50,363.14$

51. **Answers may vary.**

53. $\dfrac{x^2 - 2x - 15}{2x^2 - 9x - 5} = \dfrac{(x - 5)(x + 3)}{(2x + 1)(x - 5)} = \dfrac{x + 3}{2x + 1}$

55. $\dfrac{(3 - x)(x + 3)}{-x^2 + 9} = \dfrac{(3 - x)(x + 3)}{9 - x^2} = \dfrac{(3 - x)(x + 3)}{(3 + x)(3 - x)} = 1$

Exercise 9.2 (page 640)

1. annuities

3. payments, interest

5. regular deposit, periodic r
frequency of compoundin
number of years

7. $P = 100, i = \frac{0.06}{1}, k = 1, n = 10$

$$FV = \frac{P\left[(1+i)^{kn} - 1\right]}{i}$$

$$= \frac{100\left[\left(1 + \frac{0.06}{1}\right)^{1 \cdot 10} - 1\right]}{\frac{0.06}{1}}$$

$$\approx \$1318.08$$

9. $P = 100, i = \frac{0.06}{1}, k = 1, n = 3$

$$FV = \frac{P\left[(1+i)^{kn} - 1\right]}{i}$$

$$= \frac{100\left[\left(1 + \frac{0.06}{1}\right)^{1 \cdot 3} - 1\right]}{\frac{0.06}{1}}$$

$$\approx \$318.36$$

11. $P = 100, i = \frac{0.04}{1}, k = 1, n = 10$

$$FV = \frac{P\left[(1+i)^{kn} - 1\right]}{i}$$

$$= \frac{100\left[\left(1 + \frac{0.04}{1}\right)^{1 \cdot 10} - 1\right]}{\frac{0.04}{1}}$$

$$\approx \$1200.61$$

13. $P = 100, i = \frac{0.095}{1}, k = 1, n = 10$

$$FV = \frac{P\left[(1+i)^{kn} - 1\right]}{i}$$

$$= \frac{100\left[\left(1 + \frac{0.095}{1}\right)^{1 \cdot 10} - 1\right]}{\frac{0.095}{1}}$$

$$\approx \$1556.03$$

15. $P = 100, i = \frac{0.08}{2}, k = 2, n = 15$

$$FV = \frac{P\left[(1+i)^{kn} - 1\right]}{i}$$

$$= \frac{100\left[\left(1 + \frac{0.08}{2}\right)^{2 \cdot 15} - 1\right]}{\frac{0.08}{2}}$$

$$\approx \$5608.49$$

17. $P = 100, i = \frac{0.08}{12}, k = 12, n = 15$

$$FV = \frac{P\left[(1+i)^{kn} - 1\right]}{i}$$

$$= \frac{100\left[\left(1 + \frac{0.08}{12}\right)^{12 \cdot 15} - 1\right]}{\frac{0.08}{12}}$$

$$\approx \$34,603.82$$

19. $FV = 20,000; \ i = \frac{0.04}{1}, k = 1, n = 10$

$$P = \frac{FVi}{(1+i)^{kn} - 1}$$

$$= \frac{20,000\left(\frac{0.04}{1}\right)}{\left(1 + \frac{0.04}{1}\right)^{1 \cdot 10} - 1}$$

$$\approx \$1665.81$$

21. $FV = 20,000; \ i = \frac{0.09}{2}, k = 2, n = 10$

$$P = \frac{FVi}{(1+i)^{kn} - 1}$$

$$= \frac{20,000\left(\frac{0.09}{2}\right)}{\left(1 + \frac{0.09}{2}\right)^{2 \cdot 10} - 1}$$

$$\approx \$637.52$$

23. $P = 200, i = \frac{0.06}{12}, k = 12, n = 1$

$$FV = \frac{P\left[(1+i)^{kn} - 1\right]}{i}$$

$$= \frac{200\left[\left(1 + \frac{0.06}{12}\right)^{12 \cdot 1} - 1\right]}{\frac{0.06}{12}}$$

$$\approx \$2467.11$$

25. $P = 135,000; \ i = \frac{0.087}{12}, k = 12, n = 20$

$$FV = \frac{P\left[(1+i)^{kn} - 1\right]}{i}$$

$$= \frac{135,000\left[\left(1 + \frac{0.087}{12}\right)^{12 \cdot 20} - 1\right]}{\frac{0.087}{12}}$$

$$\approx \$86,803,923.58$$

27. $FV = 750{,}000;\ i = \frac{0.0975}{12},\ k = 12,\ n = 2.5$

$$P = \frac{FVi}{(1+i)^{kn} - 1}$$

$$= \frac{750{,}000\left(\frac{0.0975}{12}\right)}{\left(1 + \frac{0.0975}{12}\right)^{12 \cdot 2.5} - 1}$$

$$\approx \$22{,}177.71$$

29.

Bank A	Bank B
$FV = 10{,}000;\ i = \frac{0.055}{1},\ k = 1,\ n = 20$	$FV = 10{,}000;\ i = \frac{0.0535}{12},\ k = 12,\ n = 20$

Bank A:
$$P = \frac{FVi}{(1+i)^{kn} - 1}$$

$$= \frac{10{,}000\left(\frac{0.055}{1}\right)}{\left(1 + \frac{0.055}{1}\right)^{1 \cdot 20} - 1}$$

$$\approx \$286.79 \text{ per year}$$

Bank B:
$$P = \frac{FVi}{(1+i)^{kn} - 1}$$

$$= \frac{10{,}000\left(\frac{0.0535}{12}\right)}{\left(1 + \frac{0.0535}{12}\right)^{12 \cdot 20} - 1}$$

$$\approx \$23.36 \text{ per month, or } \$280.32 \text{ per year}$$

Bank B requires the lower annual contributions.

31.

First 15 years	Next 15 years
$P = 100,\ i = \frac{0.08}{12},\ k = 12,\ n = 15$	$P = 34{,}603.82,\ i = \frac{0.08}{12},\ k = 12,\ n = 15$

First 15 years:
$$FV = \frac{P\left[(1+i)^{kn} - 1\right]}{i}$$

$$= \frac{100\left[\left(1 + \frac{0.08}{12}\right)^{12 \cdot 15} - 1\right]}{\frac{0.08}{12}}$$

$$\approx \$34{,}603.82$$

Next 15 years:
$$FV = P(1+i)^{kn}$$

$$= 34{,}603.82\left(1 + \frac{0.08}{12}\right)^{12 \cdot 15} \approx \$114{,}432.12$$

33.

Last 5 years	First 5 years
$FV = 13{,}500;\ i = \frac{0.09}{1},\ k = 1,\ n = 5$	$FV = 8774.07;\ i = \frac{0.09}{1},\ k = 1,\ n = 5$

Last 5 years:
$$PV = FV(1+i)^{-kn}$$

$$= 13{,}500\left(1 + \frac{0.09}{1}\right)^{-1(5)}$$

$$\approx \$8774.07$$

First 5 years:
$$P = \frac{FVi}{(1+i)^{kn} - 1}$$

$$= \frac{8774.07\left(\frac{0.09}{1}\right)}{\left(1 + \frac{0.09}{1}\right)^{1 \cdot 5} - 1}$$

$$\approx \$1466.08 \text{ per year}$$

35.
$$\frac{2(5x - 12)}{x} = 8$$
$$2(5x - 12) = 8x$$
$$10x - 24 = 8x$$
$$2x = 24$$
$$x = 12$$

37.
$$\sqrt{2x + 3} = 3$$
$$\left(\sqrt{2x + 3}\right)^2 = 3^2$$
$$2x + 3 = 9$$
$$2x = 6$$
$$x = 3$$

Exercise 9.3 (page 646)

1. present value **3.** promissory note **5.** amortizing

7. $P = 3500, k = 1, i = \dfrac{0.0525}{1}, n = 25$

$$PV = \frac{P\left[1 - (1+i)^{-kn}\right]}{i}$$

$$= \frac{3500\left[1 - \left(1 + \frac{0.0525}{1}\right)^{-1 \cdot 25}\right]}{\frac{0.0525}{1}}$$

$$\approx \$48{,}116.14$$

9. $A = 25{,}000; \; k = 12, i = \dfrac{0.12}{12}, n = 15$

$$P = \frac{Ai}{1 - (1+i)^{-kn}}$$

$$= \frac{25{,}000\left(\frac{0.12}{12}\right)}{1 - \left(1 + \frac{0.12}{12}\right)^{-12 \cdot 15}} \approx \$300.04$$

11. $P = 700, k = 4, i = \dfrac{0.0625}{4}, n = 15$

$$PV = \frac{P\left[1 - (1+i)^{-kn}\right]}{i}$$

$$= \frac{700\left[1 - \left(1 + \frac{0.0625}{4}\right)^{-4 \cdot 15}\right]}{\frac{0.0625}{4}}$$

$$\approx \$27{,}128.43$$

13. $P = 12{,}000; \; k = 1, i = \dfrac{0.085}{1}, n = 15$

$$PV = \frac{P\left[1 - (1+i)^{-kn}\right]}{i}$$

$$= \frac{12{,}000\left[1 - \left(1 + \frac{0.085}{1}\right)^{-1 \cdot 15}\right]}{\frac{0.085}{1}}$$

$$\approx \$99{,}650.84$$

15. $A = 21{,}700; \; k = 12, i = \dfrac{0.084}{12}, n = 4$

$$P = \frac{Ai}{1 - (1+i)^{-kn}} = \frac{21{,}700\left(\frac{0.084}{12}\right)}{1 - \left(1 + \frac{0.084}{12}\right)^{-12 \cdot 4}} \approx \$533.84$$

17.

15-year	20-year
$A = 130{,}000; \; k = 12, i = \dfrac{0.12}{12}, n = 15$	$A = 130{,}000; \; k = 12, i = \dfrac{0.11}{12}, n = 20$

$$P = \frac{Ai}{1 - (1+i)^{-kn}}$$

$$= \frac{130{,}000\left(\frac{0.12}{12}\right)}{1 - \left(1 + \frac{0.12}{12}\right)^{-12 \cdot 15}} \approx \$1560.22$$

$$P = \frac{Ai}{1 - (1+i)^{-kn}}$$

$$= \frac{130{,}000\left(\frac{0.11}{12}\right)}{1 - \left(1 + \frac{0.11}{12}\right)^{-12 \cdot 20}} \approx \$1341.84$$

19.

From 55 to 80	From 20 to 55
$P = 5000; \; k = 12, i = \frac{0.0875}{12}, n = 25$	$FV = 608{,}166.24; \; i = \frac{0.0875}{12}, k = 12, n = 35$

$$PV = \frac{P\left[1 - (1+i)^{-kn}\right]}{i}$$

$$= \frac{5000\left[1 - \left(1 + \frac{0.0875}{12}\right)^{-12 \cdot 25}\right]}{\frac{0.0875}{12}}$$

$$\approx \$608{,}166.24$$

$$P = \frac{FVi}{(1+i)^{kn} - 1}$$

$$= \frac{608{,}166.24\left(\frac{0.0875}{12}\right)}{\left(1 + \frac{0.0875}{12}\right)^{12 \cdot 35} - 1}$$

$$\approx \$220.13$$

21.

First ten years	Next ten years (2 parts)

$$FV = \frac{500\left[\left(1 + \frac{0.0725}{4}\right)^{4\cdot 10} - 1\right]}{\frac{0.0725}{4}}$$

$$\approx \$29{,}003.32$$

$$FV = \frac{1500\left[\left(1 + \frac{0.0725}{4}\right)^{4\cdot 10} - 1\right]}{\frac{0.0725}{4}}$$

$$\approx \$87{,}009.96$$

$$FV = 29{,}003.32\left(1 + \frac{0.0725}{4}\right)^{4(10)} \approx \$59{,}496.55$$

$$\text{Total} = \$87{,}009.96 + \$59{,}496.55 = \$146{,}506.51$$

23. $\dfrac{6\sqrt{30}}{3\sqrt{5}} = \dfrac{6}{3}\cdot\dfrac{\sqrt{30}}{\sqrt{5}} = 2\sqrt{\dfrac{30}{5}} = 2\sqrt{6}$

25. $3\sqrt{5x} + 5\sqrt{20x} = 3\sqrt{5x} + 5\sqrt{4}\sqrt{5x} = 3\sqrt{5x} + 10\sqrt{5x} = 13\sqrt{5x}$

Chapter 9 Summary (page 649)

1. $FV = 2000\left(1 + \dfrac{0.09}{1}\right)^{1\cdot 5} \approx \$3077.25$

2. $FV = 2350\left(1 + \dfrac{0.14}{360}\right)^{360(1/6)} \approx \$2405.47$

3. $FV = 2000\left(1 + \dfrac{0.076}{4}\right)^{4\cdot 16} \approx \$6670.80$

4. BigBank: $R = \left(1 + \dfrac{0.063}{4}\right)^4 - 1 = 0.0645$ BestBank: $R = \left(1 + \dfrac{0.0621}{365}\right)^{365} - 1 = 0.0641$

BigBank has the better return.

5. $PV = 7900\left(1 + \dfrac{0.0575}{2}\right)^{-2\cdot 6}$

$\approx \$5622.23$

6. $FV = \dfrac{500\left[\left(1 + \frac{0.05}{1}\right)^{1\cdot 13} - 1\right]}{\frac{0.05}{1}}$

$\approx \$8856.49$

7. $FV = \dfrac{150\left[\left(1 + \frac{0.08}{12}\right)^{12\cdot 20} - 1\right]}{\frac{0.08}{12}}$

$\approx \$88{,}353.06$

8. $P = \dfrac{40{,}700\left(\frac{0.075}{12}\right)}{\left(1 + \frac{0.075}{12}\right)^{12\cdot 7} - 1} \approx \$369.89$

9. $PV = \dfrac{250\left[1 - \left(1 + \frac{0.065}{2}\right)^{-2\cdot 20}\right]}{\frac{0.065}{2}}$

$\approx \$5552.11$

10. $PV = \dfrac{50{,}000\left[1 - \left(1 + \frac{0.096}{1}\right)^{-1\cdot 20}\right]}{\frac{0.096}{1}}$

$\approx \$437{,}563.50$

11. $P = \dfrac{150{,}500\left(\frac{0.1075}{12}\right)}{1 - \left(1 + \frac{0.1075}{12}\right)^{-12\cdot 15}} \approx \$1687.03$; Total paid $= 1687.03 \cdot 180 = \$303{,}665.40$

12. $P = \dfrac{150{,}500\left(\frac{0.1075}{12}\right)}{1 - \left(1 + \frac{0.1075}{12}\right)^{-12\cdot30}} \approx \$1404.89$; Total paid $= 1404.89 \cdot 360 = \$505{,}760.40$

Chapter 9 Test (page 651)

1. compound   **2.** periodic   **3.** effective   **4.** annual

5. annuities   **6.** sinking fund   **7.** present value   **8.** amortizing

9. $FV = 1300\left(1 + \frac{0.05}{1}\right)^{1\cdot10} \approx \$2117.56$ **10.** $FV = 1300\left(1 + \frac{0.05}{12}\right)^{12\cdot10} \approx \$2141.11$

11. $R = \left(1 + \frac{0.05}{12}\right)^{12} - 1 = 0.05116$
$ = 5.116\%$ **12.** $PV = 5000\left(1 + \frac{0.07}{4}\right)^{-4\cdot10} \approx \$2498.00$

13. $FV = \dfrac{700\left[\left(1 + \frac{0.073}{12}\right)^{12\cdot5} - 1\right]}{\frac{0.073}{12}}$
$ \approx \$50{,}506.10$ **14.** $P = \dfrac{8000\left(\frac{0.065}{12}\right)}{\left(1 + \frac{0.065}{12}\right)^{12\cdot5} - 1} \approx \$113.20$

15. $PV = \dfrac{1000\left[1 - \left(1 + \frac{0.068}{12}\right)^{-12\cdot15}\right]}{\frac{0.068}{12}}$
$ \approx \$112{,}652.71$ **16.** $P = \dfrac{90{,}000\left(\frac{0.0895}{12}\right)}{1 - \left(1 + \frac{0.0895}{12}\right)^{-12\cdot15}} \approx \$910.16$

Cumulative Review Exercises (page 652)

1. $\begin{cases} 2x + y = 8 \\ x - 2y = -1 \end{cases}$ **2.** $\begin{cases} 3x = -y + 2 \\ y + x - 4 = -2x \end{cases}$

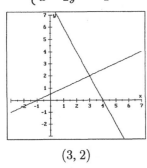

$(3, 2)$

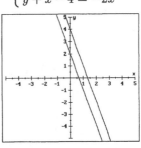

no solution, inconsistent system

3.
$$\begin{array}{l} 5x = 3y + 12 \Rightarrow 5x - 3y = 12 \Rightarrow \quad 5x - 3y = 12 \\ \underline{2x - 3y = 3} \quad\;\; \Rightarrow \underline{2x - 3y = 3} \Rightarrow \underline{-2x + 3y = -3} \\ \hphantom{2x - 3y = 3 \Rightarrow 2x - 3y = 3 \Rightarrow} \begin{array}{rcl} 3x &=& 9 \\ x &=& 3 \end{array} \end{array}$$

$\begin{array}{l} 2x - 3y = 3 \\ 2(3) - 3y = 3 \\ -3y = -3 \\ y = 1 \end{array}$

Solution:
$\boxed{x = 3,\, y = 1}$

4. $\begin{bmatrix} x \\ y \\ z \end{bmatrix} = \begin{bmatrix} 2 & 1 & -1 \\ 1 & -1 & 1 \\ 1 & 1 & -3 \end{bmatrix}^{-1} \begin{bmatrix} 7 \\ 2 \\ 2 \end{bmatrix} = \begin{bmatrix} 3 \\ 2 \\ 1 \end{bmatrix}$

5. $\begin{bmatrix} 2 & 1 & -1 & | & 0 \\ 1 & -1 & 1 & | & 3 \\ 1 & 1 & -3 & | & -5 \end{bmatrix} \Rightarrow \begin{bmatrix} 1 & 1 & -3 & | & -5 \\ 1 & -1 & 1 & | & 3 \\ 2 & 1 & -1 & | & 0 \end{bmatrix} \Rightarrow \begin{bmatrix} 1 & 1 & -3 & | & -5 \\ 0 & -2 & 4 & | & 8 \\ 0 & -1 & 5 & | & 10 \end{bmatrix} \Rightarrow$

$\qquad\qquad\qquad\qquad\qquad\qquad R_1 \Leftrightarrow R_3 \qquad\qquad\qquad -R_1 + R_2 \Rightarrow R_2$

$\qquad\qquad\qquad\qquad\qquad\qquad\qquad\qquad\qquad\qquad\quad -2R_1 + R_3 \Rightarrow R_3$

$\begin{bmatrix} 1 & 0 & -1 & | & -1 \\ 0 & 1 & -2 & | & -4 \\ 0 & 0 & 3 & | & 6 \end{bmatrix} \Rightarrow \begin{bmatrix} 1 & 0 & 0 & | & 1 \\ 0 & 1 & 0 & | & 0 \\ 0 & 0 & 1 & | & 2 \end{bmatrix}$ Solution: $\boxed{x = 1, y = 0, z = 2}$

$\quad \frac{1}{2}R_2 + R_1 \Rightarrow R_1 \qquad \frac{1}{3}R_3 + R_1 \Rightarrow R_1$

$\quad -\frac{1}{2}R_2 + R_3 \Rightarrow R_3 \qquad \frac{2}{3}R_3 + R_2 \Rightarrow R_2$

$\qquad\quad -\frac{1}{2}R_2 \Rightarrow R_2 \qquad\qquad \frac{1}{3}R_3 \Rightarrow R_3$

6. $\begin{bmatrix} 2 & -2 & 3 & 1 & | & 2 \\ 1 & 1 & 1 & 1 & | & 5 \\ -1 & 2 & -3 & 2 & | & 2 \\ 1 & 1 & 2 & -1 & | & 4 \end{bmatrix} \Rightarrow \begin{bmatrix} 2 & -2 & 3 & 1 & | & 2 \\ 0 & -4 & 1 & -1 & | & -8 \\ 0 & 2 & -3 & 5 & | & 6 \\ 0 & -4 & -1 & 3 & | & -6 \end{bmatrix} \Rightarrow \begin{bmatrix} -4 & 0 & -5 & -3 & | & -12 \\ 0 & -4 & 1 & -1 & | & -8 \\ 0 & 0 & -5 & 9 & | & 4 \\ 0 & 0 & 2 & -4 & | & -2 \end{bmatrix}$

$\qquad\qquad\qquad\qquad\qquad\qquad -2R_2 + R_1 \Rightarrow R_2 \qquad\qquad -2R_1 + R_2 \Rightarrow R_1$

$\qquad\qquad\qquad\qquad\qquad\qquad 2R_3 + R_1 \Rightarrow R_3 \qquad\qquad 2R_3 + R_2 \Rightarrow R_3$

$\qquad\qquad\qquad\qquad\qquad\qquad -2R_4 + R_1 \Rightarrow R_4 \qquad\qquad -R_4 + R_2 \Rightarrow R_4$

$\begin{bmatrix} -4 & 0 & -5 & -3 & | & -12 \\ 0 & -4 & 1 & -1 & | & -8 \\ 0 & 0 & -5 & 9 & | & 4 \\ 0 & 0 & 1 & -2 & | & -1 \end{bmatrix} \Rightarrow \begin{bmatrix} 4 & 0 & 0 & 12 & | & 16 \\ 0 & -20 & 0 & 4 & | & -36 \\ 0 & 0 & -5 & 9 & | & 4 \\ 0 & 0 & 0 & -1 & | & -1 \end{bmatrix} \Rightarrow \begin{bmatrix} 1 & 0 & 0 & 3 & | & 4 \\ 0 & -5 & 0 & 1 & | & -9 \\ 0 & 0 & -5 & 9 & | & 4 \\ 0 & 0 & 0 & 1 & | & 1 \end{bmatrix}$

$\qquad\quad \frac{1}{2}R_4 \Rightarrow R_4 \qquad\qquad\qquad -R_1 + R_3 \Rightarrow R_1 \qquad\qquad \frac{1}{4}R_1 \Rightarrow R_1$

$\qquad\qquad\qquad\qquad\qquad\qquad 5R_2 + R_3 \Rightarrow R_2 \qquad\qquad \frac{1}{4}R_2 \Rightarrow R_2$

$\qquad\qquad\qquad\qquad\qquad\qquad 5R_4 + R_3 \Rightarrow R_4 \qquad\qquad -R_4 \Rightarrow R_4$

$\Rightarrow \begin{bmatrix} 1 & 0 & 0 & 0 & | & 1 \\ 0 & -5 & 0 & 0 & | & -10 \\ 0 & 0 & -5 & 0 & | & -5 \\ 0 & 0 & 0 & 1 & | & 1 \end{bmatrix} \Rightarrow \begin{bmatrix} 1 & 0 & 0 & 0 & | & 1 \\ 0 & 1 & 0 & 0 & | & 2 \\ 0 & 0 & 1 & 0 & | & 1 \\ 0 & 0 & 0 & 1 & | & 1 \end{bmatrix}$ Solution:

$\qquad\quad -3R_4 + R_1 \Rightarrow R_1 \qquad\qquad -\frac{1}{5}R_2 \Rightarrow R_2 \qquad\qquad \boxed{x = 1, y = 2, z = 1, t = 1}$

$\qquad\quad -R_4 + R_2 \Rightarrow R_2 \qquad\qquad -\frac{1}{5}R_3 \Rightarrow R_3$

$\qquad\quad -9R_4 + R_3 \Rightarrow R_3$

7. $\begin{bmatrix} 2 & 1 \\ 1 & 4 \end{bmatrix} + \begin{bmatrix} -1 & 2 \\ 2 & 3 \end{bmatrix} = \begin{bmatrix} 1 & 3 \\ 3 & 7 \end{bmatrix}$

8. $\begin{bmatrix} -1 & 2 \\ 2 & 3 \end{bmatrix} - \begin{bmatrix} 2 & 1 \\ 1 & 4 \end{bmatrix} = \begin{bmatrix} -3 & 1 \\ 1 & -1 \end{bmatrix}$

9. $\begin{bmatrix} 2 & 1 \\ 1 & 4 \end{bmatrix} \begin{bmatrix} 2 & 0 & -1 \\ -1 & 2 & 2 \end{bmatrix} = \begin{bmatrix} 3 & 2 & 0 \\ -2 & 8 & 7 \end{bmatrix}$

10. $\begin{bmatrix} -1 & 2 \\ 2 & 3 \end{bmatrix}^2 + 2\begin{bmatrix} 2 & 1 \\ 1 & 4 \end{bmatrix} = \begin{bmatrix} -1 & 2 \\ 2 & 3 \end{bmatrix}\begin{bmatrix} -1 & 2 \\ 2 & 3 \end{bmatrix} + \begin{bmatrix} 4 & 2 \\ 2 & 8 \end{bmatrix} = \begin{bmatrix} 5 & 4 \\ 4 & 13 \end{bmatrix} + \begin{bmatrix} 4 & 2 \\ 2 & 8 \end{bmatrix} = \begin{bmatrix} 9 & 6 \\ 6 & 21 \end{bmatrix}$

11. $\begin{bmatrix} 2 & 6 & | & 1 & 0 \\ 2 & 4 & | & 0 & 1 \end{bmatrix} \Rightarrow \underset{\frac{1}{2}R_1 \Rightarrow R_1}{\begin{bmatrix} 1 & 3 & | & \frac{1}{2} & 0 \\ 2 & 4 & | & 0 & 1 \end{bmatrix}} \Rightarrow \underset{-2R_1 + R_2 \Rightarrow R_2}{\begin{bmatrix} 1 & 3 & | & \frac{1}{2} & 0 \\ 0 & -2 & | & -1 & 1 \end{bmatrix}} \Rightarrow$

$\underset{-\frac{1}{2}R_2 \Rightarrow R_2}{\begin{bmatrix} 1 & 3 & | & \frac{1}{2} & 0 \\ 0 & 1 & | & \frac{1}{2} & -\frac{1}{2} \end{bmatrix}} \Rightarrow \underset{-3R_2 + R_1 \Rightarrow R_1}{\begin{bmatrix} 1 & 0 & | & -1 & \frac{3}{2} \\ 0 & 1 & | & \frac{1}{2} & -\frac{1}{2} \end{bmatrix}} \Rightarrow$ Inverse: $\begin{bmatrix} -1 & \frac{3}{2} \\ \frac{1}{2} & -\frac{1}{2} \end{bmatrix}$

12. $\begin{bmatrix} 1 & -1 & 1 & | & 1 & 0 & 0 \\ 1 & 4 & 0 & | & 0 & 1 & 0 \\ 2 & 4 & 1 & | & 0 & 0 & 1 \end{bmatrix} \Rightarrow \underset{\substack{-R_1 + R_2 \Rightarrow R_2 \\ -2R_2 + R_3 \Rightarrow R_3}}{\begin{bmatrix} 1 & -1 & 1 & | & 1 & 0 & 0 \\ 0 & 5 & -1 & | & -1 & 1 & 0 \\ 0 & 6 & -1 & | & -2 & 0 & 1 \end{bmatrix}} \Rightarrow \underset{-R_2 + R_3 \Rightarrow R_2}{\begin{bmatrix} 1 & -1 & 1 & | & 1 & 0 & 0 \\ 0 & 1 & 0 & | & -1 & -1 & 1 \\ 0 & 6 & -1 & | & -2 & 0 & 1 \end{bmatrix}}$

$\Rightarrow \underset{\substack{R_2 + R_1 \Rightarrow R_1 \\ -6R_2 + R_3 \Rightarrow R_3}}{\begin{bmatrix} 1 & 0 & 1 & | & 0 & -1 & 1 \\ 0 & 1 & 0 & | & -1 & -1 & 1 \\ 0 & 0 & -1 & | & 4 & 6 & -5 \end{bmatrix}} \Rightarrow \underset{\substack{R_3 + R_1 \Rightarrow R_1 \\ -R_3 \Rightarrow R_3}}{\begin{bmatrix} 1 & 0 & 0 & | & 4 & 5 & -4 \\ 0 & 1 & 0 & | & -1 & -1 & 1 \\ 0 & 0 & 1 & | & -4 & -6 & 5 \end{bmatrix}} \Rightarrow \text{Inv} = \begin{bmatrix} 4 & 5 & -4 \\ -1 & -1 & 1 \\ -4 & -6 & 5 \end{bmatrix}$

13. $\begin{vmatrix} -3 & 5 \\ 4 & 7 \end{vmatrix} = (-3)(7) - (5)(4) = -21 - 20 = -41$

14. $\begin{vmatrix} 2 & -3 & 2 \\ 0 & 1 & -1 \\ 1 & -2 & 1 \end{vmatrix} = 2\begin{vmatrix} 1 & -1 \\ -2 & 1 \end{vmatrix} - (-3)\begin{vmatrix} 0 & -1 \\ 1 & 1 \end{vmatrix} + 2\begin{vmatrix} 0 & 1 \\ 1 & -2 \end{vmatrix}$

$= 2(-1) + 3(1) + 2(-1) = -2 + 3 - 2 = -1$

15. $x = \dfrac{\begin{vmatrix} 11 & 3 \\ 24 & 5 \end{vmatrix}}{\begin{vmatrix} 4 & 3 \\ -2 & 5 \end{vmatrix}}$

16. $y = \dfrac{\begin{vmatrix} 4 & 11 \\ -2 & 24 \end{vmatrix}}{\begin{vmatrix} 4 & 3 \\ -2 & 5 \end{vmatrix}}$

17. $\dfrac{-x+1}{(x+1)(x+2)} = \dfrac{A}{x+1} + \dfrac{B}{x+2}$

$= \dfrac{A(x+2)}{(x+1)(x+2)} + \dfrac{B(x+1)}{(x+1)(x+2)}$

$= \dfrac{Ax + 2A + Bx + B}{(x+1)(x+2)}$

$= \dfrac{(A+B)x + (2A+B)}{(x+1)(x+2)}$

$\begin{cases} A + B = -1 \\ 2A + B = 1 \end{cases} \Rightarrow A = 2, B = -3$

$\dfrac{-x+1}{(x+1)(x+2)} = \dfrac{2}{x+1} - \dfrac{3}{x+2}$

18.

$$\frac{x-4}{(2x-5)^2} = \frac{A}{2x-5} + \frac{B}{(2x-5)^2}$$
$$= \frac{A(2x-5)}{(2x-5)^2} + \frac{B}{(2x-5)^2}$$
$$= \frac{2Ax-5A+B}{(2x-5)^2}$$
$$= \frac{2Ax+(-5A+B)}{(2x-5)^2}$$

$$\begin{cases} 2A = 1 \\ -5A+B = -4 \end{cases} \Rightarrow \begin{array}{l} A = \frac{1}{2} \\ B = -\frac{3}{2} \end{array}$$

$$\frac{x-4}{(2x-5)^2} = \frac{\frac{1}{2}}{2x-5} - \frac{\frac{3}{2}}{(2x-5)^2}$$

19. $y \le 2x+6$

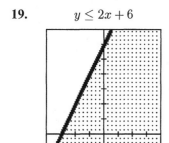

20. $\begin{cases} 2x+3y \ge 6 \\ 2x-3y \le 6 \end{cases}$

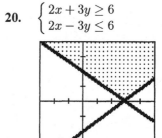

21. $(x-h)^2 + (y-k)^2 = r^2$
$(x-0)^2 + (y-0)^2 = 4^2$
$x^2 + y^2 = 16$

22. $(x-h)^2 + (y-k)^2 = r^2$
$(x-2)^2 + (y+3)^2 = 11^2$
$(x-2)^2 + (y+3)^2 = 121$

23. $x^2 + y^2 - 4y = 12$
$x^2 + y^2 - 4y + 4 = 12 + 4$
$(x-0)^2 + (y-2)^2 = 4^2$
$C(0,2), r = 4$

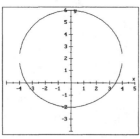

24. $x^2 - 2y - 2x = -7$
$x^2 - 2x + 1 = 2y - 7 + 1$
$(x-1)^2 = 2(y-3)$
$V(1,3)$

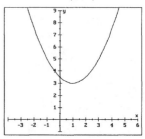

25.
$$x^2 + 4y^2 + 2x = 3$$
$$x^2 + 2x + 1 + 4y^2 = 3 + 1$$
$$(x + 1)^2 + 4y^2 = 4$$
$$\frac{(x + 1)^2}{4} + \frac{y^2}{1} = 1$$

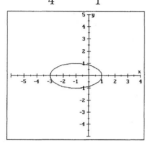

26.
$$x^2 - 9y^2 - 4x = 5$$
$$x^2 - 4x + 4 - 9y^2 = 5 + 4$$
$$(x - 2)^2 - 9y^2 = 9$$
$$\frac{(x - 2)^2}{9} - \frac{y^2}{1} = 1$$

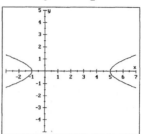

27. $a = 6, b = 4$
$$\frac{x^2}{36} + \frac{y^2}{16} = 1$$

28. $a = 5, c = 2$
$$b^2 = a^2 - c^2 = 25 - 4 = 21$$
$$\frac{(x - 2)^2}{21} + \frac{(y - 3)^2}{25} = 1$$

29. $a = 2, c = 3$
$$b^2 = c^2 - a^2 = 9 - 4 = 5$$
$$\frac{x^2}{4} - \frac{y^2}{5} = 1$$

30. $2a = 6, 2b = 6 \Rightarrow a = 3, b = 3$
$$\frac{(y - 4)^2}{9} - \frac{(x - 2)^2}{9} = 1$$

31. The 2nd term will involve $(2y)^1$.
$$\frac{8!}{7!1!}x^7(2y)^1 = 16x^7y$$

32. The 6th term will involve $(2y)^5$.
$$\frac{8!}{3!5!}x^3(2y)^5 = 1792x^3y^5$$

33. $\displaystyle\sum_{k=1}^{5} 2 = 5(2) = 10$

34. $\displaystyle\sum_{k=2}^{6}(3k + 1) = 3\sum_{k=2}^{6}k + \sum_{k=2}^{6}1 = 3(2 + 3 + 4 + 5 + 6) + 5(1) = 60 + 5 = 65$

35. $a = -2, d = 3$
$$l = a + (n - 1)d = -2 + 5(3) = 13$$
$$S = \frac{n(a + l)}{2} = \frac{6(-2 + 13)}{2} = 33$$

36. $a = \frac{1}{9}, r = 3, n = 6$
$$S = \frac{a - ar^n}{1 - r} = \frac{\frac{1}{9} - \frac{1}{9}(3)^6}{1 - 3}$$
$$= \frac{-\frac{728}{9}}{-2} = \frac{364}{9}$$

37. $P(8, 4) = \dfrac{8!}{(8 - 4)!} = \dfrac{8!}{4!} = 1680$

38. $P(24, 0) = \dfrac{24!}{(24 - 0)!} = \dfrac{24!}{24!} = 1$

39. $C(12, 10) = \dfrac{12!}{10!2!} = 66$

40. $P(4, 4) \cdot C(6, 6) = 24 \cdot 1 = 24$

41. $4! \cdot 6! = 17{,}280$

42. $\dbinom{12}{4} = \dfrac{12!}{4!8!} = 495$

43. $\dfrac{2}{36} = \dfrac{1}{18}$

44. $\dfrac{\binom{26}{5}}{\binom{52}{5}} = \dfrac{65{,}780}{2{,}598{,}960} = \dfrac{253}{9996}$

45. $(0.6)(0.8) = 0.48$

46. $1 - (0.3 + 0.4) = 1 - 0.7 = 0.3$

47.

Check $n = 1$: $3(1) + 1 \stackrel{?}{=} \dfrac{1(3(1) + 5)}{2}$ True for $n = 1$

$$4 = 4$$

Assume for $n = k$ and show for $n = k + 1$:

$$4 + 7 + 10 + \cdots + (3k + 1) = \dfrac{k(3k + 5)}{2}$$

$$\boxed{4 + 7 + 10 + \cdots + (3k + 1)} + 3(k + 1) + 1 = \boxed{\dfrac{k(3k + 5)}{2}} + 3(k + 1) + 1$$

$$4 + 7 + 10 + \cdots + (3(k + 1) + 1) = \dfrac{k(3k + 5)}{2} + \dfrac{2 \cdot (3(k + 1) + 1)}{2}$$

$$4 + 7 + 10 + \cdots + (3(k + 1) + 1) = \dfrac{3k^2 + 5k + 6k + 6 + 2}{2}$$

$$4 + 7 + 10 + \cdots + (3(k + 1) + 1) = \dfrac{3k^2 + 11k + 8}{2}$$

$$4 + 7 + 10 + \cdots + (3(k + 1) + 1) = \dfrac{(k + 1)(3k + 8)}{2}$$

$$4 + 7 + 10 + \cdots + (3(k + 1) + 1) = \dfrac{(k + 1)(3(k + 1) + 5)}{2}$$

Since this is what results when $n = k + 1$ in the formula, we have shown that the formula works for $n = k + 1$ if it works for $n = k$.

48. Expected winnings $= \dfrac{4}{52}(100) + \dfrac{4}{52}(10) + \dfrac{44}{52}(0) \approx \$8.46 \Leftarrow$ fair price

49. $PV = 10{,}000\left(1 + \dfrac{0.085}{1}\right)^{-1 \cdot 12}$
$\approx \$3757.02$

50. $P = \dfrac{110{,}000\left(\frac{0.0875}{12}\right)}{1 - \left(1 + \frac{0.0875}{12}\right)^{-12 \cdot 20}} \approx \$972.08$